AF334544

Semiconductor-Based Sensors

Semiconductor-Based Sensors

Editors

Fan Ren | Stephen J Pearton
University of Florida, USA

NEW JERSEY · LONDON · SINGAPORE · BEIJING · SHANGHAI · HONG KONG · TAIPEI · CHENNAI · TOKYO

Published by

World Scientific Publishing Co. Pte. Ltd.

5 Toh Tuck Link, Singapore 596224

USA office: 27 Warren Street, Suite 401-402, Hackensack, NJ 07601

UK office: 57 Shelton Street, Covent Garden, London WC2H 9HE

British Library Cataloguing-in-Publication Data
A catalogue record for this book is available from the British Library.

SEMICONDUCTOR-BASED SENSORS

ISBN 978-981-3146-72-3

Desk Editor: Rhaimie Wahap

Typeset by Stallion Press
Email: enquiries@stallionpress.com

Printed in Singapore

Contents

Preface

Chemical sensors have gained in importance in the past decade for applications that include homeland security, medical and environmental monitoring and also food safety. A desirable goal is the ability to simultaneously analyze a wide variety of environmental and biological gases and liquids in the field and to be able to selectively detect a target analyte with high specificity and sensitivity. The goal is to realize real-time, portable and inexpensive chemical and biological sensors and to use these as handheld gas, environmental pollutant, exhaled breath, saliva, urine, or blood monitors with wireless capability. In the medical area, frequent screening can catch the early development of diseases, reduce the suffering of patients due to late diagnoses, and lower the medical cost. For example, a 96 % survival rate has been predicted in breast cancer patients if the frequency of screening is every three months. This frequency cannot be achieved with current methods of mammography due to high cost to the patient and invasiveness (radiation). In the area of detection of medical biomarkers, many different methods, including enzyme-linked immunsorbent assay (ELISA), particle-based flow cytometric assays, electrochemical measurements based on impedance and capacitance, electrical measurement of microcantilever resonant frequency change, and conductance measurement of semiconductor nanostructures, gas chromatography (GC), ion chromatography, high density peptide arrays, laser scanning quantitiative analysis, chemiluminescence, selected ion flow tube (SIFT), nanomechanical cantilevers, bead-based suspension microarrays, magnetic biosensors and mass spectrometry (MS) have been employed. Depending on the sample condition, these methods may show variable results in terms of sensitivity for some applications and may not meet the requirements for a handheld biosensor.

While the techniques mentioned above show excellent performance under lab conditions, there is also a need for small, handheld sensors with wireless connectivity that have fast response. For medical sensing applications, disease diagnosis by detecting specific biomarkers (functional or structural abnormal enzymes, low molecular weight proteins, or antigen) in blood, urine, saliva, or tissue samples has been established. Most of the techniques mentioned earlier such as ELISA possesses a major limitation in that only one analyte is measured at a time. Particle-based assays allow for multiple detection by using multiple beads but the whole detection process is generally longer than 2 hours, which is not practical for in-office or bedside detection.

Semiconductor-based sensors can be fabricated using mature techniques from the Si chip industry and/or novel nanotechnology approaches. Silicon based sensors are still the dominant component of the semiconductor segment due to their low cost, reproducible and controllable electronic response. However, these sensors are not suited for operation in harsh environments, for instance, high temperature, high pressure or corrosive ambients. Si will be etched by some of the acidic or basic aqueous solutions encountered in biological sensing. By sharp contrast, materials like GaN are not etched by any acid or base at temperatures below a few hundred degrees and are alternative options to supplement silicon in these applications because of their high temperature/high power capability, high electron saturation velocity and simple integration with existing GaN-based UV light-emitting diode, UV detectors and wireless communication chips. ISFETs serve as platforms to fabricate a diversity of chemical and biosensors used in medical diagnostics, agriculture, food processing, pharmaceutical and chemical industries, besides their main application in pH sensing. Notable examples are chemical sensors for alkali ions, lead, mercury, ammonium, fluoride, phosphate and nitrate ions, and biosensors for glucose, urea, creatinine and triglycerides. Diabetes is a leading cause of death and disability, with over 135 million people. One of three Americans born in 2000 will develop Type II diabetes during their lifetime, and there were over 20 million Americans diagnosed with diabetes (2015).The population of diabetics is large and growing. Although the frequent monitoring of glucose levels is strongly encouraged by health professionals most of the glucose testing products currently on the market are uncomfortable for the user and dissatisfaction is high. Most current glucose monitors are cheap and the main recurring cost are the strips of plastic that are placed in the meter to read the blood glucose level (test strip market is currently $2B). Given the ever increasing incidence of diabetes in both the United States and abroad the market for diabetes testing and supplies is large and growing.

Like the market for diabetes testing, the market for breast cancer testing is also highly promising. Although numerous testing alternatives exist the market size is huge and growing – in 2015 the testing market in the United States alone was worth well in excess of $1 billion – and the most common diagnostic methods involve some level of discomfort and/or exposure to radiation giving it the same kind of patient dissatisfaction found among diabetes.

In this book, we have brought together experts in a wide variety of sensors to provide a state-of-the-art summary for those already involved in these fields or just entering them. Teyu Kao and Jenshan Lin discuss non-contact vital sign and vibration detections using Doppler radar systems. Compared to other detection methods such as laser-based and interferometers sensors, Doppler radar generally has simple architecture implemented by discrete components on printed circuit board or integrated circuit, making it a low-cost and low-power solution.

E.S. McLamor and collaborators discuss biomimetic fractal nanometals as a transducer layer in electrochemical bio-sensing to enhance signal acquisition. In electrochemical sensing, nanomaterials are used to increase the active surface area of the sensor surface, where chemical interactions are monitored by recording electrical signals. The increase in sensor surface area facilitates targeted adsorption of biorecognition agents (e.g., proteins, DNA) and also enhances transduction of signal via charge transfer, surface redox reactions, or changes in charge transfer resistance/capacitance.

Sichen Zhan and co-workers detail the status of carbon nanodots, which display excellent optical properties, biocompatibility, water solubility, photo-stability, and ease of cost-effective synthesis and thus have gained tremendous attention in a wide spectrum of chemical and biomedical applications, ranging from sensing to bioimaging, fluorescent ink and phototherapy. In recent years, fluorescent C-dots have been regarded as a promising candidate to replace the conventional semiconductor quantum dots in various applications.

Yu-Lin Wang and collaborators discuss rapid detection of biotoxin and pathogen, and quick identification of ligand-receptor binding affinity using AlGaN/GaN high electron mobility transistors for accurate, affordable, point of care health care sensor platforms. There is also a strong need for the fast detection of biotoxins and pathogens. Biotoxins are possibly used as biological weapons by terrorists because they are usually much cheaper than traditional weapons, and they cause large mortality if they are dispersed to public facilities such as airports, rail stations, malls, or museums. A field-deployable, reliable, and sensitive biosensor is needed in fields for routinely monitoring biotoxins in real-time to ensure the security. The deployable sensor has to be able to sustain in a harsh environment for a considerable period of time. It is also preferred that the sensor has a small size and a lightweight so it could be made as a portable device. A portable device is convenient to inspect foods, drinks or suspected objects instantly. Finally, the sensor has to be very cheap so that it can be deployed in multiple sites in a facility to collect complete information of possible contaminated regions with dispersed toxins. The fast detection of pathogens is also critical for natural or farm-raised aquatic animals such as fish and shellfish. Some aquatic animals have very high economic value in the market. However, the population of aquatic animals is subject to the appearance of pathogens in freshwater or in sea. Pathogen infection results in sickness of aquatic animals, which may eventually lead to the death of these animals

The stability and Reliability of III-Nitride Based Biosensors is detailed by Nathaniel Rohrbaugh and co-workers. Specific biosensor device functionality, specificity, selectivity and operation are discussed on a material by material basis. Additionally, a background into the materials themselves is provided from a stability perspective with regard to solution exposure, doping, and interactions with biological systems.

GaN-Based Hydrogen Sensors are summarized by Kwang Hyeon Baik and Soohwan Jang. Interest in hydrogen as an alternative energy source and a viable commercial carrier has grown significantly in recent years. It is an emission-free fuel that when reacted with oxygen can produce heat combustion of 141.9 kJ/g. By comparison, only 47.0 and 45.0 kJ/g is yielded on average from gasoline and diesel, respectively. Electrochemical cells and combustion in internal engines are used to transport hydrogen-generated energy to power vehicles and electric devices. Ironically, most commercial hydrogen is produced through steam reforming from hydrocarbons which originate from fossil fuels. In addition, this process creates unwanted carbon dioxide and carbon monoxide bi-products. In response to these concerns, researchers have examined more efficient and economical means of producing hydrogen. These include radiolysis and water splitting through photo-electrochemical, photocatalytic and photo-biological processes.

Geonyeop Lee and collaborators discuss the state-of-the-art in graphene-based chemical sensors. Graphene possesses a large specific surface area ($2630 \ m^2 \cdot g^{-1}$), which indicates that, theoretically, all atoms of graphene can contribute to its detection capacity. The crystal lattice of graphene allows a low level of Johnson noise ($1/f$), which, in turn, makes the electrical conductivity responsive to small changes in carrier concentrations. These properties have made graphene a promising candidate for next generation sensors with ultrahigh sensitivity and responsivity. Due to its excellent mechanical and electrical properties under a mechanical strain, graphene sensors can be integrated in various areas including flexible and wearable devices. In this chapter, the authors focus on graphene as a sensing material.

Electronic micro-sensors for metabolite detection based on conductivity change of polyaniline is summarized by Yu-Lin Wang et al., including detection of hydrogen peroxide, sensing mechanisms, Cholesterol sensors and hydroxyl radical sensors

C.F. Lo et al. summarize the status of ZnO Nanorod Based Sensors. With proper surface functionalization with these nanorods, AlGaN/GaN HEMT sensors have been used for different sensing applications, such as gas sensing for detecting hydrogen, carbon monoxide, carbon dioxide, and ammonium, as well as liquid sensing for detecting proteins, pH values of aqueous solutions, lactic acid, breast cancer in saliva, DNA, kidney injury molecules, prostate cancer, glucose, chloride ion, and mercury ions

The production of scalable nanomanufactured broadband antireflection coatings on semiconductors is discussed by Pratik Kothary et al. Traditional nanomanufacturing technologies for scalably producing periodic subwavelength moth-eye AR coatings suffer from low throughput and high cost. In this chapter, the authors review two scalable nanomanufacturing platforms based on a microfabrication-compatible spin-coating technique and a continuous Langmuir-Blodgett (LB) approach for large-scale production of broadband moth-eye AR gratings on a variety of technologically important semiconductor wafers, such as crystalline silicon (both single-crystalline and multicrystalline), GaAs, and GaSb.

The breath biomarker detection by chemical sensors is summarized by J. Huang etal. Biomarkers find many applications in disease diagnosis and health status monitoring, either as diagnostic tools for a disease; as a tool for staging of disease or classification of the extent of the disease; or as an indicators of disease prognosis; and as diagnostic tool for the prediction and monitoring of the clinical response to an intervention. Commonly used biomarkers include the concentration of glucose in the blood for diabetes monitoring, the concentration of cholesterol in blood for cardiovascular disease, the blood pressure for stoke, the tidal volume for pulmonary disease, to name a few.

The status of Gallium Nitride microelectronics for high-temperature environments is the subject of a chapter by Debbie G. Senesky et al., including physical sensing, strain sensing, pressure sensing, optical and radiating sensors and high temperature transistors and contacts.

The emerging nanotechnology for strain gauge sensors is the subject of a chapter by Sheng-Po Fang et al., including use of nanomaterials, nanocomposites such as carbon nanofiber nanocomposites and electro-spun nanofibers as well as their detection mechanisms.

Overall, this book provides a comprehensive summary of the status of emerging sensor technologies and provides a framework for future advances in the field.

Fan Ren and Steve Pearton
Gainesville, FL
March 2016.

CHAPTER 1

60-GHz CMOS Micro-radar System-in-package for Noncontact and Noninvasive Measurement of Human Vital Signs and Vibrations

Teyu Kao, Jenshan Lin

Department of Electrical and Computer Engineering
University of Florida
1064 Center Drive, NEB 559, Gainesville, FL 32611-6130

1.1. Introduction

1.1.1. Doppler radar for vital sign and vibration detection

Non-contact vital sign and vibration detections using Doppler radar system have drawn intensive interests over the past decades. Significant progress has been made on improving detection range and accuracy, reducing system size and cost, and exploring various detection targets and applications [1]-[5]. Compared to other detection methods such as laser-based [6] and interferometer [7] sensors, Doppler radar generally has simple architecture implemented by discrete components on printed circuit board (PCB) or integrated circuit (IC), making it a low-cost and low-power solution. Depending on the frequency and power of electromagnetic wave transmitted by the radar, it was also found effective in special environments such as low visibility and through-wall search and detections. A typical Doppler radar system for non-contact detections is shown in Figure 1 where an un-modulated signal $T(t)$ is transmitted with the amplitude normalized to unity:

$$T(t) = \cos(2\pi f t + \phi_{vco}) \tag{1}$$

where f and ϕ_{vco} are the frequency and phase noise of the voltage controlled oscillator (VCO). $T(t)$ is reflected and phase modulated by the target

1

displacement $x(t)$ which is chest-wall movement in this case. The baseband output $B(t)$ can be expressed as [8]:

$$B(t) \approx G_s \cdot \cos\left(\frac{4\pi x(t)}{\lambda} + \phi_t\right) \qquad (2)$$

where λ is wavelength of $T(t)$, G_s is defined as the total system gain, and ϕ_t is total residue phase accumulated in the circuit and transmission path. Assuming $x(t)$ is much smaller than λ and ϕ_t is odd multiples of $\pi/2$, the system shows approximately linear transfer function (optimal detection points). For example, as $\phi_t = -\pi/2$, $B(t)$ can be simplified by small angle approximation:

$$B(t) \approx G_s \frac{4\pi x(t)}{\lambda}. \qquad (3)$$

As $x(t) \ll \lambda$, the baseband output $B(t)$ is proportional to the target displacement $x(t)$, and it can be sampled by an analog-to-digital convertor (ADC) and performed Fast Fourier Transform (FFT) to obtain frequency domain information such as heartbeat and respiration rates.

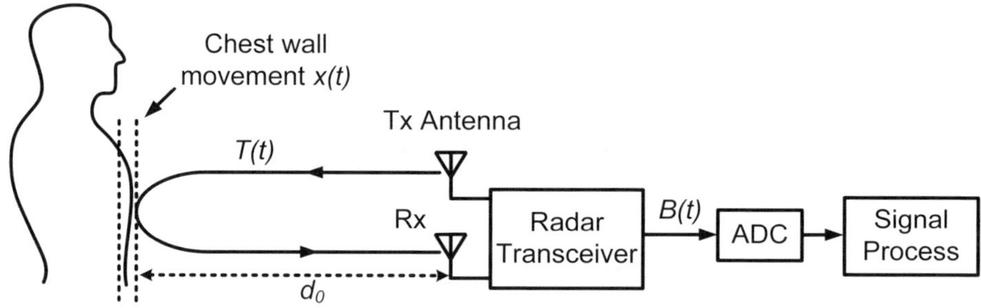

Figure 1. A typical vital sign detection using Doppler radar system.

1.1.2. 60-GHz radar implementation on CMOS

Many successful human heartbeat and respiration detections by Doppler radar systems have been demonstrated at the frequency range from a few to hundreds of GHz [9] [10]. The increase in radar frequency improves the sensitivity to small movement and reduces the circuit component size due to smaller wavelength. As shown in equation (3), the shorter λ in the denominator provides a higher "system demodulation gain" to distinguish small displacement at a longer distance. For instance, a 228-GHz heterodyne radar system reports the detection of vital sign at 50 m away [10]. Ideally the linear system response can track arbitrary movement of $x(t)$ as long as λ is much greater than $x(t)$. Higher radar frequency potentially enables the detection of very small vibration of insects, acoustic devices, micro-electro-mechanical systems (MEMS), and factory equipment for different applications.

The total system gain Gs in equation (2) is usually determined by antenna gains, power loss during the reflection and propagation (distance), and transceiver circuits. In radar design, moving the operating frequency up to millimeter-wave (mm-wave) range theoretically achieves larger radar received power by less antenna area. Generally the antenna gain G increases with frequency for the same antenna effective area A_e [11]:

$$G = \frac{4\pi A_e}{\lambda^2}. \tag{4}$$

For a simple flat-plate reflector perpendicular to the line of sight (LOS) at far-field, the radar cross section σ is approximately [12]:

$$\sigma \approx \frac{4\pi T^2}{\lambda^2} \tag{5}$$

where T is the actual area of the plate. Radar range equation can be used to estimate the received power under far-filed condition for simple analysis:

$$\frac{P_r}{P_t} = \sigma \frac{G_t G_r}{4\pi} \left[\frac{\lambda}{4\pi R^2} \right]^2 \tag{6}$$

where P_t and P_r are the transmitted and received power. G_t and G_r are the gain of transmitter (Tx) and receiver (Rx) antennas, and R is the distance between the target and radar. Plugging equations (4) and (5) into equation

(6), the received power can be estimated by:

$$P_r = P_t \frac{A_t A_r T^2}{\lambda^4 R^4} \tag{7}$$

where A_t and A_r are the effective area of Tx and Rx antennas. The equation shows apparent advantage of short λ if air absorption is negligible in a few meters range. Comparing 6-GHz and 60-GHz radar systems, for example, P_r at 60 GHz is theoretically 10^2 times higher than that at 6 GHz even if 1/10 area is used for Tx and Rx antennas.

Given the booming growth of modern wireless technology, sub-6-GHz bands are increasingly crowded by various protocols such as Bluetooth, Wi-Fi, and LTE. Next generation of cellular network (5G), for example, is expected to move to mm-wave frequencies for wider available bandwidth and less interference. Similarly the researches on Doppler radar for biomedical, security, and other daily applications have been exploring higher frequencies such as K_a band [3] [13] and 60-GHz unlicensed band [14] for higher detection resolution (shorter λ) and less chance of interfering with other wireless products. Design challenges including loss, noise, and transistor performances at higher frequencies need to be overcome by more accurate electromagnetic modeling, proper circuit architecture, and compact system integration.

Doppler radar system can be implemented by discrete components on PCB [3] or instrument level for testing and proof of concepts [9]. Compared to the board level implementation, SoC (system-on-chip) or SiP (system-in-package) are desired in terms of cost and system integration. Traditionally for IC operating above tens of GHz, III-V compound semiconductor (GaAs) and SiGe heterojunction bipolar transistors (HBT) process are adopted for its superior high-frequency performance. For example, the unity-current-gain frequency (f_T) / power-gain cutoff frequency (f_{MAX}) of a HBT can reach around 200/300 GHz in 130-nm process. Several single-chip Doppler radar systems are implemented on SiGe based processes to achieve high conversion gain and better noise performance [5] [15].

Nowadays the economies-of-scale of complementary-symmetry metal-oxide-semiconductor (CMOS) technology with the ability to integrate other circuits on the same die makes it a very low-cost and appealing IC platform. Single-chip Doppler radar transceivers on CMOS have been successfully developed for non-contact vital sign detection [2] [16] [17] in sub-6-GHz frequencies. Attributing to the improved transistor performance in scaled CMOS, f_T / f_{MAX} reaches about 120/150 GHz in 90-nm process and 200/300 GHz in 45-nm process, and the same trend is expected in more advanced CMOS technologies [18]. Increasing number of works on mm-wave CMOS circuits and systems have been reported for wireless communication, radar, and imaging [19]-[21]. The work in this chapter introduces the first CMOS Doppler radar chip targeting 60-GHz unlicensed band, and it is fully integrated with antennas to demonstrate a low-cost SiP. System architecture, component design considerations, and experiment results will be presented in the following sessions.

1.2. CMOS Radar Chip Design and Fabrication

1.2.1. System consideration

Doppler radar is known to have optimal and null detection points. As explained in equation (3), the system response is at optimal detection point (linear) if $x(t)$ is much smaller than λ and ϕ_t is odd multiples of $\pi/2$. When ϕ_t in equation (2) is even multiples of $\pi/2$, $B(t)$ is at the null detection point and generates higher order harmonics on spectrum. The alternating optimal and null points occurs every $\lambda/8$ as ϕ_t varies with the distance d_0 [8], and it becomes more problematic when λ is small. Double-sideband architecture [17] or frequency [13] tuning is possible to alleviate the null point issue, but it also becomes impractical if λ is very small. Complex signal demodulation (CSD) [4] was found to be an effective method to solve the issue, as in-phase (I) and quadrature-phase (Q) baseband outputs are generated and the CSD signal is constructed in real time:

$$S(t) = I(t) + jQ(t) = \cos\left(\frac{4\pi x(t)}{\lambda} + \phi_t\right) + j\sin\left(\frac{4\pi x(t)}{\lambda} + \phi_t\right)$$

$$= e^{j\left[\frac{4\pi x(t)}{\lambda}\right]} \cdot e^{j\phi_t} \tag{8}$$

Since $\exp(j\phi_t)$ has a constant magnitude, the method eliminates the effect of total residue phase ϕ_t while applying FFT to obtain output spectrum.

The I/Q generation requires careful system-level planning in mm-wave design. It can be realized by a RF quadrature VCO in a direct-conversion system [19]; however, distributing the high-frequency, large local oscillator (LO) signals is a main concern in terms of loss, power consumption, and I/Q mismatch [22]. For 60-GHz radar transceiver for example, I/Q separation at an intermediate frequency (IF = 6 GHz) by a ring oscillator was shown to be a simple and power-efficient topology [23]. The ring oscillator, which does not require any inductor and capacitor, is very compact and is able to drive large IF passive mixers with low flicker noise. Especially in CMOS technology, flicker noise is a major factor affecting Rx performance since the vital sign outputs is very close to dc (0-2 Hz for human heartbeat and respiration). The flicker noise can be represented by [24]:

$$\overline{v_n}^2 = \frac{K_{fk}}{C_{ox}WL} \cdot \frac{1}{f} \tag{9}$$

where K_{fk} is a process-dependent constant, W and L are gate width and length of a transistor, and C_{ox} is the total gate capacitance. Consequently in vital sign detection system, the last mixer stage at IF which down-converts signal to baseband normally adopts passive mixers with large transistor sizes to minimize the flicker noise presented at the outputs.

Dealing with harmonics and intermodulation is another major challenge for 60-GHz vital sign detection. For human heartbeat and respiration, $x(t)$ in equation (2) can be modeled as:

$$x(t) = x_r(t) + x_h(t) = m_r \sin(2\pi f_r t) + m_h \sin(2\pi f_h t) \tag{10}$$

where m_r and m_h are the amplitudes and f_r and f_h are the frequencies of the respiration and heartbeat. First, human m_r is typically 1-5 mm which is comparable to the wavelength at 60GHz. The system response is not linear since the small angle approximation in equation (3) is no longer valid. This introduces multiple respiration harmonics on spectrum which increases detection difficulty. Second, human heartbeat m_h is usually one order of magnitude smaller than m_r, making it easily be overwhelmed by respiration harmonics on spectrum. Figure 2 shows a simulated comparison between 6-GHz and 60-GHz detection results after CSD. In Figure 2 (B), the relatively small heartbeat peak is blocked by the harmonics of respiration even without the presence of system and environmental noise. It should be noticed in the nonlinear system, the magnitudes on spectrum (RR and HR) are not proportional to m_r and m_h. They are determined by the two-tone intermodulation analysis using Bessel function [25]. Certain m_r shows very small magnitude on spectrum which is not able to be distinguished. As a result, signal processing techniques [26] and algorithm [23] were developed to ensure accurate 60-GHz Doppler radar detection.

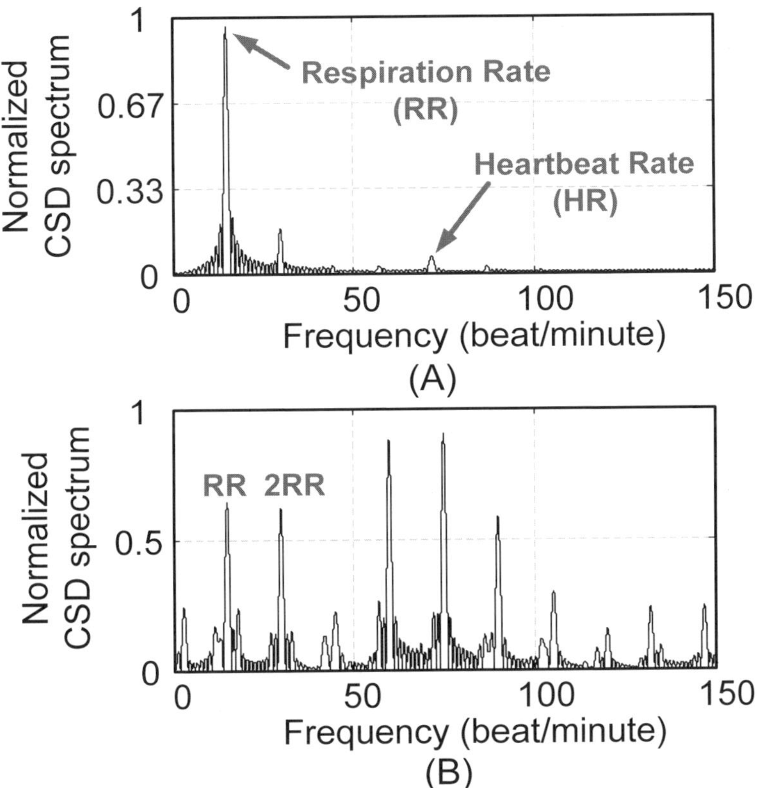

Figure 2. Simulated comparison of vital sign detection using (A) 6-GHz and (B) 60-GHz Doppler radar system.

Figure 3 shows an overview of the system including the CMOS radar chip and RT/duroid 5870 laminate. RF pads on-chip and the metal traces on the laminate are designed for mm-wave flip-chip integration. The

Tx containing two VCOs, an up-conversion mixer, a balun, and a driver is designed to transmit an un-modulated, 60-GHz continuous wave (CW) signal through the Tx antenna. The signal is reflected and phase modulated by a small vibration of the target due to Doppler effect, and then received by the Rx antenna. For the Rx, the weak received signal is amplified by a 60-GHz LNA and down-converted by the same VCOs. Since the same VCOs are used for Tx and Rx, the range correlation effect [8] results in significant reduction of VCO phase noise at baseband outputs, and thus free-running VCOs can be adopted in the system.

For the chip floor planning, the active RF mixers are designed to be very close to RF VCO, and the antennas are placed closely on the same side of the chip to minimize loss and power consumption of 54-GHz LO distribution. It also prevents the dc/baseband connections from interfering with antennas. The G(ground)-S(signal)-G-S-G transition achieves impedance match and provides enough Tx/Rx isolation to reduce dc offset at baseband. By using the same 150-μm pitch and 50-Ω impedance interface, on-wafer (for chip) and on-board (for antennas) probing measurements can be conducted separately. The technique enables the antennas and flip-chip transition to be designed at the measured optimal frequency of transceiver chip. It ensures the fabrication yield against the modeling error and under-estimation of parasitics which are very common at mm-wave frequencies.

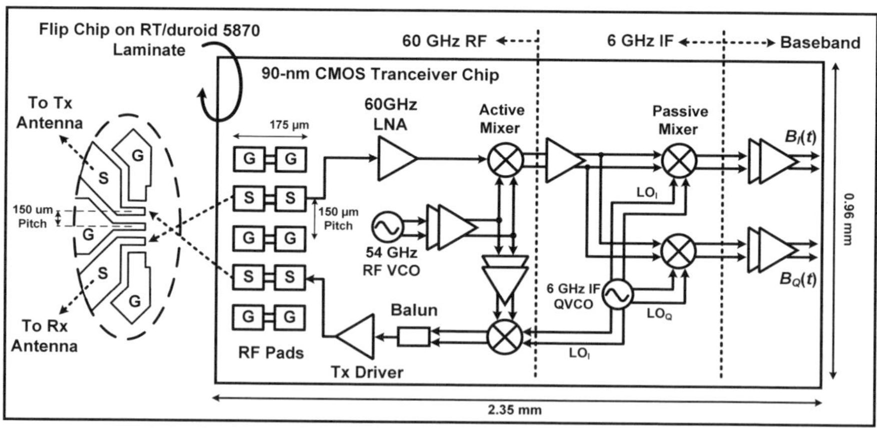

Figure 3. Top view of 60-GHz CMOS micro-radar system. (Not in scale)

1.2.2. Receiver RF circuit fabrication

Passive circuit components were used throughout the chip for various impedance transformation needs. Transmission line theory is one of the methods to design the passive components to have very well-control EM field distribution and characteristic impedance. It generally occupies larger area due to low inductance/capacitance per unite length. From Smith Chart point of view, impedance match using transmission lines usually requires to travel through a long circular path. The other approach is to model the passive components by lumped resistor (R), inductors (L), and capacitors (C), and it is important to confirm that the structure of interest is much smaller than λ. At 60-GHz frequency range, λ is estimated to be around 1.5-2.5 mm, depending on the relative dielectric constant on chip. For an inductor around 150 um long, for example, it is safe to model it by lumped elements using double-π model [27] which can be easily included in active circuit simulation. The models reveal physical meanings such as loss and parasitics and provide more insights to designers than s-parameter boxes do. Besides, passive components modeled by lumped elements is normally very compact due to high inductance/capacitance per unite length. It require accurate 3-D EM simulation to be individually optimized.

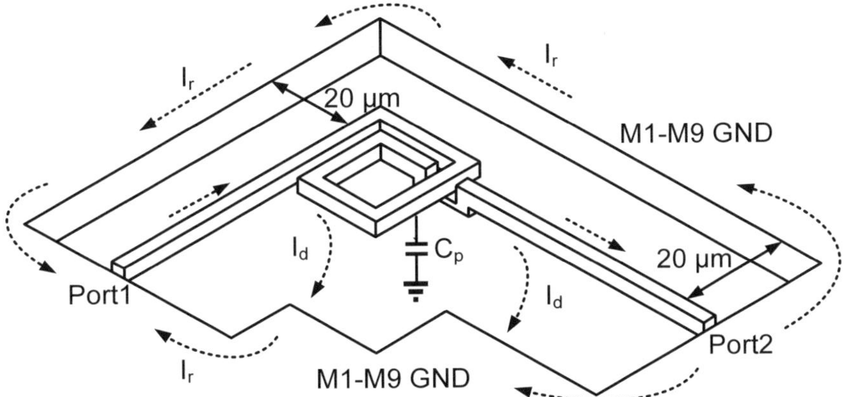

Figure 4. Simulation of a 1.5-turn, 95-pH inductor along with the surrounding ground plane. The return current (I_r) and displacement current (I_d) illustrate the EM interactions between the ground plane and the inductor.

The 3-D EM simulators such as Ansys HFSS are extensively used for mm-wave design. Ideally it is desired to include all 3-D structures of interest in a single simulation, so that the EM interaction among the components can be captured. However, it is sometimes not feasible due to computational limitation. To simulate a single component, a rule of thumb is to include the surrounding ground (gnd) shape, so that a confined EM field distribution and return current path can be obtained to mimic the real chip situation. Figure 4 shows a 95-pH inductor at 60 GHz. As excited from port 1, the return current I_r seeks the lowest impedance path back to the source and contributes negative mutual inductance. The loss mechanism includes conductor loss due to current crowding [28], induced Eddy current (H-field magnetic coupling) on substrate and nearby conductors, and displacement current I_d (E-field capacitive coupling) flowing to nearby ground. In the 90 nm CMOS process, the inductor is implemented on the top metal layer (M9) to reduce the loss to substrate. If a ground plane on bottom metal (M1) is placed underneath the inductor, the inductance density drops and parasitic capacitance (C_p) increases, which degrade the quality factor (Q) [27] of the inductor. Thick ground wall stacking from p-type substrate all the way to M9 is used to minimize the parasitic resistance and inductance on the chip ground. Figure 5 (A) shows a die photo of the L-shaped inductor and Figure 5 (B) presents a circular inductor which is useful in a situation that port 1 and port 2 are very close to each other. This configuration can also be used as a high-Q differential inductor for VCO resonant tank if the center tap is applied.

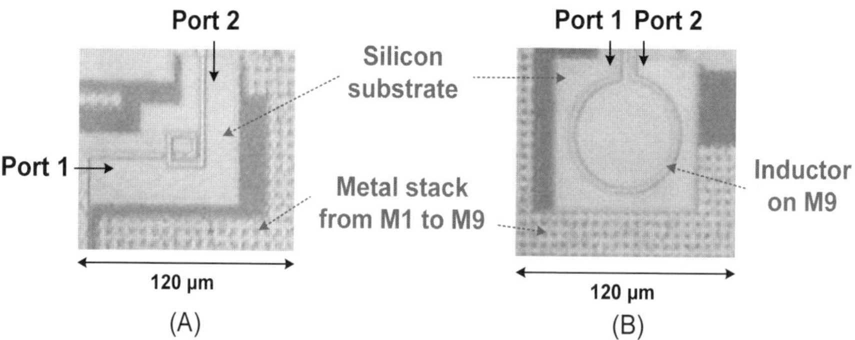

Figure 5. Microphotograph of the on-chip inductors used in the 60-GHz front-end. (A) 95-pH, L-shaped inductor (simulated Q ≈ 18). (B) 128-pH circular inductor (simulated Q ≈ 19.5).

Building a high-gain LNA at the first stage of receiver is critical to the radar Rx sensitivity. In Figure 6 V_{in} coming from the single-ended Rx patch antenna, the Rx front-end is composed of a 60-GHz five-stage cascode LNA, a single-balanced mixer, and a 54-GHz VCO. Transistors were biased at the current density around 0.2 mA/μm for the optimal f_{MAX} and noise performance [29], and extra parasitic capacitance across the gate, drain, and source due to interconnects was estimated by a 3-port extraction in HFSS and included in the circuits simulation [30]. L_{g1} and L_{s1} are used to match the impedance to 50-Ω flip-chip transition. Stage 2 - 4 are identical while inter-stage conjugate matching was achieved by designing the values of C_{g2} to C_{g5}. In this manner the 95-pH inductor in Figure 4 can be reused in $L_{p1} - L_{p4}$ and $L_{d1} - L_{d4}$ which greatly reduce the custom layout time.

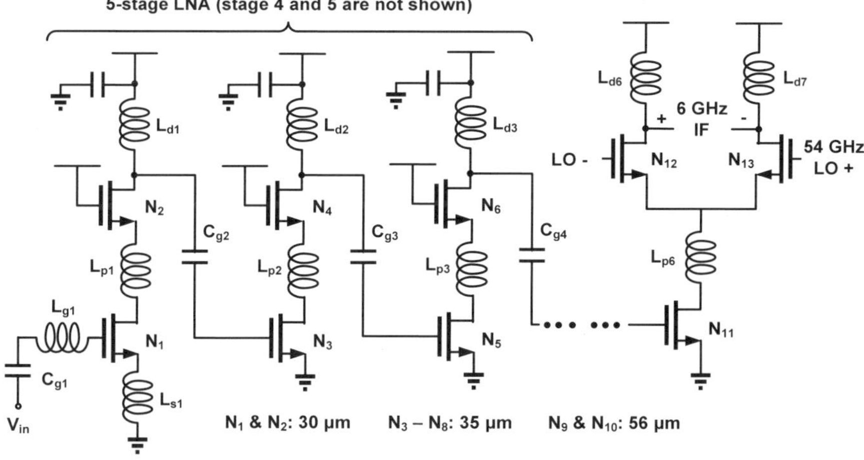

Figure 6. Rx front-end (60 GHz to 6 GHz) including the 5-stage LNA, single-balanced mixer, and 54-GHz LOs (VCO and LO distribution details are shown in Figure 8).

As presented in Figure 7, the 1.5-turn, L-shape inductors are compactly arranged to make best use of the area, and the surrounding gnd walls help to maintain low coupling level between adjacent inductors. The 5-stage cascode layout consumes a total area of 0.16 mm². Simulation shows the LNA provides a 38-dB gain and 5.2-dB NF at 60 GHz while consuming 38 mA at a 1.2 V power supply (Vdd).

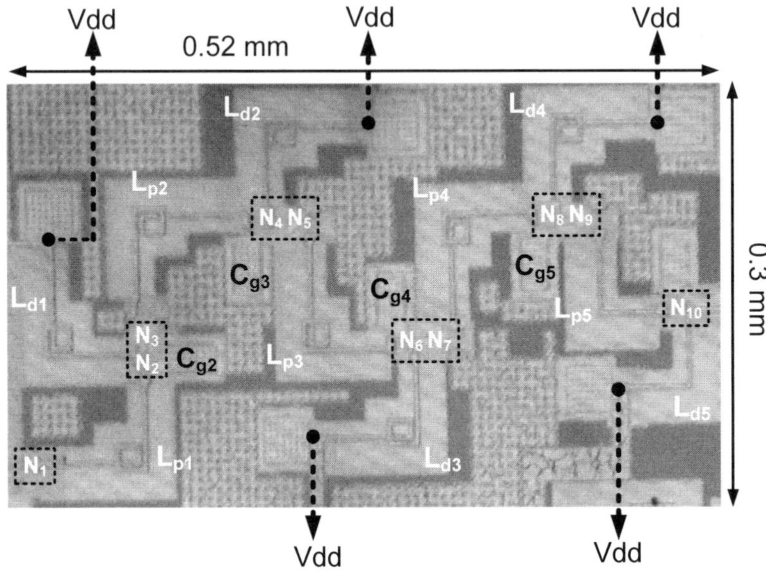

Figure 7. 5-stage LNA layout arrangement and easy access to the power supply (Vdd).

The single-balanced mixer down-converts the 60-GHz signal to 6-GHz IF and also serves as a single-ended to differential conversion for the following stages. The 54-GHz LO feed-through is far away from the down-converted signal on the spectrum and also greatly attenuated by the load resonant tank of the mixer. Figure 8 shows the 54-GHz VCO tuned by accumulation-mode varactors [31] with LO buffers to drive up- and down-convert mixers. The fabricated chip is presented in Figure 9. Simulated tuning range covers from 51.6 to 54.9 GHz, and the phase noise is -101 dBc/Hz (at 1-MHz offset) at 54 GHz. The VCO core consumes a total power of 18 mW.

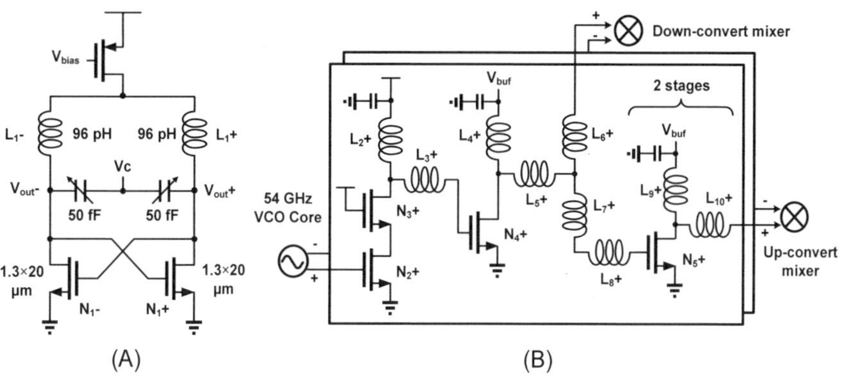

Figure 8. 54-GHz RF LO generation and distribution. (A) 54-GHz VCO core. (B) Differential LO distribution buffer network to up- and down-convert mixers.

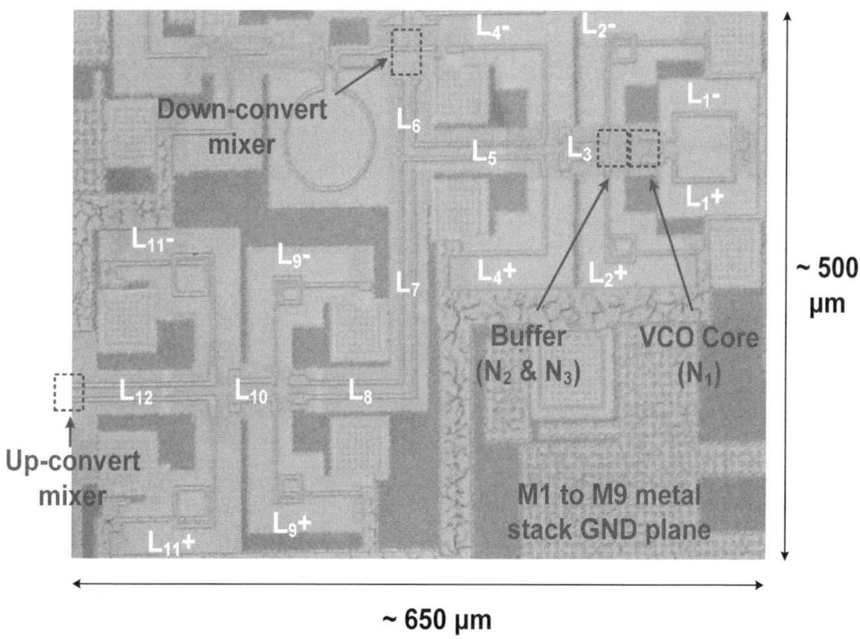

Figure 9. 54-GHz LO distribution network from RF VCO to up- and down-convert mixers.

At a total power consumption of 110 mW, the receiver front-end is designed to provide a single-ended conversion gain of 38 dB when one of the IF outputs terminated by 50-Ω. Simulated input 1-dB compression/ point (P_{1dB}) is at -44 dBm which is much higher than the expected radar received power.

1.2.3. Transmitter RF circuit fabrication

As explained in the previous session, Tx shares the same RF and IF VCOs as Rx to utilize radar range correlation effect [8] so that no phase-locked-loop is needed to reduce phase noise. Figure 10 shows the Tx front-end design. A double-balanced mixer is used to accommodate the differential signals from the RF and IF VCOs and convert the 60-GHz outputs to single-ended by a balun load. A three-stage tuned amplifier following the balun is used to drive the Tx patch antenna. The lump-element-modeled transformer balun in Figure 11 occupies relatively small area. The primary inductor (L_p) is implemented on the high aluminum pad layer to reduce the loss to substrate, while M9 is used for the secondary inductor (L_s) to form a vertically coupled structure. The vertical transformer usually shows less magnetic flux leakage and higher coupling coefficient (k) compared to a planar coupled one. The transformer balun provides the conversion between differential impedance Z_s and single-ended impedance Z_p [32]:

$$\frac{Z_s}{Z_p} = \frac{n^2}{1} = \frac{L_s}{L_p}. \tag{11}$$

Here n is conventionally defined as the turn ratio. Adjusting the value of L_p and L_s achieves impedance match. Based on EM simulation, the differential to single-ended insertion loss of the passive balun is around 5 dB at 60 GHz.

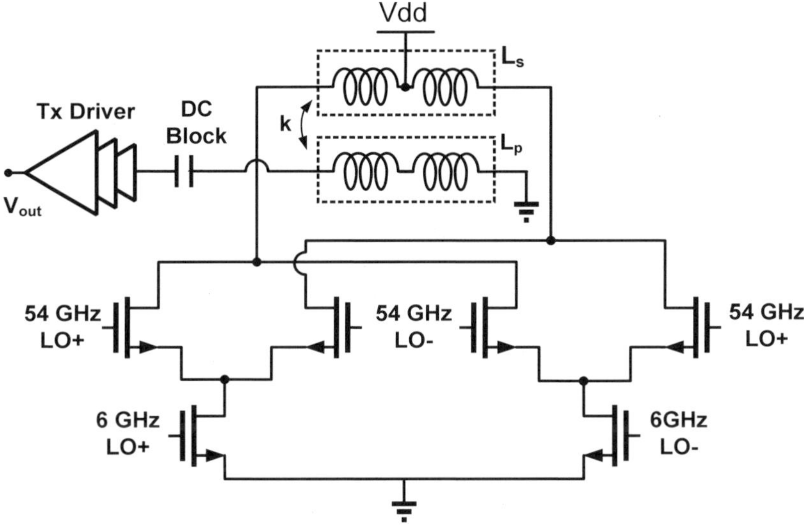

Figure 10. Tx front-end (6 GHz to 60 GHz) using the double-balanced up-convert mixer, transformer balun, and three-stage driver at 60 GHz.

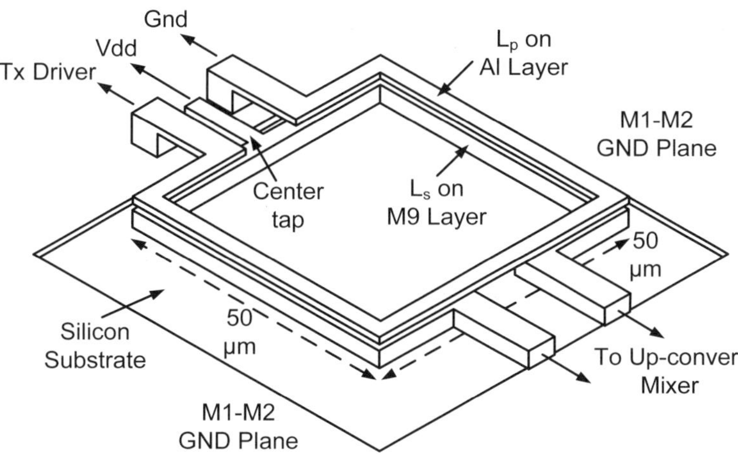

Figure 11. Transformer balun structure on chip for differential to single-ended conversion.

Since Doppler radar transmitting non-modulated signal as shown in equation (1) and no information is contained in amplitude, the three-stage TX driver can operate in the non-linear region to maximize the transmitted power. In the simulation, the power level from IF VCO is -10.7 dBm (differential) at 6 GHz, and it is up-converted to 60 GHz with the power level boosted to 2 dBm (single-ended) by the mixer and driver. The conversion gain from IF to RF is 12.7 dB, and the Tx front-end consumes 35-mW power and 0.13-mm^2 area, which are both less than 10% of the overall system consumption.

1.2.4. IF and baseband

The I/Q separation is performed at IF stage as the quadrature ring VCO drives two passive mixers for low flicker noise at baseband outputs in Figure 12. The two-stage ring VCO is able to generate four output phases ($0°$, $90°$, $180°$, and $270°$) [33]. The differential I and Q LO signals with two-stage buffers are able to drive large passive mixers, while I LO signals are also needed by the up-convert mixer in the Tx front-end. The buffers (A_I and A_Q) have separate power supply (V_I and V_Q) to compensate possible I/Q amplitude mismatch due to the different loads. The two passive mixers with large transistor sizes (75 μm) avoid transconductor stage and dc bias current to minimize the flicker noise at the baseband outputs. The simulated phase noise is -83 dBc/Hz (at 1-MHz offset) when the oscillating frequency is at 6.4 GHz, and total power consumption of the IF stage is 99 mW.

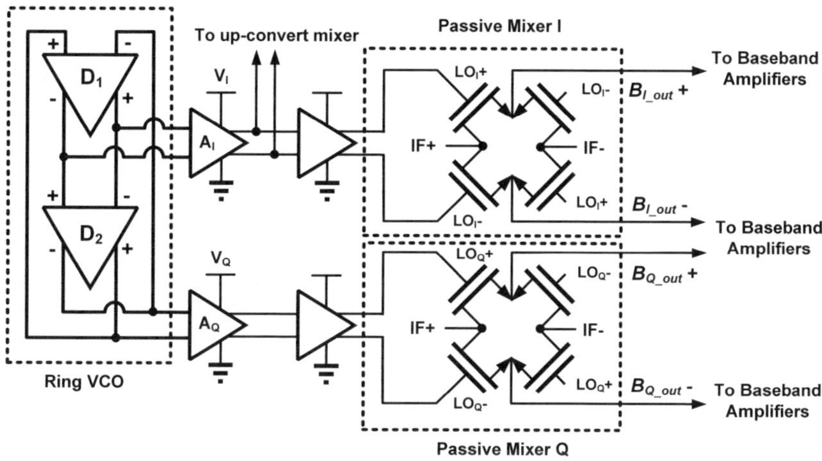

Figure 12. IF stage (6 GHz to dc) including the quadrature ring VCO, IF LO buffers, and passive mixers.

The final radar chip in 90 nm CMOS is shown in Figure 13. The 60-GHz RF front-end is on the left and the down-converted baseband I/Q outputs are on the right. The area of the 60 GHz RF core is 0.73 mm^2. It can be seen that baseband amplifiers are also integrated on chip. They provide 30-dB gain so that the outputs can be directly sampled by the oscilloscope or ADC. All the dc biases are supplied from the dc pads on the top and bottom sides and do not interfere with the RF antennas.

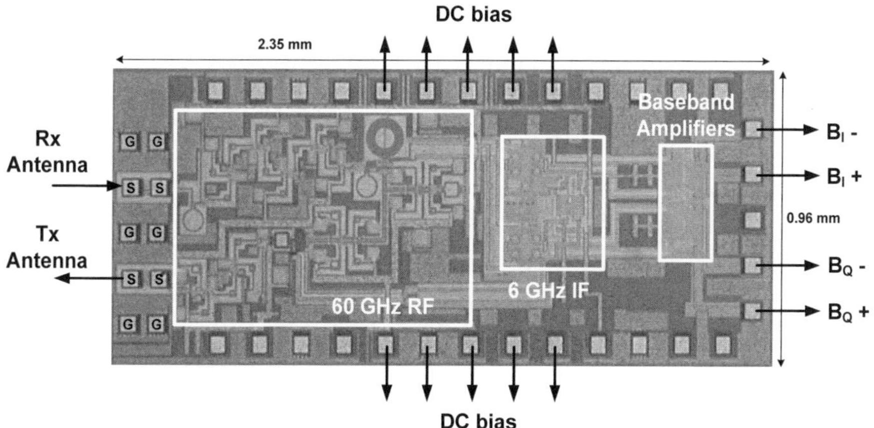

Figure 13. Microphotograph of the 60-GHz CMOS micro-radar chip.

1.3. Millimeter-Wave System in Package Design and Fabrication

1.3.1. Flip-chip packaging

To enable radars operating at higher frequency, a transition structure between the transceiver chip and antennas is designed to target low insertion loss, 50-Ω impedance match, and high isolation between Tx and Rx. On-chip antennas [15] can be used to have minimum transition loss, as the antenna size is usually small enough around 60 GHz. However, they generally suffer from low radiation efficiency due to silicon substrate and thus limited patch antenna gain. Another drawback is narrow antenna bandwidth which is caused by thin dielectric thickness on chip [34]. To use PCB antennas with superior performance compared to on-chip ones, custom mm-wave packaging techniques such as [20] successfully built the supporting structure to horizontally align the chip and antenna, which greatly reduce the parasitic inductance.

Flip-chip packaging is widely used in current IC industry not only for high-density digital chips but also for RF interconnection. As illustrated in Figure 14 (A), traditional bonding wire could be several hundreds of micrometers and the high parasitic inductance in nano-Henry range is usually not acceptable for mm-wave applications. Bumping and flip-chip attachment in Figure 14 (B) have been proven to be a simple and low-cost approach. Once the chip is flipped on the PCB, the bonding is established by the re-flow of solder bumps under heat and pressure. The series inductance of the transition is usually in 100-200 pH range, and the insertion loss is typically less than 1dB at 60 GHz.

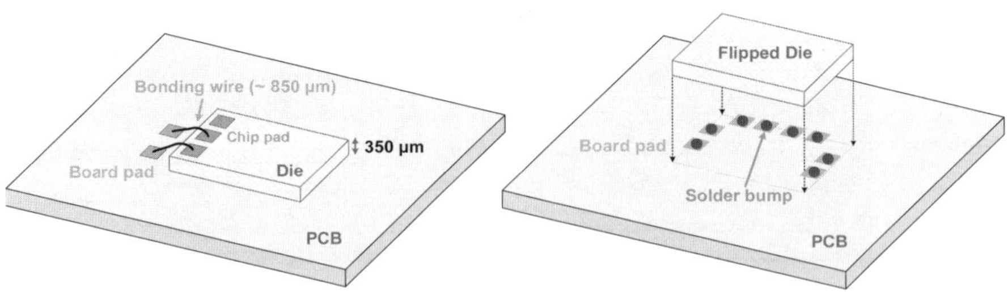

Figure 14. (A) Typical wire-bonding packaging configuration. The parasitic inductance of the wire is usually in nano-Henry range at around 10 GHz. (B) Flip-chip attachment.

The flip-chip transition design on the RT/duriod 5870 laminate is described in Figure 15. Two bumps are used on each RF signal pad to reduce the series inductance and resistance, and similarly three pairs of bumps are placed on RF ground pads. Since the width of 50-Ω feed-line on PCB is around 350 μm and the pitch of chip pads is only 150 μm, the taper structure is designed for the smooth transition. The increased inductance due to the taper structure is capacitively compensated by controlling the length d_1 and d_2 to maintain 50-Ω impedance match. The thin RF laminate (127 μm) is chosen to reduce surface wave loss at the cost of antenna bandwidth [34]. Actually for Doppler radar application, the antenna does not require large bandwidth since single frequency is transmitted and the limited bandwidth also helps to remove unwanted harmonics and images. Using thin PCB substrate also helps to narrow the width of 50-Ω transmission lines for easier flip-chip transition design and reduces the spurious radiation. A thick FR4 PCB is needed underneath to support the soft RT/duriod 5870 laminate.

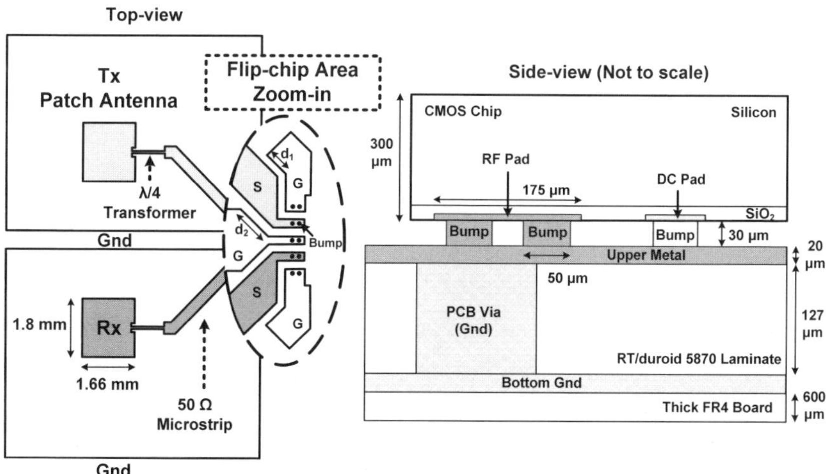

Figure 15. Flip-chip transition design between the 60-GHz CMOS radar chip and PCB patch antennas on RT/duroid 5870 laminate.

Microphotograph in Figure 16 shows the actual flip-chip area on the laminate. In the flip-chip process, two solder bumps are placed on each

RF copper trace (G-S-G-S-G), and one solder bump is placed on each dc bias trace. All the copper traces are partially covered by the solder mask to control the reflow of solder when the heat and pressure are applied. It should be noticed that a through hole under the chip is needed to maintain the EM characteristics in air.

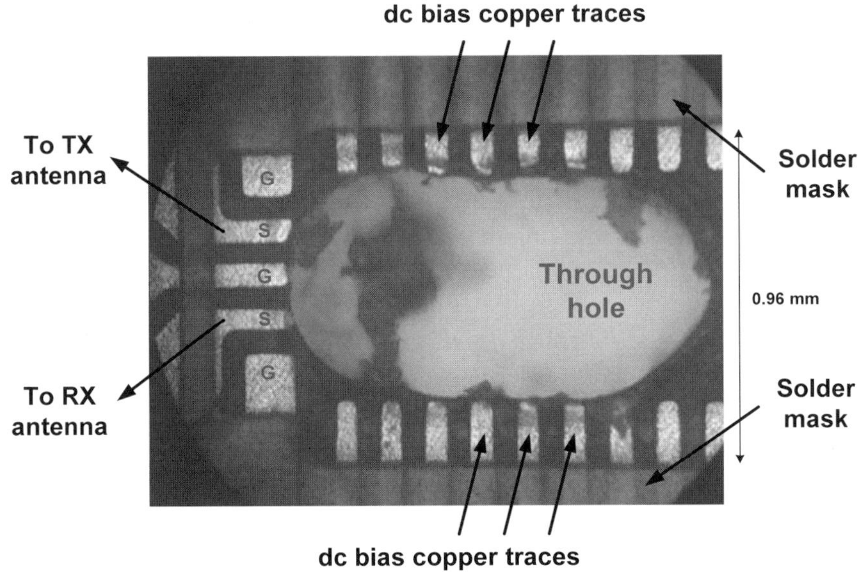

Figure 16. Microphotograph of the flip-chip area on RT/duriod 5870 surface. The solder mask in dark green is deigned to control the reflow of solder bumps during the flip-chip process. The hole penetrates through the RT/duriod 5870 and FR4 supporting board.

1.3.2. Patch antenna

The Tx and Rx path antennas were designed at the optimal transceiver operating frequency (55GHz) which will be shown in the next session. Figure 17 (A) shows the simulation setup of antennas after the chip is flipped on the laminate. The ports are placed on the chip as port 1 is for Tx and port 2 is for Rx. By using the orthogonal antenna feed-lines and G-S-G-S-G arrangement, the isolation between port 1 and port 2 reaches -34 dB at 55 GHz in Figure 17 (B) even the Tx and Rx antennas are placed closely to each other. Minimizing the un-modulated signal directly coupled from Tx to Rx reduces the dc offset at baseband outputs. The simulated gain of the single patch antenna is around 5 dBi as shown in Figure 18.

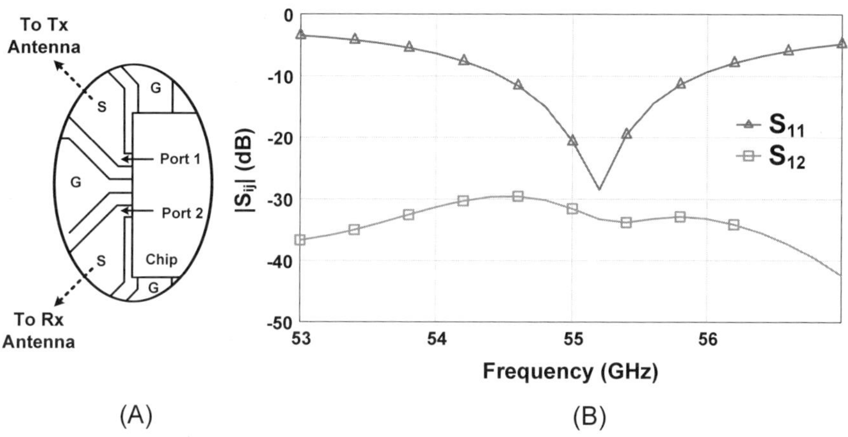

Figure 17. (A) Antenna simulation setup (B) Patch antenna s-parameter with flip-chip packaging. S11 shows the input matching and S12 represents the isolation between Tx and Rx ports.

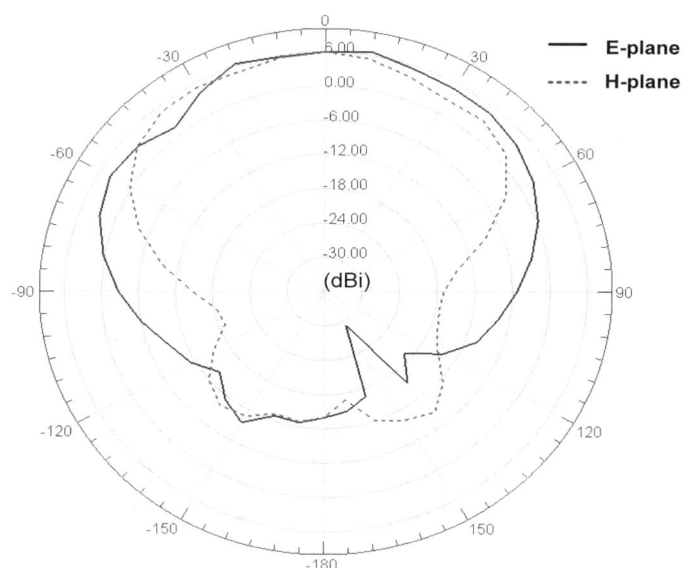

Figure 18. Simulated patch antenna pattern after the flip-chip packaging.

1.4. Radar System Test

1.4.1.   Receiver Front-end Measurement

The gain of 60-GHz receiver front-end is one of the key factors impacting total receiver noise figure and sensitivity. In practice, it is common to see the optimal operating frequency deviates from original simulation due to component modeling limitation and inaccuracy at mm-wave frequencies. Consequently it is critical to perform on-die measurement to characterize RF front-end as the first step of system verifications. Figure 19 shows the simulated and measured down-conversion (60 GHz to 6 GHz) gains of the receiver front-end. A separate test structure which only contains the 60-GHz RF core (LNA, RF mixer, and RF VCO) was fabricated on die for the measurement purpose. A G-S-G probe was landed at LNA input and another G-S-G probe was placed at one of the differential mixer outputs, while the other mixer output was terminated by a 50-Ω on-die resistor. Several DC probes were used to provide power supply, DC bias, and VCO control voltages. The input power at the LNA probe was set at a low level (-60 dBm) to emulate the weak reflected signal received by the Rx antenna and avoid gain compression.

At a total 190 mW power consumption (1.2 V power supply), the Rx front-end provides more than 30 dB down-conversion gain from 52 GHz to 56.5 GHz. The peak down-conversion gain is 36 dB with the RF VCO operating at 48.4 GHz. The tuning range of the RF VCO is from 48 to 51 GHz. In the first pass, the reduced peak gain and shifted peak frequency from 60 GHz to 55 GHz are possibly due to several reasons. For example, the modeling of active and passive devices might under-estimates the parasitic capacitance and inductance. Also the current return path assumed in the EM simulation setup (Figure 4) is different from the complex situation on the actual chip, causing the discrepancy in inductor values.

The measurement reveals that the optimal operating frequency is centered around 55GHz, at which the flip-chip transition and antennas can be designed accordingly. The input P1dB of the receiver front-end is measured at -42 dBm which is much higher than the expected received radar signal.

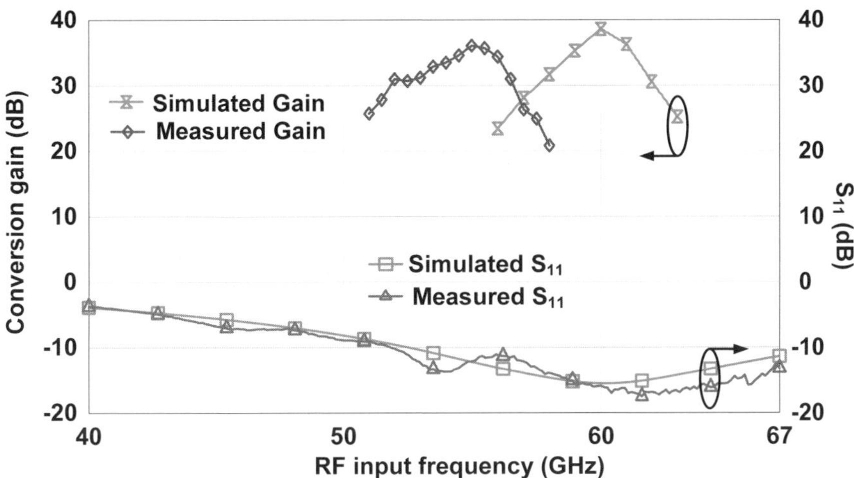

Figure 19. Measured down-conversion gain (60 GHz to 6 GHz) versus RF input frequency. The RF input power was set at -60 dBm to emulate the weak reflected radar signal.

1.4.2. Transmitted Power Test

Once the optimal Rx operating frequency is measured, the Tx frequency is identical since the same VCOs are used for up- and down-convert. To verify that the transmitted power is near expected level, the following test measures the power level (P_t) transmitted by Tx patch antenna. Figure 20 shows the micro-radar system-in-package including the CMOS transceiver chip, two patch antennas, and dc bias through the blue wires. Bypass capacitors with large capacitance value (22 µF) are used on the PCB to suppress the power supply noise. The weight of the system shown in the photo is less than 10 gram (0.3 ounce) and can be easily pasted to an upright cardboard facing the target in the experiments. Figure 21 and Figure 22 show the experimental setup including a horn antenna with a gain (Gr) of 23 dBi, waveguide-to-coaxial-cable adapters, a 50 − 75 GHz down-convert mixers, and E4448A spectrum analyzer. The spectrum analyzer provides built-in LO to down-convert the 55 GHz signal to IF.

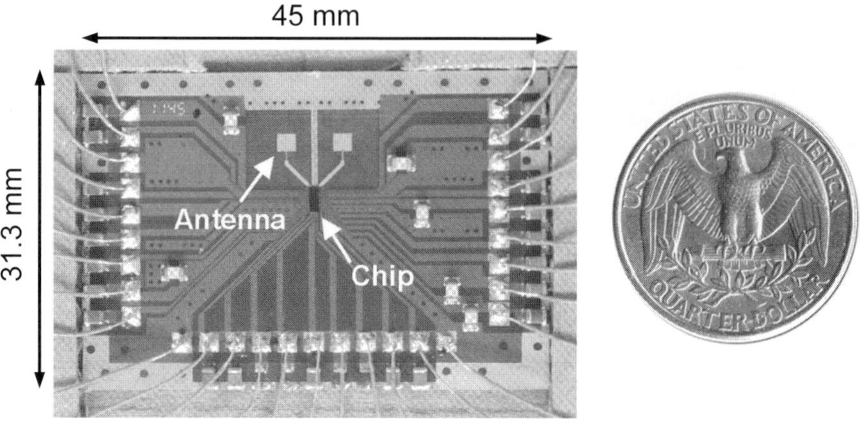

Figure 20. The final system assembly of the 60-GHz micro-radar system-in-package including the CMOS transceiver chip, two PCB patch antennas for Tx and Rx, and dc biasing wires.

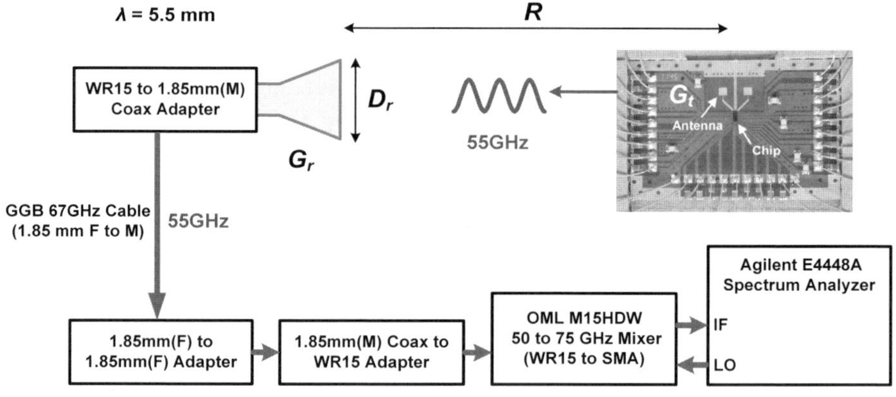

Figure 21. Experimental setup for the TX output power of CMOS transceiver chip.

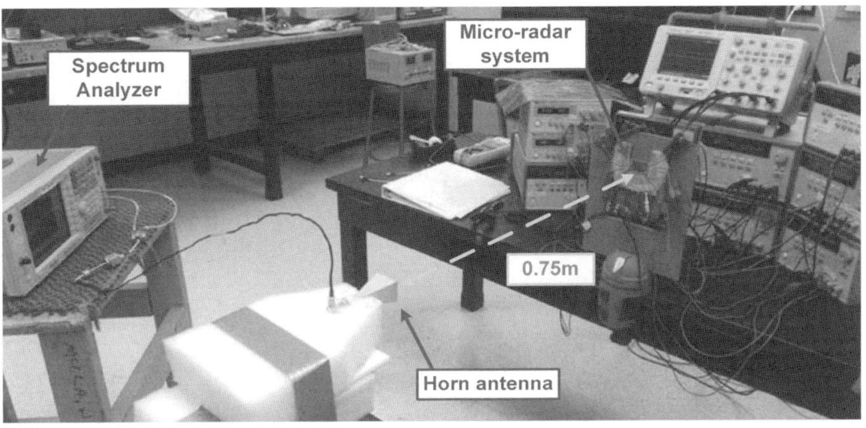

Figure 22. Photo of the experimental setup to test TX transmitted power of CMOS micro-radar chip. The distance R is chosen to be 0.75 m to satisfied far-field condition.

A 55-GHz power source was used to calibrate the loss of the down-convert path. Without the horn antenna, a known power level was sent through the path so that the total loss (L) of the adapters, cable, and mixer was measured to be 42 dB. Besides, the experiment setup needs to maintain far-field condition, so that Friis transmission equation can be used to calculate the transmitted power. The far-field condition is summarized as follows [35]:

$$R > \frac{2D^2}{\lambda} \qquad (12.a)$$

$$R \gg D \qquad (12.b)$$

$$R \gg \lambda \qquad (12.c)$$

where D is the maximum dimension of the antennas. For the patch antenna, D is around 1.8 mm as shown in Figure 3, and the horn antenna dimension is 40 mm. Based on equations (12), R was chosen to be 0.75 m to ensure the far-field condition at 55GHz. It should be noticed that in many of the indoor, short-distance vital sign detection applications, the distance often results in the near-field condition due to the relatively large reflector (human body). Friss transmission equations (13) or radar range equation (6) are normally not applicable to estimate the parameters such as detection distance and reflected power. It mainly relies on experiments to characterize the system.

Figure 23 shows a screenshot of the received power P_r = -82.24 dBm on spectrum analyzer. The resolution bandwidth was lowered from default 3 MHz to 100 KHz to reduce the noise floor. This helps to distinguish the small received signal on the spectrum at a cost of longer refresh time. Friis transmission equation including the down-convert loss L can be used to calculate the transmitted power P_t:

$$P_r = P_t + G_r + G_t - L + 20 \cdot \log(\frac{\lambda}{4\pi R}). \tag{13}$$

The transmitted power P_t is estimated to be about -2.56 dBm assuming the Tx patch antenna gain G_t = 4 dBi. The experiments verifies that the transmitter of CMOS micro-radar chip works properly and is ready to be used for the vibration and vital sign detection tests.

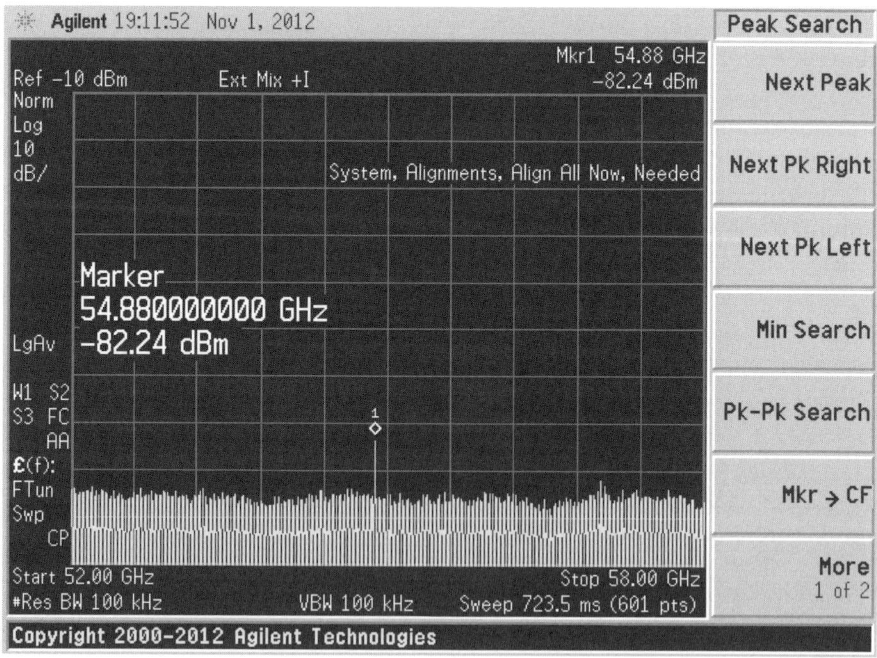

Figure 23. Received power P_r = -82.24 dBm on the spectrum analyzer.

1.4.3. Mechanical vibration detection

To verify the I/Q separation by the quadrature receiver, a 0.15 m-by-0.15 m flat metal plate was attached to a Zaber T-LA60A-S actuator and placed at a distance $D = 0.3$ m from the radar. A 1-Hz mechanical vibration was generated by the actuator to test the I and Q operations. In Figure 24 (A), the differential Q channels are near the optimal detection point as the baseband output waveforms (B_Q+ and B_Q-) show the successful detection and the vibration frequency (1 Hz) can be easily read from the time axis. On the other hand, the differential I channels (B_I+ and B_I-) are near the null detection point. The output waveforms near null detection point have smaller amplitude and do not contain the fundamental tone of the vibration. Figure 24 (B) shows the opposite scenario at a slightly different D which changes the total residue phase ϕ_t. This result indicates the I/Q generation of IF quadrature VCO with passive mixers are working properly, and thus the CSD in equation (8) can be used to ensure robust detections.

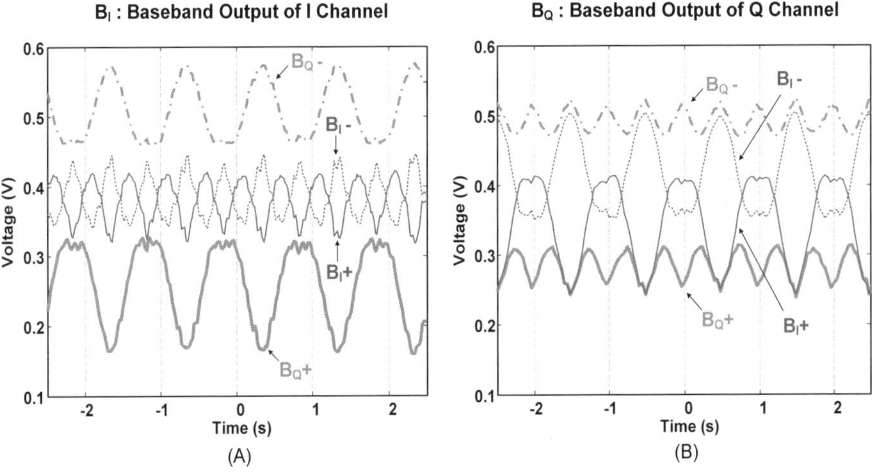

Figure 24. (A) Q is near optimum detection point and I is near null detection point. (B) I is near optimum detection point and Q is near null detection point.

1.4.4. Human vital sign detection

In the experiment, the CMOS micro-radar system was pasted to a vertically placed cardboard facing the human target. The heartbeat signals in Figure 25 (A) were directly sampled by the oscilloscope as the person sitting on a chair with his chest wall 0.3 m in front of the radar. The breath was held to avoid the blocking from the harmonics of respiration signal as mentioned previously. The result shows the robustness of the quadrature architecture to avoid the null detection points, which occur every 1.25 mm ($\lambda/4$) in detection distance. At the time t = 0 ~ 7 s, Q channel was around the optimal point while I channel was near the null point. However, the target entered the null point of Q channel after t = 9 s due to some slight body movement, and I channel started to take over the detection. Figure 25 (B) shows the CSD spectrum of detected signal and the heartbeat rate at 69 beats/minute (BPM), which agrees with the result of human counting. The Doppler frequency shift due to body movement also shows up in the spectrum. The total power consumption of the micro-radar SiP is 377 mW at 1.2 V power supply. More experiment data of vital sign and mechanical vibration detection can be found in [14] and [23].

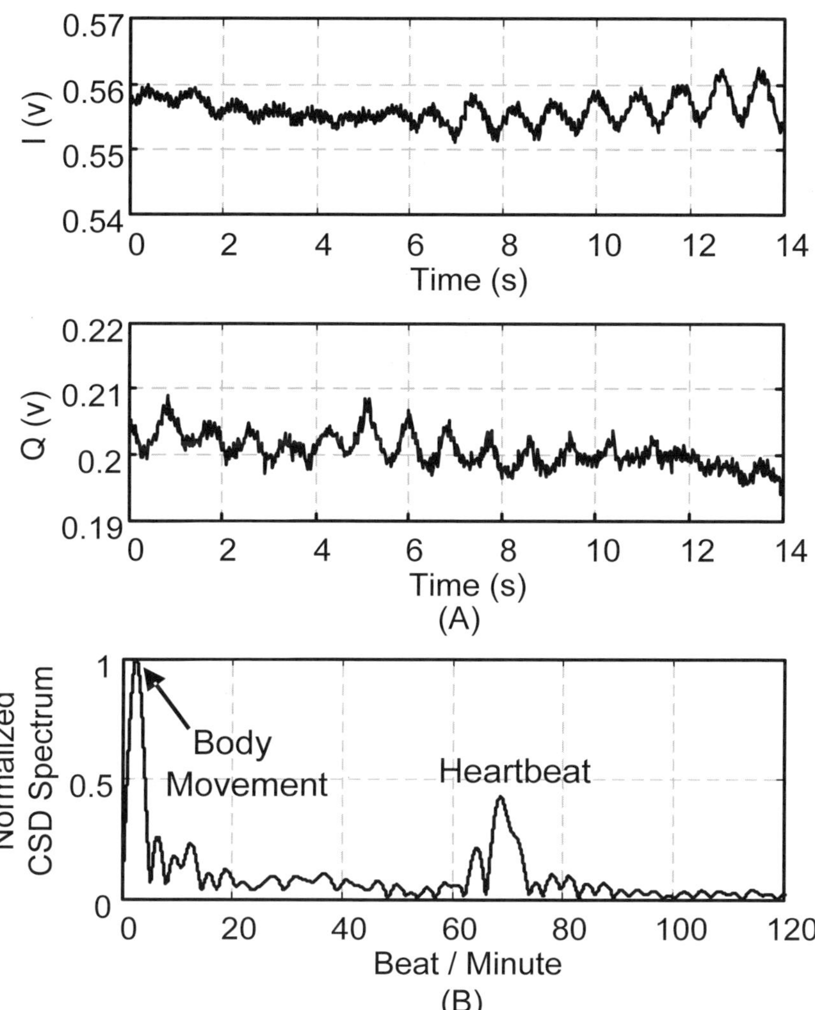

Figure 25. Heartbeat detection using the 60-GHz radar when the target holds the breath at 0.3 m away. (A) The baseband output signals (B) Normalized CSD spectrum showing accurate heartbeat rate at 69 BPM.

1.5. Conclusion

Operating at a short wavelength of 5 mm, the 60-GHz single-tone CW radar is able to detect small vibrations of 0.02 mm from 30 cm away or 0.2 mm from 2 m away [14], and is shown to detect human heartbeats while the subject is holding the breadth. Because of the large displacement of respiratory movement and the small displacement of heartbeat movement result in signals of large dynamic range, proper signal processing is needed to simultaneously measure both the respiration and heartbeat [23], [26]. When both the respiration and heartbeat displacements are very small, the small wavelength of 60-GHz radar is a perfect match and can potentially become a very useful tool in biomedical research. It was shown in [36] that the 60-GHz radar can measure the respiration and heart rates of a laboratory rat and monitor the variations over time. In addition, the nonlinear Doppler phase demodulation method enables the simultaneous measurement of respiratory and heartbeat displacements. [36] Demonstrated for the first time, the simultaneous measurement of both respiration and heart rates and their displacements in an animal, without using any contact sensor and wire.

Reference

1. A. D. Droitcour, V. M. Lubecke, J. Lin, and O. Boric-Lubecke, "A microwave radio for Doppler radar sensing of vital signs," IEEE MTT-S Int. Microw. Symp. Digest, pp. 176-178, May, 2001.

2. C. Li, X.Yu, C. Lee, L. Ran, and J. Lin, "High-sensitivity software configurable 5.8 GHz radar sensor receiver chip in 0.13 μm CMOS for non-contact vital sign detection," IEEE Trans. Microw. Theory Techn., vol. 58, no 5, pp. 1410-1419, May 2010.

3. C. Li, and J. Lin, "Non-contact measurement of periodic movements by a 22-40 GHz radar sensor using nonlinear phase modulation," IEEE MTT-S Int. Microw. Symp. Digest, pp. 579-582, June 2007.

4. C. Li, and J. Lin, "Complex signal demodulation and random body movement cancellation techniques for non-contact vital sign detection," IEEE MTT-S International Microwave Symposium, pp. 567-570, Atlanta, June, 2008.

5. S. Nicolson, P. Chevalier, A. Chanter, B. Sautreuil and S. P. Voinigescu "A 77–79 GHz Doppler radar transceiver in silicon", IEEE Compound Semiconduct. Integr. Circuits Symp., pp.1, 2007

6. K. Abe, K. Otsuka, and J.-Y. Ko "Self-mixing laser Doppler vibrometry with high optical sensitivity: Application to real-time sound reproduction," New J. Phys., vol. 5, no. 1, pp. 8.1-8.9, 2003.

7. S. Kim and C. Nguyen "On the development of a multifunction millimeter-wave sensor for displacement sensing and low-velocity measurement", IEEE Trans. Microw. Theory Techn., vol. 52, no. 52, pp.2503 -2512, Nov. 2004

8. A. D. Droitcour, O. Boric-Lubecke, V. M. Lubecke, J. Lin, and G. T. A. Kovac, "Range correlation and I/Q performance benefits in single-chip silicon Doppler radars for noncontact cardiopulmonary monitoring," IEEE Trans. Microw. Theory Techn., vol. 52, no.3, pp. 838-848, Mar. 2004.

9. H. Chuang, H. Kuo, F. Lin, T. Huang, C. Kuo, and Y. Ou, "60-GHz millimeter-wave life detection system (MLDS) for noncontact human vital-signal monitoring," IEEE Sensors Journal, vol. 12, issue 3, pp. 602-609, March 2012.

10. D. T. Petkie, C. Benton, and E. Bryan, "Millimeter wave radar for remote measurement of vital signs," IEEE Radar Conference, pp.1-3, May 2009

11. C. Balanis, Antenna Theory Analysis and Design. Hoboken, NJ: John Wiley & Sons, Inc., 2005, ch. 2, pp. 110-112.

12. E. Knott, Radar Handbook. New York, NY: McGraw Hill Company, 2008, ch. 14, pp. 10.

13. Y. Xiao, J. Lin, O. Boric-Lubecke, and V. M. Lubecke, "Frequency tuning technique for remote detection of heartbeat and respiration using low-power double-sideband transmission in Ka-band," IEEE Trans. Microw. Theory Techn., vol. 54, pp. 2023-2032, May 2006.

14. T. Kao, A. Chen, Y. Yan, T. Shen, and J. Lin, "A flip-chippackaged and fully-integrated 60 GHz CMOS micro-radar sensor for heartbeat and mechanical vibration detections," in Proc. IEEE Radio Frequency Integr. Circuits Symp., Jun. 2012, pp. 443–446.

15. T. Yao, L. Tchoketch-Kebir, O. Yuryevich, M. Q. Gordon, and S. P. Voinigescu "65 GHz Doppler sensor with on-chip antenna in 0.18 um SiGe BiCMOS", IEEE MTT-S Int. Microw. Symp. Dig., pp. 1493-1496, June 2006

16. A. D. Droitcour, O. Boric-Lubecke, V. M. Lubecke, and J. Lin, "0.25µm CMOS and BiCMOS single chip direct conversion Doppler radars for remote sensing of vital signs," IEEE Int. Solid State Circuits Conference, Digest of Technical Papers, pp. 348-349, Feb. 2002.

17. C. Li, Y. Xiao, and J. Lin, "A 5 GHz double-sideband radar sensor chip in 0.18 µm CMOS for non-contact vital sign detection," IEEE Microw. and Wireless Components Letters, vol. 18, issue 7, pp. 495-496, 2008

18. W. Sansen "Analog IC Design in Nanometer CMOS Technologies," Proceedings of VLSI Design 2009: Improving productivity through higher abstraction, The 22nd International Conference on VLSI Design, New Delhi, India, Jan. 2009.

19.   E. Laskin, M. Khanpour, S. T. Nicolson, A. Tomkins, P. Garcia, A. Cathelin, D. Belot, and S. P. Voinigescu, "Nanoscale CMOS transceiver design in the 90-170 GHz range," IEEE Trans. Microw. Theory Techn., vol. 57, pp. 3477-3490, Dec. 2009.

20.   J. Lee, Y. Chen, and Y. Huang, "A low-power low-cost fully-integrated 60-GHz transceiver system with OOK modulation and on-board antenna assembly," IEEE J. Solid State Circuits, vol. 45, no. 2, pp.264-275, Feb. 2010.

21.   A. Tang, Q. J. Gu, and M.-C. F. Chang, "CMOS receivers for active and passive mm-wave imaging," IEEE Communication Magazine, vol. 49, no. 10, Oct. 2011.

22.   B. Razavi, "Design of millimeter-wave CMOS radios: A tutorial," IEEE Transactions on Circuits and Systems I: Regular Papers, vol.56, no.1, pp.4-16, Jan. 2009

23.   T. Kao, Y. Yan, T. Shen, A. Chen, and J. Lin, "Design and analysis of a 60-GHz CMOS Doppler micro-radar system-in-package for vital sign and vibration detection," IEEE Trans. Microw. Theory Techn., vol. 61, no. 4, pp. 1649–1659, Apr. 2013.

24.   B. Razavi, Design of Analog CMOS Integrated Circuits. New York, NY: McGraw Hill Company, 2001, ch. 7, pp. 215

25.   Y. Yan, L. Cattafesta, C. Li, J. Lin, "Analysis of detection methods and realization of a real-time monitoring RF vibrometer," IEEE Trans. on Microw. Theory Techn., vol. 59, no. 12, pp. 3556-3566, Dec 2011.

26.   T. Kao and J. Lin, "Vital Sign Detection Using 60-GHz Doppler Radar System," IEEE International Wireless Symposium, Apr. 2013, pp.1–4.

27.   T. O. Diskson, M. A. LaCroix, S. Boret, D. Gloria, R. Beekens, and S. P. Voinigescu, "30-100 GHz inductors and transformers for millimeter-wave (Bi)CMOS integrated circuits," IEEE Trans. Microw. Theory Techn., vol. 53, no. 1, pp.123-134, Jan. 2005.

28.   Y. Cao, R. A. Groves, X. Huang, N. D. Zamdmer, J. O. Plouchart, R. A. Wachnik, T. J. King, and C. Hu, "Frequency-independent equivalent circuit model for no-chip spiral inductors," IEEE J. Solid-State Circuits, vol. 38, no. 3, pp. 419-426, Mar. 2003.

29.   T. Yao, M. Q. Gordon, K. K. W. Tang, Kenneth H. K. Yau, M. Yang, P. Schvan, and S. P. Voinigescu, "Algorithmic design of CMOS LNAs and PAs for 60 GHz Radio," IEEE J. Solid-State Circuits, vol. 42, no. 5, pp. 1044-1057, May 2007.

30.   C. K. Lian and B. Razavi, "Systematic transistor and inductor modeling for millimeter-wave design," IEEE J. Solid-State Circuits, pp. 450-457, Feb. 2009

31.   C. Cao and K. K. O, "Millimeter-wave voltage-controlled oscillators in 0.13-m CMOS technology," IEEE J. Solid State Circuits, vol. 41, no. 6, pp. 1297-1304, June 2006

32.   J. R. Long, "Monolithic transformers for silicon RF IC design" IEEE J. of Solid-State Circuits., vol. 35, no.9, pp. 1368-1382, Sept. 2000.

33.   W. Yan and H. Luong, "A 900-MHz CMOS low-phase-noise voltage-controlled ring oscillator," IEEE Trans. Circuits Syst. II: Analog Digital Signal Process., vol. 48, no. 2, pp.216-221, Feb. 2001

34.   D. M. Pozar, "Considerations for millimeter wave printed antennas," IEEE Trans. Antennas Propagat., vol AP-31, pp. 740-747, Sept. 1983.

35.   W. Stutzman, and G. Thiele, Antenna Theory and Design. Hoboken, NJ: John Wiley & Sons, Inc., 2005, ch. 1, pp. 30.

36.   T.-Y. Huang, J. Lin, and L. Hayward, "Non-invasive measurement of laboratory rat's cardiorespiratory movement using a 60-GHz radar and nonlinear Doppler phase modulation," in 2005 IEEE International Microwave Workshop Series on RF and Wireless Technologies for Biomedical and Healthcare Applications (IMWS-Bio 2015), Sept. 2015.

2.2. Fractal Patterns Ontogeny

Any fractal pattern is proven to be the result of a transport optimization where the energy represents a functional component that is minimized for the transport of system fluxes [35,36]. Yet, any system structure observed in nature is an optimal or suboptimal structure optimizing system function, or it has the tendency to reach an optimal state ideally. In nature this tendency toward optimality is attributed to an evolutionary pathway between order and "chaos" where order is played out by clear optimization rules and "chaos" is attributable to unpredictable events disturbing the search toward optimal states [35,36]. In a sense, this natural evolution can be seen in the same optic of the "blind watchmaker" [37].

Much recent work in nanofractal patterning has focused on the biomimetic design of nanomaterials used as the sensor transducer layers [38-41], although some work has focused on pacture of targets such as cancer cells based on the unique fractal nature of the morphological structure of the target [16, 17, 19]. The premise for this work is that the rules of nature can be used to build sensing technology that optimizes the accuracy of measurements by overlapping the assembled sensing nanomaterials to transportation pathways of biological fluxes to be measured. Resulting patterns can be dominating in one or multiple dimensions, but typically assembled sensing nanomaterials are developed with one or two dominating dimensions. The dimensions of assembled sensing nanomaterials do not need to match the dimensions of the sensed support necessarily which depends on the need to perform point or surface scale measurement.

Some of the properties of spanning structures have, indeed, biological implications as shown by previous literature [35,36]. For instance, we consider typical allometric scaling relationships, that is, power laws withscaling exponent, relating measures of metabolic rates of living organisms with their body mass. This is typically done by analyzing the general features of spanning networks serving an assigned volume. Such allometric scaling relationships can be used for designing sensing technology that, for instance, relates bio-sensing mass to metabolic fluxes to measure.

Fractal nanometal structures in the current literature have names that resemble living organisms. The impetus behind this nomenclature is that natural systems have evolved to be highly efficient heat/mass/photon transport systems that result to be fractals. Based on this concept, a number of research groups in the last decade have developed methods for fabrication of biomimetic fractal nanostructures to be used as a transducer layer in biosensing. A schematic of the basic concept behind some of these hybrid bionanomaterials is shown in Figure 1. Biorecognition agents (e.g., proteins, DNA, etc.) are adsorbed to the metal surface or encapsulated in hydrogels, forming a bioactive fractal structure. The enhanced electron transport in the underlying metal nanofractal layer improves sensor reponse time, sensitivity, and limit of detection.

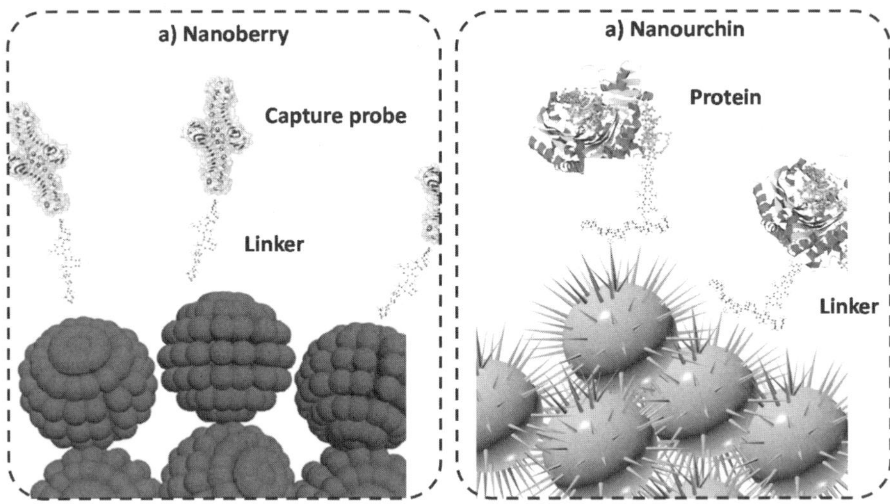

Figure 1. Conceptual schematic of fractal nanometals used in electrochemical biosensing. (a) Schematic for metal nanoberry structure used as a transducer layer in biosensing. A generic capture probe (aptamer) is shown linked on the metal surface. (b) Schematic for metal nanourchin structure used as a transducer layer. A generic protein (xanthine oxidase) is shown linked to the metal surface.

2.3. Synthesis of Fractal Nanometals

In general, high metallic precursor concentration is favorable for the formation of dendritic metallic nanostructures (e.g., flowers), while low precursor concentration results in formation of morphologies such as coral, urchin, berry, and cauliflower [31]. Although there is indeed a relationship between metal precursor concentration and morphology, there are many other factors that play a pivotal role in structure formation by affecting metal growth kinetics. It is generally true that relatively slow growth kinetics favor formation of anisotropic particles. The synthesis factors that alter growth kinetics include synthesis method, substrate type (for surface deposition), metal precursor concentration, reducing agent concentration, temperature, and mixing, among others.

Flower-like nanostructures have been synthesized using gold [39], nickel-copper-iron alloys [28], tin dioxide [40], and platinum [38]. Nanostructures with similar features to cauliflower or broccoli have been synthesized with copper [29], zinc oxide [30], and iron oxide [27]. Other nanostructures such as raspberries, urchins, and confeito have also been synthesized [41, 42]. Many of these nanostructures were formed in the suspended phase, which is a disadvantage in electrochemical biosensing, where surface chemistry at a solid conductive surface is transduced into output signal. In addition, some of the aforementioned materials are super-hydrophobic [26], and have use in energy conversion systems [27-29], optoelectronics [30] and imaging [31]. However, it is a disadvantage for use in biosensing where some degree of hydrophilicity is needed to interface with biological materials. Two of the most common methods for nanofractal synthesis are briefly described below.

2.3.1. Suspended materials

In general, suspended nanomaterials have been produced with two major synthesis strategies. The first strategy uses a "top-down" approach, and relies on production of nanoscale materials from a bulk material using physical fabrication methods [43, 44]. The second uses a "bottom-up" approach. This approach is based on the production of nanomaterials from the organized assembly of atoms. The top-down approach is cost effective, facile, and less time consuming than the bottom up approach; there are many current examples of large-scale production of top down nanometerials for industrial use [45]. However, with top down synthesis methods it is very difficult to control purity, shape and distribution of nanoparticles. Conversely, the size, shape and distribution of nanoparticles can be controlled with the bottom-up approach [46,47]. This method is therefore preferred for production of high quality nanoparticles in solution (homogeneous size distribution and morphology) for biomedical and bioanalytical applications. Currently, the main drawback of this approach is the production of low quantities of nanoparticles.

The three most common synthesis techniques within the class of bottom-up strategies includes: i) co-precipitation [48-50], ii) solvothermal [51-54] and iii) thermal decomposition [55-58]. These methods can be used to synthesize high quality nanoparticles from one individual component or by using a multi-component material as the initiator. The formation of uniform and monodisperse suspended particles was reported by LaMer and coworkers as early as the 1940's. They demonstrated that the suspended materials with a narrow size distribution can be prepared with short nucleation and slow growth [59,60]. The LaMer model was originally developed for sulfur hydrosols and oil aerosol microparticles, but this concept can also be applied to nanoparticle synthesis. According to the Lamer model, the synthesis of nanoparticles can be explained in three major successive steps: nucleation, growth process and formation. Recent reviews on the synthesis of suspended nanoparticles describe these methods in detail [61-64].

2.3.2. Nanometal deposition

Common methods for nanometal deposition include physical vapor deposition (i.e., electron beam evaporation, thermal evaporation, and sputtering) [5, 65], atomic layer deposition [66,67], chemical vapor deposition, galvanic displacement [68,69], and electrodeposition [70]. These methods have become industry standards in fabrication of sensors and biosensors, particularly for fabrication of biochips on silicon or glass substrates.

Metals and metal oxides nanostructures can be formed on various surfaces using physical vapor deposition (PVD). PVD typically involves heating metals under very high temperature and pressure through a variety of techniques including electron beam evaporation, thermal evaporation, and molecular beam epitaxy (MBE), to grow a variety of nano structures. This nanostructures can be grown in a wide variety of morphologies including needle-like wires or whiskers to berry-like or spherical structures. The concomitance of high surface area and small size (e.g., quantum confinement) significantly enhances the sensitivity and signal-to-noise ratios of sensors/biosensors comprised of transduction layers developed through CVD layers. Furthermore PVD grown materials often provide high hardness, wear and corrosion resistance, low friction, and specific optical or electrical properties (Helmersson et al., 2006) that are well suited for a variety of sensor/biosensor applications. Some of the more commonly formed fractal-like materials grown through PVD are ZnO nanowires. For example,. Kong et al [71] developed UV emitting ZnO nano wires using physical vapor deposition while Lyu et al. [72] demonstrated fabrication of a high quality single-crystalline ZnO nanowire array on the Al_2O_3 substrate using simple PVD process at low growth temperature. Wang et. al [73] also synthesized well aligned ZnO wires without any additives and catalysts. In the realm of metallic nanowires, Richter et.al [74] fabricated ultrahigh strength single crystalline Cu nanowhiskers by controlling defect and flaw densities through molecular beam epitaxy (MBE). Au metallic nanostructures were also developed on chemically modified graphene via PVD methods by Pandey et.al [75]. Provezano and Holtz [76] also developed metal and metal oxides nano structures (e.g. Copper-niobium, silver-nickel, copper-aluminium and nanostructured palladium) using relatively inexpensive to produce, amenable to near-net shape forming, and scalable to industrial production techniques via inert-gas condensation PVD method at the US

Naval Research Laboratory (NRL). This system created nano composite samples using a combination of co-evaporation or co-sputtering to produce nanosized powders. Finally, a wide variety of sensors have been developed from PVD grown nanostructures. For example, Kumar et. al [77] developed NO gas sensor on tellurium nanotubes, grown on bare and silver/gold nanoparticle deposited, silicon substrates. Yang et.al [78] detected ammonia at room temperature using micro/nano MoO_3 using PVD. Park et. al [79] prepared tin oxide films by physical vapor deposition of Sn for detection of CH_2Cl_2 in O_2.

Atomic layer deposition, or atomic layer epitaxy, can deposit a monolayer or less of metals and metal oxides onto substrates by combing two or more different vapor-phase reactants. A wide variety of metals including tungsten [80,81], ruthenium [82], and platinum [82] as well as copper, cobalt, iron, and nickel [83]. Recent reports have also shown the ability to deposit a variety of noble metals and their oxides including (ruthenium, platinum, gold, silver, palladium, iridium) [66]. Atomic layer deposition can be used to coat layers on nanowires/nanoneedle or other fractal-like materials to improve their functionality in biosensing scenarios. For example, Al_2O_3, TiO_2, and SiO_2 coatings of 3-5nm in thickness were deposited onto gallium nitride nanowires via ALD to transform the nanowires into surfaces that are subsequently more amenable to protein binding for subsequent biosensing scenarios [83]. Interestingly ALD has been used to form nanoparticles or nanoberries at the defect sites of carbon nanomaterials, such as as graphene, where the dangling bonds or functional groups at thes defect sites interact with with ALD precursors to afford selective nanoparticle growth This leads to an interesting and simple way to decorate and visualize defects in graphene [84]. Such an ALD technique has been used to grow Ni nanoberries along multi-walled carbon nanotubes. These CNT-Ni nanocomposites provide high surface area and electrochemical activity for subsequent non-enzymatic glucose sensing [85].

Chemical vapor deposition (CVD) can also be used to grow nanostructures of various size, shape, and density. During CVD nanostructure synthesis, vaporized chemical precursors are added in to the reaction chamber and adsorb onto surfaces maintained at high temperature [86]. These chemical precursors such as methane and ethane can be used to synthesize a variety nanostructures or be used to develop highly conductive materials that can act as the subsequent electrical support for nanostructures. For example, both carbon nanotubes and graphene have been developed, via CVD methods, to create single-walled carbon nanotubes and multilayered graphene to facilitate the growth of platinum berries and needles [7,38]. Likewise Claussen et.al [87] designed a novel brush-like electrode based on carbon nanotube (CNT) nano-yarn fiber for electrochemical biosensor applications and others have used similar techniques for growing vertically aligned carbon nanotubes [88]. Huang et.al [89] also used CVD to develop highly oriented, multiwalled carbon nanotubes on polished polycrystal and additionally to grow single crystal Ni substrates by plasma enhanced hot filament CVD at low temperature. Oxide nanostructures have also been developed via CVD processes for a variety of sensor applications. For example, Wu and Liu [90] used CVD growth to grow well aligned ZnO nanorods at low temperature. Ansari et al. [91] fabricated glucose sensor using nano-baskets of $SnO(2)$ on anodized aluminum oxide pores using plasma enhanced chemical vapor deposition (PECVD). Wu et.al [92] developed ZnO nano-rods ozone sensor using CVD process.

Electrodeposition persists as a competitive technique due to the relatively low cost, simplicity and high deposited electroactive surface area for noble metals. Kloke et al. [93] and others reviewed some standard methods for controlling nanometal (platinum) morphology and thickness during electrodeposition. One of the keys to electrodeposition of nanometals is the control of the deposited particle sizes, i.e. the smaller the deposited particle sizes, the higher the electroactive surface area and thus higher electrochemical sensor performance. The control of nanoparticle growth during electrodeposition can be controlled by regulating temperature, precursor concentration, exogenous reducing agents or capping agents, with one of the primary goals being the increase of nucleation rates which in turn leads to smaller deposited nanoparticles [93].

Recently, pulsed electrodeposition (PED) has gained much attention which consists of short (typically <1s) bursts of plating potential applied rather than a continuous applied potential. This technique allows for more uniformly distributed and smaller particle nanometal films without additional chemicals and controls needing to be exhibited over the plating system [94-96]. Another technique, sonoelectrodeposition (SED), has also been used for exhibiting controls over electrodeposition. Sonoelectro-deposition is performed by applying constant ultrasonic stimulation to the plating solution during electrodeposition which causes cavitation of gas bubbles that form on the surface during plating resulting in microjetting and microstreaming effects increasing the rate of metal mass transfer to the surface and reduction of the diffusion layer [97]. This technique has been widely adopted in the production of nanometal suspensions and has been demonstrated for a variety of metals including copper and platinum [98-101].

2.4. Fractal Nanometal Morphology

To date, the two most common morphologies are nanoflowers and nanopetals, followed by other morphologies including structures which are similar to marine corals, urchins, raspberries, and cauliflowers. Other nanostructures such as confeito have also been synthesized, but are not considered here in detail [31]. Each of the morphologies listed avove is briefly summarized. Examples for sensing and biosensing are provided in the section titled *"Recent Applications of Nanofractal Metals in Biosensing."*

2.4.1. Nanoflowers and Nanopetals

Nanoflower structures (i.e., dendritic nanometals) have been prepared from a wide array of metals and metal oxides (e.g., gold, platinum, zinc oxide). These nanostructures have broad applications, including: use as capacitors in lithium batteries [102,103], waste treatment [104-107], catalysis of volatile organics [108], development of dye-sensitized solar cells [109-111], contrast agents in fluorescent assays [112], and enzymatic processing in proteomics [113]. Nanometals with petal morphologies are another common structure in the current literature, and have applications in energy storage/production [114,115], heavy metal removal [116], and fluorescence signal enhancement [117]. Electron micrographs of nanogold flower [42] and petal structures [118] from the recent literature are shown in Figure 2. The nanogold flowers by Ujihara and Imae [42] had dendritic structures (confeito-like) ranging from tens to hundreds of nanometers. The gold nanopetals synthesized by Liu et al (2013) have edge features in the 800-900 nm range, with a petal thickness in the tens of nanometers.

Figure 2. Nanoflower and nanopetal structures synthesized with gold precursor. a) Nanoflower dendritic structure (Courtesy of Ujihara and Imae, 2013)[42]. b) Nanopetal structure (courtesy of Liu et al., 2013)[118].

2.4.2. Nanocorals

To date, most nanocoral structures have been developed using surface deposition of noble metals (e.g., gold, platinum, palladium), metal oxides (e.g., zinc oxide, copper oxide) and chalcogenides (e.g., cadmium sulphoselenide) onto a conductive surface. These structures have been used as materials for supercapacitors [119], photovoltaic cells [120], solar hydrogen generators [121], chemical reactors for the reduction of environmtal contaminants [122] , and thin film semiconductors [123]. Some examples of nanocoral structures can be seen in Figure 3. Xu et al. deposited a gold nanocoral layer on a glassy carbon electrode with dendritic features of approximately 5 nm [124]. Vanegas et al. developed platinum nanocoral structures that resemble a marine microbial thrombolite using sonoelectrodeposition [8]. The surface was composed of relatively homogenous 20 nm platinum nanospheres formed on either graphene oxide or multiwalled carbon nanotubes.

Figure 3. Nanocoral structures electrodeposited onto metal electrodes. a) nanocoral gold structure resembling a marine sponge (Courtest of Xu et al.)[124]. b) platinum nanocoral bed similar to a marine thrombolite (Courtesy of Vanegas et al .)[8].

2.4.3. Nanourchins

Urchin-like nanostructures are increasing in popularity, and have been developed using metal oxides (e.g., zinc oxide, nickel oxide) noble metals (e.g., palladium, gold-palladium alloy), and composites of chalcogenides and metal oxide (e.g., CdS/ZnO). Nanourchins have use in solar cells [125], hydrogen generation [126], degradation of environmental pollutants [127], and catalysis of solvent oxidation/combustion [128,129]. Figure 4 shows some recent examples of ZnO nanourchins using hydrothermal processing. Zheng et al. developed a thin film of ZnO nanourchins with sparse nanorods extending from the core [125]. The urchins had a spherical core that was approximately 500 nm in diameter, decorated with a base of ZnO nanorods that were 15 to 30 nm in diameter and 50 to 250 nm in length. Umar et al. developed ZnO nanourchins with dense nanoneedles formed along the spherical core with a mean diameter of 2 µm [127]. The diameter of the nanoneedles at the tip was approximately 50 nm, and the diameter at the base was significantly larger (approximately 200 nm). Claussen and coworkers have developed Pt nanourchins on both carbon nanotubes and paper (cellulose)-based materials via a facile chemical reduction of the metal salt chloroplatinic acid (H_2PtCl_6) with formic acid (HCOOH) [5,130]. The velocity of the Pt nanourching growth mechanism and thus the size, density, and morphopology of the nanourchins is controlled by changing the pH of the metal salt solution. In effect, the lower the pH, the slower the reaction velocity and accordingly the higher density and more needle-like the individual needles of the Pt nanourchins appear. For example a low solution pH of 1.75 yields dense nanourchins with diameters of 3-5 nm while higher pH values (pH 4.5) yields larger jagged-like spikes ranging in 10-30 nm in diameter [5].

Figure 4. Zinc oxide nanourchin structures fabricated using hydrothermal processes. a) ZnO nanourchins grown in a 2D array with a spherical core of 500 nm and sparse nanorods that were 15 to 30 nm in diameter and 50 to 250 nm in length (Courtesy of Zheng et al.)[125]. b) ZnO nanourchins with dense nanoneedles along the spherical core; nanoneedles were 50 nm in diameter at the tip (courtesy of Umar et al.)[127].

2.4.4. Nanoberries

Nanoberries (strawberries, raspberries, mulberries) have been synthesized using noble metals as the precursor. Most berry nanostructures have been fabricated with platinum, palladium, and gold [131,132]. Nanoberry structures have diverse applications as components of lithium batteries [133], nanoscale electronics [134], photovoltaic cells [131], and electrochemical sensing [135]. Exampes of berry-like nanofractal structures are shown in Figure 5. Das et al. developed gold nanoraspberry structures using NAD^+-mediated electrodeposition [135]. The raspberry aggregates were approximately 100 nm in diameter, with drupelet-like fractal structures as small as 2-5 nm. Simonet developed platinum nanostructures similar to mulberries using cathodic electrodeposition in the presence of tetraalkylammonium salts. The mulberry structures (500 to 900 nm) contained drupelet-like fractal structures with an approximate size of 100 nm [136].

Figure 5. Raspberry and mulberry nanostructures synthesized using noble metals. a) Gold nanoraspberry structures developed with electrodeposition using nicotinamide adenine dinucleotide (NAD^+) as an oxidizing agent (Courtesy of Das et al.)[135]; b) Platinum nanomulberries formed using cathodic electrodeposition in the presence of tetraalkylammonium salts (Courtesy of Simonet) [136].

2.4.5. Nanocauliflowers

Nanometal structures with morphology similar to cauliflower or broccoli have been synthesized with gold [137], copper [29], zinc oxide [30], iron oxide [27], and cadmium sulfide [138]. These nanomaterials have applications in hydrogen fuel cells [30], solar cells [138,139], biofuel cells [137], and light emitting diodes [30]. Examples of metal nanocauliflower and nanobroccoli fractal structures are shown in Figure 6. Kong et al. developed a nanocauliflower structure with copper using electroless deposition on indium tin oxide (ITO)[29]. The isolated nanostructures resemble peduncles approximately 1 μm in diameter, with evident sepal-like structures that have feature sizes from 20 to 200 nm. Liu et al. developed gold nanobroccoli structures on commercial screen printed electrodes using electrodeposition in a nanogold precursor solution[39]. The broccoli-like nanostructures had curd-like structures that were approximately 2-5 μm in diameter, with peduncle and sepal structures as small as 100 nm.

Figure 6. a) Nanocauliflower and nanobroccoli metal structures. Copper nanocauliflowers developed with electroless deposition on ITO electrodes (courtesy of Kong et al.)[29]; b) Gold nanobroccoli formed on commercial screen printed electrodes using electrodeposition (courtesy of Liu et al.)[39].

2.4.6. Nanoroots and Nanobranches

Root-like structures have been synthetized using nanomaterials like ZnO; Co, and SnO_2. These structures have been mostly fabricated *via* electrospinning. Aplications of metal nanoroots include sensing, biosensing, and advanced tissue regeneration [39, 140,141]. Nanobranches have also been prepared from metals and metal oxides such as SnO_2, Au, TiO_2, NiO, Co_3O_4 and ZnO. Methods such as vapor transport with calcination, seeded growth in the presence of cationic surfactants, solution-phase deposition, hydrothermal routes, chemical bath deposition, and cathodic electrodeposition have been applied to obtain nanobranches. Applications listed in the literature include: supercapacitors, optical data storage, submicrometer metallic barcodes, resonant optical antennas, chemical and biological sensing, biological imaging, photothermal therapy, dye-sensitized solar cells, and photoelectrolysis [142-145]. SEM micrographs of ZnO nanoroots and TiO2/NiO nanobranches are shown in Figure 7.

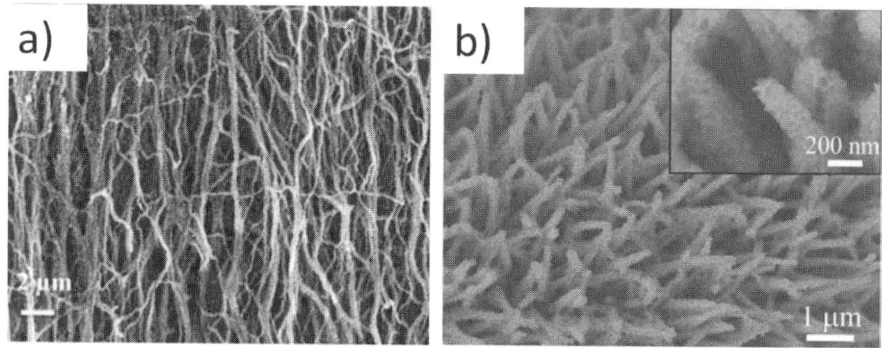

Figure 7. Root and branch nanostructures synthesized using metal oxides. A) ZnO nanoroot structures fabricated by electrospinning and subsequent calcination at 500 ^{0}C (Courtesy of Ali et al.)[140]; b) TiO2/NiO nanobranches grown by a combination of atomic layer deposition and chemical bath deposition (Courtesy of Xia et al.)[143].

2.5. Recent Applications of Nanofractal Metals in Sensing

2.5.1. Nanoflowers and Nanopetals

Nanoflower structures have been used in gas sensing [146-148], but to date, there are only a few examples of nanoflowers used in biosensing. These include graphene/platinum/PEDOT(poly(3,4-ethylenedioxythio-phene)) nanopetal structures for enzymatic glucose sensing [7], carbon nanotube/platinum nanoflowers for glucose sensing [149], and hierarchical nanoflowers for hydrogen peroxide sensing [150]. Nanopetals have also been used in gas sensing [151,152], and in detection of dissolved silver ions [153]. Although nanopetals have not been applied extensively in biosensing with the use of a biorecognition agent, Tong et al. developed a non-enzymatic glucose sensor that showed excellent sensitivity and response time toward glucose in buffer, but this approach has not yet been pusblished in a complex sample such as blood or serum[154].

2.5.2. Nanocorals

A wide variety of nanocoral structures have been developed that could potentially be used in electrochemical biosensing, but to date there are only a few demonstrations of this approach. Most recently, Xu et al. developed gold nanocoral layer for peroxide sensing based on electrochemiluminescence based on the luminol system [124]. The nanocoral structures were deposited on a glassy carbon electrode, and had dendritic features with a diameter of approximately 5 nm. Cheng et al formed gold quasi-one-dimensional nanocoral structures on a glass carbon electrode for enzyme-free glucose sensing [155]. Similar to the enzyme-free nanopetal device by Tong et al. [156], the gold nanocoral structure had a relatively high sensitivity, wide sensing range, and a low detection limit in buffer but has not yet been validated in a biological sample.

2.5.3. Nanourchins

Rajkumar et al. developed palladium nanourchins decorated on multiwalled carbon nanotubes for the detetion of hydrazine or H_2O_2[157]. Chaturvedi et al. developed platinum nanourchins on a graphene-nanoceria film for driving dismutase reactions in a biosensor designed for electrochemical detection of superoxide anion produced by xanthine oxidase[158]. Zhou et al. developed a non-enzymatic sensor for detection of glucose using gold, but had similar issues as previously discussed for nanocoral and nanopetal enzyme-free sensors[159]. Zhou et al. palladium nanourchins for use in amperometric sensing of hydrogen peroxide[160]. Claussen et al developed platinum nanourchins on both cellulose and graphene-based substrates [5,130]. This platinum nanowire urchin structures displayed were capable of lowering the activation energy for the decomposition of H_2O_2 to 26.9 to 29.5 kJ mol^{-1} which amounts to approximately 20% decrease in similar Pt nanostructured catalysts and 50% or greater reduction as compared to conventional catalyst materials (e.g., ferrihydrates, iron oxides) [161-163].

2.5.4. Nanoberries

Das et al. recently developed nanoraspberries composed of gold with a nicotinamide adenine dinucleotide backbone for sensing of methanol[135] Kinoshita et al. developed gold-polyaniline nanoraspberries and studied the formation of dendrite-like rod structures with a tunable lengthscale for applications in nanoelectronics and biosensing, although no specific application or demonstration was shown [134]. Shiigi et al. used antibody-tagged gold-polyanaline 5 nm nanoraspberry structures for the targeted detection of *Pseudomonas aeruginosa* and *Escherichia coli* based on surface enhanced raman spectroscopy after selective binding of surface proteins [164].

2.5.5. Nanocauliflowers

Liu et al. developed gold nanodendrite structures as a transducer layer for glucose biosensors [39]. Li et al. developed a gold and copper oxide (Au/CuO) alloy nanocomposite with nanocauliflower morphology used for nonenzymatic glucose detection [102].

2.5.6. Nanoroots and Nanobranches

Although only a few aplications of dendritic nanoroots and nanobranches have been reported in the literature, these fractal nanostructures show promising potential for revolutionizing the field of sensing technologies. For instance, Sharma et al. fabricated root-like nanostructures by electrospining a composite of SnO_2 with polyaniline [141]. The nanoroots were used for hydrogen gas sensing at low temperature (nearly at room temperature), which is a major improvement over previous metal-oxide based sensors that require high operating temperatures (typically above 200°C), which may pave the way for future development of biosensors using the dendritic nanostructure. Another proof-of-concept demonstration was shown by Ali et al.[140]. This group developed an inmunosensor based on anti-epidermal growth factor receptor conjugated with mesoporous ZnO nanoroots. The inmunosensor was designed for the label-free detection of breast cancer biomarkers. This device showed excellent impedimetric performance, being about an order of magnitude more sensitive than the best demonstrated in the current literature, and about three orders of magnitude better than the ELISA standard for breast cancer detection.

2.6. Conclusions: Challenges and Future Directions

Fractal patterns found in nature are the result of both mechanistic and random events that have occurred throughout the course of biological evolution. Such patterns possess a clear effect on the energy efficiency of living systems. Mass, heat, photon, and charge transport can be optimized by fractal shapes. Recent advancements in engineering and technology are taking advantage of these ontogenetic principles. Particularly, biomimetic design of nanomaterial-based transducers is currently a growing trend in the field of biosensing, where electrochemical performance depend directly on the transport dynamics that take place at the sensor-sample interface. By imprinting fractal patterns on the surface of a sensor, many groups have shown that it is possible to maximize the biosensor performance while minimizing the net amount of nanomaterials required for the deposition process; which is an economic saving in terms of materials, complexity, and time required for sensor fabrication.

Metal nanomaterials such as platinum, palladium, nickel, and gold have been used to create a variety of biomimetic fractal structures. The most common techniques used for synthesis of this type of structures are electrodeposition, chemical vapor deposition, and sputtering. During fabrication, several factors (e.g. precursor concentration, temperature, reducing agent, etc.) influence the end morphology of the nanometal assembly. To date, biomimetic structures like nanoflowers, nanopetals, nanourchins, nanoberries, nanocauliflowers, and nanoroots have been successfully produced by different research groups. Some proof of concept applications in the field of biosensing include gas-pahse monitoring, sensing of small molecules (e.g. glucose, methanol, hydrogen peroxide, and superoxide anion), quantification of metal ions, and detection of bacteria.

For determining the applications of a given fractal nanostructure, it is imperative to consider its particular physicochemical properties in relation with the interfacial interactions that can potentially affect the overall behavior of the system. For instance, some of the aforementioned nanostructures are super-hydrophobic and even toxic, which currently limits biological applications to *ex vivo* analysis. In terms of future directions, morphology-programmed and controlled synthesis with a predictive model based on relationships between fractal domains and physicochemical properties are current engineering needs that once addressed will greatly advance the field of nanotechnology. Additionally, development of new materials that possess the same electrical properties without imparting ecological or acute toxic effects are greatly desired.

Acknowledgments

The authors greatly acknowledge the following funding agencies for supporting research in our labs that contributed to the work reviewed here: National Scence Foundation: Nanobiosensors Program (No. 1511953; Gomes and McLamore); the Colombian Agencies Colfuturo y Colciencias (No. 001375882; Vanegas); the UF Early Career Award (USDA Hatch Project No. FLA-ABE-005062; McLamore); the NASA Florida Space Grant Consortium (No. 00091356; McLamore); the Roy J Carver Charitable Trust Foundation (Claussen)

References

1. Xia, F. N., Perebeinos, V., Lin, Y. M., Wu, Y. Q., and Avouris, P., 2011, The origins and limits of metal-graphene junction resistance. *Nature Nanotechnology, 6*(3), 179-184.
2. Shipway, A. N., Katz, E., and Willner, I., 2000, Nanoparticle arrays on surfaces for electronic, optical, and sensor applications. *Chemphyschem, 1*(1), 18-52.
3. Vairavapandian, D., Vichchulada, P., and Lay, M. D., 2008, Preparation and modification of carbon nanotubes: Review of recent advances and applications in catalysis and sensing. *Analytica Chimica Acta, 626*(2), 119-129.
4. Kloke, A., von Stetten, F., Zengerle, R., and Kerzenmacher, S., 2011, Strategies for the Fabrication of Porous Platinum Electrodes. *Advanced Materials, 23*(43), 4976-5008.
5. Claussen, J. C., Daniele, M. A., Geder, J., Pruessner, M., Makinen, A. J., Melde, B. J., Twigg, M., Verbarg, J. M., and Medintz, I. L., 2014, Platinum-Paper Micromotors: An Urchin-like Nanohybrid Catalyst for Green Monopropellant Bubble-Thrusters. *Acs Applied Materials & Interfaces, 6*(20), 17837-17847.
6. Claussen, J. C., Franklin, A. D., ul Haque, A., Porterfield, D. M., and Fisher, T. S., 2009, Electrochemical Biosensor of Nanocube-Augmented Carbon Nanotube Networks. *Acs Nano, 3*(1), 37-44.
7. Claussen, J. C., Kumar, A., Jaroch, D. B., Khawaja, M. H., Hibbard, A. B., Porterfield, D. M., and Fisher, T. S., 2012, Nanostructuring Platinum Nanoparticles on Multilayered Graphene Petal Nanosheets for Electrochemical Biosensing. *Advanced Functional Materials, 22*(16), 3399-3405.
8. Vanegas, D. C., Taguchi, M., Chaturvedi, P., Burrs, S., Tan, M., Yamaguchi, H., and McLamore, E. S., 2014, A comparative study of carbon-platinum hybrid nanostructure architecture for amperometric biosensing. *Analyst, 139*(3), 660-667.
9. Hezard, T., Fajerwerg, K., Evrard, D., Colliere, V., Behra, P., and Gros, P., 2012, Gold nanoparticles electrodeposited on glassy carbon using cyclic voltammetry: Application to Hg(II) trace analysis. *Journal of Electroanalytical Chemistry, 664*, 46-52.
10. Jin, G. P., Ding, Y. F., and Zheng, P. P., 2007, Electrodeposition of nickel nanoparticles on functional MWCNT surfaces for ethanol oxidation. *Journal of Power Sources, 166*(1), 80-86.
11. Plowman, B. J., Bhargava, S. K., and O'Mullane, A. P., 2011, Electrochemical fabrication of metallic nanostructured electrodes for electroanalytical applications. *Analyst, 136*(24), 5107-5119.
12. Tsai, W. C., Wan, C. C., and Wang, Y. Y., 2003, Frequency effect of pulse plating on the uniformity of copper deposition in plated through holes. *Journal of the Electrochemical Society, 150*(5), C267-C272.

13. McLamore, E. S., Porterfield, D. M., and Banks, M. K., 2009, Non-Invasive Self-Referencing Electrochemical Sensors for Quantifying Real-Time Biofilm Analyte Flux. *Biotechnology and Bioengineering, 102*(3), 791-799.

14. Carraro, C., Maboudian, R., and Magagnin, L., 2007, Metallization and nanostructuring of semiconductor surfaces by galvanic displacement processes. *Surface Science Reports, 62*(12), 499-525.

15. Shahrokhian, S., Salimian, R., and Rastgar, S., 2014, Pd-Au nanoparticle decorated carbon nanotube as a sensing layer on the surface of glassy carbon electrode for electrochemical determination of ceftazidime. *Materials Science & Engineering C-Materials for Biological Applications, 34*, 318-325.

16. Traversi, C., Bianciardi, G., Tasciotti, A., Berni, E., Nuti, E., Luzi, P., and Tosi, G. M., 2008, Fractal analysis of fluoroangiographic patterns in anterior ischaemic optic neuropathy and optic neuritis: a pilot study. *Clinical and Experimental Ophthalmology, 36*(4), 323-328.

17. Bizzarri, M., Giuliani, A., Cucina, A., D'Anselmi, F., Soto, A. M., and Sonnenschein, C., 2011, Fractal analysis in a systems biology approach to cancer. *Seminars in Cancer Biology, 21*(3), 175-182.

18. Zhang, P. C., and Wang, S. T., 2014, Designing Fractal Nanostructured Biointerfaces for Biomedical Applications. *Chemphyschem, 15*(8), 1550-1561.

19. Zhang, P. C., Chen, L., Xu, T. L., Liu, H. L., Liu, X. L., Meng, J. X., Yang, G., Jiang, L., and Wang, S. T., 2013, Programmable Fractal Nanostructured Interfaces for Specific Recognition and Electrochemical Release of Cancer Cells. *Advanced Materials, 25*(26), 3566-3570.

20. He, L. L., Kim, N. J., Li, H., Hu, Z. Q., and Lin, M. S., 2008, Use of a Fractal-like Gold Nanostructure in Surface-Enhanced Raman Spectroscopy for Detection of Selected Food Contaminants. *Journal of Agricultural and Food Chemistry, 56*(21), 9843-9847.

21. Ariga, K., Yamauchi, Y., Ji, Q. M., Yonamine, Y., and Hill, J. P., 2014, Research Update: Mesoporous sensor nanoarchitectonics. *Apl Materials, 2*(3)

22. Fernandez, Y. D., Sun, L. L., Gschneidtner, T., and Moth-Poulsen, K., 2014, Research Update: Progress in synthesis of nanoparticle dimers by self-assembly. *Apl Materials, 2*(1), 010702.

23. Sawada, Y., Dougherty, A., Gollub, J.P., 1986. Dendritic and Fractal Patterns in Electrolytic Metal Deposits. Phys. Rev. Lett. 56, 1260.

24. Kane, R.S., Takayama, S., Ostuni, E., Ingber, D.E., Whitesides, G.M., 1999. Patterning proteins and cells using soft lithography. Biomaterials 20, 2363–2376. doi:10.1016/S0142-9612(99)00165-9

25. Sims, D.W., Southall, E.J., Humphries, N.E., Hays, G.C., Bradshaw, C.J. a, Pitchford, J.W., James, A., Ahmed, M.Z., Brierley, A.S., Hindell, M. a, Morritt, D., Musyl, M.K., Righton, D., Shepard, E.L.C., Wearmouth, V.J., Wilson, R.P., Witt, M.J.,

Metcalfe, J.D., 2008. Scaling laws of marine predator search behaviour. Nature 451, 1098–1102. doi:10.1038/nature06518

26. Roach P, Shirtcliffe N, Newton M (2008) Progess in superhydrophobic surface development. Soft Matter 4:224–240

27. Warren, S. C., Voitchovsky, K., Dotan, H., Leroy, C. M., Cornuz, M., Stellacci, F., Hebert, C., Rothschild, A., and Gratzel, M., 2013, Identifying champion nanostructures for solar water-splitting. *Nature Materials, 12*(9), 842-849.

28. Giz, M. J., Marengo, M. C., Ticianelli, E. A., and Gonzalez, E. R., 2003, Electrochemical and physical characterization of Ni-Cu-Fe alloy for chlor-alkali hydrogen cathodes. *Ecletica Quimica, 28*(2), 21-28.

29. Kong, C. C., Sun, S. D., Zhang, J., Zhao, H. D., Song, X. P., and Yang, Z. M., 2012, Nanocube-aggregated cauliflower-like copper hierarchical architectures: synthesis, growth mechanism and electrocatalytic activity. *Crystengcomm, 14*(18), 5737-5740.

30. Izaki, M., Watanabe, M., Aritomo, H., Yamaguchi, I., Asahina, S., Shinagawa, T., Chigane, M., Inaba, M., and Tasaka, A., 2008, Zinc oxide nano-cauliflower array with room temperature ultraviolet light emission. *Crystal Growth & Design, 8*(4), 1418-1421.

31. Wang, L., Imura, M., and Yamauchi, Y., 2012, Tailored synthesis of various Au nanoarchitectures with branched shapes. *Crystengcomm, 14*(22), 7594-7599.

32. Shu Chen ,†‡ Ying Li ,§ Chen Guo ,*†‡ Jing Wang ,† Junhe Ma ,†‡ Xiangfeng Liang ,†‡ Liang-Rong Yang ,†‡ and Hui-Zhou Liu *†‡ Temperature-Responsive Magnetite/PEO–PPO–PEO Block Copolymer Nanoparticles for Controlled Drug Targeting Delivery. Langmuir, 2007, 23 (25), pp 12669–12676

33. Y. Koc b, A. J. de Mello b, G. McHale a, M. I. Newton a, P. Roach a and N. J. Shirtcliffe *a Nano-scale superhydrophobicity: suppression of protein adsorption and promotion of flow-induced detachment† DOI: 10.1039/B716509A (Paper) Lab Chip, 2008, 8, 582-586

34. Y. Wu, F. Simonovsky, B. Ratner and T. Horbett, *J. Biomed. Mater. Res. A,* 2005, **74**(4), 722–738.

35. Jayanth R. Banavara, Todd J. Cookeb, Andrea Rinaldoc,Journal of Statistical Physics, July 2001, Volume 104, Issue 1, pp 1-48. First online: Scaling, Optimality, and Landscape Evolution.

36. Jayanth R. Banavara, Todd J. Cookeb, Andrea Rinaldoc, and Amos Maritane Journal of Statistical Physics, July 2001, Volume 104, Issue 1, pp 1-48.

37. Dawkins, R. (1988) The Blind Watchmaker, Penguin Books, England.

38. Yude Wang, Igor Djerdj, Bernd Smarsly, and Markus Antonietti,Antimony-Doped SnO2 Nanopowders with High Crystallinity for Lithium-Ion Battery Electrode Chem. Mater., 2009, 21 (14), pp 3202–3209.

39. Wang, J., Yuan, R., Chai, Y., Li, W., Fu, P., Min, L., 2010. Using flowerlike polymer-copper nanostructure composite and novel organic-inorganic hybrid material to

construct an amperometric biosensor for hydrogen peroxide. Colloids Surf. B. Biointerfaces 75, 425–31. doi:10.1016/j.colsurfb.2009.09.015.

40. Jiajia Ning, Quanqin Dai, Tao Jiang, Kangkang Men, Donghua Liu, Ningru Xiao, Chenyuan Li‖, Dongmei Li, Bingbing Liu, Bo Zou, Guangtian Zou‡ and William W. Yu, Facile Synthesis of Tin Oxide Nanoflowers: A Potential High-Capacity Lithium-Ion-Storage Material Langmuir, 2009, 25 (3), pp 1818–1821.

41. J. Wang, B. Ding, J. Yub, M. Wang, Engineering biomimetic superhydrophobic surfaces of electrospun nanomaterials Xianfeng Nano Today Volume 6, Issue 5, October 2011, Pages 510–530.

42. Wang, X., Jiao, L., Sheng, K., Li, C., Dai, L., Shi, G., 2013. Solution-processable graphene nanomeshes with controlled pore structures. Sci. Rep. 3, 1996. doi:10.1038/srep01996.

43. Herzer, G.,1989, Grain structure and magnetism of nanocrystalline ferromagnet. *IEEE Transactions on Magnetics*, 25, 3327–3329.

44. 44. Skorvánek, I., O'Handley, R. C., 1995, Fine-particle magnetism in nanocrystalline Fe-CuNb-Si-B at elevated temperatures. *Journal of Magnetism and Magnetic Materials,* 140-144, 467–468.

45. Jayanth R. Banavara,1, Todd J. Cookeb,1, Andrea Rinaldoc,d,1,2, and Amos Maritane,1 Form, function, and evolution of living organisms vol. 111 no. 9 > Jayanth R. Banavar, 3332–3337, doi: 10.1073/pnas.1401336111.

46. Cheon, J., Kang, N.-J., Lee, S.-M., Lee, J.-H., Yoon, J.-H., Oh, S. J., 2004, Shape evolution of single-crystalline iron oxide nanocrystals. *Journal of the American Chemical Society*, 126, 1950–1951.

47. Lu, A.-H., Salabas, E. L., Schüth, F., 2007, Magnetic nanoparticles: synthesis, protection, functionalization, and application. *Angewandte Chemie International Edition*, 46, 1222–1244.

48. Kang, Y. S., Risbud, S., Rabolt, J. F., Stroeve, P., 1996, Synthesis and characterization of nanometer-size Fe_3O_4 and γ-Fe_2O_3 particles. *Chemistry of Materials*, 8, 2209–2211.

49. Kuo, P. C., Tsai, T. S., 1989, New approaches to the synthesis of acicular α-FeOOH and cobalt-modified iron oxide particles. *Journal of Applied Physics*, 65, 4349–4356.

50. Hong, C.-Y., Jang, I. J., Horng, H. E., Hsu, C. J., Yao, Y. D., Yang, H. C., 1997, Magnetochromatic effects in magnetic fluid thin films. *Journal of Applied Physics*, 81, 4275–4277.

51. Lu, J., Jiao, X., Chen, D., Li, W., 2009, Solvothermal synthesis and characterization of Fe_3O_4 and γ-Fe_2O_3 nanoplates. *Journal of Physical Chemistry* C, 113, 4012–4017.

52. Zhang, W., Yang, Z., Liu, Y., Tang, S., Han, X., Chen, M., 2004, Controlled synthesis of Mn_3O_4 nanocrystallites and MnOOH nanorods by a solvothermal method. *Journal of Crystal Growth*, 263, 394–399.

53. Rath, C., Sahu, K., Anand, S., Date, S., Mishra, N., Das, R., 1999, Preparation and characterization of nanosize Mn–Zn ferrite. *Journal of Magnetism and Magnetic Materials*, 202, 77–84.

54. Baruwati, B., Nadagouda, M. N., Varma, R. S., 2008, Bulk synthesis of monodisperse ferrite nanoparticles at water–organic interfaces under conventional and microwave hydrothermal treatment and their surface functionalization. *Journal of Physical Chemistry C*, 112, 18399–18404.

55. Puntes, V. F., Krishnan, K. M., Alivisatos, P., 2001, Synthesis, self-assembly, and magnetic behavior of a two-dimensional superlattice of single-crystal ε-Co nanoparticles. *Applied Physics Letters*, 78, 2187–2189.

56. Stoeva, S., Klabunde, K. J., Sorensen, C.M., Dragieva, I., 2002, Gram-scale synthesis of monodisperse gold colloids by the solvated metal atom dispersion method and digestive ripening and their organization into two- and three-dimensional structures. *Journal of the American Chemical Society*, 124, 2305–2311.

57. Dinega, D. P., Bawendi, M., 1999, A Solution-phase chemical approach to a new crystal structure of cobalt. *Angewandte Chemie International Edition*, 38, 1788–1791.

58. Redl, F. X., Black, C. T., Papaefthymiou, G. C., Sandstrom, R. L., Yin, M., Zeng, H., Murray, C. B., O'Brien, S. P., 2004, Magnetic, electronic, and structural characterization of nonstoichiometric iron oxides at the nanoscale. *Journal of the American Chemical Society*, 126, 14583–14599.

59. LaMer, V. K., Dinegar, R. H., 1950, Theory, production and mechanism of formation of monodispersed hydrosols. *Journal of the American Chemical Society*, 72, 4847–4854.

60. Sugimoto, T., 2001, Monodispersed particles, 1st ed.; Elsevier: Amsterdam, New York.

61. Li, C., Chen, T., Ocsoy, I., Zhu, G., Yasun, E., You, M., Wu, C, Zheng, J., Song, E., Huang, C. Z., Tan., Tan, 2014, Gold-coated Fe_3O_4 nanoroses with five unique functions for cancer cell targeting, imaging, and therapy. *Advanced Functional Materials*, 24, 1772–1780.

62. Ocsoy, I., Gulbakan, B., Shukoor, M. I., Xiong, X., Chen, T., Powell, D. H., Tan, W., 2013, Aptamer-conjugated multifunctional nanoflowers as a platform for targeting, capture, and detection in laser desorption ionization mass spectrometry. *ACS Nano*, 7 (1), 417–427.

63. Chen, T., Öçsoy, I., Yuan, Q., Wang, R., You, M., Zhao, Z., Song, E., Zhang, X., Tan., 2012, One-step facile surface engineering of hydrophobic nanocrystals with designer molecular recognition. *Journal of the American Chemical Society*, 134 (32), 13164–13167.

64. Schladt, T. D., Shukoor, M. I., Schneider, K., Tahir, M. N., Natalio, F., Ament, I., Becker, J., Jochum, F. D., Weber, S., Khler, O., Theato, P., Schreiber, L. M., Snnichsen, C., Schrder, H. C., Mller, W. E. G., Tremel, W. D., 2010, Au@MnO nanoflowers: hybrid nanocomposites for selective dual functionalization and imaging. *Angewandte Chemie International Edition*, 49, 3976 –3980.

65. Suen, J. C., Hengenius, J. B., Wickner, M. M., Fisher, T. S., Umulis, D. M., and Porterfield, D. M., 2011, Effects of Carbon Nanotube-Tethered Nanosphere Density on Amperometric Biosensing: Simulation and Experiment. *Journal of Physical Chemistry C, 115*(43), 20896-20904.

66. Liu, H. C., Tsai, C. C., and Wang, G. J., 2013, Glucose biosensors based on a gold nanodendrite modified screen-printed electrode. *Nanotechnology, 24*(21), 215101.

67. Ujihara, M., and Imae, T., 2013, Versatile one-pot synthesis of confeito-like Au nanoparticles and their surface-enhanced Raman scattering effect. *Colloids and Surfaces a-Physicochemical and Engineering Aspects, 436*, 380-385.

68. Zhang, M., Gao, B., Vanegas, D. C., McLamore, E. S., Fang, J., Liu, L., Wu, L., and Chen, H., 2014, Simple approach for large-scale production of reduced graphene oxide films. *Chemical Engineering Journal, 243*, 340-346.

69. Taylor, R., Coulombe, S., Otanicar, T., Phelan, P., Gunawan, A., Lv, W., Rosengarten, G., Prasher, R., and Tyagi, H., 2013, Small particles, big impacts: A review of the diverse applications of nanofluids. *Journal of Applied Physics, 113*(1).

70. Ghorbani, H. R., 2014, A Review of Methods for Synthesis of Al Nanoparticles. *Oriental Journal of Chemistry, 30*(4), 1941-1949.

71. Kong, Y. C., Yu, D. P., Zhang, B., Fang, W., and Feng, S. Q. (2001). Ultraviolet-emitting ZnO nanowires synthesized by a physical vapor deposition approach. *Applied Physics Letters* **78**, 407-409.

72. Lyu, S. C., Zhang, Y., Lee, C. J., Ruh, H., and Lee, H. J. (2003). Low-temperature growth of ZnO nanowire array by a simple physical vapor-deposition method. *Chemistry of Materials* **15**, 3294-3299.

73. Wang, L. S., Zhang, X. Z., Zhao, S. Q., Zhou, G. Y., Zhou, Y. L., and Qi, J. J. (2005). Synthesis of well-aligned ZnO nanowires by simple physical vapor deposition on c-oriented ZnO thin films without catalysts or additives. *Applied Physics Letters* **86**.

74. Richter, G., Hillerich, K., Gianola, D. S., Moenig, R., Kraft, O., and Volkert, C. A. (2009). Ultrahigh Strength Single Crystalline Nanowhiskers Grown by Physical Vapor Deposition. *Nano Letters* **9**, 3048-3052.

75. Pandey, P. A., Bell, G. R., Rourke, J. P., Sanchez, A. M., Elkin, M. D., Hickey, B. J., and Wilson, N. R. (2011). Physical Vapor Deposition of Metal Nanoparticles on Chemically Modified Graphene: Observations on Metal-Graphene Interactions. *Small* **7**, 3202-3210.

76. Provenzano, V., and Holtz, R. L. (1997). Nanostructured metals produced by physical vapour deposition or structural and gas-reactive applications. *Philosophical magazine. B* **76**, 593-604.

77. Kumar, V., Sen, S., Sharma, M., Muthe, K. P., Jagannath, Gaur, N. K., and Gupta, S. K. (2009). Tellurium Nano-Structure Based NO Gas Sensor. *Journal of Nanoscience and Nanotechnology* **9**, 5278-5282.

78. Yang, S., Liu, Y., Jin, W., Qi, Y., Zakharova, G. S., and Chen, W. (2015). Controlled Synthesis of Micro/Nano MoO3 by Physical Vapor Deposition and Its Gas Sensing Properties to NH3 Gas at Room Temperature. *Ferroelectrics* **477**, 112-120.

79. Park, S. H., Son, Y. C., Willis, W. S., Suib, S. L., and Creasy, K. E. (1998). Tin oxide films made by physical vapor deposition thermal oxidation and spray pyrolysis. *Chemistry of Materials* **10**, 2389-2398.

80. Elam, J. W., Nelson, C. E., Grubbs, R. K., and George, S. M., 2001, Kinetics of the WF6 and Si2H6 surface reactions during tungsten atomic layer deposition. *Surface Science, 479*(1-3), 121-135.

81. Klaus, J. W., Ferro, S. J., and George, S. M., 2000, Atomic layer deposition of tungsten using sequential surface chemistry with a sacrificial stripping reaction. *Thin Solid Films, 360*(1-2), 145-153

82. Aaltonen, T., Alen, P., Ritala, M., and Leskela, M., 2003, Ruthenium thin films grown by atomic layer deposition. *Chemical Vapor Deposition, 9*(1), 45-49.

83. Lim, B. S., Rahtu, A., and Gordon, R. G., 2003, Atomic layer deposition of transition metals. *Nature Materials, 2*(11), 749-754.

84. Gutes, A., Laboriante, I., Carraro, C., and Maboudian, R., 2010, Palladium nanostructures from galvanic displacement as hydrogen peroxide sensor. *Sensors and Actuators B-Chemical, 147*(2), 681-686.

85. Coleman, E. J., and Co, A. C., 2014, Galvanic displacement of Pt on nanoporous copper: An alternative synthetic route for obtaining robust and reliable oxygen reduction activity. *Journal of Catalysis, 316*, 191-200.

86. Burda, C., *et al.* (2005) Chemistry and properties of nanocrystals of different shapes, *Chemical Reviews*, **105**, 1025-1102.

87. Claussen, J.C., *et al.* (2012) Nanostructuring Platinum Nanoparticles on Multilayered Graphene Petal Nanosheets for Electrochemical Biosensing, *Advanced Functional Materials*, **22**, 3399-3405.

88. Chhowalla, M., *et al.* (2001) Growth process conditions of vertically aligned carbon nanotubes using plasma enhanced chemical vapor deposition, *Journal of Applied Physics*, **90**, 5308-5317.

89. Huang, Z.P., *et al.* (1998) Growth of highly oriented carbon nanotubes by plasma-enhanced hot filament chemical vapor deposition, *Applied Physics Letters*, **73**, 3845-3847.

90. Wu, J.J. and Liu, S.C. (2002) Low-temperature growth of well-aligned ZnO nanorods by chemical vapor deposition, *Advanced Materials*, **14**, 215-+.

91. Ansari, S.G., *et al.* (2008) Glucose sensor based on nano-baskets of tin oxide templated in porous alumina by plasma enhanced CVD, *Biosensors & Bioelectronics*, **23**, 1838-1842.

92. Wu, R.J., *et al.* (2008) Application of Nanostructure ZnO for Room Working Temperature Ozone Sensor, *Sensor Letters*, **6**, 800-802.

93. Kloke, A., Von Stetten, F., Zengerle, R., Kerzenmacher, S., 2011. Strategies for the fabrication of porous platinum electrodes. Adv. Mater. 23, 4976–5008. doi:10.1002/adma.201102182.

94. Electrochimica Acta 53 (2008) 3313–3322 Review article Pulse and pulse reverse plating—Conceptual, advantages and applications M.S. Chandrasekar, Malathy Pushpavanam.

95. X. Chen, N. Li, K. Eckhard, L. Stoica, W. Xia, J. Assmann, M. Muhler and W. Schuhmann, *Electrochem. commun.*, 2007, **9**, 1348–1354.

96. H. Natter and R. Hempelmann, *Electrochim. Acta*, 2003, **49**, 51–61.

97. V. Sáez and T. J. Mason, *Molecules*, 2009, **14**, 4284–99.

98. I. Haas, S. Shanmugam and A. Gedanken, *J. Phys. Chem. B*, 2006, **110**, 16947–52.

99. Herdman, R.D.; Pearson, T.; Long, E.; Gardner, A. Controlling the hardness of electrodeposited copper coatings by variation of current profile. US Pat. 7329334, 2008.

100. W.-C. Tsai, C.-C. Wan and Y.-Y. Wang, *J. Electrochem. Soc.*, 2003, **150**, C267.

101. H. Heli, N. Sattarahmady, R. Dehdari Vais and A. R. Mehdizadeh, *Sensors Actuators B Chem.*, 2014, **192**, 310–316.

102. Li, Z. Z., Xin, Y. M., Zhang, Z. H., Wu, H. J., and Wang, P., 2015, Rational design of binder-free noble metal/metal oxide arrays with nanocauliflower structure for wide linear range nonenzymatic glucose detection. *Scientific Reports, 5*

103. Chen, J. S., Zhu, T., Hu, Q. H., Gao, J. J., Su, F. B., Qiao, S. Z., and Lou, X. W., 2010, Shape-Controlled Synthesis of Cobalt-based Nanocubes, Nanodiscs, and Nanoflowers and Their Comparative Lithium-Storage Properties. *Acs Applied Materials & Interfaces, 2*(12), 3628-3635.

104. Liu, S., Tian, J. Q., Wang, L., Luo, Y. L., and Sun, X. P., 2012, One-pot synthesis of CuO nanoflower-decorated reduced graphene oxide and its application to photocatalytic degradation of dyes. *Catalysis Science & Technology, 2*(2), 339-344.

105. Srivastava, V., Sharma, Y. C., and Sillanpaa, M., 2015, Green synthesis of magnesium oxide nanoflower and its application for the removal of divalent metallic species from synthetic wastewater. *Ceramics International, 41*(5), 6702-6709.

106. Wu, J. M., and Qi, B., 2007, Low-temperature growth of a nitrogen-doped titania nanoflower film and its ability to assist photodegradation of rhodamine B in water. *Journal of Physical Chemistry C, 111*(2), 666-673.

107. Wang, J., Zhu, J. Z., Zhou, X. X., Du, Y. Y., Huang, W. M., Liu, J. J., Zhang, W. Q., Shi, J. L., and Chen, H. R., 2015, Nanoflower-like weak crystallization manganese oxide for efficient removal of low-concentration NO at room temperature. *Journal of Materials Chemistry A, 3*(14), 7631-7638.

108. Yan, Q. Y., Li, X. Y., Zhao, Q. D., and Chen, G. H., 2012, Shape-controlled fabrication of the porous Co3O4 nanoflower clusters for efficient catalytic oxidation of gaseous toluene. *Journal of Hazardous Materials, 209*, 385-391.

109. Dou, X. C., Sabba, D., Mathews, N., Wong, L. H., Lam, Y. M., and Mhaisalkar, S., 2011, Hydrothermal Synthesis of High Electron Mobility Zn-doped SnO2 Nanoflowers as Photoanode Material for Efficient Dye-Sensitized Solar Cells. *Chemistry of Materials, 23*(17), 3938-3945.

110. Lui, G., Liao, J. Y., Duan, A. S., Zhang, Z. S., Fowler, M., and Yu, A. P., 2013, Graphene-wrapped hierarchical TiO2 nanoflower composites with enhanced photocatalytic performance. *Journal of Materials Chemistry A, 1*(39), 12255-12262.

111. Jiang, C. Y., Sun, X. W., Lo, G. Q., Kwong, D. L., and Wang, J. X., 2007, Improved dye-sensitized solar cells with a ZnO-nanoflower photoanode. *Applied Physics Letters, 90*(26), 263501.

112. Wang, T., Costan, J., Centeno, A., Pang, J. S., Darvill, D., Ryan, M. P., and Xie, F., 2015, Broadband enhanced fluorescence using zinc-oxide nanoflower arrays. *Journal of Materials Chemistry C, 3*(11), 2656-2663.

113. Yin, Y. Q., Xiao, Y., Lin, G., Xiao, Q., Lin, Z., and Cai, Z. W., 2015, An enzyme-inorganic hybrid nanoflower based immobilized enzyme reactor with enhanced enzymatic activity. *Journal of Materials Chemistry B, 3*(11), 2295-2300.

114. Hsu, Y. K., Yu, C. H., Chen, Y. C., and Lin, Y. G., 2013, Fabrication of coral-like Cu2O nanoelectrode for solar hydrogen generation. *Journal of Power Sources, 242,* 541-547.

115. Zhang, Y. X., Li, F., and Huang, M., 2013, One-step hydrothermal synthesis of hierarchical MnO2-coated CuO flower-like nanostructures with enhanced electrochemical properties for supercapacitor. *Materials Letters, 112,* 203-206.

116. Liang, H. F., Xu, B. B., and Wang, Z. C., 2013, Self-assembled 3D flower-like alpha-Fe2O3 microstructures and their superior capability for heavy metal ion removal. *Materials Chemistry and Physics, 141*(2-3), 727-734.

117. Fu, C. C., Ossato, G., Long, M., Digman, M. A., Gopinathan, A., Lee, L. P., Gratton, E., and Khine, M., 2010, Bimetallic nanopetals for thousand-fold fluorescence enhancements. *Applied Physics Letters, 97*(20), 203101.

118. Synthesis and Ethanol Sensing Properties of Novel Hierarchical Sn3O4 Nanoflowers Xingyang Li, Fengping Wang, Jianhai Tu, Hidayat Ullah Shah, Jianling Hu, Yan Li, Yanzhen Lu, and Mei Xu, Journal of Nanomaterials Volume 2015, Article ID 980170.

119. A facile, one-pot synthesis of highly branched Au nanocorals and their enhanced electrocatalytic activity for ethanol oxidation Zhenyuan Liu, Gengtao Fu,†a Yawen Tang, Dongmei Sun,a Yu Chen*ab and Tianhong Lua Cite this: CrystEngComm, 2014, 16, 8576.

120. Borysiewicz, M. A., Wojciechowski, T., Dynowska, E., Wielgus, M., Bar, J., Wojtowicz, T., Kaminska, E., and Piotrowska, A., 2014, Nanocoral ZnO films fabricated on flexible poly(vinyl chloride) using a carrier substrate. *Thin Solid Films, 550,* 145-148.

121. Fabrication of coral-like Cu2O nanoelectrode for solar hydrogen generation, Yu-Kuei Hsua, , , Chun-Hao Yua, Ying-Chu Chenb, Yan-Gu Linc Journal of Power Sources, Volume 242, 15 November 2013, Pages 541–547.

122. Effects of Biomineralization Peptide Topology on the Structure and Catalytic Activity of Pd Nanomaterials Jose Isagani B. Janairo, Tatsuya Sakaguchi, Kenji Hara, Atsushi Fukuoka and Kazuyasu Sakaguchi, Chem. Commun., 2014,50, 9259-9262ii

123. Khot, K. V., Mali, S. S., Pawar, N. B., Kharade, R. R., Mane, R. M., Kondalkar, V. V., Patil, P. B., Patil, P. S., Hong, C. K., Kim, J. H., Heo, J., and Bhosale, P. N., 2014, Development of nanocoral-like Cd(SSe) thin films using an arrested precipitation technique and their application. *New Journal of Chemistry, 38*(12), 5964-5974.

124. Xu, M., Qi, W. J., Zhang, L., Lai, J. P., Aziz-ur-Rehman, Majeed, S., and Xu, G. B., 2014, New synthesis of gold nanocorals using a diazonium compound, and their application to an electrochemiluminescent assay of hydrogen peroxide. *Microchimica Acta, 181*(7-8), 737-742.

125. Zheng, Y. Z., Ding, H. Y., Liu, Y., Tao, X., Cao, G. Z., and Chen, J. F., 2014, In situ hydrothermal growth of hierarchical ZnO nanourchin for high-efficiency dye-sensitized solar cells. *Journal of Power Sources, 254*, 153-160.

126. Barpuzary, D., Khan, Z., Vinothkumar, N., De, M., and Qureshi, M., 2012, Hierarchically Grown Urchinlike CdS@ZnO and CdS@Al2O3 Heteroarrays for Efficient Visible-Light-Driven Photocatalytic Hydrogen Generation. *Journal of Physical Chemistry C, 116*(1), 150-156.

127. Umar, A., Akhtar, M. S., Al-Hajry, A., Al-Assiri, M. S., Dar, G. N., and Islam, M. S., 2015, Enhanced photocatalytic degradation of harmful dye and phenyl hydrazine chemical sensing using ZnO nanourchins. *Chemical Engineering Journal, 262*, 588-596.

128. Bai, G. M., Dai, H. X., Deng, J. G., Liu, Y. X., and Ji, K. M., 2012, Porous NiO nanoflowers and nanourchins: Highly active catalysts for toluene combustion. *Catalysis Communications, 27*, 148-153.

129. You, H., Zhang, F., Liu, Z., and Fang, J., 2014, Free-Standing Pt–Au Hollow Nanourchins with Enhanced Activity and Stability for Catalytic Methanol Oxidation. *ACS Catalysis, 4*(9), 2829-2835.

130. Marr, K. M., Chen, B. L., Mootz, E. J., Geder, J., Pruessner, M., Melde, B. J., Vanfleet, R. R., Medintz, I. L., Iverson, B. D., and Claussen, J. C., 2015, High Aspect Ratio Carbon Nanotube Membranes Decorated with Pt Nanoparticle Urchins for Micro Underwater Vehicle Propulsion via H2O2 Decomposition. *Acs Nano, 9*(8), 7791-7803.

131. Gao, P. C., Ma, H. M., Yan, T., Fan, D. W., Hu, L. H., Du, B., and Wei, Q., 2014, Mulberry-like gold nanospheres supported on graphene nanosheets: one-pot synthesis, characterization and photoelectrochemical property. *New Journal of Chemistry, 38*(7), 3166-3171.

132. Shiigi, H., Yamamoto, Y., Yoshi, N., Nakao, H., and Nagaoka, T., 2006, One-step preparation of positively-charged gold nanoraspberry. *Chemical Communications* (41), 4288-4290.

133. Zhang, J. Q., Li, D., Zhu, Y. M., Chen, M. M., An, M. Z., Yang, P. X., and Wang, P., 2015, Properties and electrochemical behaviors of AuPt alloys prepared by direct-current electrodeposition for lithium air batteries. *Electrochimica Acta, 151*, 415-422.

134. Kinoshita, T., Murakami, H., Muranaka, Y., Yamamoto, Y., Nishino, T., Shiigi, H., and Nagaoka, T., 2014, Tracking the Growth of Tadpole-shaped Aggregates by Scanning Electron Microscopy. *Analytical Sciences, 30*(3), 319-322.

135. Das, A. K., Layek, R. K., Kim, N. H., Samdani, J., Kang, M. C., and Lee, J. H., 2015, Nicotinamide adenine dinucleotide assisted shape-controlled synthesis of catalytically active raspberry-like gold nanostructures. *Electrochimica Acta, 151*, 195-202.

136. Simonet, J., 2006, The Platinised Platinum Interface Under Cathodic Polarisation morphology changes under the influence of tetralkylammonium salts in super-dry solution. *Platinum Metals Review, 50*(4), 180-193.

137. Sattayasamitsathit, S., O'Mahony, A. M., Xiao, X. Y., Brozik, S. M., Washburn, C. M., Wheeler, D. R., Gao, W., Minteer, S., Cha, J., Burckel, D. B., Polsky, R., and Wang, J., 2012, Highly ordered tailored three-dimensional hierarchical nano/microporous gold-carbon architectures. *Journal of Materials Chemistry, 22*(24), 11950-11956.

138. Wilson, K. C., Manikandan, E., Ahamed, M. B., and Mwakikunga, B. W., 2014, Nanocauliflower like structure of CdS thin film for solar cell photovoltaic applications: In situ tin doping by chemical bath deposition technique. *Journal of Alloys and Compounds, 585*, 555-560.

139. Céline M. Leroy *a, Alexandra E. Maegli b, Kevin Sivula a, Takashi Hisatomi a, Nicolas Xanthopoulos c, Eugenio H. Otal b, Songhak Yoon b, Anke Weidenkaff b, Rosendo Sanjines d and Michaël Grätzel a LaTiO2N/In2O3 photoanodes with improved performance for solar water splitting†‡, Chem. Commun., 2012, 48, 820-822.

140. Ali, M. A., Mondal, K., Singh, C., Malhotra, B. D., and Sharma, A., 2015, Anti-epidermal growth factor receptor conjugated mesoporous zinc oxide nanofibers for breast cancer diagnostics. *Nanoscale, 7*(16), 7234-7245.

141. Sharma, H. J., Sonwane, N. D., and Kondawar, S. B., 2015, Electrospun SnO2/Polyaniline composite nanofibers based low temperature hydrogen gas sensor. *Fibers and Polymers, 16*(7), 1527-1532.

142. Sun, S. D., Zhang, X. Z., Sun, Y. X., Zhang, J., Yang, S. C., Song, X. P., and Yang, Z. M., 2013, A facile strategy for the synthesis of hierarchical CuO nanourchins and their application as non-enzymatic glucose sensors. *Rsc Advances, 3*(33), 13712-13719 (2014).

143. Xia, X. H., Zeng, Z. Y., Li, X. L., Zhang, Y. Q., Tu, J. P., Fan, N. C., Zhang, H., and Fan, H. J., 2013, Fabrication of metal oxide nanobranches on atomic-layer-deposited TiO2 nanotube arrays and their application in energy storage. *Nanoscale, 5*(13), 6040-6047.

144. Organic solar cells with solution-processed graphene transparent electrodes Junbo Wu,1 Héctor A. Becerril,2 Zhenan Bao,2 Zunfeng Liu,3 Yongsheng Chen,3 and Peter Peumans4, *Applied Physics Letters, 92*, 263302 2008

145. Firooz, A. A., Mahjoub, A. R., and Khodadadi, A. A., 2009, Highly sensitive CO and ethanol nanoflower-like SnO2 sensor among various morphologies obtained by using single and mixed ionic surfactant templates. *Sensors and Actuators B-Chemical, 141*(1), 89-96.

146. Organic solar cells with solution-processed graphene transparent electrodes Junbo Wu,1 Héctor A. Becerril,2 Zhenan Bao,2 Zunfeng Liu,3 Yongsheng Chen,3 and Peter Peumans4, *Applied Physics Letters, 92*, 263302 2008.

147. Acharyya, D., and Bhattacharyya, P., 2015, An efficient BTX sensor based on ZnO nanoflowers grown by CBD method. *Solid-State Electronics, 106*, 18-26.

148. Jiang, X. H., Ma, S. Y., Li, W. Q., Wang, T. T., Jin, W. X., Luo, J., Cheng, L., Mao, Y. Z., and Zhang, M., 2015, Synthesis of hierarchical ZnO nanostructure assembled by nanorods and their performance for gas sensing. *Materials Letters, 142*, 299-303.

149. Su, L. A., Jia, W. Z., Zhang, L. C., Beacham, C., Zhang, H., and Lei, Y., 2010, Facile Synthesis of a Platinum Nanoflower Monolayer on a Single-Walled Carbon Nanotube Membrane and Its Application in Glucose Detection. *Journal of Physical Chemistry C, 114*(42), 18121-18125.

150. Enhanced electrocatalytic reduction and highly sensitive nonenzymatic detection of hydrogen peroxide using platinum hierarchical nanoflowers, Sensors and Actuators B: Chemical, Volume 192, 1 March 2014, Pages 310–316.

151. Sonker, R. K., Sabhajeet, S. R., Singh, S., and Yadav, B. C., 2015, Synthesis of ZnO nanopetals and its application as NO2 gas sensor. *Materials Letters, 152*, 189-191.

152. Tiwari, J. N., Tiwari, R. N., and Lin, K. L., 2010, Synthesis of Pt Nanopetals on Highly Ordered Silicon Nanocones for Enhanced Methanol Electrooxidation Activity. *Acs Applied Materials & Interfaces, 2*(8), 2231-2237.

153. H2 Sensing Properties of Pd Modified WO3–Fe2O3 Nanostructured Composite Films Prepared by Amorphous W–Fe Dealloying, Gao, Wubin; Gao, Wubin; Wu, Gengjiu; Wu, Gengjiu; Ling, Yunhan; Ling, Yunhan; Sun, Jialin; Sun, Jialin, Journal of Nanoscience and Nanotechnology, Volume 13, Number 2, February 2013, pp. 1190-1193(4).

154. Amin, G., Asif, M. H., Zainelabdin, A., Zaman, S., Nur, O., and Willander, M., 2012, CuO Nanopetals Based ElecTiwaritrochemical Sensor for Selective Ag+ Measurements. *Sensor Letters, 10*(3-4), 754-759.

155. Tong, G. X., Liu, F. T., Wu, W. H., Shen, J. P., Hu, X., and Liang, Y., 2012, Polymorphous alpha- and beta-Ni(OH)(2) complex architectures: morphological and phasal evolution mechanisms and enhanced catalytic activity as non-enzymatic glucose sensors. *Crystengcomm, 14*(18), 5963-5973.

156. Cheng, T. M., Huang, T. K., Lin, H. K., Tung, S. P., Chen, Y. L., Lee, C. Y., and Chiu, H. T., 2010, (110)-Exposed Gold Nanocoral Electrode as Low Onset Potential Selective Glucose Sensor. *Acs Applied Materials & Interfaces, 2*(10), 2773-2780.

157. Rajkumar, M., Hong, C. P., and Chen, S. M., 2013, Electrochemical Synthesis of Palladium Nano Urchins Decorated Multi Walled Carbon Nanotubes for Electrocatalytic Oxidation of Hydrazine and Reduction of Hydrogen peroxide. *International Journal of Electrochemical Science, 8*(4), 5262-5274.

158. Chaturvedi, P., D.C. Vanegas, S.L. Burrs, M. Taguchi, P. Sharma, E.S. McLamore (2014). A nanoceria-platinum-graphene nanocomposite for electrochemical biosensors. Biosensors & Bioelectronics, 58: 179-185.

159. Ultrasensitive non-enzymatic glucose sensor based on three-dimensional network of ZnO-CuO hierarchical nanocomposites by electrospinning, Chunyang Zhou, Lin Xu, Jian Song, Ruiqing Xing, Sai Xu, Dali Liu & Hongwei Song, Scientific Reports 4, Article number: 7382 (2014)doi:10.1038/srep07382.

160. Novel Dendritic Palladium Nanostructure and Its Application in Biosensing, Ping Zhou , Zhihui Dai ,* Min Fang , Xiaohua Huang , and Jianchun Bao, Jiangfeng Gong, J. Phys. Chem. C, 2007, 111 (34), pp 12609–12616.

161. Hasnat, M. A., Rahman, M. M., Borhanuddin, S. M., Siddiqua, A., Bahadur, N. M., and Karim, M. R., 2010, Efficient hydrogen peroxide decomposition on bimetallic Pt–Pd surfaces. *Catalysis Communications, 12*(4), 286-291.

162. Huang, C. P., Huang, Y. F., Cheng, H. P., and Huang, Y. H., 2009, Kinetic study of an immobilized iron oxide for catalytic degradation of azo dye reactive black B with catalytic decomposition of hydrogen peroxide. *Catalysis Communications, 10*(5), 561-566.

163. Ma, Y., Meng, S., Qin, M., Liu, H., and Wei, Y., 2012, New insight on kinetics of catalytic decomposition of hydrogen peroxide on ferrihydrite: Based on the preparation procedures of ferrihydrite. *Journal of Physics and Chemistry of Solids, 73*(1), 30-34.

164. Shiigi, H., Kinoshita, T., Fukuda, M., Le, D. Q., Nishino, T., and Nagaoka, T., 2015, Nanoantennas as Biomarkers for Bacterial Detection. *Analytical Chemistry, 87*(7), 4042-4046.

Chapter 3

Carbon Nanodots for Sensor Applications

Sichen Zhang[*], Xiangcheng Sun[†], Yupeng Wu[‡], Yu Lei[*,¶]

[*] *Department of Biomedical Engineering, University of Connecticut, 260 Glenbrook Road, Storrs, CT 06269, United States*

[†] *Department of Chemistry and Chemical Biology, Cornell University, Ithaca, NY 14853, United States*

[‡] *Department of Architecture and Built Environment, Faculty of Engineering, University of Nottingham, University Park, Nottingham, NG7 2RD, UK*

[¶] *Department of Chemical and Biomolecular Engineering, University of Connecticut, 191 Auditorium Road, Storrs, CT 06269, United States*

3.1. Introduction

Carbon nanomaterials including carbon nanotubes, graphene sheets, nanodiamonds, fullerenes and fluorescent carbon nanodots (carbon nanoparticles or carbon quantum dots) have attracted much research interest in recent years. Especially, carbon nanodots (C-dots), because of their excellent optical properties, biocompatibility, water solubility, photo-stability, and ease of cost-effective synthesis, have gained tremendous attention in a wide spectrum of chemical and biomedical applications, ranging from sensing, bioimaging, phototherapy, catalysis to energy-related applications. In recent years, fluorescent C-dots have been regarded as a promising candidate to replace the conventional semiconductor quantum dots in various applications [1].

Fluorescent C-dots were coincidentally discovered and produced in the process of purification of single-wall carbon nanotubes derived from candle soot by Xu et al [2]. This finding of fluorescent C-dots initiated tremendous research interests on this type of novel nanoscale carbon materials around the world. However, C-dots typically suffer from low fluorescence quantum yield in the early development stage. With the improvement of synthetic methods and doping of different elements into C-dots, the fluorescent photoluminescence (PL) properties are greatly improved and thus suitable for a wide range of applications. This short book chapter aims at summarizing recent research progress in fluorescent C-dots and emphasizes on the preparation methods of C-dots and their various applications. Starting with an introduction of different preparation methods of C-dots and the strategies that are employed to tune the fluorescence property of C-dots, this book chapter discusses various applications of C-dots in bioimaging [3], biosensing [4, 5], biomedicine [6], heavy metal ions sensing [7, 8], explosive detection [9, 10], light harvesting[11], and finally closes with a conclusion and future trends. Through this chapter, we would like to display a general picture of current studies in this field and to inspire the research in the improvement of C-dots' properties with more diverse applications.

3.2. Preparation Methods of Carbon Nanodots

Carbon dots can be prepared by a number of methods but they all generally fall into two broad categories: bottom-up and top-down approaches. In bottom-up approaches, the C-dots are obtained from assembling molecular precursors under a range of different reaction conditions, including hydrothermal, microwave-assisted, ultrasonic, acid dehydration, and pyrolytic condition. C-dots could also be prepared through "top down" approaches from large cluster of carbon materials, such as carbon nanotubes, graphite, carbon soot and so on through arc-charge, laser ablation and chemical oxidation. The preparation methods of C-dots will be discussed in the subsequent sections.

3.2.1. Hydrothermal Method

Hydrothermal method has been widely employed in the synthesis of nanostructured materials. Recently, Cui et al. [12] reported the fabrication of N-doped carbon dots using ammonium citrate, ammonium hydroxide and hydrogen peroxide as precursors under a hydrothermal condition of 180°C for 2 h. The as-prepared N-doped C-dots showed strong green fluorescence with quantum yield of 15.7%. During the synthesis, ammonium hydroxide served as nitrogen resource on one hand; on the other hand, the release of NH_3 from ammonium hydroxide during the hydrothermal process resulted in high-pressured alkaline environment, which favored the decomposition of ammonium citrate to form C-dots. The role of H_2O_2 in the synthesis was to shift the emission band of C-dots from blue to green emission due to its hydroxyl group. It was interesting to find that the effect of H_2O_2 in this hydrothermal method was different from that of previous reports in which H_2O_2 was usually employed as the fluorescence quencher [13, 14]

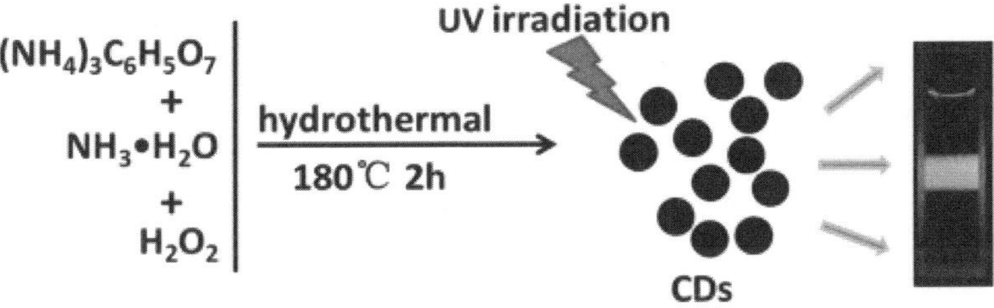

Figure 1. Hydrothermal synthesis of green fluorescent carbon nanodots using ammonium citrate, ammonium hydroxide and hydroxide peroxide as precursors. Reprinted with permission from Y. Cui, C. Zhang, L. Sun, Z. Hu and X. Liu, Particle & Particle Systems Characterization, 2015, 32, 542-546. Copyright 2015 WILEY-VCH.

3.2.2. Microwave-assisted Method

Due to its simplicity, low cost, fast reaction rate and the ability to generate a variety of C-dots in a single step, microwave-assisted method has also become popular in the synthesis of fluorescent carbon nanodots with diverse optical properties. In general, microwave heating of the precursor solution resulted in the high temperature and/or high pressure environment to trigger the decomposition/reaction of precursors, thus forming C-dots. Recently, Qu and co-workers [15] synthesized the dual emission fluorescent carbon nanodots, which simply used citric acid and urea as precursors. Under microwave-assisted conditions of 4-5 mins in a domestic 750 W microwave oven, the C-dots composed of C, H, N and O and hydroxyl, carboxyl and amine groups, which endowed the as-prepared C-dots with high water solubility, photostability and unique fluorescence property, were synthesized. However, the use of multiple precursors resulted in the difficulty to understand the synthetic mechanism. Very recently, our group synthesized blue fluorescent carbon dots using a single precursor (ammonium citrate dibasic) and a home-use microwave oven in less than one minute [16]. It was believed that the fluorescent carbon nanoparticles were formed due to polymerization of the precursor ammonium citrate dibasic followed by partial carbonization of the resulting polymer. It was presumably assumed that ammonium citrate dibasic was condensed and inter-molecularly dehydrated to form oligomers when the reaction temperature was relatively low. Further increasing the temperature due to microwave heating produced a burst of high concentration of nucleation, which grew uniformly and was further carbonized to form noncrystalline carbon nanoparticles with hydrophilic functional groups on the surface. Both the ammonium and citrate groups in ammonium citrate dibasic contributed to the formation of fluorescent carbon nanoparticles. Control experiments under similar conditions but using citric acid or ammonium carbonate respectively were conducted, and after microwave heating we observed little fluorescence under UV lamp. Following this study, the same group also synthesized N,P co-doped CDs with an average diameter of 8.1 nm (Figure 2) and dual fluorescence emission (Figure 3) and ultrahigh quantum yield through microwave-assisted approach [9].

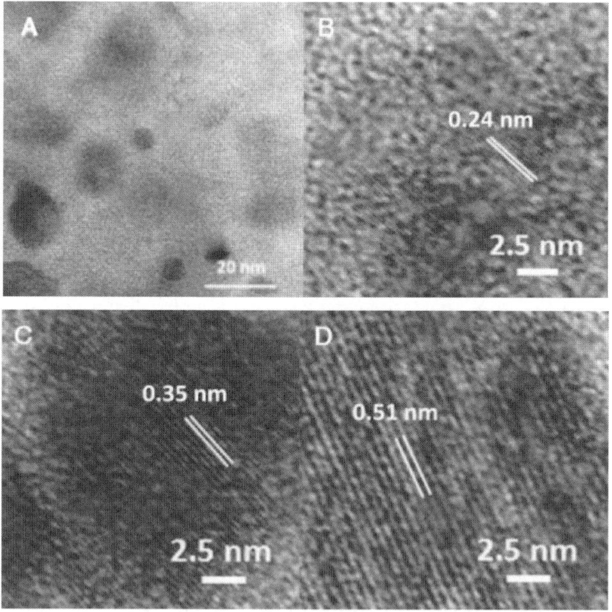

Figure 2. (A) TEM graph of the N,P-doped C-dots; (B-D) HRTEM images and lattice structures of different N,P-doped C-dots. Reprinted with permission from X. Sun, C. Brückner and Y. Lei, Nanoscale, 2015, **7**, 17278-17282. Copyright 2015 RSC.

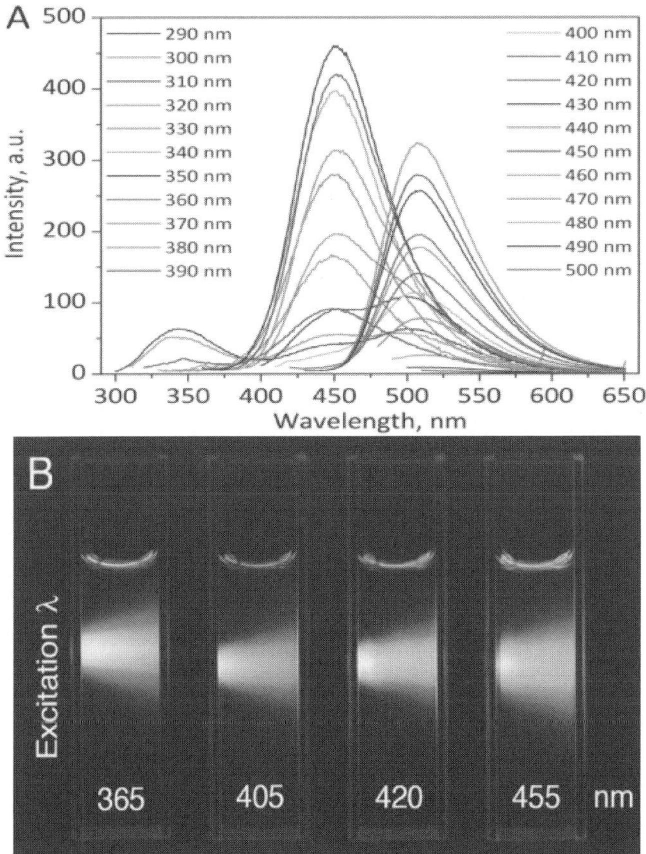

Figure 3. (A) Emission spectra (in water) of N,P-CDs at the various excitation wavelengths indicated. (B) Photographs of diluted N,P-CDs solutions (in water; quartz cuvettes) excited with fibre-optics-coupled LEDs of the wavelengths indicated, showing their excitation dependent emission properties. Reprinted with permission from X. Sun, C. Brückner and Y. Lei, Nanoscale, 2015, **7**, 17278-17282. Copyright 2015 RSC.

3.2.3. Pyrolytic Method

Pan et al. [17] and Zhou et al.[18] developed a pyrolytic method to prepare C-dots. In brief, EDTA-2Na·2H$_2$O was used as the precursor and filled in a quartz boat, and then thrusted into a quartz tube which was calcined at 400°C for 2 hours under nitrogen environment. Because of the incomplete pyrolysis of the precursor, H, N, O and Na were incorporated into the C-dots. The obtained black powder was then dissolved in acetone and centrifuged for 15 mins to remove the unreacted compounds. TEM characterization showed that the C-dots were spherical with the mean size of 3.8 nm, while the FTIR spectra in Figure 4 revealed that the peaks at around 1470 cm^{-1} and 1647 cm^{-1} were assigned to carboxylic group, and the peaks appeared at 3346 cm^{-1} was attributed to the presence of O-H bond vibration. The surface groups not only promoted hydrophilicity of the C-dots, but also enhanced their PL property.

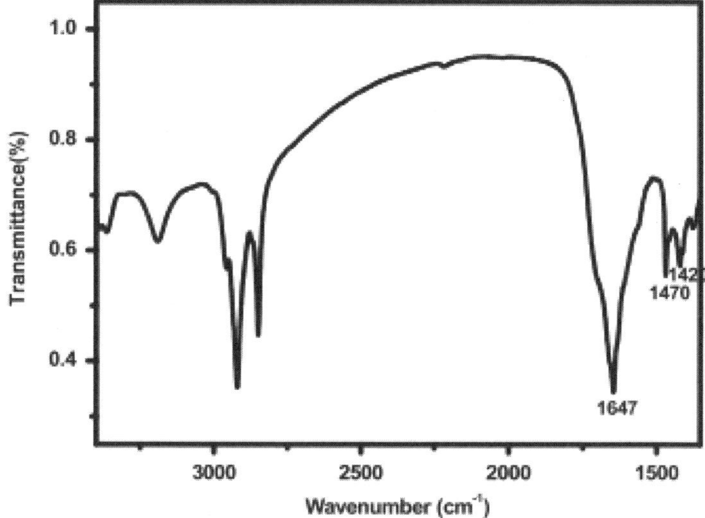

Figure 4. Fourier transform infrared (FTIR) image of C-dots for functional group analysis. Reprinted with permission form L. Zhou, Y. Lin, Z. Huang, J. Ren and X. Qu, Chemical communications, 2012, 48, 1147-1149. Copyright 2012 RSC.

3.2.4. Acid Dehydration Method

Citric acid, phosphate, sulfuric acid and nitrate have already been employed for the preparation of fluorescent carbon nanodots. Wang et al.[11] reported the synthesis of nitrogen-rich fluorescent C-dots utilizing melamine and glycerol as precursors, and citric acid, H_2SO_4 or H-ZSM-5 as acid catalyst (Figure 5). Specifically in this work, melamine was dissolved in glycerol and subsequently heated up to 270 °C with nitrogen flowing. Citric acid was quickly added into the high temperature solution and maintained at 270 °C for 15 mins and then cooled down to room temperature to obtain the dark brown solution of C-dots. Following by centrifugation and dialysis procedure, pure C-dots solution was obtained. Three fractions of C-dots with different diameters were obtained by using two dialysis membranes with molecular weight cut-off (MWCO) 3500 and MWCO 1000, respectively. The higher oxygen content accompanied with lower carbon percentage for small diameter fraction of C-dots suggested that the fraction of C-dots with larger diameter has underwent further dehydration and condensation. According to the FTIR spectra, the appearance of new surface functional groups and similar surface structure of the C-dots indicated the C-dots were produced through one spot reaction between melamine and glycerol. The effect of citric acid serving as catalyst was also investigated. The quantum yield of the C-dots obtained without the use of citric acid was about 50% lower than those which underwent acid dehydration reaction.

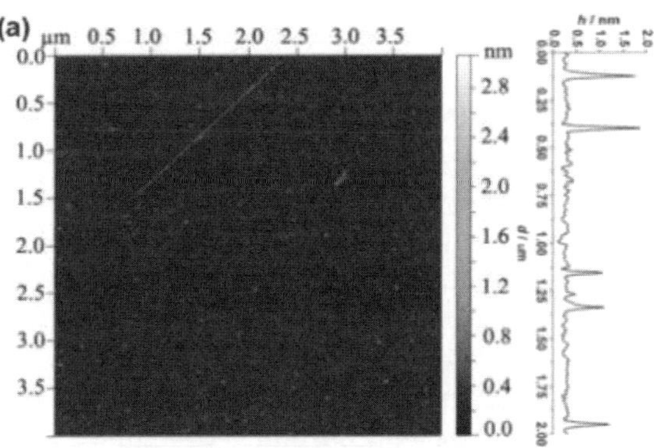

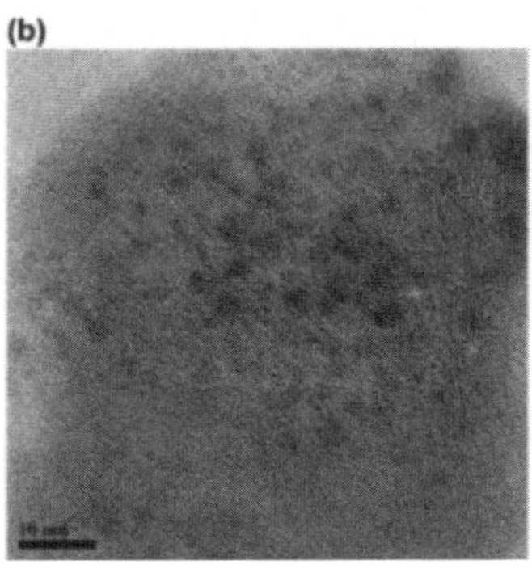

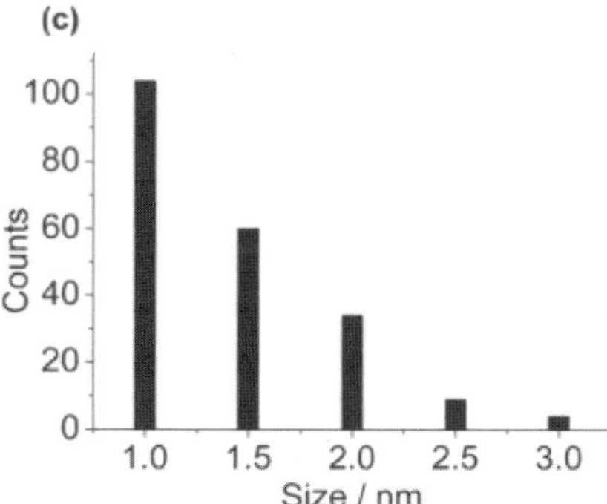

Figure 5, AFM (a) and TEM (b) images of C-dots, and the size distribution of C-dots (c) based on AFM image analysis. Reprinted with permission from C. Wang, X. Wu, X. Li, W. Wang, L. Wang, M. Gu and Q. Li, Journal of Materials Chemistry, 2012, 22, 15522-15525. Copy right 2012 RSC.

3.2.5. Arc-discharge Method

Fluorescent carbon nanodots were firstly reported by Xu's group[2] when they purified single-walled carbon nanotubes (SWNTs) from arc-discharge soot. Three discriminate substances, which were long nanotubes, tubular material and fluorescent carbon material (Figure 6), were obtained in the process of purification of arc-discharge soot suspension through electrophoretic method. The separated and purified fluorescent carbon materials were further segregated based on the different color of fluorescence emission under 366 nm excitation. The molecule weight of the fluorescent carbon materials could be correlated to the emission wavelength of fluorescent materials, and the nominal molecular weights of the green-blue, yellow and orange were 3000-10000, 10000-30000, 30000-50000, respectively.

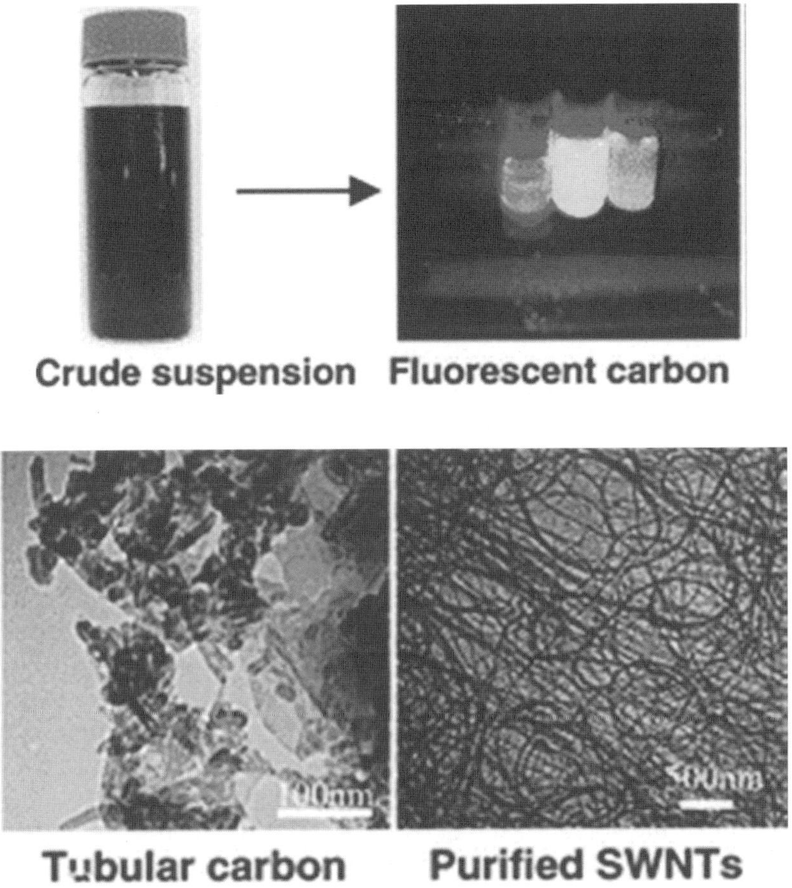

Figure 6. Fluorescent C-dots, tubular carbon and purified SWNTs prepared through arc-discharge method. Reference with permission. Reprinted with permission form X. Xu, R. Ray, Y. Gu, H. J. Ploehn, L. Gearheart, K. Raker and W. A. Scrivens, Journal of the American Chemical Society, 2004, 126, 12736-12737. Copyright 2004 ACS publications.

Carboxylic groups were functionalized on the as-prepared carbon nanomaterials which were proved by FTIR characterization. In conjunction with electrophoresis-based separation, arc-discharge method offers a facile way to prepare C-dots.

3.2.6. Laser Ablation

Adapting a top-down approach, Hu et al.[19] reported a novel one-step synthesis of fluorescent carbon nanodots, utilizing laser irradiation of carbon powders in organic solvent such as diamine hydrate, diethanolamine, and poly(ethylene glycol) (PEG200N). Taking PEG_{200N} as an example of a representative solvent (Figure 7), graphite powders were dispersed in PEG_{200N} followed by simultaneous treatment of laser irradiation and ultrasonication for 2 hours. Laser irradiation was conducted with a wavelength of 1.064 μm and power density of 6.0×10^6 W cm^{-2} while ultrasound was employed to facilitate the movement of carbon particles. The black supernatant (Sample A) was obtained from the suspension via centrifugation. In order to understand the relationship between surface functional groups and PL of fluorescent carbon nanodots, both the suspension of graphite powders in DI water using the same procedure described above and the suspension of graphite powders in DI water followed by subsequently heating in the mixture of perchloric acid ($HClO_4$) and PEG_{200N} (Sample B) were used as a control group (Figure 8). The C-dots obtained from the supernatant of graphite powders dispersed in PEG_{200N} and the mixture of $HClO_4$ and PEG_{200N} with graphite powder had emission peak of 490 nm, while that synthesized in DI water did not show any PL emission. In addition, the functional group such as carboxylate group appeared on the fluorescent C-dots, but was not presented in non-fluorescent materials. These results not only indicated the effect of the solvent used on the synthesis of fluorescent C-dots but also revealed that the surface functional groups played an important role in the origin of the fluorescence. Further study of the effect of the carbon source and organic solvent on the fluorescence properties of carbon dots concluded that the PL property of C-dots was not obviously influenced by the type of carbon source in the same organic solvent. However, the emission peaks of the C-dots obtained in different solvents (PEG_{200N}, diamine hydrate and diethanolamine) were different due to the formation of different surface functional groups. Furthermore, the PL peak of C-dots shifted to a longer wavelength with increasing of PEG molecular weights was also proposed.

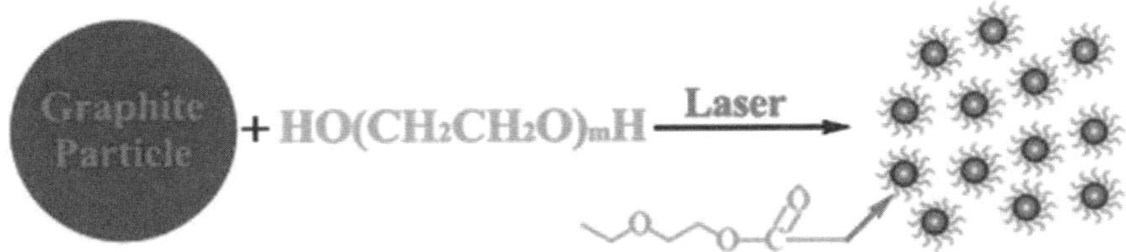

Figure 7. Schematic map of the one-step synthesis of luminescent C-dots in PEG200N solvent. Reprinted with permission from S.-L. Hu, K.-Y. Niu, J. Sun, J. Yang, N.-Q. Zhao and X.-W. Du, Journal of Materials Chemistry, 2009, 19, 484-488. Copyright 2009 RSC.

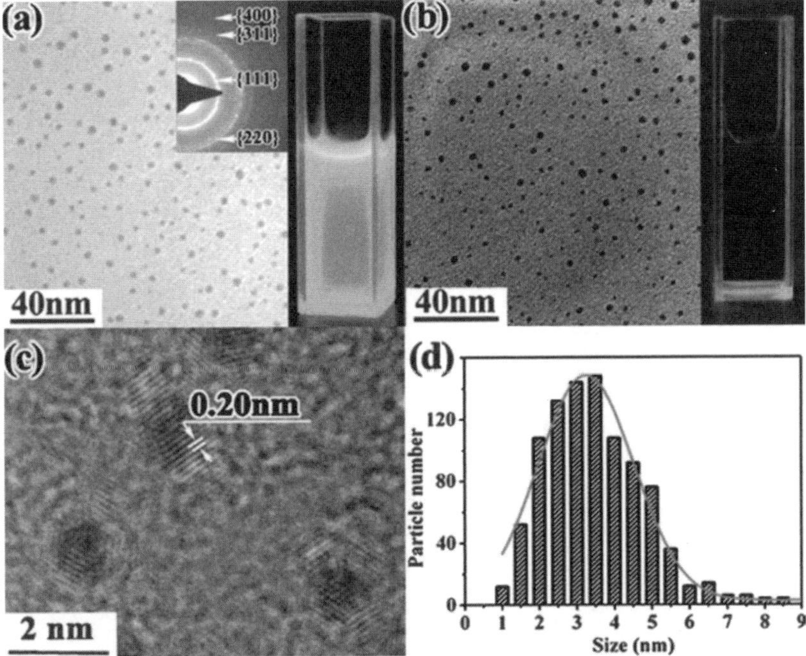

Figure 8. Representative TEM images of the C-dots and direct photographs of suspensions under a 365 nm UV lamp. (a) TEM image of the C-dots in sample A and direct photograph of sample A under a 365 nm UV lamp, the inset is the SAED pattern of the C-dots; (b) TEM image of the C-dots in sample B and direct photograph of sample B under a 365 nm UV lamp; (c) HRTEM image of the C-dots in sample A; (d) the size distribution of the C-dots in sample A. Reprinted with permission from S.-L. Hu, K.-Y. Niu, J. Sun, J. Yang, N.-Q. Zhao and X.-W. Du, Journal of Materials Chemistry, 2009, 19, 484-488. Copyright 2009 RSC.

3.2.7. Chemical Oxidation

Utilizing candle soot, a common carbon source, with an oxidative acid treatment, Liu et al.[20] prepared multi-color fluorescent C-dots (Figure 9). The candle soot was treated with 5 M HNO_3 and refluxed for 12 hours, and then the supernatant was separated from the suspension by centrifugation. The pure fluorescent C-dots were separated from the neutralized candle-soot dispersion through denaturing PAGE gel electrophoresis. The fluorescent carbon nanodots separated in the PAGE gel showed distinguished colorful bands under 321 nm UV light. The different C-dots were collected from the gel and dissolved in DI water. The PL spectra showed that the emission peaks of different C-dots were ranged from 415 nm to 615 nm under the same UV light. However, the quantum yields of the as-synthesized C-dots were lower than 1%, relatively poorer than other literature reports. In addition, the PL properties of the C-dots showed strong pH-dependency, and the fluorescence intensity could decrease by 40-89% if the pH value of carbon dots solution changed from neutral to acidic or basic.

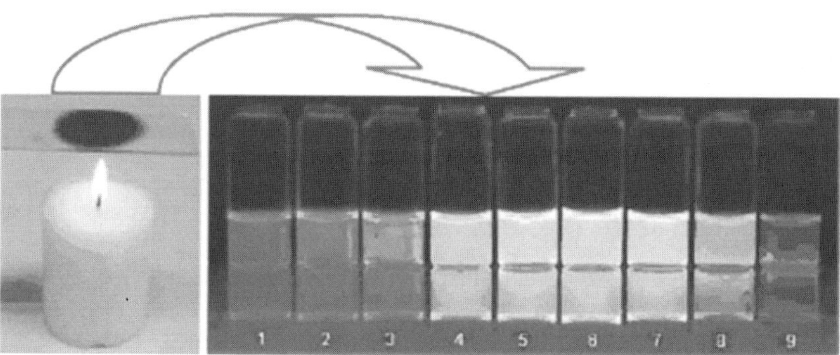

Figure 9. Water-soluble, multicolor fluorescent carbon nanoparticles are prepared by refluxing candle soot with nitric acid. Reprinted with permission from H. Liu, T. Ye and C. Mao, Angewandte Chemie International Edition, 2007, 46, 6473-6475. Copyright 2007 WILEY-VCH.

3.3. Applications of C-dots

On the basis of the unique fluorescent properties, good aqueous solubility, low toxicity and photostability of C-dots, we will advance the discussion about the applications of C-dots in various fields, namely, bioimaging, nanoplatform therapy, metal ions and biologically meaningful molecule detection, fluorescent ink, explosives detection and light harvesting.

3.3.1. Bio-imaging

Fluorescent C-dots meet the prerequisites for bio-application, which typically requires materials with low toxicity, biocompatibility and photo-stability. Generally, the longer the emission wavelength of fluorescent carbon nanodots is, the better for bioimaging, in order to eliminate the effects of the autofluorescence in cells and tissue as well as the potential damage to the cells and tissue caused by short wavelength light.

Recently Cui et al.[12] reported on the one-step synthesis of green fluorescent C-dots, which could be applied for the bio-imaging of live human HeLa cancer cells without extra surface modification. The excitation wavelength-dependent properties of the carbon dots were revealed by the fluorescence emission spectra, and the emission peaks located from 500-580 nm with the variation of excitation wavelength. The cytotoxicity of the C-dots was analyzed by treating the HeLa cells with different dose of as-synthesized C-dots ranging from 0-2 mg/mL. Cell viability was not obviously decreased with the increase of the C-dots concentrations. 98% cells were viable after 24 hours incubation even though they were treated with high concentration of C-dots (2 mg/mL). After direct treatment of HeLa cells with the as-prepared C-dots, HeLa cells showed strong green fluorescence emission under blue light excitation (Figure 10), indicating that C-dots were an excellent fluorescent material for cell imaging.

Figure 10. Bright field and green fluorescence microphotograph of HeLa cells treated with as-synthesized C-dots. Reprinted with permission from Y. Cui, C. Zhang, L. Sun, Z. Hu and X. Liu, Particle & Particle Systems Characterization, 2015, 32, 542-546. Copyright 2015 WILEY-VCH.

Many other research groups have also been exploiting the applications of C-dots in bio-imaging. For instance, Gong et al.[21] reported a facile and green synthetic route, which utilized pumpkin as the source under low-temperature (90 °C) hydrothermal treatment in the presence of H_3PO_4, to prepare phosphorous and nitrogen co-doped C-dots with the yellow fluorescence emission (Figure 11). The as-synthesized C-dots were demonstrated with the capabilities for cell imaging and real-time tracking of the pH variation in cells. Recently Shi et al.[3] also prepared C-dots with dual-emission using naked oats through pyrolysis and microwave-assisted treatment. The biocompatible C-dots had the potential to be applied in bio-imaging of living cells. Jiang et al.[22] used hydrothermal method to produce a yellow fluorescent C-dots from 1,2,4-triaminobenzene. The as-prepared C-dots not only were applied in cell imaging, but also displayed bifunctional sensing property, which can be used for monitoring Ag^+ in aqueous solution and Cys in human plasma sensing.

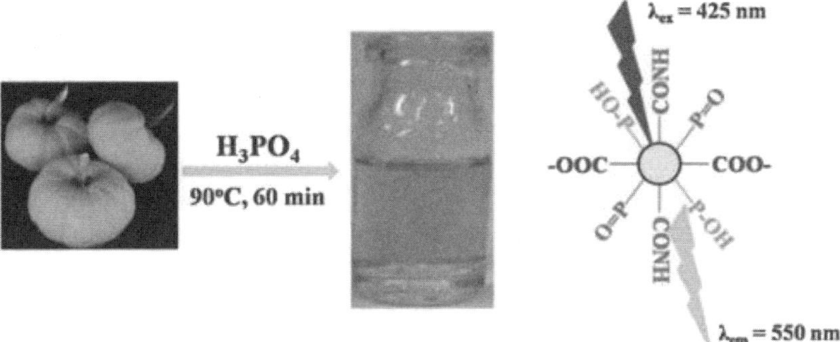

Figure 11. Scheme of C-dots synthetic route and the proposed structure of as-synthesized C-dots derived from pumpkin. Reprinted with permission from X. Gong, W. Lu, Y. Liu, Z. Li, S. Shuang, C. Dong and M. M. Choi, Journal of Materials Chemistry B, 2015, 3, 6813-6819. Copyright 2015 RSC.

3.3.2. Biomolecular Sensing

Biologically meaningful molecules, such as proteins, phytic acid or other small molecules in cells, play important roles in the physiological process. The abnormal level of these molecules in cells could lead to diseases. Consequently, there is an urgent need to develop selective and sensitive probes for monitoring concentration variation of these molecules.

Mercapto biomolecules including cysteine (Cys), homocysteine (Hcy), and glutathione (GSH) were regarded as important indexes for human health due to their important roles in cellular antioxidant defense systems and maintaining biological redox homeostasis. Recently a C-dots based ratiometric fluorescence probe for mercapto biomolecules sensing was developed by Lan's group[23]. The C-dots with 30% quantum yield were prepared by employing melamine and trisodium citrate as precursors through a microwave-assisted hydrothermal method. The sensing probe for biothiols was designed by first adding 2 mM Hg^{2+} into C-dots solution in an appropriate ratio, and thus the blue fluorescence of C-dots was fully quenched by the addition of mercury solution due to electrostatic interaction between the surface functional groups (ammonia, hydroxyl group and carboxylic group) on C-dots and Hg^{2+} (Figure 12). The aforementioned non-fluorescent C-dots/Hg^{2+} complex system could be applied for sensing biothiols via the tunable fluorescence of C-dots. The quenched fluorescence by Hg^{2+} could be rapidly recovered by addition of the mercapto biomolecules in 10 seconds because the strong reaction between Hg^{2+} and thiol group, thus resulting in the fluorescence "turn-on" of C-dots. The biocompatibility of C-dots/Hg^{2+} complex was also investigated by treating the A549 cells with different concentrations of C-dots/Hg^{2+} complex prepared using C-dots in the range of 0-500 µg/mL of C-dos and 10 µM Hg^{2+}. The viability of the A549 cells did not show an obvious decrease in comparison to that of the control group (without addition of C-dots/Hg^{2+}), indicating that the low toxicity of designed C-dots/Hg^{2+} complex and its high potential for biothiols sensing *in vivo*. When living A549 cells were incubated with C-dots/Hg^{2+}, no fluorescence signal was observed. By further adding Cys into the culture media, the Cys was uptaken by cells and interacted with Hg^{2+}, thus turning on blue fluorescence of C-dots. The limit of detection (LOD) of the fluorescence "turn-on" probe for biothiols was 15 nM. The probe also showed high selectivity against other common metal ions and amino acids without thiol group, indicating that the as-prepared C-dots offer an excellent sensing platform for sensitive and selective detection of mercapto thiols.

Figure 12. Schematic diagram showing the fabrication process of the C-dots/Hg^{2+} sensor. The fluorescence of the sensor can be turned on by addition of Cys. Reprinted with permission from M. Lan, J. Zhang, Y.-S. Chui, H. Wang, Q. Yang, X. Zhu, H. Wei, W. Liu, J. Ge and P. Wang, Journal of Materials Chemistry B, 2015, 3, 127-134. Copyright 2015 RSC.

Several other fluorescence "turn off" or "turn-on" systems for biologically meaningful molecules based on the tunable PL property of C-dots have also been reported in recent years. Gao et al.[4] reported a C-dot based fluorescent probe for phytic acid sensing. In this study, the C-dots were prepared using citric acid and lysine as carbon source and surface passivation reagent, respectively, under hydrothermal treatment in autoclave. The obtained C-dots emitting bright blue fluorescence was fully quenched by ferric ions to form the C-dots/Fe^{3+} complex, which could serve as the sensing probe. The fluorescence of the biocompatible and low toxic C-dots/Fe^{3+} complex would be recovered in the presence of phytic acid, which can be correlated to the concentration of phytic acid (Figure 13). The C-dots/Fe^{3+} based probe could detect phytic acid as low as 0.36 µM with high sensitivity and selectivity. Recently Wang et al.[24] investigated a class of N,S- co-doped fluorescent C-dots, which employed ammonium persulfate, glucose, and ethylenediamine as precursors and prepared under hydrothermal conditions. The obtained C-dots were able to detect the methotrexate in human serum with the detection limit of 0.33 nM based on fluorescence quenching property. All these sensing application opens a new research direction for C-dots.

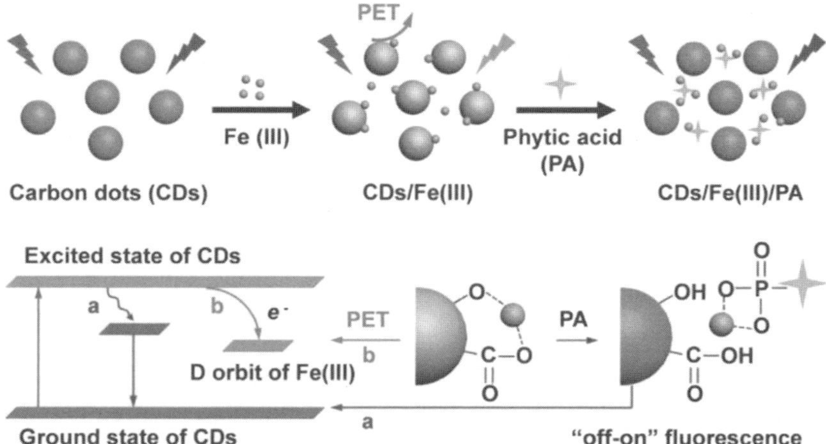

Figure 13. Schematic illustration of fluorescence response of C-dots in the presence of Fe (III) ions and Fe (III)/PA. Mode a: photo-induced fluorescence emission yielding bright blue luminescence. Mode b: photo-induced electron transfer (PET) in which an electron in the excited state enters the unfilled d orbit of Fe, leading to fluorescence quenching of the C-dots. Reprinted with permission from Z. Gao, L. Wang, R. Su, R. Huang, W. Qi and Z. He, Biosensors and Bioelectronics, 2015, 70, 232-238. Copyright 2015 ELSEVIER.

3.3.3. Phototherapy

Phototherapy has received specific attention in recent years, due to its minimal invasiveness and the realization of localized treatment. Generally, there are two categories of phototherapy for cancers treatment, namely photodynamic therapy (PDT) and photothermal therapy (PTT). In PDT treatment, photosensitizers (PS) are used as PDT agents to react with the oxygen in tissue under laser irradiation, thus generating reactive oxygen species (ROS) to kill cancer cells, while in PTT treatment, PS are excited by infrared wavelength light and release vibration energy. Consequently, the heat produced by the excited PS could kill the target cells.

Ge et al.[6] reported a novel fluorescent C-dots based nanomedicine, which combined the therapy function of PDT and PTT for cancer treatment. The precursors of C-dots were synthesized from the thiophene benzoic acid through a series of reactions, forming polythiophene phenylpropionic acid. The obtained dark red polythiophene phenyl-propionic acid polymers were then dissolved in alkaline solution to produce C-dots through hydrothermal method. The as-synthesized C-dots emitted bright red fluorescence, with the emission wavelengths ranging from 640 to 680 nm. The cytotoxicity of the C-dots and its application in PDT/PTT were analyzed *in vitro* by incubating B16-F0 skin cancer cells with as-synthesized C-dots followed by laser irradiation under different powder densities (Figure 14). There was no cell death observed in the presence of 200 µg/mL of C-dots or in the absence of C-dots but under laser irradiation (0.1 or 2 W cm^{-2}), indicating the low toxicity of the as-prepared C-dots and the negligible killing effect of laser-produced heat on the viability of cancer cells. However, in the presence of C-dots and under irradiation of 635 nm laser with low power density (0.1 W cm^{-2}), a portion of cells died due to the sole effect of PDT. More interestingly, in the presence of C-dots and under irradiation of 635 nm laser with high power density (2 W cm^{-2}), all cancer cells were killed, accompanying the increase of temperature in the incubated cells solution up to 50 °C within 5 mins. This result strongly supported the killing effect to cancer cells can be ascribed to the synergistic effects of PDT and PTT when higher power density laser was used. The *in vivo* therapy efficacy of C-dots was further investigated using HeLa-bearing nude tumor mice model. After the injection of 100 µL of 2 mg/mL C-dots dissolved in PBS, the cancer cells were partially destroyed and slightly delayed increase of the tumor size was observed under lower power laser irradiation (0.1 W cm^{-2}), which was consistent with *in vitro* result. This result also indicated that PDT therapy itself was insufficient to kill cancer cells. Alternatively, after localized treatment using higher power laser irradiation (2 W cm^{-2}), all cancer cells were damaged. The therapy efficacy of PDT was evaluated to be 27% using the ROS ($^1O_2$) QY analysis, and the PTT conversion efficiency could reach 36.2 %. Further histological analysis of the treated mice showed that there was no damage observed in main organs. All these features suggest that the as-synthesized C-dots hold great potential in cancer phototherapy.

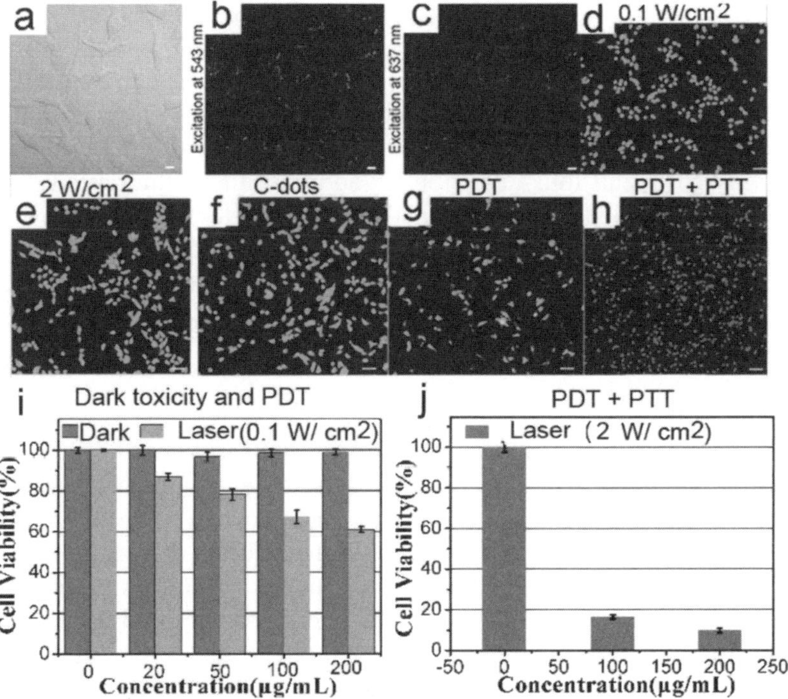

Figure 14. *In vitro* imaging and PDT/PTT. Confocal microscopy images of B16-F0 skin cancer cells incubated with C-dots (20 µg/mL) for 4 h at bright field a), Ex = 543 nm b), and Ex = 637 nm c). Scale bar = 10 µm. Fluorescence images of calcein AM/PI-stained B16-F0 skin cancer cells incubated with various media: d) laser only (0.1 W cm^{-2}), e) laser only (2 W cm^{-2}), f) C-dots only (200 µg/mL), g) 200 µg/mL of C-dots + laser (635 nm, 0.1 W cm^{-2}, PDT), and h) 200 µg/mL of C-dots + laser (635 nm, 2 W cm^{-2}, PDT/PTT). Scale bar = 100 µm. Relative viability of B16-F0 skin cancer cells incubated with various concentrations of C-dots under dark or irradiation by 635 nm laser at power densities of i) 0.1 W cm^{-2} (PDT) and j) 2 W cm^{-2} (PDT/PTT) for 10 min. Reprinted with permission from J. Ge, Q. Jia, W. Liu, M. Lan, B. Zhou, L. Guo, H. Zhou, H. Zhang, Y. Wang and Y. Gu, Advanced Healthcare Materials, 2015. Copyright 2015 WILEY-VCH.

3.3.4. Heavy Metal Ions Detection

Heavy metal ions play important and essential roles in many fields. Tiny amount of the metal ions change in the human body will affect physiological activities. In addition, some heavy metal ions are highly toxic to cells and tissues. Therefore, in recent years, researchers have engineering C-dots through surface functionalization during synthesis and then developed fluorescent C-dots based probes for selective and sensitive metal ions detection. Surface functionalization of the C-dots has been realized during the C-dots synthesis through different doping strategies (e.g., nitrogen, sulfide or phosphor doping). The electron-donating elements not only enhance the PL of C-dots, but also endow them the capability for metal ions detection. In addition, those fluorescent C-dots could possess good selectivity and selectivity for metal ions due to abundant carboxyl groups, hydroxyl groups and nitrogen groups on the surface of C-dots. The fluorescence of the C-dots can be quenched through the electron transfer process between the functionalized C-dots and the electron-deficient heavy metals ions (e.g. Hg^{2+}, Fe^{3+}, and Cu^{2+}) even at very low concentration.[25-28]

Gong et al.[21] reported the nitrogen-doped C-dots for ferric ions detection. Chitosan, acetic acid and 1, 2-ethylenediamine (EDA) were utilized as carbon and nitrogen sources under pyrolysis in the domestic microwave oven. The dark brown solution of the C-dots emitted strong blue fluorescence under UV-light and the maximum PL intensity of emission peak was at around 440 nm. The sensitivity and selectivity of C-dots for Fe^{3+} were demonstrated by titrating the as-prepared C-dots aqueous solution (0.10 mg/mL) with different Fe^{3+} concentrations and other common metal ions at 10 mM), respectively. The PL intensity of the C-dots was not obviously changed except for slightly decreased by Fe^{2+}, Cu^{2+}, Pb^{2+}, Hg^{2+}, Ag^{2+}, Cd^{2+} and Cr^{3+} ions, compared with that was fully quenched by Fe^{3+} (Figure 15). The detection limit for Fe^{3+} ion could reach as low as 10 ppb. The biocompatibility of the as-synthesized C-dots was investigated through standard MTT assay, and the C-dots also showed good performance for Fe^{3+} sensing in cells.

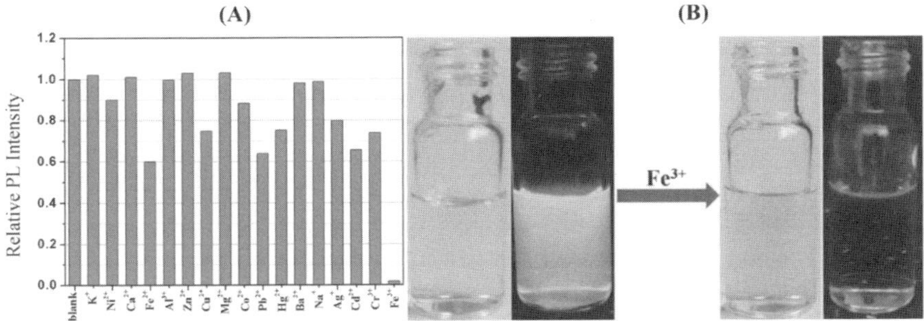

Figure 15. (A) The relative change in PL intensity of N-C-dots (0.10 mg/mL) after reacting with different metal ions (10 mM). (B) Photographic images of N-C-dots (0.10 mg/mL) before and after adding Fe^{3+} (280 ppm) under daylight (left panel) and UV irradiation (right panel). Reprinted with permission from X. Gong, W. Lu, M. C. Paau, Q. Hu, X. Wu, S. Shuang, C. Dong and M. M. Choi, Analytica chimica acta, 2015, 861, 74-84. Copyright 2015 ELSEVIER.

Recently Wen et al.[29] also prepared C-dots using cotton through pyrolysis and microwave methods. The as-prepared blue fluorescent C-dots showed high selectivity and sensitivity toward Fe^{3+} detection with a limit of detection as low as 2.7×10^{-8} M. Other C-dots based probes for Fe^{3+} sensing have also been reported in literatures. For instance, Jiang et al.[7] reported N-doped C-dots that was produced from citric acid and isoleucine using hydrothermal method. The as-synthesized C-dots showed high selectivity and sensitivity for Fe^{3+} compared with other competitive metal ions. Beyond Fe^{3+} detection, it is also important to detect Hg^{2+} ions due to its high toxicity and frequent occurrence in water system. Li et al.[30] demonstrated that their synthesized C-dots possessed high selectivity against other interference ions and the limit of detection for Hg^{2+} was as low as 10 nM. Lan et al.[31] also reported the C-dots based ratiometric fluorescent sensor for Hg^{2+} detection in buffer solution and living cells. Additionally, Liu et al.[32] reported a green route to synthesize C-dots under hydrothermal treatment of grass. The obtained N-doped C-dots exhibited high sensitivity for Cu^{2+} ions detection. Jiang et al.[22] reported the yellow emissive C-dots synthesized from 1,2,4-Triaminobenzene hydrothermally and its application in Ag^+ sensing. Furthermore, Qu et al.[33] investigated the possibility to differentiate multiple metal ions using dual-emission C-dots and the results showed that the prepared C-dots could successfully distinguish Fe^{3+}, Bi^{3+} and Al^{3+} due to the fluorescence quenching in different wavelength regions.

3.3.5. Fluorescent Ink

Fluorescent C-dots used as fluorescent ink was rarely reported in previous literatures because of the limitations such as low quantum yield of fluorescent C-dots and self-quenching of high dose of C-dots via aggregation formation.

Recently Qu's group[15] reported dual fluorescence emission of C-dots, which can be used as fluorescence ink. The fluorescent C-dots were synthesized using citric acid and urea as precursors through one-step microwave-assisted method. The obtained brown fluorescent C-dots solution showed excitation-wavelength-dependent PL property. The emission peaks located from 420 to 570 nm, which were responsible to blue, green and red emission colors. In addition, when the C-dots aqueous solution coated on the commercially available lens paper, the excitation-wavelength-dependent PL property could also be maintained, and the maximum fluorescence intensity of the C-dots coated on the paper was enhanced by 40%, indicating that the as-synthesized C-dots can be used as fluorescent ink. As shown in Figure 16, characters "395" were written on the lens paper with C-dots solution and then closed the "circles" of 395 with commercial green fluorescent ink to result in characters "888". Under blue UV-light irradiation, the green fluorescent number "888" appeared on the paper. Under green UV-light, the green fluorescent ink was invisible and only "395" written with C-dots solution was observed with red fluorescence. Based on the dual-fluorescence-emission property, the integrated system combining C-dots and commercial green fluorescent ink may provide a new route in counterfeiting and information encryption.

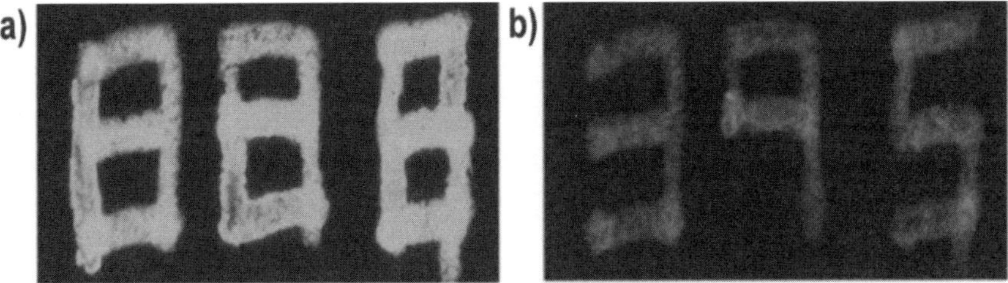

Figure 16. Fluorescent characters on commercially available lens paper under different excited UV-light. a) Characters 888 written by commercial green fluorescent ink and as-synthesized C-dots under blue UV-light; b) just characters 395 labeled with C-dots solution showed red fluorescence under green UV-light. Reprinted with permission from S. Qu, X. Wang, Q. Lu, X. Liu and L. Wang, Angewandte Chemie, 2012, 124, 12381-12384. Copyright 2012 WILEY-VCH.

The fluorescent C-dots with high quantum yield were also utilized as fluorescent ink for patterning and painting purpose. Wen et al.[29] reported the fluorescent C-dots derived from cotton for multicolor patterning. The obtained C-dots aqueous solution was drawn on commercially available papers without UV fluorescence background. The C-dots labeled area on papers showed blue and green fluorescent patterns under 365 and 470 nm excitation, respectively. In addition, the bright blue and green patterns still could be observed after several months, demonstrating the excellent photostability of C-dots. Wang et al[34] also developed a simple and large-scale preparation route of C-dots. The excitation-wavelength-dependent PL behavior enabled the as-synthesized C-dots to be used as printing inks. The as-prepared C-dots aqueous sample was filled in a commercial inkjet printer and the printed words and images showed green and red fluorescence under 455 and 523 nm excitation, respectively, as shown in Figure 17. Its application as fluorescent ink was also demonstrated using other substrates such as silk fibers and silk fabric.

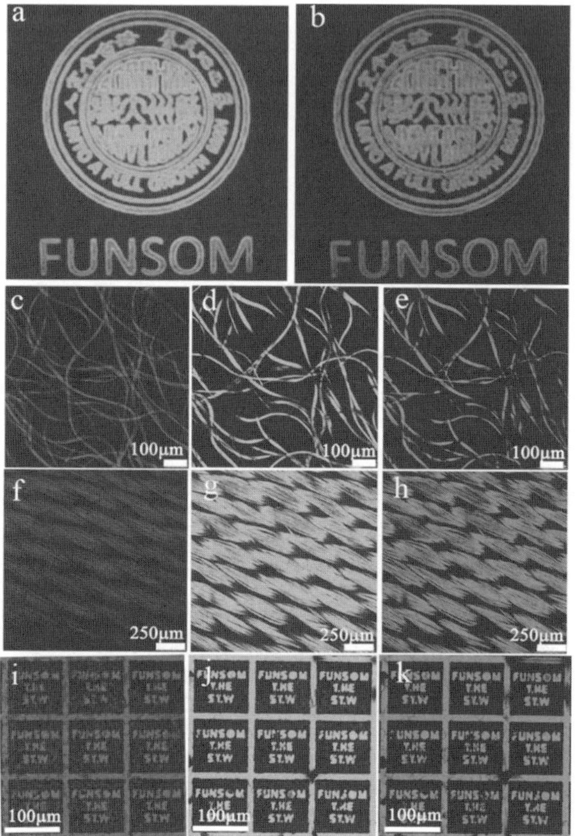

Figure 17. (a,b) Fluorescent images on commercially purchased paper stained with the C-dot ink captured by the CRI Maestro *in vivo* imaging system with excitation at 455 nm, emission at 500–600 nm, and exposure time 30 ms (a) and excitation at 523 nm, emission at 560–650 nm, and exposure time 75 ms (b). (c–e) Fluorescence images of the C-dot-stained silk fibers: (c) λ excitation = 405 nm; emission band pass: 420–480 nm; (d) λ excitation = 488 nm; emission band pass: 500–580 nm; (e) λ excitation = 543 nm; emission band pass: 580–660 nm. (f–h) Fluorescence images of the C-dots-coated silk fabric: (f) λ excitation = 405 nm; emission band pass: 420–480 nm; (g) λ excitation = 488 nm; emission band pass: 500–580 nm; (h) λ excitation = 543 nm; emission band pass: 580–660 nm. (i–k) C-dot ink is drop-cast on a hydrophilic photolithography pattern: (i) λ excitation = 405 nm; emission band pass: 420–480 nm; (j) λ excitation = 488 nm; emission band pass: 500–580 nm; (k) λ excitation = 543 nm; emission band pass: 580–660 nm. All the pictures are imaged with a laser-scanning confocal microscope (LSCM). Reprinted with permission from J. Wang, F. Peng, Y. Lu, Y. Zhong, S. Wang, M. Xu, X. Ji, Y. Su, L. Liao and Y. He, Advanced Optical Materials, 2015, **3**, 103-111. Copyright 2014 WILEY-VCH.

3.3.6. Explosives Detection

Instant, sensitive and selective detection of explosives has gained increasing attention in recent years because of the implications in homeland security, global demining and environmental safety. Although trinitrotoluene (TNT) is the most widely used military explosives, picric acid (PA, also known as 2,4,6-trinitrophenol) is regarded as an even more violent explosive than the well-known TNT. Therefore, PA has been widely used in the preparation of lethal weapons, manufacture of rocket fuels, dyes, fireworks and matches. In addition, the wide use of PA could contaminate our environment and thus PA has been recognized as an environmental pollutant. Consequently, it is highly demanded to develop novel sensory materials for highly sensitive and selective picric acid detection. Among all sensing technologies, fluorescence-based explosive detection possesses the advantages of simple operation, cost effectiveness, quick response and high sensitivity.

Recently our group developed an ultrafast synthetic route to produce fluorescent C-dots by employing ammonium citrate dibasic (ACD) as single precursor which was further applied for picric acid (PA) sensing.[16] The obtain C-dots solution showed excitation-wavelength-independent behavior and its blue fluorescence emission could be sensitively and selectively quenched by picric acid explosive. Upon addition of different concentrations of PA solutions in the range of 0-100 µM (Figure 18), the fluorescent intensity of the C-dots solution was gradually quenched, while other nitro-explosives (TNT, DNT, 2-NT, DNB and NM) affected the fluorescence intensity insignificantly, indicating good selectivity in PA detection. The sensing mechanism was systematically investigated and the synergistic effect of its low molecular orbitals, the presence of fluorescence resonance energy transfer as well as acid-base interactions between picric acid and fluorescent C-dots was believed to contribute to the excellent sensing performance to picric acid. This study demonstrated a simple way to prepare highly fluorescent C-dots and also broadened the application of C-dots.

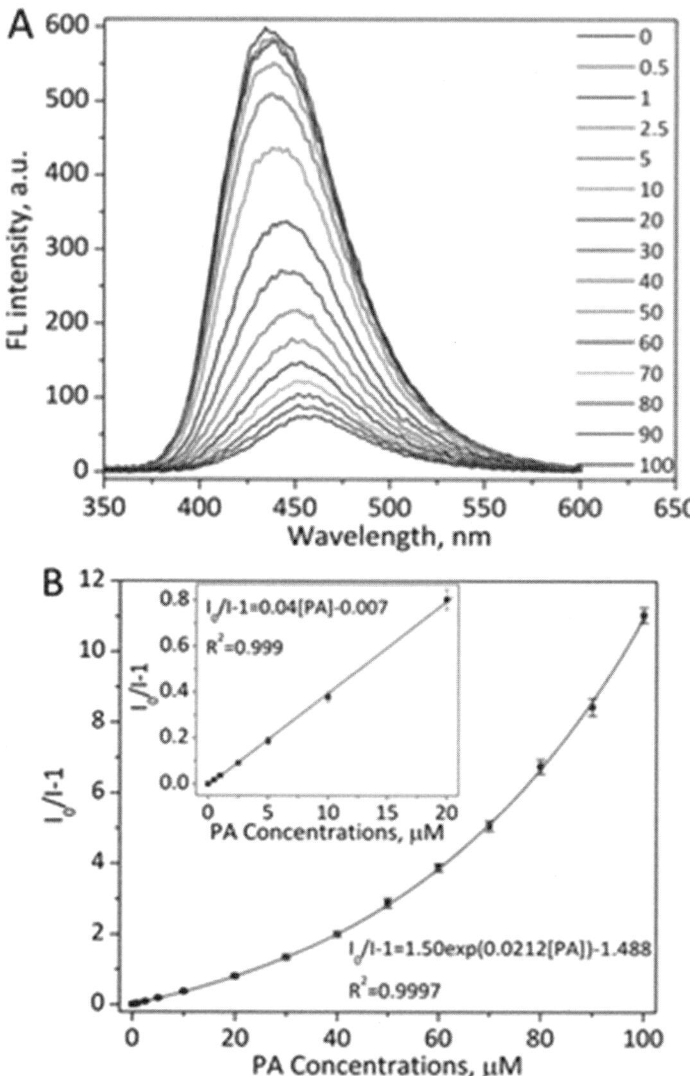

Figure 18. (A) Fluorescence spectra of C-dots upon addition of various PA concentrations (0-100 µM) in water. (B) Fitting curve of the nonlinear Stern-Volmer plot for PA by exponential quenching equation, Inset shows the fitting curve of the linearity of Stern-Volmer plot for PA of 0-20 µM. Reprinted with permission form X. Sun, J. He, Y. Meng, L. Zhang, S. Zhang, X. Ma, S. Dey, J. Zhao and Y. Lei, Journal of Materials Chemistry A, 2016. 4, 4161–4171. Copyright 2016 RSC.

Obtaining fluorescence emission with different colors has been considered as a key requirement for successfully utilizing carbon dots in various applications. So far most of the carbon dots emit blue color under ultraviolet excitation. Although the excitation-dependent fluorescence emission has been reported, a significant decrease in fluorescent intensity was also observed with changing of excitation wavelength. Researchers are trying to develop synthetic methods to shift the maximum emission light from blue, to green, yellow, and red in order to broaden the applications of C-dots. However, most of current synthesis endured issues such as complex procedures, toxic components or low quantum yield. It is still difficult to synthesize C-dots through a facile strategy with different fluorescence emission and high quantum yield in one single material. The emission of C-dots could be tuned by controlling the dehydration reaction, chemical manipulation, doping of other elements. Nitrogen is by far the most prominent doping candidate into carbon to obtain fluorescent C-dots. Doping the carbon materials with other heteroatoms such as boron (B), and sulfur (S) has also been reported. In another work in our group, we reports a novel carbon dots with high quantum yields and dual fluorescence emission[9]. The C-dots were synthesized using nitrogen and phosphorous co-doped carbon sources through one simple and fast microwave-assisted method. The optical results suggested that more than one type of excitation energy trapped on the surface of C-dots and the C-dots demonstrated excitation-dependent and pH-independent emission properties. Especially, under different excitation wavelengths, the C-dots showed blue and green emission, the quantum yield of which are 0.51 and 0.38, respectively. The nitro-explosives detection through the quenching of different emission were investigated, in which picric acid quenched both blue and green emission, while TNT only quenched green emission to some extent (Figure 19). We expect that the as-prepared C-dots provide new insights into the development of a ratiometric sensor to distinguish different explosives because of their dual fluorescence emission.

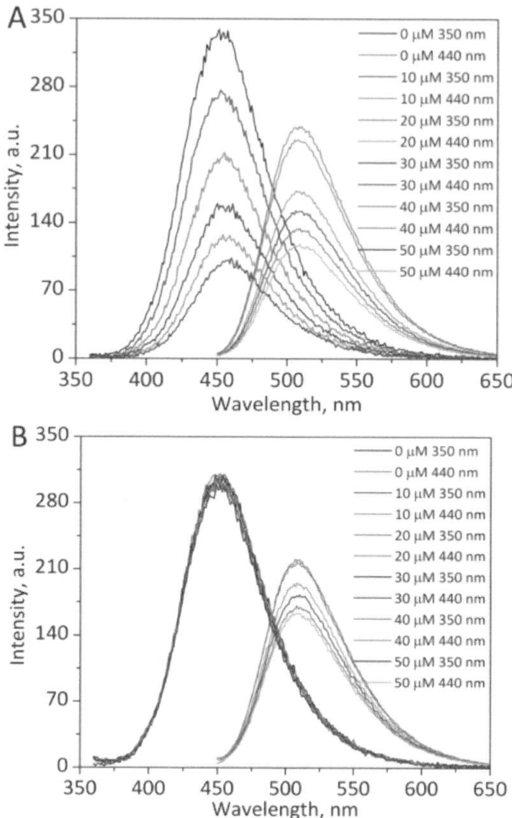

Figure 19 Explosives detection using diluted C-dots solutions with incremental addition (0-50 μM) of (A) picric acid and (B) TNT, respectively. Reprinted with permission form X. Sun, C. Brückner and Y. Lei, Nanoscale, 2015, **7**, 17278-17282. Copyright 2015 RSC.

Additionally, Krishna et al.[35] introduced boron and nitrogen co-doped C-dots (BN-C-dots) for selective and sensitive detection of picric acid (Figure 20). The BN-C-dots with strong photoluminescence and a quantum yield of ~14.2% was produced by a hydrothermal method using a mixture of sucrose, boric acid, and urea as the precursor. Its application for selective and sensitive detection of picric acid (PA) was realized through remarkable photoluminescence signal quenching upon the interaction between PA and C-dots. The good selectivity was also demonstrated with insignificant interference from other common explosives. PA, as low as 10 nM, can be detected in pH 7 solution at room temperature.

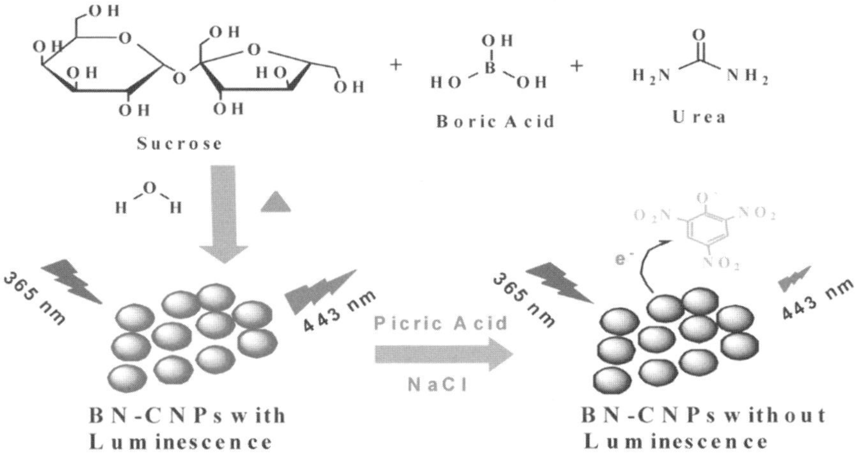

Figure 20. The synthesis of BN-doped C-dots and its application for picric acid detection. Reprinted with permission from H. K. Sadhanala and K. K. Nanda, The Journal of Physical Chemistry C, 2015. Copyright 2015 ACS publications.

3.3.7. Light Harvesting

As a new application, C-dots were also employed in light harvesting[11]. Nitrogen-enriched C-dots with diameter of 1 to 3 nm and fluorescent quantum yield of up to 22% were produced through a one-pot reaction between melamine and glycerol at an elevated temperature. The N-rich C-dots were then applied to sensitize TiO_2 based photoanode to study the conversion of near IR irradiation to photocurrent (Figure 21). The results indicated that the as-prepared N-doped C-dots displayed comparable performance as other conventional sensitizers such as CdSe and InAs quantum dots at their early development stage. As IR accounts for over 50% of solar energy, the developed C-dots hold great promise in light-harvesting application and it can be used as a complementary material to other photosensitizers, thus maximizing energy conversion efficiency.

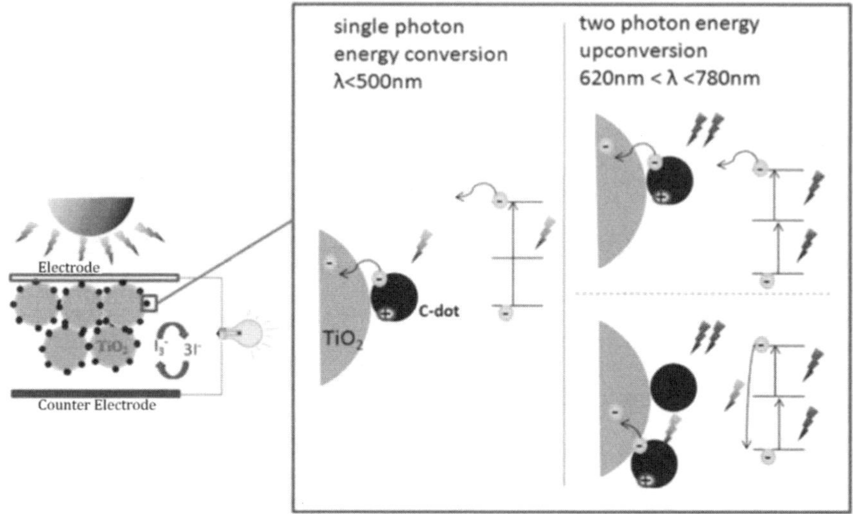

Figure 21. Working mechanisms of nitrogen-enriched C-dots as light harvesting agents for the TiO$_2$ based photoanode. Reprinted with permission from C. Wang, X. Wu, X. Li, W. Wang, L. Wang, M. Gu and Q. Li, Journal of Materials Chemistry, 2012, 22, 15522-15525. Copyright 2012 RSC.

3.4. Conclusion

In this book chapter, we have briefly introduced a range of synthetic methods (Bottom-up and Top-down) to prepare C-dots, followed with an extensive discussion with an emphasis on the applications of C-dots. Owing to the low toxicity and PL stability, C-dots have been regarded as promising materials in biosensing, bioimaging, nanomedicine and many other fields. In the early development stage of C-dots, most of C-dots emit in the blue range under ultraviolet excitation. In order to expand its application for cell imaging and *in vivo* applications, many efforts are underway to develop novel synthetic methods for C-dots with emission in the green, yellow, or red range. Although early success has been achieved, many synthesis reported are complex, involve toxic components, or the resulting C-dots suffer from low emission quantum yields.[36-41] Therefore, to freely tune the fluorescence emission of C-dots has been considered as a key requirement in order to broaden the application of C-dots, but remains elusive.[37] Various strategies including control of the condensation reaction conditions, chemical manipulations, or doping of other elements into C-dots have been exploited to generate C-dots with tunable emission.[37, 42, 43] Nitrogen (N) is by far the most promin-ent doping candidate but other elements such as boron (B), sulfur (S) and phosphorus (P) were also used in combination with nitrogen to further enhance the emission tunability.[8, 43-47] Excitation-dependent fluorescence emissions of C-dots were also reported, but generally there is a significant decrease in the fluorescence intensity upon modulation of the excitation wavelength.[48-50]

Although various applications have been exploited using C-dots, including bioimaging, nanoplatform therapy, heavy metal ions detection, fluorescent ink, explosive detection and light harvesting, its wider application as a class of fluorescent materials was greatly hurdled compared to other common fluorescent molecules, due to the limited information and understanding of fundamentals in material structure-property-function relationship. The preparation of C-dots with unique fluorescence property and its applications in the past is basically on a basis of trial-and-error, which is lack of systematic study and cannot generate a full picture. Therefore, there are two overall future research trends in developing unique C-dots and exploiting its applications: 1) increased efforts in fundamental studies for better understanding of the relation between C-dots structure and the optical properties. Most of the C-dots have so far been developed empirically. The structure and properties of C-dots should be characterized carefully and modeling and simulation can be involved to establish guidelines for the design of C-dots and develop reliable criteria for predicting the most useful C-dots for individual applications; and 2) New or improved C-dots synthesis with full controllability of C-dots properties is required. Currently many optical and electronic properties of C-dots have not been well controlled because of the unpredictable synthesis process of C-dots, resulting in the unpredictable chemical structure of C-dots. Although C-dots synthesis using inexpensive, recycling source and easily obtained natural materials, such as food waste[51], milk[34], cotton[29] and soy milk[52] as precursors have gained much attention as green synthetic route in recent years, the complicity of the precursors could cause more obstacles in understanding the relationship between material structure and material properties. In this regard, materials simulation is a very powerful tool to understand the correlation of material structure and properties, however, it is barely employed in the design/selection of C-dots up to date. With the establishment of more and more accurate materials models to reveal the structure-property-function relationship, a computation method to project the application of various functional materials or composites is expected to play a critical role in future research of C-dots.

We believe that with further research efforts in C-dots, functional C-dots with improved and controllable optical properties will endow C-dots unprecedented power in various applications and also serve as an alternative for traditional organic fluorescent dye and inorganic semiconductor quantum dots.

Acknowledgement

The authors are grateful for the financial support from the National Science Foundation and the University of Connecticut Prototype Fund.

References

1. B. A. Kairdolf, A. M. Smith, T. H. Stokes, M. D. Wang, A. N. Young and S. Nie, *Annual review of analytical chemistry (Palo Alto, Calif.)*, 2013, **6**, 143.
2. X. Xu, R. Ray, Y. Gu, H. J. Ploehn, L. Gearheart, K. Raker and W. A. Scrivens, *Journal of the American Chemical Society*, 2004, **126**, 12736-12737.
3. L. Shi, X. Li, Y. Li, X. Wen, J. Li, M. M. Choi, C. Dong and S. Shuang, *Sensors and Actuators B: Chemical*, 2015, **210**, 533-541.
4. Z. Gao, L. Wang, R. Su, R. Huang, W. Qi and Z. He, *Biosensors and Bioelectronics*, 2015, **70**, 232-238.
5. C. Yu, X. Li, F. Zeng, F. Zheng and S. Wu, *Chemical Communications*, 2013, **49**, 403-405.
6. J. Ge, Q. Jia, W. Liu, M. Lan, B. Zhou, L. Guo, H. Zhou, H. Zhang, Y. Wang and Y. Gu, *Advanced healthcare materials*, 2015.
7. Y. Jiang, Q. Han, C. Jin, J. Zhang and B. Wang, *Materials Letters*, 2015, **141**, 366-368.
8. C. Wang, D. Sun, K. Zhuo, H. Zhang and J. Wang, *RSC Advances*, 2014, **4**, 54060-54065.
9. X. Sun, C. Brückner and Y. Lei, *Nanoscale*, 2015, **7**, 17278-17282.

10.   Y. Wang, A. La, C. Brückner and Y. Lei, *Chemical Communications*, 2012, **48**, 9903-9905.

11.   C. Wang, X. Wu, X. Li, W. Wang, L. Wang, M. Gu and Q. Li, *Journal of Materials Chemistry*, 2012, **22**, 15522-15525.

12.   Y. Cui, C. Zhang, L. Sun, Z. Hu and X. Liu, *Particle & Particle Systems Characterization*, 2015, **32**, 542-546.

13.   S. Liu, J. Tian, L. Wang, Y. Luo and X. Sun, *RSC Advances*, 2012, **2**, 411-413.

14.   H. Peng and J. Travas-Sejdic, *Chemistry of Materials*, 2009, **21**, 5563-5565.

15.   S. Qu, X. Wang, Q. Lu, X. Liu and L. Wang, *Angewandte Chemie*, 2012, **124**, 12381-12384.

16.   X. Sun, J. He, Y. Meng, L. Zhang, S. Zhang, X. Ma, S. Dey, J. Zhao and Y. Lei, *Journal of Materials Chemistry A*, 2015. 2016, 4, 4161–4171.

17.   D. Pan, J. Zhang, Z. Li, C. Wu, X. Yan and M. Wu, *Chemical Communications*, 2010, **46**, 3681-3683.

18.   L. Zhou, Y. Lin, Z. Huang, J. Ren and X. Qu, *Chemical communications*, 2012, **48**, 1147-1149.

19.   S.-L. Hu, K.-Y. Niu, J. Sun, J. Yang, N.-Q. Zhao and X.-W. Du, *Journal of Materials Chemistry*, 2009, **19**, 484-488.

20.   H. Liu, T. Ye and C. Mao, *Angewandte Chemie International Edition*, 2007, **46**, 6473-6475.

21.   X. Gong, W. Lu, M. C. Paau, Q. Hu, X. Wu, S. Shuang, C. Dong and M. M. Choi, *Analytica chimica acta*, 2015, **861**, 74-84.

22.   K. Jiang, S. Sun, L. Zhang, Y. Wang, C. Cai and H. Lin, *ACS applied materials & interfaces*, 2015, **7**, 23231-23238.

23.   M. Lan, J. Zhang, Y.-S. Chui, H. Wang, Q. Yang, X. Zhu, H. Wei, W. Liu, J. Ge and P. Wang, *Journal of Materials Chemistry B*, 2015, **3**, 127-134.

24.   W. Wang, Y.-C. Lu, H. Huang, A.-J. Wang, J.-R. Chen and J.-J. Feng, *Biosensors and Bioelectronics*, 2015, **64**, 517-522.

25.   Y. Guo, L. Zhang, S. Zhang, Y. Yang, X. Chen and M. Zhang, *Biosensors and Bioelectronics*, 2015, **63**, 61-71.

26.   Y. Guo, Z. Wang, H. Shao and X. Jiang, *Carbon*, 2013, **52**, 583-589.

27.   Y. Dong, R. Wang, G. Li, C. Chen, Y. Chi and G. Chen, *Analytical chemistry*, 2012, **84**, 6220-6224.

28.   B. Cao, C. Yuan, B. Liu, C. Jiang, G. Guan and M.-Y. Han, *Analytica chimica acta*, 2013, **786**, 146-152.

29.   X. Wen, L. Shi, G. Wen, Y. Li, C. Dong, J. Yang and S. Shuang, *Sensors and Actuators B: Chemical*, 2015, **221**, 769-776.

30.   H. Li, J. Zhai, J. Tian, Y. Luo and X. Sun, *Biosensors and Bioelectronics*, 2011, **26**, 4656-4660.

31.   M. Lan, J. Zhang, Y.-S. Chui, P. Wang, X. Chen, C.-S. Lee, H.-L. Kwong and W. Zhang, *ACS applied materials & interfaces*, 2014, **6**, 21270-21278.

32.   S. Liu, J. Tian, L. Wang, Y. Zhang, X. Qin, Y. Luo, A. M. Asiri, A. O. Al-Youbi and X. Sun, *Advanced Materials*, 2012, **24**, 2037-2041.

33.   F. Qu, S. Wang, D. Liu and J. You, *RSC Advances*, 2015, **5**, 82570-82575.

34.   J. Wang, F. Peng, Y. Lu, Y. Zhong, S. Wang, M. Xu, X. Ji, Y. Su, L. Liao and Y. He, *Advanced Optical Materials*, 2015, **3**, 103-111.

35.   H. K. Sadhanala and K. K. Nanda, *The Journal of Physical Chemistry C*, 2015.

36.   Y. Fang, S. Guo, D. Li, C. Zhu, W. Ren, S. Dong and E. Wang, *ACS Nano*, 2012, **6**, 400-409.

37.   S. Hu, A. Trinchi, P. Atkin and I. Cole, *Angew. Chem. Int. Ed.*, 2015, **54**, 2970-2974.

38.   S. K. Bhunia, A. Saha, A. R. Maity, S. C. Ray and N. R. Jana, *Sci. Rep.*, 2013, **3**, 1473.

39.   S. Qu, X. Wang, Q. Lu, X. Liu and L. Wang, *Angew. Chem. Int. Ed.*, 2012, **51**, 12215-12218.

40.   J. Zhou, P. Lin, J. Ma, X. Shan, H. Feng, C. Chen, J. Chen and Z. Qian, *RSC Advances*, 2013, **3**, 9625-9628.

41.   S. Zhu, J. Zhang, C. Qiao, S. Tang, Y. Li, W. Yuan, B. Li, L. Tian, F. Liu, R. Hu, H. Gao, H. Wei, H. Zhang, H. Sun and B. Yang, *Chem. Commun.*, 2011, **47**, 6858-6860.

42.   S. N. Baker and G. A. Baker, *Angew. Chem. Int. Ed.*, 2010, **49**, 6726-6744.

43.   Y. Dong, H. Pang, H. B. Yang, C. Guo, J. Shao, Y. Chi, C. M. Li and T. Yu, *Angew. Chem. Int. Ed.*, 2013, **52**, 7800-7804.

44.   L. Zhang, Z.-Y. Zhang, R.-P. Liang, Y.-H. Li and J.-D. Qiu, *Anal. Chem.*, 2014, **86**, 4423-4430.

45.   S. Jahan, F. Mansoor, S. Naz, J. Lei and S. Kanwal, *Anal. Chem.*, 2013, **85**, 10232-10239.

46.   H. Ding, J.-S. Wei and H.-M. Xiong, *Nanoscale*, 2014, **6**, 13817-13823.

47.   S. Chandra, D. Laha, A. Pramanik, A. Ray Chowdhuri, P. Karmakar and S. K. Sahu, *Luminescence*, 2015, DOI: 10.1002/bio.2927, n/a-n/a.

48.   M. J. Krysmann, A. Kelarakis, P. Dallas and E. P. Giannelis, *J. Am. Chem. Soc.*, 2012, **134**, 747-750.

49.   S. Zhu, Q. Meng, L. Wang, J. Zhang, Y. Song, H. Jin, K. Zhang, H. Sun, H. Wang and B. Yang, *Angew. Chem. Int. Ed.*, 2013, **52**, 3953-3957.

50.   S. H. Jin, D. H. Kim, G. H. Jun, S. H. Hong and S. Jeon, *ACS Nano*, 2013, **7**, 1239-1245.

51.   S. Y. Park, H. U. Lee, E. S. Park, S. C. Lee, J.-W. Lee, S. W. Jeong, C. H. Kim, Y.-C. Lee, Y. S. Huh and J. Lee, *ACS applied materials & interfaces*, 2014, **6**, 3365-3370.

52.   C. Zhu, J. Zhai and S. Dong, *Chemical Communications*, 2012, **48**, 9367-9369.

Chapter 4

Rapid Detection of Biotoxin and Pathogen, and Quick Identification of Ligand-Receptor Binding Affinity Using AlGaN/GaN High Electron Mobility Transistors

Yu-Lin Wang, Fan Ren†, Stephen Pearton‡,
Chih-Cheng Huang [1], You-Ren Hsu [1]*

** Institute of Nanoengineering and Microsystems, National Tsing Hua
University, Hsin Chu, Taiwan 30013*
*† Department of Chemical Engineering, University of Florida, Gainesville,
FL32611*
*‡ Department of Materials Science and Engineering, University of Florida,
Gainesville, FL32611*

4.1. Introduction

There is a strong need for the fast detection of biotoxins and pathogens in fields. Biotoxins are possibly used as biological weapons by terrorists because biotoxins are usually much cheaper than traditional weapons, and they cause a large mortality if they are dispersed to public facilities such as airports, rail stations, malls, or museums. Currently, biotoxins can be diagnosed in labs, but it takes a long time to process, which is not practical to response to the events in time in fields. A field-deployable, reliable, and sensitive biosensor is needed in fields for routinely monitoring biotoxins in real-time to ensure the security. The deployable sensor has to be able to sustain in a harsh environment for a considerable period of time. It is also preferred that the sensor has a small size and a lightweight so it could be made as a portable device. A portable device is convenient to inspect foods, drinks or suspected objects instantly. Finally, the sensor has to be very cheap so that it can be deployed in multiple sites in a facility to collect complete information of possible contaminated regions with dispersed toxins and take proper actions to response to it accordingly.

103

The fast detection of pathogens in fields is critical for natural or farm-raised aquatic animals such as fish and shellfish. Some aquatic animals have very high economic value in market. However, the population of aquatic animals is subject to the appearance of pathogens in freshwater or in sea. Pathogen infection results in sickness of aquatic animals, which may eventually lead to the death of these animals. Pathogen infection does not only lead to a huge mortality of aquatic animals resulting in a large economic loss, but also may break the natural equilibrium of the eco system. A field-deployed biosensor is needed to monitor the population of pathogens in real time in the environment. Once the pathogen is detected, or the population is over the criteria, necessary actions can be actuated by an auto-response system. On the other hand, for most aquatic products, they need to be screened for certain pathogens before they are sold to the market, or exported to other countries and vice versa. The fast detection of the pathogen can largely shorten the process time and therefore can lower the mortality rate of the aquatic animals in trade.

AlGaN/GaN high electron mobility transistors (HEMTs) have been successfully demonstrated for detection of gases, chemicals, and bio-molecules[1-15]. The AlGaN/GaN HEMT sensors have shown high sensitivity due to the high concentration of two-dimensional electron gas induced by the piezoelectric polarization of the strained AlGaN layer and the spontaneous polarization[16-18]. The gate region of the HEMTs can be modified or immobilized with various sensing materials to interact with analytes, which result in potential change on gate area leading to the current change of transistors. Gallium nitride (GaN) is thermally stable because of the wide-bandgap property. Devices made of GaN can operate up to 500 °C. GaN is also chemically stable. Most acids and bases cannot etch GaN in regular condition. Only molten NaOH or KOH is known to be able to etch GaN. Therefore GaN-based sensors are particularly suitable for detection with salt, ions, chemicals, or biomolecules in liquid. The AlGaN/GaN HEMT-based sensors are excellent candidates for detection under harsh environments such as in buffered solution, serum, or in sea. In addition to the good thermal and chemical stability, the AlGaN/GaN HEMT has advantages such as real-time detection, fast response, high sensitivity, small size, light-weight, low cost and good reliability. These advantages ensure the AlGaN/GaN HEMT-based sensors to be field-deployed, sensitive, quickly responsive, portable, reliable, and cheap devices, which are perfect for the fast detection of biotoxins and pathogens in fields.

The AlGaN/GaN HEMTs have shown promise to bio-sensing applications. In fact, they are not only can be used as biosensors to detect various biomolecules, but can also be used to study the binding affinity of ligand-receptor complexes. The ligand-receptor pairs may include antigen-antibody, DNA-DNA, protein-DNA, protein-RNA, protein-protein, drug-protein, or substrate/inhibitor-enzyme systems. On the gate region of the HEMT sensors, receptors are usually immobilized to detect the ligand concentration in the bulk solution. The sensing mechanism is through the ligand-receptor affinity, which can be quantitatively expressed as equilibrium constant (affinity constants) or dissociation constants. Besides, a receptor may bind with more than one ligand, and each ligand may have a unique dissociation constant with the receptor. With a quantitative analysis of the signals from the sensors, we are able to reveal how many binding sites are on the receptor for certain ligands and what the dissociation constants are for ligand-receptor complexes. If the receptor can bind with different types of ligands, on either the same or the different binding site, the sensor can be used to study the competition or cooperativity between these different ligands. Due to the fast response and low price of the AlGaN/GaN HEMT sensors, the analysis can be quickly done and the information about the ligand-receptor binding affinity can be obtained with very low expenses. The need to study the ligand-receptor binding affinity is increasing due to its importance in drug development[19], ligand selection[20], antibody[21] and nucleic acid development[22], and exploring the detailed mechanism of biomolecular interactions[19-22]. Conventional methods to study ligand-receptor interactions include enzyme-linked immunosorbent assay (ELISA)[23], isothermal titration calorimetry (ITC)[24], ultra-violet/visible light (UV/VIS) spectrum[25], and surface plasmon resonance (SPR)[26]. Typically, ELISA and UV/VIS spectrum require labeling probes. ITC usually needs a large amount of sample. The commercial equipment (BIAcore) utilizing SPR technology has been used to characterize binding affinity of ligand-receptor complexes[19]. However, the equipment is expensive and therefore the cost for research is high. In sharp contrast, the AlGaN/GaN HEMT sensors can achieve very good performances as mentioned previously while in the meantime keep a very low cost for research.

To investigate the number of binding sites and the dissociation constants of a receptor, the surface coverage ratio is introduced in one-binding-site model or multiple-binding-site model. The surface coverage ratio means the ratio of the number of ligand-bound receptors to the total number of receptors on the sensor. The ratio is related to the bulk concentration of the ligand, which is associated with the magnitude of the signals from the sensor. Dissociation constants are extracted by fitting experimental results into the binding-site models, and the number of binding sites on the receptors is obtained. The limit-of-detection (LOD) of the sensor is not only observed from the experimental results, but also from the surface coverage ratio calculated in the experimental-fitted binding-site model. The analysis uncovers a truth that the LOD is not only depending on the performance of the transistors, but also directly related to the dissociation constants, which put an intrinsic limitation in any kind of sensor. The ranges of the ligand concentrations that can be detected largely depend on how many dissociation constants in the system and how these dissociation constants distribute. It turns out that the idea of sensitivity of a sensor is composed of the information of dissociation constants.

In this chapter, the operation principles and the fabrication of the AlGaN/GaN HEMTs are shown first. Fast detections of biotoxins and pathogens are illustrated as botulinum toxin and Perkinsus Marinus, respectively. The long-term stability of the botulinum toxin sensor is also shown. The Perkinsus Marinus was detected in sea waters that infected clams lived in. Antibody-antigen and protein-DNA binding pairs are illustrated to study the binding affinity with the theoretical binding-site models and the experimental results. One-binding-site mode and multiple-binding-site model are derived, discussed, and fitted with experimental results. The analysis of the signals from AlGaN/GaN HEMT sensors using the binding-site models demonstrate that the AlGaN/GaN HEMTs cannot only be used as biosensors, but also can be used to study the biological binding affinity of ligand-receptor pairs.

4.2. Operation Principles

The AlGaN/GaN HEMTs are depletion mode n-channel field-effect transistors (FETs). The HEMT sensors shown in this chapter are operated under the linear region of conventional FETs. The drain current can be expressed as,

$$I_D \cong (\frac{W\mu_n C}{L})(V_G - V_T - \frac{1}{2}V_D)V_D \qquad (1)$$

where W and L represent the width and length of the gate region. C is the capacitance of the dielectric. μ_n is the electron mobility. V_G, V_T, and V_D are the gate voltage, the threshold voltage and the drain voltage, respectively. Because this device is a depletion mode FET, without applying any gate bias the device is always "on" under a constant drain bias. This means the threshold voltage is negative and the term $(V_G - V_T)$ is always larger than zero. Once ligands specifically bind with the receptors immobilized on the gate region, the binding will cause the surface potential change, which means the gate voltage change. The gate voltage change can then lead to the drain current change, which is the signal for the sensors. Many biomolecules bring charges, and it is why they can change the surface potential when they bind on surface. Even though some molecules do not bring net charges, they may still cause surface potential change due to the polarity induced on surface by the molecules attached. The more the ligand molecules bind with the receptors, the larger the gate voltage change, resulting in larger drain current change. The magnitude of drain current change depends on the transconductance of the FETs, shown as,

$$g_m = (\frac{\partial I_D}{\partial V_g})_{V_D} \qquad (2)$$

As a FET has a larger transconductance, higher drain current can be induced at a certain ligand concentration. Therefore, a FET is expected to have a large g_m to create a large signal.

4.3. Sensor Fabrication

The HEMT structures consisted of a 3 μm thick undoped GaN buffer, 30Å thick $Al_{0.3}Ga_{0.7}N$ spacer, and 220Å thick Si-doped $Al_{0.3}Ga_{0.7}N$ cap layer. The epi-layers were grown by metal organic chemical vapor deposition (MOCVD) on thick GaN buffers produced on Si or sapphire substrates. Mesa isolation was performed with an Inductively Coupled Plasma (ICP) etching with Cl_2/Ar based discharges at –90 V dc self-bias, ICP power of 300 W at 2 MHz and a process pressure of 5 mTorr. 10 × 50 μm^2 Ohmic contacts separated with gaps of 5 μm consisted of e-beam deposited Ti/Al/Pt/Au patterned by lift-off and annealed at 850 °C, 45 sec under flowing N_2. 5nm or 10nm gold were deposited on the gate region was patterned by lift-off process. 400-nm-thick 4% polymethyl methacrylate (PMMA) or photoresist were used to encapsulate the source/drain regions, with only the gate region open to allow the liquid solutions to cross the surface. Thiolated sensing elements such as antibody or DNA were immobilized on the gate region through the covalent boding between the thiol group and gold. The chemical linker with bi-functional groups, the thiol and the carboxyl groups, was used to link the antibody to the gate surface. The carboxyl group can form amide bonds with the amine group of the antibody. The DNA was thiolated after synthesized. The source-drain current-voltage characteristics were measured at 25°C using an Agilent 4156C parameter analyzer with the gate region exposed.

4.4. Rapid Detection of Biotoxins and Pathogens using AlGaN/GaN HEMTs

4.4.1. Detection of Biotoxin: Botulinum toxin detection

Biological weapons are particularly attractive tools for terror because biological agents are available and easy to manufacture, small amounts are required to cause large-scale effects, and attacks can easily overwhelm existing medical resources, even in the US. Targets can range from public places such as amusement parks and stadiums to high profile people, as was seen in the anthrax attacks. Reliable detection of biological agents in the field and in real time has proved to be a fundamental challenge. Clostridium botulinum neurotoxins are the most deadly toxins known to man and are listed as a NIAID—Category A agent for bioterrorism potential. The lethal dose in unvaccinated humans is estimated at 1ng/kg[27]. Conventional methods of detection involve the use of HPLC, mass spectrometry and colorimetric ELISAs; but these are impractical because such tests can only be carried out at centralized locations, and are too slow to be of practical value in the field. However, these methods still tend to be the methods of choice in current detection of toxins. For example, the standard test for botulinum toxin detection is the 'mouse assay', which relies on the death of mice as an indicator of toxin presence[28]. Clearly, such methods are slow and impractical in the field. A fast detection of the toxin in field is therefore highly needed. The sensors made of AlGaN/GaN high electron mobility transistors can achieve fast detection of botulinum in real-time with a low cost.

Figure 1 shows a schematic device cross sectional view with thioglycolic acid followed by botulinum antibody coating. The Au surface was functionalized with a specific bi-functional molecule, thioglycolic acid. A self-assembled monolayer of thioglycolic acid, $HSCH_2COOH$, an organic compound and containing both a thiol (mercaptan) and a carboxylic acid functional group, on the Au surface in the gate area through strong interaction between gold and the thiol-group of the thioglycolic acid. The devices were first placed in the oxygen plasma chamber and then submerged in 1 mM aqueous solution of thioglycolic acid at 4 °C. This resulted in binding of the thioglycolic acid to the Au surface in the gate area with the COOH groups available for further chemical linking of other functional groups. X-Ray Photoelectron Spectroscopy and electrical measurements confirming a high surface coverage and Au-S bonding formation on the GaN surface have been previously published[12]. The device was incubated in a phosphate buffered saline (PBS) solution of 200 μg/ml botulinum polyclonal rabbit antibody for 18 hours before real time measurement of botulinum toxin.

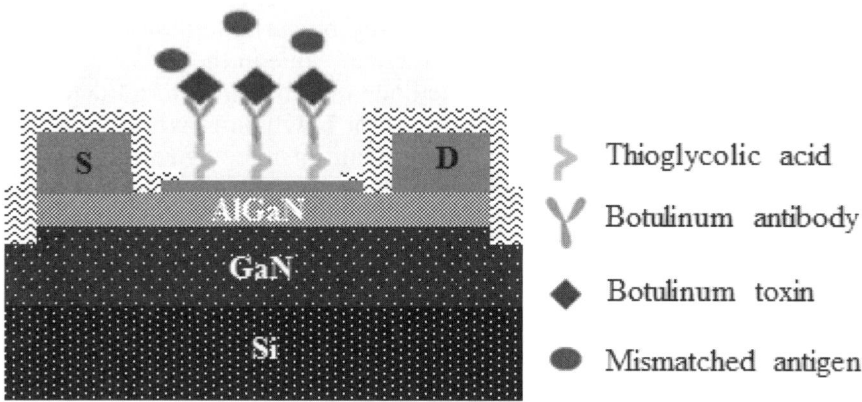

Figure 1. Cross-sectional schematic of the AlGaN/GaN HEMT-based botulinum sensor.

After incubation with a PBS buffered solution containing botulinum antibody at a concentration of 200 µg/ml, the device surface was thoroughly rinsed off with PBS and dried by a nitrogen blower. The source and drain current from the HEMT were measured before and after the sensor was exposed to 100 ng/ml of botulinum toxin at a constant drain bias voltage of 500 mV, as shown in Figure 2 (top). Any slight changes in the ambient of the HEMT affect the surface charges on the AlGaN/GaN. These changes in the surface charge are transduced into a change in the concentration of the 2DEG in the AlGaN/GaN HEMTs, leading to the slight decrease in the conductance for the device after exposure to botulinum toxin.

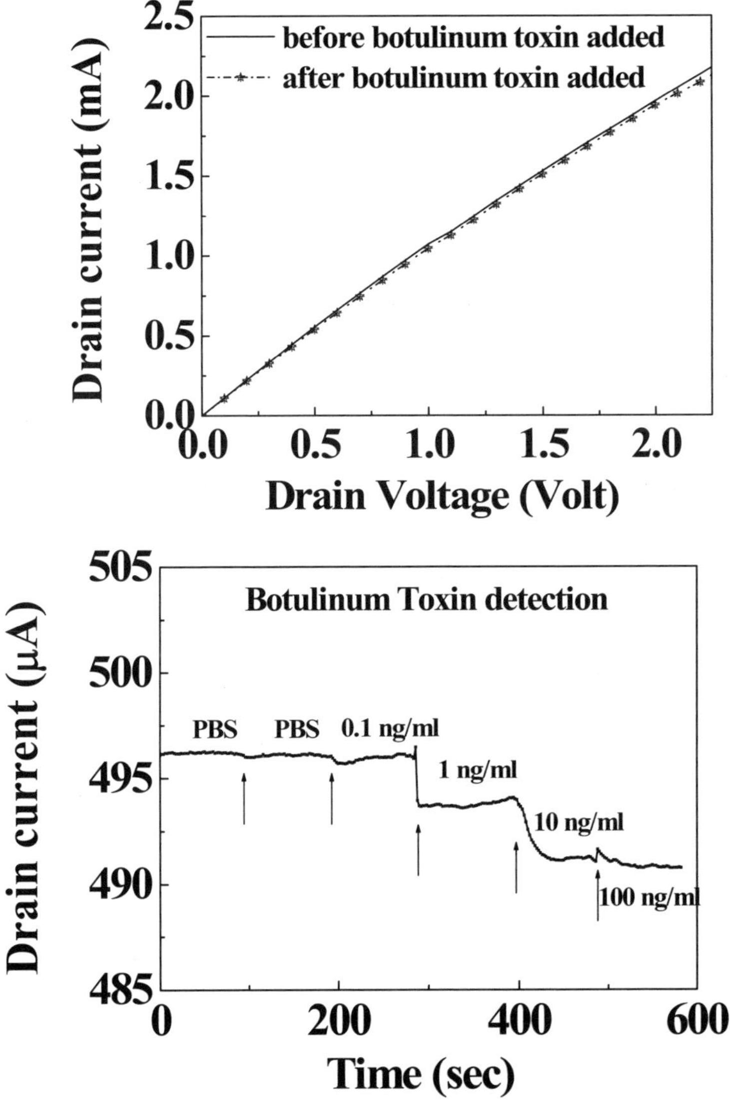

Figure 2. (Left) Current-Voltage characteristics of AlGaN/GaN HEMT sensor before and after exposure to 100ng/ml botulinum toxin. (Right) Drain current of an AlGaN/GaN HEMT versus time for botulinum toxin from 0.1 ng/ml to 100 ng/ml.

Figure 2 (bottom) shows a real-time botulinum toxin detection in PBS buffer solution using the drain current change with constant bias of 500 mV. No current change can be seen with the addition of buffer solution around 100 seconds, showing the specificity and stability of the device. In clear contrast, the current change showed a rapid response in less than 5 seconds when target 1 ng/ml botulinum toxin was added to the surface. The abrupt current change due to the exposure of botulinum toxin in a buffer solution was stabilized after the botulinum toxin thoroughly diffused into the buffer solution. Different concentrations (from 0.1 ng/ml to 100 ng/ml) of the exposed target botulinum toxin in a buffer solution were detected. Above the concentration of 10ng/ml of the botulinum toxin, the sensor becomes saturated. The fast response of the sensor to botulinum toxin makes the sensor become a perfect botulinum toxin alarm. The experiment at each concentration was repeated four times to calculate the standard deviation of source-drain current response, as shown in Figure 3. The limit of detection of this device was below 1 ng/ml of botulinum toxin in PBS buffer solution. The result shows that the sensor was very sensitive to the botulinum toxin.

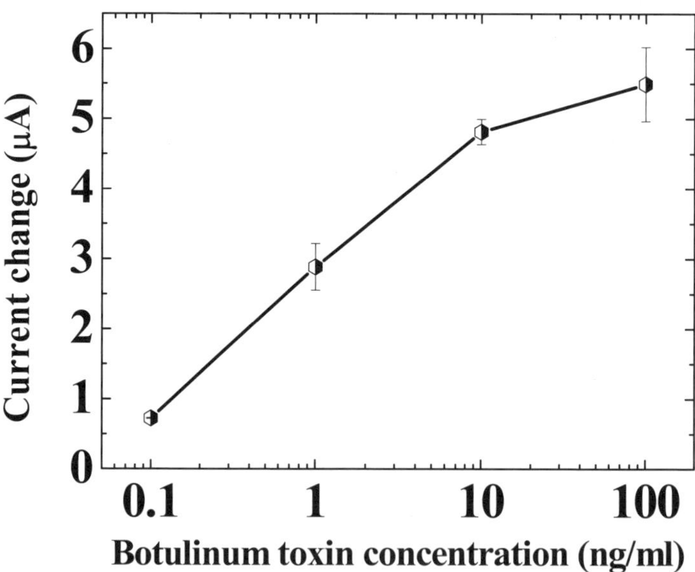

Figure 3. The drain current change versus the botulinum toxin concentration.

Besides fast response and reproducibility, the long-term stability is essentially important for the field-deployed biosensor chips. To date, there are only few reports on the stability of antibody-based immunosensors[29-32]. Here the long-term stability of the botulinum sensors was investigated. A fresh sensor was used to detect botulinum toxin of different concentrations (from 0.1 ng/ml to 30 ng/ml). After the first detection, the sensor was rinsed thoroughly with a PBS buffer solution (pH=6, 10mM) to wash off the toxin. The procedures of toxin detection and sensor surface reactivation were repeated for five times. The sensor showed good detection repeatability as previously described. Then the sensor was stored at 4°C in a refrigerator and then used to detect botulinum every three months. Figure 4 shows the real time detection of botulinum toxin at constant bias of 500 mV for a fresh sensor (top) and a sensor stored at 4 °C in a refrigerator for nine months (bottom). This result demonstrated that after 9-month storage, the sensor still could detect the toxin and could be reactivated right after the test with PBS buffer solution rinse. This indicated that the toxin can be completely washed away from the antibodies.

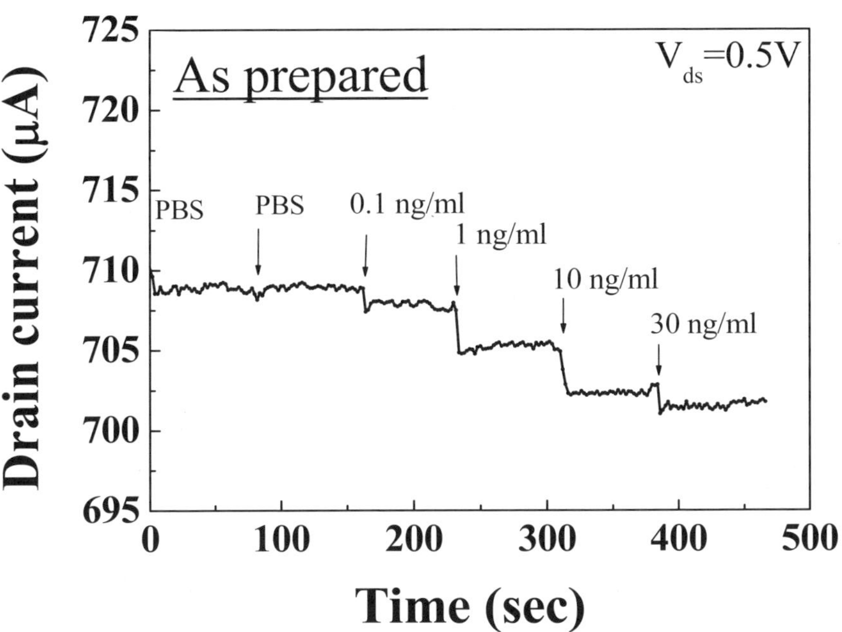

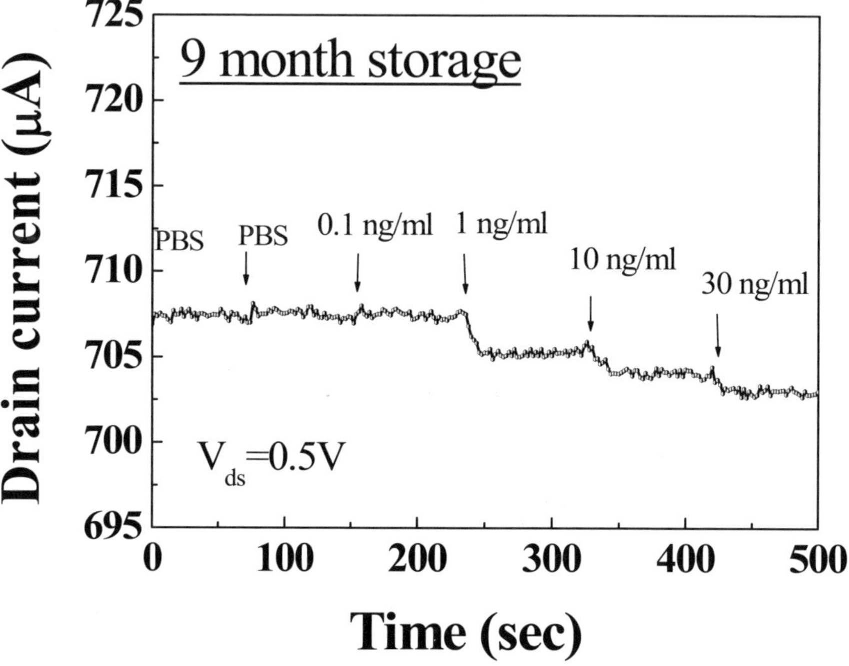

Figure 4. Real time detection of botulinum toxin with a fresh sensor (top) and the recycled 9-month-stored sensor (bottom), respectively.

However, it was obvious that the sensitivity decreased after 9 months of storage. The decrease of the sensitivity after 9 months of storage was not caused by the remaining of the surface-bound toxin-antibody complex, but was rather due to the decrease of antibody activity [29-32]. Another important finding was that the response time of the 9-month-stored sensor increased from 5 seconds of the fresh sensor to around 10 seconds, when target toxins were exposed to the sensor. The longer response time may be also due to the decreased number of highly active sites on the antibodies after long-term storage. This corresponds to the lower sensitivity of the sensor. Figure 5 shows a photograph of the packaged sensor placed in a Petri dish for long-term storage. PBS buffer solution was dropped on the active region of the sensor and the Petri dish as well.

The Petri dish was then covered and sealed in order to keep the antibodies on the sensor in a PBS environment.

Figure 5. Packaged botulinum sensor placed in a Petri dish with PBS buffer solution on the sensor and the dish for long-term storage.

Figure 6 shows the current changes of the sensors tested after different storage times with the same toxin concentration of 10 ng/ml. After 3, 6 and 9 months of storage, the current change drops 2%, 12% and 28%, respectively. Within 3 months of storage, this sensor showed almost the same sensitivity as when it was fresh. Although, after 6 and 9 months of storage, the sensor would need to be re-calibrated for the purpose of quantitative determination of toxin concentrations, there is no need for recalibration for the use as the first responder of the detection for the presence or absence of the toxin.

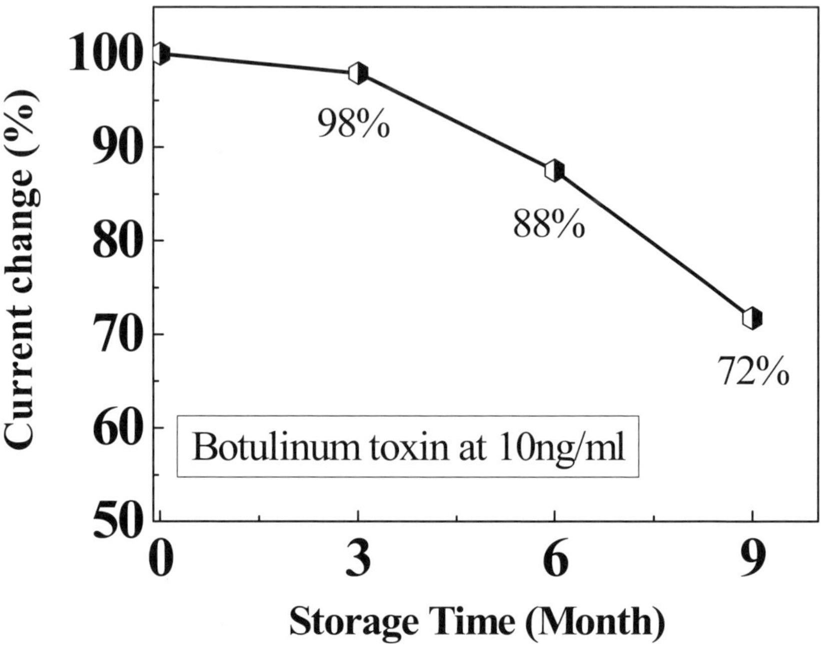

Figure 6. The drain-source current change percentages of the initial, 3, 6 and 9 month stored sensors. The current change in the initial test is defined as 100%.

In summary, through a chemical modification, the Au-gated AlGaN/GaN HEMT can be functionalized for the detection of botulinum toxin with fast response (response time less than 5 seconds) and a low limit of detection (less than 1 ng/ml). The long-term stability study shows that the sensor still keeps 70% of the original sensitivity. This electronic detection of biomolecules is a significant step towards a field-deployed sensor chip, which can also be integrated with a commercially available wireless transmitter to realize a real-time, fast response and high sensitivity botulinum toxin detector.

4.4.2. Detection of Pathogen: Perkinsus Marinus Detection

A protozoan pathogen of oysters, Perkinsus marinus (P. marinus), is highly prevalent along the east coast of the United States. Widespread

116

mortality caused by Perkinsus species infections in both natural and farm-raised oyster populations, resulting in severe economic losses for the shellfish industry and detrimental effects on the environment[33-37]. Currently, the standard diagnostic method for Perkinsus species infections has been fluid thioglycollate medium (FTM) assay detection[38]. However, this method of detection takes several days. The polymerase chain reaction (PCR)-based technique is also used to diagnose Perkinsus, but it is quite expensive and time-consuming, and requires exquisite controls to assure specificity and accuracy[39]. Clearly, such methods are slow and impractical in this age of global trade that requires rapid detection of such pathogens. AlGaN/GaN HEMT sensors have shown the capability of the fast response to biotoxin. In this section, quick detection of P. marinus in seawater can be achieved with the antibody-immobilized AlGaN/GaN HEMTs.

Figure 7 shows a picture of *Tridacna crocea,* which was used as the carrier for P. marinus in this study. The *Tridacna crocea* are extremely popular ornamental reef clams imported in huge numbers into the USA from the Indo-Pacific for the aquarium trade and they are known to be vulnerable to P. marinus [40, 41]. The tanks for hosting *Tridacna crocea* were set up using bedding of crushed coral reef sold commercially as "aragonite", which added a random set of minerals to help establish a biofilm. Mimic seawater was prepared by dissolving sea salt in the water and pH value, salinity and ammonia was controlled to mimic the real seawater. The infection status of the *Tridacna crocea* in this study was verified by histopathology, FTM, and polymerase chain reaction assays.

Figure 7. A picture of the clam (Tridacna crocea) which carrys the perkinsus marinus.

After immobilizing the anti-P. marinus rabbit antibody on the gate area of the AlGaN/GaN HEMT sensor, the sensor surface was thoroughly rinsed off with PBS. For the P. marinus detection, the drain current of the HEMT sensor was measured at a constant drain bias voltage of 500 mV before and after the sensor was exposed to the water from the first tank (in which the sick and dying clams were housed), as shown in Figure 8. Specific protein from decomposed P. marinus released into the seawater and bound with the antibodies bringing surface charges to the gate region of the HEMT, leading to the decrease in the conductance of the channel of the device.

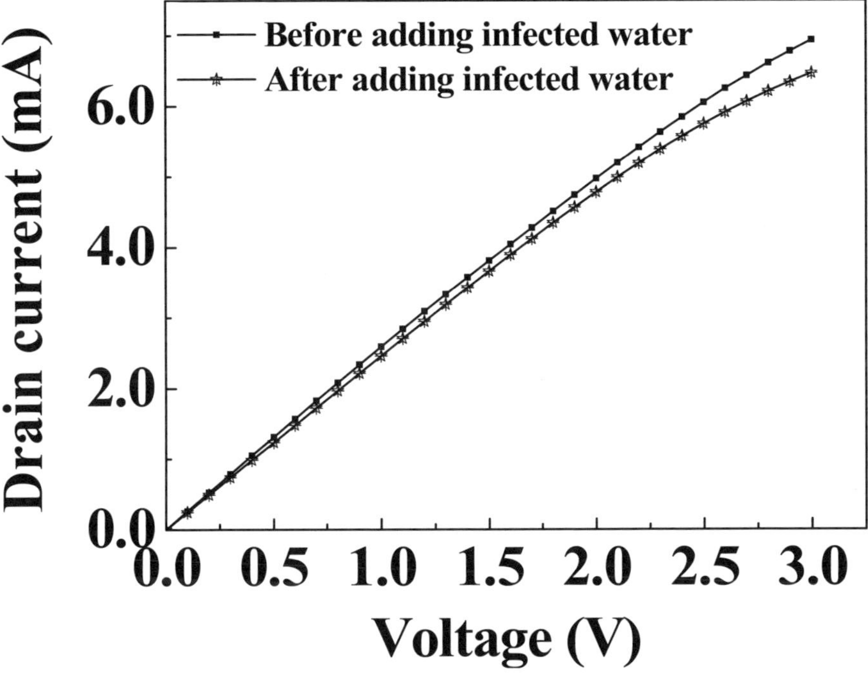

Figure 8. Drain I-V characteristics of AlGaN/GaN HEMT sensor before and after exposure to the P. marinus-infected water from tank water.

Figure 9 shows a real time P. marinus detection by the drain current change at constant drain bias of 500 mV. The current change showed a rapid response in less than 5 seconds when 2 ul of infected seawater was added to the surface of the sensor. The abrupt drain current change due to the exposure of P. marinus in a buffer solution was stabilized after the antigen thoroughly diffused into the buffer solution. Continuous 2 ul of the infected seawater added into the sensor resulted in further decreases of drain current. Obviously, the tank housed sick and dying calms releasing P. marinus organisms into the water, which subsequently shed surface antigens readily detected by the sensors. Then, the sensor was washed with PBS (pH 6.5) and used to detect the P. marinus again. The recycled sensor still shows very good sensitivity.

These results demonstrated that the fast detection of P. marinus in real-time can be achieved in just a few seconds with the AlGaN/GaN HEMT sensor.

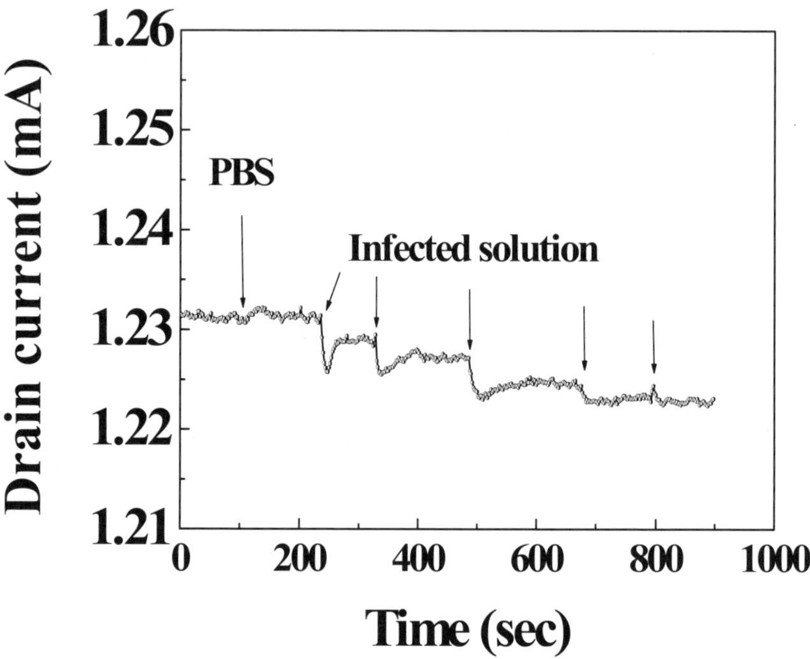

Figure 9. Real-time detection of P. marinus in an infected seawater.

4.5. Quick Identification of Ligand-receptor Binding Affinity Using AlGaN/GaN HEMTs

4.5.1. Theoretical Binding-site Models

The chemical reaction between the surface-immobilized receptors and the free ligands in bulk solution can be expressed as the following equations.

$$[R]+[L]\leftrightarrow[RL] \tag{3}$$

$$K = K_A = \frac{1}{K_D} = \frac{[RL]}{[R][L]} \tag{4}$$

where K is the equilibrium constant for formula (3), K_A is the association constant, and K_D is the dissociation constant. $[R]$ is the concentration of the unbound receptors immobilized on the sensor surface. $[L]$ is the ligand concentration in the bulk solution. $[RL]$ is the concentration of the ligand-receptor complexes on the sensor surface.

Because the total amount of receptors on the surface is limited, the maximum concentration of receptors on surface equals to $[R]_{max}$. Therefore, the total receptors including ligand-bound or unbound ones equal to the maximum receptor concentration.

$$[R]_{max} = [R]+[RL] \tag{5}$$

Here we can define a parameter, the surface coverage ratio α, expressed as,

$$\alpha \equiv \frac{[RL]}{[R]_{max}} \tag{6}$$

By combining equation (4), (5), and (6), the surface coverage ratio can be shown as,

$$\alpha \equiv \frac{[RL]}{[R]_{max}} = \frac{1}{1+\dfrac{K_D}{[L]}} \tag{7}$$

Rearranging equation (7), a linear equation can be obtained as,

$$\frac{[L]}{[RL]} = \frac{[L]}{[R]_{max}} + \frac{K_D}{[R]_{max}} \tag{8}$$

Equation (8) is the well-known Langmuir isotherm equation. Now, if we assume when ligands bind receptors, the formation of the ligand-receptor complexes bring signals from the transducer, which is the transistor in our case, then the signal P under a certain ligand concentration $[L]$ is proportional to the amount of the complexes $[RL]$ on the sensor surface. Therefore, when the receptors are saturated with the ligands, the maximum signal P_{max} is proportional to the maximum concentration of receptor $[R]_{max}$ on surface. Accordingly, the Langmuir equation can also be expressed as,

$$\frac{[L]}{P} = \frac{[L]}{P_{max}} + \frac{K_D}{P_{max}} \tag{9}$$

Equation (9) is a linear equation of ligand concentration $[L]$ in bulk solution with the slope $\dfrac{1}{P_{max}}$ and the intercept in y-axis $\dfrac{K_D}{P_{max}}$. When signals are measured at different ligand concentration, the dissociation constant can be extracted from the line fitting of equation (9). Figure 10 shows an example of the line fitting and extraction of dissociation constant from the Langmuir equation.

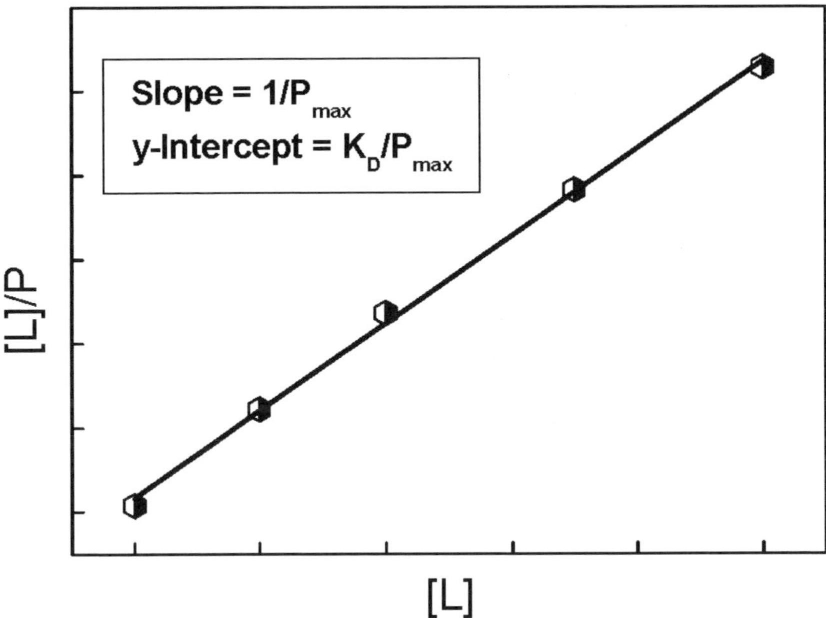

Figure 10. The line fitting of the Langmuir equation to experimental data.

The Langmuir isotherm equation is largely used to obtain dissociation constant between ligand and receptor [42-44]. For many biological receptors, they have more than one binding site for a certain ligand. However, the Langmuir equation (eqn. 8 or eqn. 9), is obvious for one-binding site receptor only. For some receptors with the unknown number of binding sites, it will be dangerous to just simply extract the dissociation constant from the Langmuir equation, because it leads to just one dissociation constant and the receptor is regarded to have one binding site only. To avoid this problem, a multiple binding site model is necessary. The multiple binding site model adopts the idea of the surface coverage ratio that we have mentioned in equation (6) and equation (7). Let's start with the one-binding site and two-binding site model using the surface coverage ratio. From equation (7), the concentration of ligand-receptor complex is given as,

$$[RL] = \frac{1}{1+\dfrac{K_D}{[L]}} \times [R]_{\max} = \alpha \times [R]_{\max} \qquad (10)$$

Since we know $[RL] \propto P$, and $[R]_{\max} \propto P_{\max}$, the equation (10) can be expressed as,

$$P = \frac{1}{1+\dfrac{K_D}{[L]}} \times P_{\max} = \alpha \times P_{\max} \qquad (11)$$

We then can measure signals at different ligand concentrations from very dilute, to medium, and to very high ligand concentration until the signal saturates. The signal saturates because almost all receptors on surface are bound with the ligand and nearly no free receptors are available for binding. The relationship among ligand concentrations, the surface coverage ratios, and the signals are illustrated as,

At $[L] \sim 0$, the surface coverage ratio $\alpha \sim 0$ and the signal $P = \alpha \times P_{\max}$.

At $[L] = K_D$, the surface coverage ratio $\alpha \sim 0.5$ and the signal $P = 0.5 P_{\max}$.

At $[L] \sim \infty$, the surface coverage ratio $\alpha \sim 1$ and the signal $P \cong P_{\max}$.

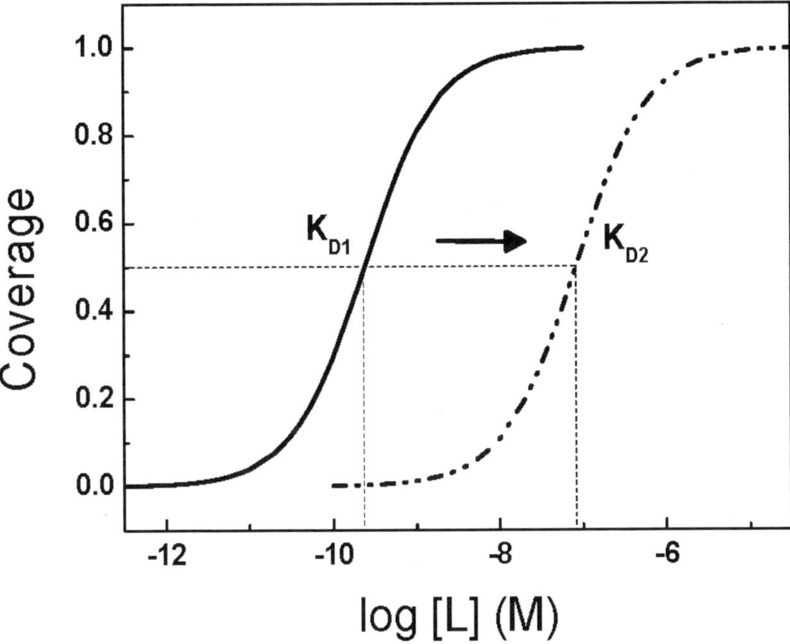

Figure 11. Ideal surface coverage curve for one-binding-site model.

The ideal surface coverage ratio as a function of the ligand concentration is shown as Figure 11. This is the typical curve for the one-binding-site model. Notice that the surface coverage ratio is only dependent on the ratio of dissociation constant to the ligand concentration. Changing the dissociation constant does not alter the shape of the curve, but shifts it laterally on the x-axis, as shown in Figure 11. When the surface coverage ratio equals to 0.5, the ligand concentration equals to the dissociation constant. This does not only work in surface-immobilized sensors, but is also applicable to any system with limited amount of receptors. In fact, it is a well-known definition in traditionally analytical chemistry.

For a one-binding-site system, by fitting experimental data points into this ideal surface coverage ratio model, the dissociation constant can be easily extracted. From equation (11), it is obvious that the experimental coverage ratio just equals to $P/P_{\max}$. When K_D equals to 10 folds or 10% of $[L]$, the surface coverage ratio goes to around 9.09% and 90.91%, respectively. Beyond 90.91% or below 9.09% of the surface coverage ratio, increasing or decreasing the concentration of ligand does not change the surface coverage ratio much. Therefore, we conclude that for a one-binding-site model, the most significant change of the surface coverage ratio is within the range of the ligand concentration between one order higher and one order lower than the value of the dissociation constant. Between 9.09% and 90.91% of the surface coverage ratios, if the ligand concentration ranges across two orders of magnitude, the receptor should have one binding site. Within this range of coverage ratio, if the experimental data points cross more than two orders, such as three or four orders, then the receptor may have two dissociation constants, or more. Here a two-binding-site model is proposed as the following. In the two-binding-site model, the chemical equations are described as below,

$$[R]_{site1} + [L] \leftrightarrow [RL]_{site1} \qquad (12)$$
$$K_{D1}$$

$$[R]_{site2} + [L] \leftrightarrow [RL]_{site2} \qquad (13)$$
$$K_{D2}$$

where the K_{D1} and K_{D2} are the dissociation constants for the two binding sites on a receptor.

For a system with more than one dissociation constants, the signals from experiments can be regarded as the sum of the signal contributed from each individual binding site, which has its own dissociation constant. Therefore, the overall signals for a two binding-site ligand-receptor complex can be expressed as the linear combination of the signal from each binding site. The signal from each binding site is proportional to the coverage ratio of each single binding site and the maximum signal of each binding site when the binding sites are saturated by ligands as well. The signal from a single binding site is simply regarded as a one-binding-site system. In other words, the two-binding-site model is mathematically composed of two one-binding-site models, as shown as followings,

$$P = P_1 + P_2 \tag{14}$$

$$P = \frac{P_{\max 1}[L]}{K_{D1} + [L]} + \frac{P_{\max 2}[L]}{K_{D2} + [L]} = \frac{P_{\max 1}}{\dfrac{K_{D1}}{[L]} + 1} + \frac{P_{\max 2}}{\dfrac{K_{D2}}{[L]} + 1} \tag{15}$$

$$P = \alpha_1 P_{\max 1} + \alpha_2 P_{\max 2} \tag{16}$$

where α_1 and α_2 are the surface coverage ratio at binding site 1 and site 2 of the receptor. $P_{\max 1}$ and $P_{\max 2}$ are the maximum (saturated) signal contributed by site 1 and site 2, respectively. If we assume that ligands bound at the two different binding sites of a receptor equally affect either the channel conductance or current of the transistor, it is quite reasonable to allow $P_{\max 1} = P_{\max 2} = P_{\max}/2$. If the two binding sites on a receptor are expected to give different maximum (saturated) signals, then the $P_{\max 1}$ and $P_{\max 2}$ can be extracted individually. For even more complex multiple-binding-site systems, such as three- or four-binding-site receptors, the models can be built up as the similar procedures as shown in the two-binding site model. These models are not only applicable to semiconductor transistor-based sensors, but also can be used in other ligand-receptor type sensors. In fact, the models also work in non-surface-type sensors. For example, ligands or receptors are not immobilized on surface but are free in solution. As long as the amount of receptors are limited in the system and the signals are proportional to the amount of the ligand-receptor complex, then these models work well either in surface-type or non-surface-type sensors. Finally, if a receptor can bind different kinds of ligands, then the models can be derived with the similar techniques we have shown in this section. In such a case, the cooperativity, which describes the competiveness among different ligands, can be obtained by fitting the model with the experimental data. The model can be applied to study the interactions among enzymes, substrates, and inhibitors, for example.

4.5.2. Protein-peptide Binding Affinity

Here we demonstrate the ligand-receptor binding affinity by using the theory we developed in the previous section. An antibody-antigen system was introduced in this study. In this study, anti-ferritin heavy chain (FHC) antibodies were immobilized on the AlGaN/GaN HEMTs for detecting the antigen (a short peptide). One-binding-site model and two-binding-site model were fitted with experimental results in the analysis of the surface coverage ratio. The surface coverage ratio is defined as the ratio of the amount of peptide-antibody complexes to that of the total antibodies immobilized on a transistor. The analysis of the surface coverage ratio reveals how many binding sites a ligand/receptor system may have and what the dissociation constants are at different binding sites for that system. The characteristics of source-drain current-voltage of the sensor was measured at 20°C using an Agilent B1500 parameter analyzer with the gate region exposed. The source-drain bias was fixed at 0.5V. The antigen, a short peptide, consists of 20 amino acids. This peptide can specifically bind to the FHC antibody. Different concentrations of the peptides were sequentially dropped on the sensor in a real-time detection.

Figure 12 shows the real time detection of the peptide at constant bias of 500 mV for the sensor. Real-time current monitoring spanned the range of target concentrations from 0.5pM to 5μM of the peptide. Upon these target concentrations of the peptide, the 50pM gave a most significant current change. The current change gradually saturated as the peptide concentration increased. Eventually, no more current change increase was observed at high concentration of the peptide, as shown in Figure 12 and Figure 13(top). Figure 13(top) and Figure 13(bottom) show the current change versus the target concentration of the peptide in linear scale and in log scale, respectively. From Figure 13(bottom), the limit of detection of this sensor was estimated about 10pM of the peptide in the buffer solution.

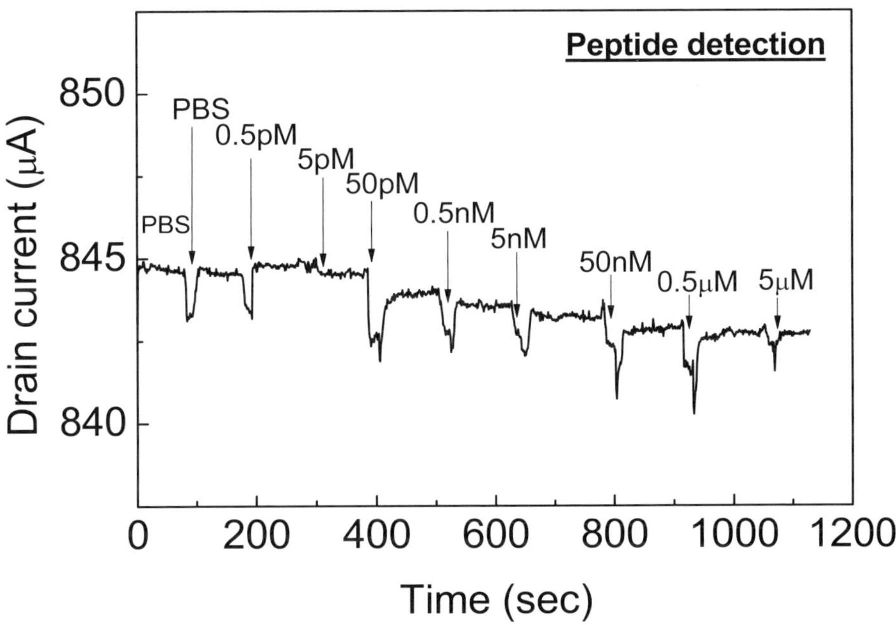

Figure 12. Real-time detection of the peptide from 0.5pM to 5μM at a constant bias of 500 mV.

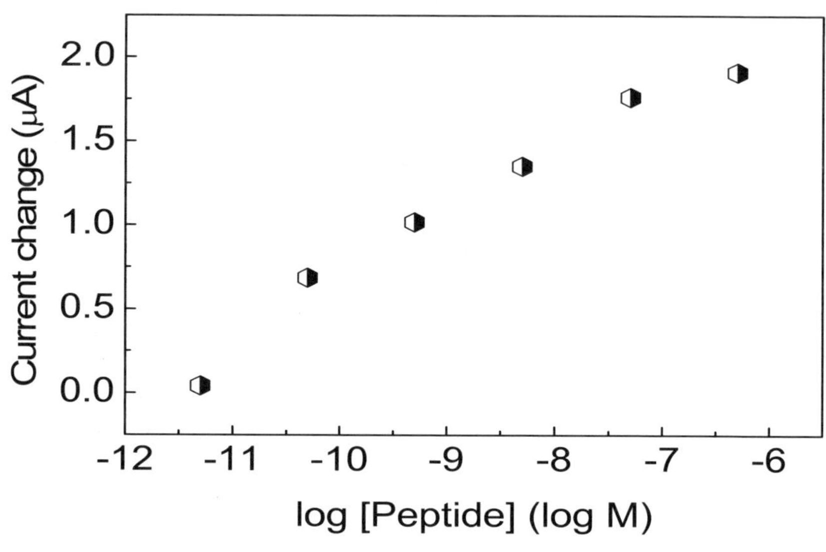

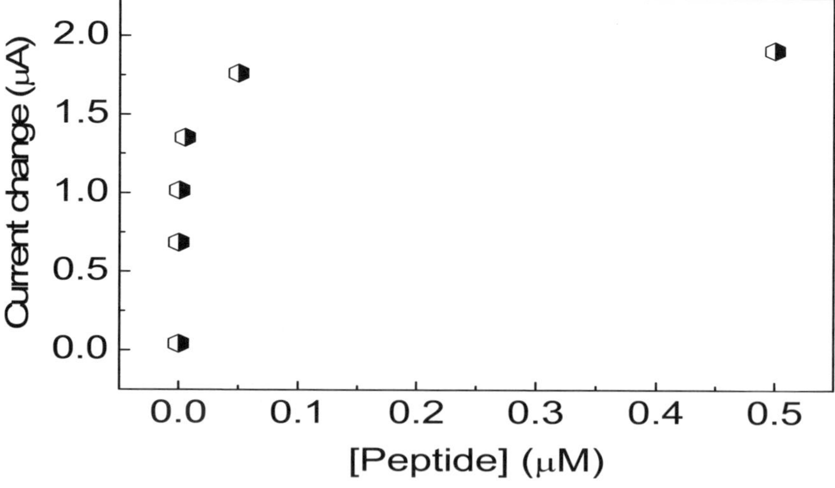

Figure 13. The current change versus the peptide target concentration (top) in linear scale and (bottom) in log scale.

In this example, the receptor is the antibody immobilized on the sensor surface, and the ligand is the antigen, which is the short peptide. The chemical reaction therefore becomes as following,

$$[Ab] + [Ag] \leftrightarrow [Ab - Ag] \tag{17}$$

$$K = K_A = \frac{1}{K_D} = \frac{[Ab - Ag]}{[Ab][Ag]} \tag{18}$$

where K is the equilibrium constant for formula (17), K_A is the association constant, and K_D is the dissociation constant. $[Ab]$ is the concentration of the unbound antibody immobilized on the sensor surface. $[Ag]$ is the antigen concentration in the bulk solution. $[Ab - Ag]$ is the concentration of the antibody-antigen complex on the sensor surface.

To estimate the dissociation constant, the Langmuir isotherm equation was used to extract the dissociation constant. Langmuir isotherm equation is much frequently used to extract the dissociation constant [42-44]. Here the current changes were introduced as signals to the Langmuir isotherm equation for calculating the dissociation constant. Using the equation (9), the Langmuir isotherm equation in this case is given as,

$$\frac{[Ag]}{\Delta I} = \frac{[Ag]}{\Delta I_{max}} + \frac{K_D}{\Delta I_{max}} \tag{19}$$

where ΔI is the current change at the bulk concentration of antigen $[Ag]$, and ΔI_{max} is the saturated current change. The dissociate constant can be extracted from the linear regression by using equation (19), as shown in Figure 14. The y-axis and the x-axis are $[Ag]/\Delta I$ and $[Ag]$, respectively. The slope and the y-intercept are $1/\Delta I_{max}$ and $K_D/\Delta I_{max}$, respectively. The line-fitting shown in Figure 14 gives a good linear regression correlation coefficient (r^2=0.99992). The extracted maximum current change was about 1.91μA, which is very close to the experimental observed one, 1.90μA. The extracted dissociation constant was obtained as 1.37×10^{-9}M, which is also in the reasonable range for most IGg antibody-antigen complexes [21, 45, 46].

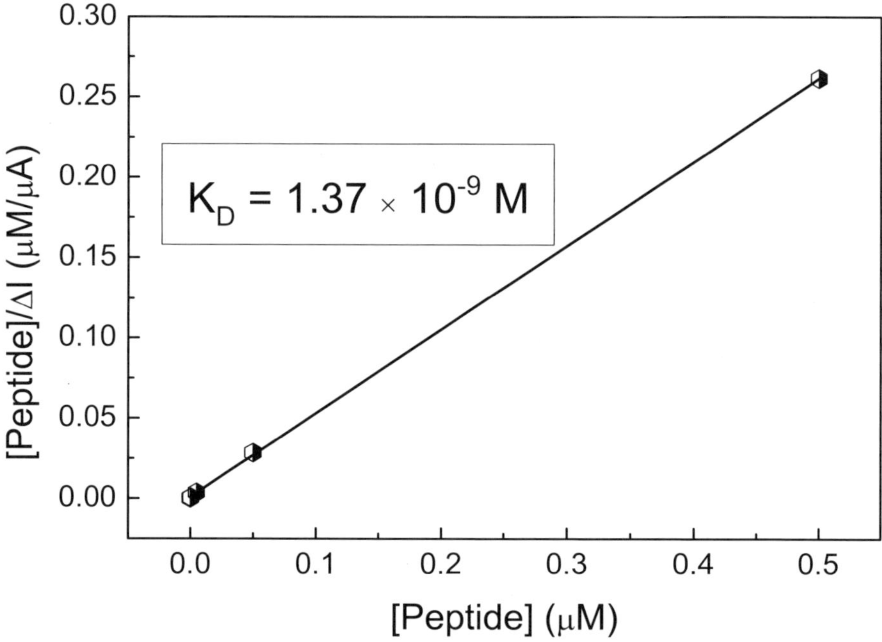

Figure 14. [Ag]/ΔI versus [Ag] and the dissociation constant extracted from the linear regression using Langmuir adsorption isotherm.

Now let's put this dissociation constant, which is extracted from the equation (19), into the equation (7) and (11) to obtain the equation (20).

$$\frac{1}{1+\dfrac{K_D}{[Ag]}} = \frac{[Ab-Ag]}{[Ab]_{max}} \equiv \alpha \tag{20}$$

The surface coverage at different antigen (peptide) concentration can be obtained from equation (20), as shown in Figure 15.

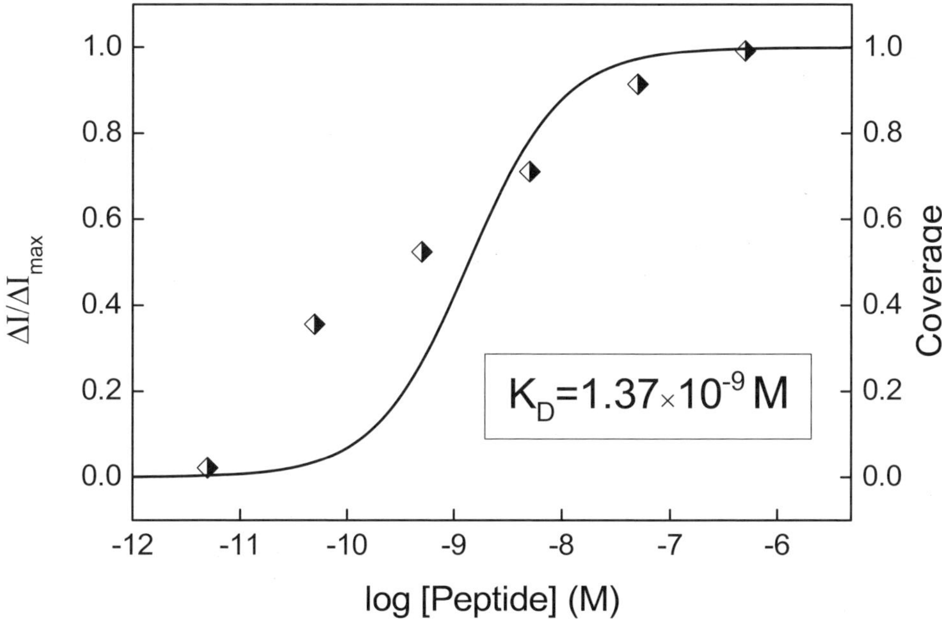

Figure 15. Surface coverage ratio (the blue curve) as a function of the antigen concentration calculated with the dissociation constant extracted from Figure 14. The left y-axis ($\Delta I/\Delta I_{max}$) from the experiment is shown as square dots versus the antigen concentration.

It is obvious that the curve of the surface coverage ratio does not match well with those experimental dots. From equation (7), it is shown that the surface coverage ratio actually depends on the ratio of dissociation constant to the concentration of antigen, that is $K_D/[Ag]$. When K_D equals to 10 folds or 10% of $[Ag]$, the surface coverage ratio goes to around 9.09% and 90.91%, respectively. Beyond 90.91% or below 9.09% of the surface coverage ratio, increasing or decreasing the concentration of antigen does not change the surface coverage ratio much. Therefore, for a one binding-site model, the most significant change of the surface coverage ratio is within the range of the antigen concentration between one order higher and one order lower than the value of the dissociation constant, as mentioned in previous section. This was also shown by experimental data in literature [47]. Thus, ideally once the dissociation constant is determined, the highest sensitivity for the sensor is only within this two-order range of antigen concentration. On the other hand, for any experimental result, the surface coverage ratio between 9.09% and 90.91% should only cover two orders of antigen concentration for an ideal one-binding site model. However, in Figure 15, it is significant that between 9.09% and 90.91% of the surface coverage, the experimental data points cover almost 4 orders of the antigen concentration. Therefore, it is not possible to fit these experimental points well into the surface coverage ratio with the one-binding site model. The wide range of the antigen concentration within that coverage ratio (9.09% ~ 90.91%) may be resulted from multiple binding-sites of the receptor. Compared with Figure 14, Figure 15 is clearly more sensitive to judge whether the model can really fit into the experimental data. From this comparison, we learn that the surface coverage ratio is more useful than the Langmuir isotherm equation to decide whether one-binding-site model is adequate for the receptor or not. The good correlation coefficient of the linear regression using the Langmuir isotherm equation may lead to a false judgment on the binding sites, or more specifically, the numbers of dissociation constants in the ligand-receptor system. Thus, the surface coverage ratio versus the antigen concentration will be used for finding out the number of binding sites and the dissociation constants. Now, we should move on to a multiple-binding-site model and see if that can better fit into the experimental results. Let's start from the two-binding-site model.

Using equation (12) and (13), in the two-binding-site model, the chemical equations are shown as below in this case,

$$[Ab]_{site1} + [Ag] \leftrightarrow [Ab - Ag]_{site1} \qquad (21)$$
$$K_{D1}$$
$$[Ab]_{site2} + [Ag] \leftrightarrow [Ab - Ag]_{site2} \qquad (22)$$
$$K_{D2}$$

where the K_{D1} and K_{D2} are the dissociation constants for the two binding sites on an antibody.

Replace the signals as current changes in equation (14), (15), and (16). Then the total current change is assumed to be the sum of the current change resulted from antibody-antigen complex at site 1 and at site 2, respectively, as shown in equation (23), (24) and (25).

$$\Delta I = \Delta I_1 + \Delta I_2 \qquad (23)$$

$$\Delta I = \frac{\Delta I_{max1}[Ag]}{K_{D1} + [Ag]} + \frac{\Delta I_{max\,2}[Ag]}{K_{D2} + [Ag]} = \frac{\Delta I_{max1}}{\dfrac{K_{D1}}{[Ag]} + 1} + \frac{\Delta I_{max\,2}}{\dfrac{K_{D2}}{[Ag]} + 1} \qquad (24)$$

$$\Delta I = \alpha_1 \Delta I_{max\,1} + \alpha_2 \Delta I_{max\,2} \qquad (25)$$

where α_1 and α_2 are the surface coverage ratio at site 1 and site 2 of the antibody. We first assume the two binding sites on an antibody equally contribute to the current change when they are saturated with the antigens. Therefore we can write the total maximum current change as $\Delta I_{max1} = \Delta I_{max2} = \Delta I_{max}/2$. At such a condition, the overall surface coverage ratio can then be shown as,

$$\alpha = \frac{\Delta I}{\Delta I_{max}} = \frac{\alpha_1 + \alpha_2}{2} \qquad (26)$$

Now we can fit this two-binding-site model into the experimental data points to find out the two dissociation constants, K_{D1} and K_{D2}.

Figure 16 shows the surface coverage ratio versus the antigen concentration. The model and the experimental data points perfectly fit together. The K_{D1} and K_{D2} are 2.723×10^{-11}M and 6.994×10^{-9} M, respectively. The correlation coefficient, r^2 is 0.9926. These two binding constants are in a reasonable range of regular antibody-antigen binding constants[21, 46]. The antigen, the peptide, consisting of only 20 amino acids is pretty small and the antibody is allowed to bind two antigens on its two binding sites. Because the two binding sites of an antibody are usually regarded as identical structures, the site 1 and site 2 in this reaction should not be recognized as two different sites. Instead, they are more preferably regarded as the binding sequence with the antigens. The difference between these two dissociation constants is probably ascribed to the stereo hindrance resulted from the first antigen-antibody complex for the second one.

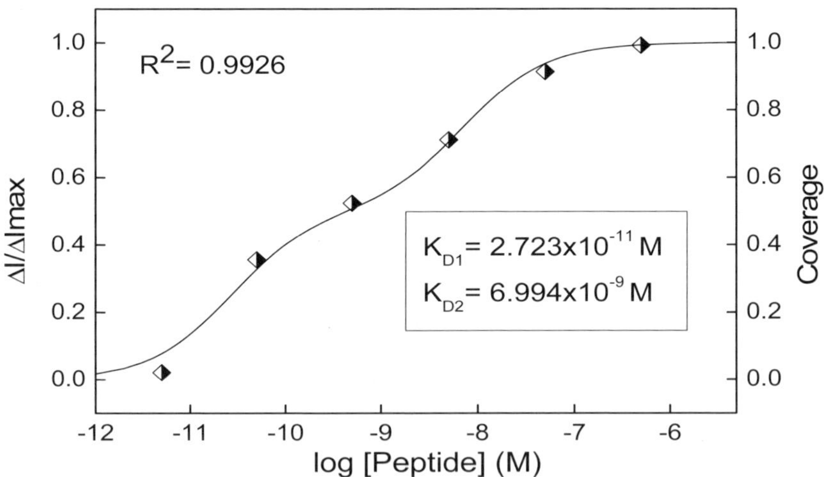

Figure 16. Surface coverage ratio as a function of the antigen concentration with equal maximum current changes for the two binding complexes on an antibody.

It may be arguing that whether the assumption of equal maximum current changes for the two binding complexes is appropriate or not. If ΔI_{max1} and ΔI_{max2} are allowed to be variables and equation (24) is used for the curve fitting (assuming ΔI_{max1} $=$c1ΔI_{max} and ΔI_{max1}) $=$ c2ΔI_{max} = (1 - c1) ΔI_{max}), we can get c1=0.551, c2=0.449, K_{D1} =3.411 $\times$ 10^{-11} M, K_{D2} = 9.590 $\times$ 10^{-9} M and r^2=0.9946. Figure 17 shows the surface coverage ratio derived from this non-equal maximum current changes for the two binding complexes. These two dissociation constants have less than one order of magnitude shift compared to the ones derived from the equal maximum current change for the two binding complexes. What happened if the high sensitivity region covers 5~6 orders of magnitude? In that case, we may further need to consider a three-binding-site model. Now when we look back at the detection limit of the sensor, which is between 5 pM~50 pM of the antigen is just consistent with the lower-half high sensitivity region of K_{D1} (between one order lower and the order of the dissociation constant). This result shows that the AlGaN/GaN HEMTs are still very sensitive in low antigen concentration. However, the detection limit of this sensor for the antigen is not only depending on the transistor, but also on the affinity of the antigen to the antibody.

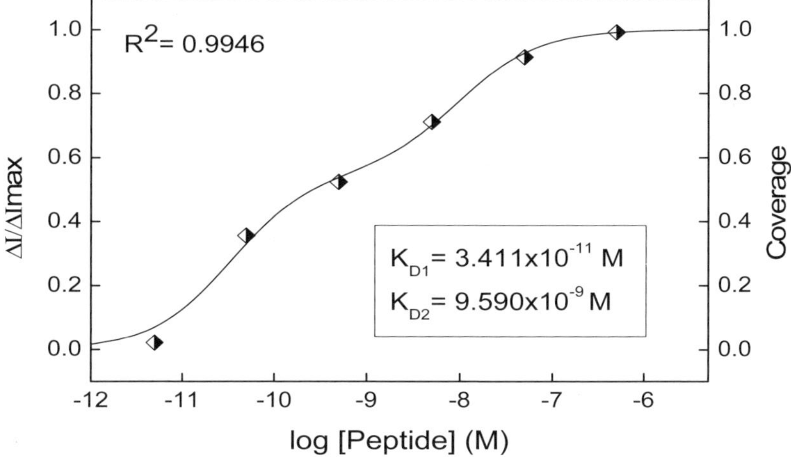

Figure 17. Surface coverage ratio as a function of the antigen concentration with non-equal maximum current changes for the two binding complexes on an antibody.

4.5.3. DNA-protein Binding Affinity

The breakout of Severe Acute Respiratory Syndrome (SARS) in 2002 caused a large mortality and almost paralyzed the world economy in 2003. More than 8000 probable infected cases and 774 deaths were reported worldwide ranging 29 countries, according to World Health Organization (WHO). This disease is caused by the SARS coronavirus (SARS-CoV). A coronavirus protein, nucleocapsid protein (N protein), encapsidates the coronavirus genomic RNA playing an important role in the virus replication. SARS-CoV N protein has been shown its capability of interacting with RNA and DNA, and therefore is known as a nucleic acid binding protein[48-52]. Investigating the nucleic acid-SARS-CoV N protein interaction can help us to explore the virus genome packaging process to construct a genome packaging model for SARS coronavirus. Electrophoretic Mobility Shift Assay (EMSA) and Filter binding assay have been widely used to study protein-nucleic acid interaction in the last 30 years [53-56]. However, these two methods require labeling of fluorescent probes or isotope elements on nucleic acids to provide signals for quantitative molecule detection. Thus, the cost for these methods is high, and the labeling may alter the binding affinity of molecules. Developing an efficient and molecule-labeling free binding assay with low cost and high sensitivity becomes a very important issue.

Here AlGaN/GaN HEMTs were immobilized with double stranded DNA (dsDNS) on gate region as the receptors for detecting N proteins. The sequences of the dsDNA are 5'-Thiol-TTTTTTTTTTTTTTTTTTTTTTTTCATCGAGGCCACGCGGAGTACGATCGAGGGTACAGTGAA -3', and 3'AAGTAGCTCCGGTGCGCCTCATGCTAGCTCCCATGTCACTT-5'). The N protein can be divided into two binding domains, the C-terminal domain (CTD) and the N-terminal domain (NTD). Both the two domains can bind with nucleic acids through their electropositive region. In this example, the N protein that we use is CTD only. The sensors were used to study the affinity between the CTD and the dsDNA. Figure 18 shows the schematics of the N protein sensors.

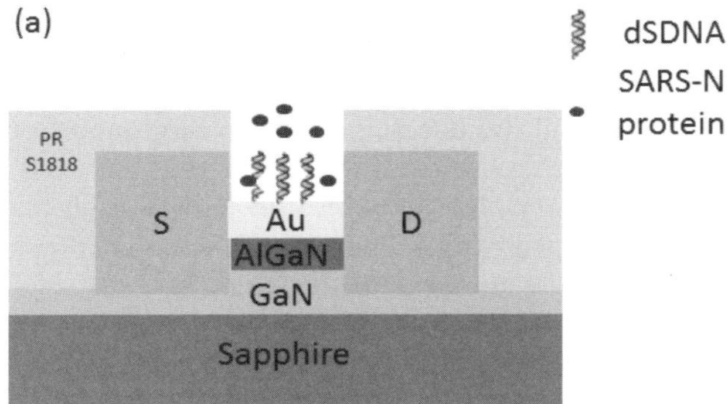

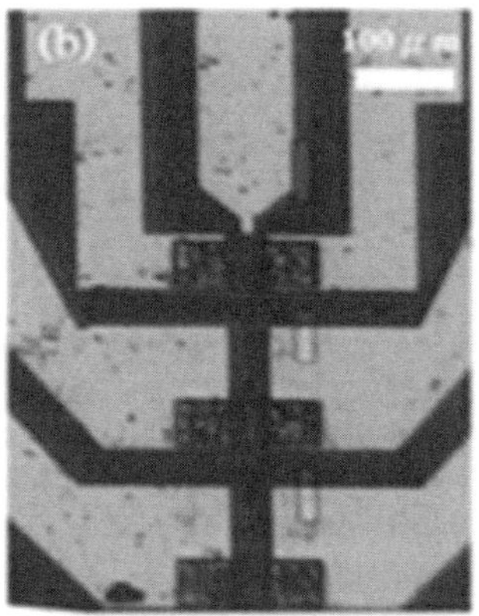

Figure 18. (a) Schematics of the N protein sensor. (b) Plan-view photography of the sensor.

Figure 19 shows the real-time detection of the N protein at constant bias of 350 mV for the sensor. Phosphate buffered saline (10 mM sodium phosphate pH 6.0, 50 mM NaCl, 1 mM EDTA) was initially dropped on the sensor. There is no net current change until the target protein concentration of 0.003nM of the protein was added. A clear current change was observed as the system reached a steady state. Real-time current monitoring spanned the range of protein concentrations from 0.003nM to 3000nM, and showed saturation as the concentration of N protein larger than 300nM. Figure 20(a) and Figure 20(b) show the current changes at different N protein-CTD concentrations for the sensor, the background and the one subtracting background from the sensor, in linear and log scale, respectively. The red line and the black line represent the current changes from the HEMT with and without DNA immobilized, respectively. The green one is the net current by subtracting the black line from the red line.

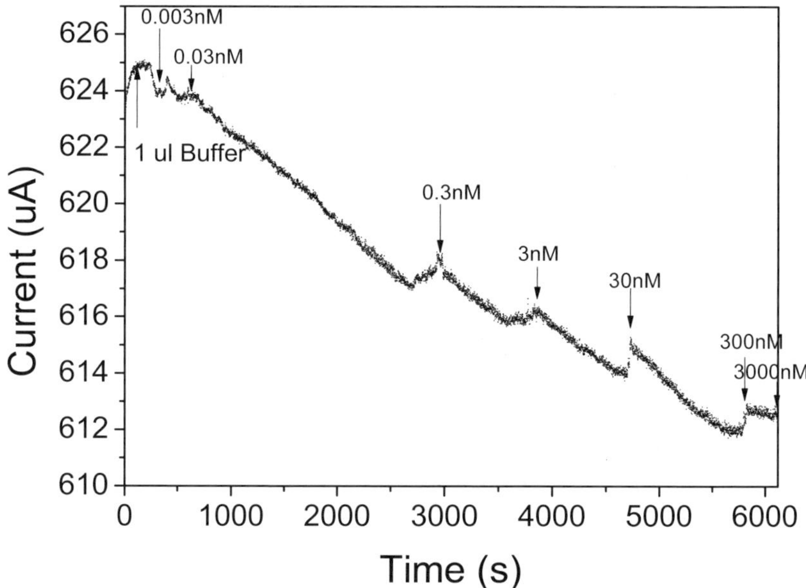

Figure 19. Real-time detection of the N protein from 0.003nM to 3000nM at constant bias of 350 mV.

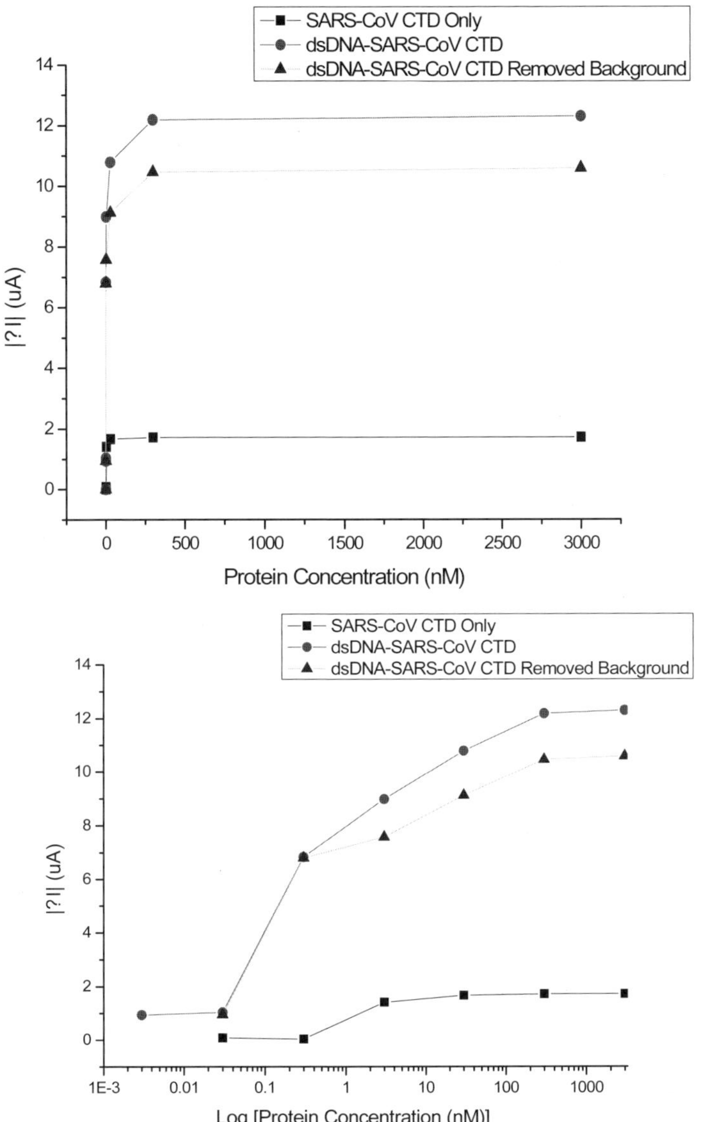

Figure 20. Current changes at different concentrations of N protein CTD in (a) linear and (b) log scale. The red line and the black line represent the current changes from the HEMT with and without DNA immobilized, respectively. The green one is the net current by subtracting the black line from the red line.

By using the binding-site model that was mentioned previously, we are able to analyze the experimental results. Figure 21 shows the surface coverage ratio at different N protein concentrations. The experimental data points nicely fit into the two-binding site model. Two dissociation constants, (K_{D1} = 0.1094 nM and K_{D2} = 51.24 nM) were extracted from the models. This result suggests that there are at least two SARS-CoV CTD binding on the dsDNA.

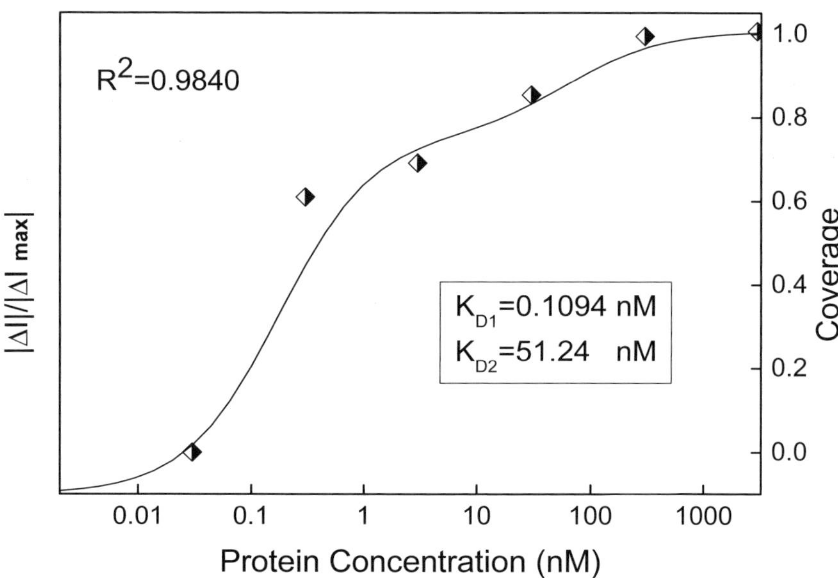

Figure 21. The surface coverage ratio at different N protein CTD concentrations.

5.6. Summary

In this chapter, we present the applications of the AlGaN/GaN HEMTs in the fast detection of a biotoxin and a pathogen with high sensitivity. The HEMTs are also shown the capability to reveal the binding affinity and binding site information of ligand-receptor complex with the help of the binding-site models. Protein-peptide and DNA-protein interactions are demonstrated using the AlGaN/GaN HEMTs and the binding-site models. These biosensors are very promising in wide applications in security, environmental protection, global trade, and even the fundamental research.

References

1. Zhang, A.P., et al., *Correlation of device performance and defects in AlGaN/GaN high-electron mobility transistors.* Journal of Electronic Materials, 2003. **32**(5): p. 388-394.
2. Ambacher, O., et al., Proc. Electrochemical Society Pennington, NJECS 02-14, 27, 2002.
3. Neuberger, R., et al., *High-electron-mobility AlGaN/GaN transistors (HEMTs) for fluid monitoring applications.* Physica Status Solidi a-Applied Research, 2001. **185**(1): p. 85-89.
4. Schalwig, J., et al., *Group-III-nitride based gas sensing devices.* Physica Status Solidi a-Applied Research, 2001. **185**(1): p. 39-45.
5. Steinhoff, G., et al., *pH response of GaN surfaces and its application for pH-sensitive field-effect transistors.* Applied Physics Letters, 2003. **83**(1): p. 177-179.
6. Eickhoff, M., et al., *Wetting behaviour of GaN surfaces with Ga- or N-face polarity.* Physica Status Solidi B-Basic Research, 2001. **228**(2): p. 519-522.
7. Kang, B.S., et al., *Electrical detection of biomaterials using AlGaN/GaN high electron mobility transistors.* Journal of Applied Physics, 2008. **104**(3).
8. Kang, B.S., et al., *Pressure-induced changes in the conductivity of AlGaN/GaN high-electron mobility-transistor membranes.* Applied Physics Letters, 2004. **85**(14): p. 2962-2964.
9. Pearton, S.J., et al., *GaN-based diodes and transistors for chemical, gas, biological and pressure sensing.* Journal of Physics-Condensed Matter, 2004. **16**(29): p. R961-R994.
10. Steinhoff, G., et al., *AlxGa1-xN - A new material system for biosensors.* Advanced Functional Materials, 2003. **13**(11): p. 841-846.
11. Steinhoff, G., et al., *Recording of cell action potentials with AlGaN/GaN field-effect transistors.* Applied Physics Letters, 2005. **86**(3).
12. Kang, B.S., et al., *Enzymatic glucose detection using ZnO nanorods on the gate region of AlGaN/GaN high electron mobility transistors.* Applied Physics Letters, 2007. **91**(25).
13. Kang, B.S., et al., *Electrical detection of immobilized proteins with ungated AlGaN/GaN high-electron-mobility transistors.* Applied Physics Letters, 2005. **87**(2).

14.   Kang, B.S., et al., *Electrical detection of deoxyribonucleic acid hybridization with AlGaN/GaN high electron mobility transistors.* Applied Physics Letters, 2006. **89**(12).

15.   Eickhoff, M., et al., *Electronics and sensors based on pyroelectric AlGaN/GaN heterostructures – Part B: Sensor applications.* physica status solidi (c), 2003. **0**(6): p. 1908-1918.

16.   Neuberger, R., et al., *Ion-induced modulation of channel currents in AlGaN/GaN high-electron-mobility transistors.* Physica Status Solidi a-Applied Research, 2001. **183**(2): p. R10-R12.

17.   Gangwani, P., et al., *Polarization dependent analysis of AlGaN/GaN HEMT for high power applications.* Solid-State Electronics, 2007. **51**(1): p. 130-135.

18.   Shen, L., et al., *High-power polarization-engineered GaN/AlGaN/GaN HEMTs without surface passivation.* Ieee Electron Device Letters, 2004. **25**(1): p. 7-9.

19.   Bertucci, C. and S. Cimitan, *Rapid screening of small ligand affinity to human serum albumin by an optical biosensor.* Journal of Pharmaceutical and Biomedical Analysis, 2003. **32**(4-5): p. 707-714.

20.   Morrill, P.R., R.B. Millington, and C.R. Lowe, *Imaging surface plasmon resonance system for screening affinity ligands.* Journal of Chromatography B-Analytical Technologies in the Biomedical and Life Sciences, 2003. **793**(2): p. 229-251.

21.   Kim, D.H., et al., *Premature antibodies with rapid reaction kinetics and their characterization for diagnostic applications.* Analytical Biochemistry, 2012. **420**(1): p. 54-60.

22.   Krusinski, T., A. Ozyhar, and P. Dobryszycki, *Dual FRET assay for detecting receptor protein interaction with DNA.* Nucleic Acids Research, 2010. **38**(9).

23.   Orosz, F. and J. Ovadi, *A simple method for the determination of dissociation constants by displacement ELISA.* Journal of Immunological Methods, 2002. **270**(2): p. 155-162.

24.   Chiad, K., et al., *Isothermal Titration Calorimetry: A Powerful Technique To Quantify Interactions in Polymer Hybrid Systems.* Macromolecules, 2009. **42**(19): p. 7545-7552.

25.   del Toro, M., et al., *Study of the interaction between the G-quadruplex-forming thrombin-binding aptamer and the porphyrin 5,10,15,20-tetrakis-(N-methyl-4-pyridyl)-21, 23H-porphyrin tetratosylate.* Analytical Biochemistry, 2008. **379**(1): p. 8-15.

26.   Homola, J., *Surface plasmon resonance sensors for detection of chemical and biological species.* Chemical Reviews, 2008. **108**(2): p. 462-493.

27. Arnon, S.S., et al., *Botulinum toxin as a biological weapon - Medical and public health management.* Jama-Journal of the American Medical Association, 2001. **285**(8): p. 1059-1070.

28. Bagramyan, K., et al., *Attomolar Detection of Botulinum Toxin Type A in Complex Biological Matrices.* Plos One, 2008. **3**(4).

29. Qiang, Z., et al., *A new potentiometric immunosensor for determination of α-fetoprotein based on improved gelatin–silver complex film.* Electrochimica Acta, 2006. **51**(18): p. 3763-3768.

30. Pohanka, M., et al., *Detection of Aflatoxins in Capsicum Spice Using an Electrochemical Immunosensor.* Analytical Letters, 2008. **41**(13): p. 2344-2353.

31. Pan, N.Y. and J.S. Shih, *Piezoelectric crystal immunosensors based on immobilized fullerene C60-antibodies.* Sensors and Actuators B-Chemical, 2004. **98**(2-3): p. 180-187.

32. Grogan, C., et al., *Characterisation of an antibody coated microcantilever as a potential immuno-based biosensor.* Biosensors & Bioelectronics, 2002. **17**(3): p. 201-207.

33. Dungan, C.F. and B.S. Roberson, *BINDING SPECIFICITIES OF MONOCLONAL AND POLYCLONAL ANTIBODIES TO THE PROTOZOAN OYSTER PATHOGEN PERKINSUS-MARINUS.* Diseases of Aquatic Organisms, 1993. **15**(1): p. 9-22.

34. Burreson, E.M. and L.M.R. Calvo, *Epizootiology of Perkinsus marinus disease of oysters in Chesapeake Bay, with emphasis on data since 1985.* Journal of Shellfish Research, 1996. **15**(1): p. 17-34.

35. Ford, S.E., *Range extension by the oyster parasite Perkinsus marinus into the northeastern United States: Response to climate change?* Journal of Shellfish Research, 1996. **15**(1): p. 45-56.

36. Villalba, A., et al., *Perkinsosis in molluscs: A review.* Aquatic Living Resources, 2004. **17**(04): p. 411-432.

37. Pecher, W.T., et al., *Assessment of the Northern Distribution Range of Selected Perkinsus Species in Eastern Oysters (Crassostrea virginica) and Hard Clams (Mercenaria mercenaria) with the Use of PCR-Based Detection Assays.* Journal of Parasitology, 2008. **94**(2): p. 410-422.

38. Ray, S.M., *A review of the culture method of detecting Dermocystidium marinum with suggested modifications and precautions.* Proc. Natl. Shellfish. Assoc., 1966. **54**: p. 55-69.

39. Reece, K.S. and C.F. Dungan, *AFS-FHS Blue Book: Suggested Procedures for Detection and Identification of Certain Finfish and Shellfish Pathogens 2005 Edition*: (American Fisheries Society, Bethesda, 2006), Chap. 5.2, pp. 1-17.

40. Sheppard, B.J. and A.C. Phillips, *Perkinsus olseni detected in Vietnamese aquacultured reef clams Tridacna crocea imported to the USA, following a mortality event.* Diseases of Aquatic Organisms, 2008. **79**(3): p. 229-235.

41. Sheppard, B.J. and C.F. Dungan, *Exotic Perkinsus sp. Protozoa in an Imported Vietnamese Ornamental Clam (Tridacna crocea) Maintained in a Home Aquarium.* Journal of Zoo and Wildlife Medicine, 2009. **40**(1): p. 140-146.

42. Wang, W.U., et al., *Label-free detection of small-molecule-protein interactions by using nanowire nanosensors.* Proceedings of the National Academy of Sciences of the United States of America, 2005. **102**(9): p. 3208-3212.

43. Lin, T.W., et al., *Label-free detection of protein-protein interactions using a calmodulin-modified nanowire transistor.* Proceedings of the National Academy of Sciences of the United States of America, 2010. **107**(3): p. 1047-1052.

44. Chen, Y.N., et al., *Electronic Detection of Lectins Using Carbohydrate-Functionalized Nanostructures: Graphene versus Carbon Nanotubes.* Acs Nano, 2012. **6**(1): p. 760-770.

45. Dragusanu, M., et al., *On-Line Bioaffinity-Electrospray Mass Spectrometry for Simultaneous Detection, Identification, and Quantification of Protein-Ligand Interactions.* Journal of the American Society for Mass Spectrometry, 2010. **21**(10): p. 1643-1648.

46. Mitsakakis, K. and E. Gizeli, *Multi-sample acoustic biosensing microsystem for protein interaction analysis.* Biosensors & Bioelectronics, 2011. **26**(11): p. 4579-4584.

47. Maehashi, K., et al., *Aptamer-Based Label-Free Immunosensors Using Carbon Nanotube Field-Effect Transistors.* Electroanalysis, 2009. **21**(11): p. 1285-1290.

48. Saikatendu, K.S., et al., *Ribonucleocapsid Formation of Severe Acute Respiratory Syndrome Coronavirus through Molecular Action of the N-Terminal Domain of N Protein.* Journal of Virology, 2007. **81**(8): p. 3913-3921.

49. Hsieh, P.-K., et al., *Assembly of Severe Acute Respiratory Syndrome Coronavirus RNA Packaging Signal into Virus-Like Particles Is Nucleocapsid Dependent.* Journal of Virology, 2005. **79**(22): p. 13848-13855.

50. Chang, C.K., et al., *Multiple Nucleic Acid Binding Sites and Intrinsic Disorder of SARS Coronavirus Nucleocapsid Protein - Implication for Ribonucleocapsid Protein Packaging.* Journal of Virology, 2008: p. JVI.02001-08.

51. Takeda, M., et al., *Solution Structure of the C-terminal Dimerization Domain of SARS Coronavirus Nucleocapsid Protein Solved by the SAIL-NMR Method.* Journal of Molecular Biology, 2008. **380**(4): p. 608-622.

52. Chen, C.Y., et al., *Structure of the SARS Coronavirus Nucleocapsid Protein RNA-binding Dimerization Domain Suggests a Mechanism for Helical*

Packaging of Viral RNA. Journal of Molecular Biology, 2007. **368**(4): p. 1075-1086.

53.   Klig, L.S., I.P. Crawford, and C. Yanofsky, *Analysis of trp repressor-operator interaction by filter binding.* Nucleic Acids Research, 1987. **15**(13): p. 5339-5351.

54.   Garner, M.M. and A. Revzin, *A gel electrophoresis method for quantifying the binding of proteins to specific DNA regions: application to components of the Escherichia coli lactose operon regulatory system.* Nucleic Acids Research, 1981. **9**(13): p. 3047-3060.

55.   Fried, M. and D.M. Crothers, *Equilibria and kinetics of lac repressor-operator interactions by polyacrylamide gel electrophoresis.* Nucleic Acids Research, 1981. **9**(23): p. 6505-6525.

56.   Fried, M.G. and M.A. Daugherty, *Electrophoretic analysis of multiple protein-DNA interactions.* ELECTROPHORESIS, 1998. **19**(8-9): p. 1247-1253.

Chapter 5

Stability and Reliability of III-Nitride Based Biosensors

Nathaniel Rohrbaugh, Ronny Kirste, Ramon Collazo, and
Albena Ivanisevic

*Department of Materials Science and Engineering
North Carolina State University
Raleigh, North Carolina, 27695*

5.1. III-Nitride Materials

Two particular materials of significant interest within recent years have been AlGaN and GaN. These two materials are commonly combined in a heterojunction structure for high electron mobility transistors (HEMT). From a biological compatibility stand point the AlGaN/GaN system has been shown to be non-toxic when put under physiological conditions as well as maintaining a chemically stable surface [1]. Furthermore cellular growth studies have revealed that AlGaN/GaN is a suitable surface for cell growth and tissue interfacing [2]. These devices are fabricated such that layered AlGaN/GaN produces a conductive channel at their interface, close to the sensing surface. It is this interface that gives this material its characteristic properties as a biosensing platform, and allows for quantification of dissolved or airborne compounds. Interactions between the surface of the AlGaN/GaN device and its environment generates changes along the surface that influence the conductive channel [3]. The electrical fields of these interactions produce conductance changes in the channel, thus these devices fall under the category of field effect transistors (FETs). One of the largest benefits of the AlGaN/GaN system is the adaptability of the surface to different modification and functionalization schemes. This is made possible by the availability of surface bonding sites and the innate chemical inactivity in the presence of both acidic and basic solutions [4]. Thus, combining the benefits of physiological stability, sensitivity to surface interactions, and wide range of surface modifications the AlGaN/GaN system is one of the most widely used III-nitride currently being used in biosensor research and development. While the AlGaN/GaN system provides an excellent example of applied III-nitrides to biosensors it represents only one area in the larger research field that is III-nitrides. AlN, GaN, InN all share similar surface properties and have been examined as possible base materials for novel biosensing platforms.

In this chapter we will be discussing III-nitrides for biosensing as a wider field of study. Although there will be a large amount of focus given to AlGaN/GaN as the most prominent III-nitride of study in literature for biosensing, other less commonly used III-nitrides will be explored as well. Specific biosensor device functionality, specificity, selectivity and operation will be discussed on a material by material basis. Additionally, a background into the materials themselves will be provided from a stability perspective with regard to solution exposure, doping, and interactions with biological systems.

The group III-Nitrides can be defined as the class of materials that encompasses all compounds formed between a III group element binding with nitrogen. This includes AlN, GaN, InN, InGaN, and AlGaN; as well as all other binary, tertiary, and quaternary combinations. Interest has greatly increased in the past decades as the need for light emitting diodes (LEDs) and other optoelectronics has risen. III-Nitrides offer both wide bandgap semiconductor materials like GaN and AlN, as well as narrow bandgap materials like InN [5-7]. It is due to this wide bandgap character that these materials have taken such an interest for optoelectronics [8]. The bandgap can be tuned depending on the alloy compositions, or other factors induced during processing such as strain [9,10]. All of these materials have a direct bandgap making them attractive for light emission and detector applications [11].

III-nitride materials may crystallize into two structures: wurtzite (hexagonal) and zinc blende (cubic). Example diagrams of both wurtzite and zinc blende can be seen in Figure 1. Wurtzite is the most common solid phase of the two, and is a form of the hexagonal crystal system with primary axes in the *a* and *c* directions [12]. Anisotropic behavior exists between the *a* and *c* directions [13]. Polarization is inherent to the wurtzite structure of the III-nitrides and can be categorized as spontaneous and piezoelectric. Spontaneous polarization comes from electric fields generated between alternating layers of N and III group metal in the material and the non-centro symmetry of the hexagonal crystal structure [14].

Zinc blende structures have similar alternating layers to wurtzite but lacks this spontaneous polarization due to the lack of non-centro symmetry. Piezoelectric polarization arises from stresses on the III-nitride material [14]. Typical heteroepitaxy used for these materials results in strain; therefore both polarization effects will always be present in III-nitride deposited layers.

III-nitrides films are typically grown in the $+c$ orientation of the crystal with the surface being the (0001) plane. Due to the non-centrosymmetric nature of the hexagonal structure, two different orientations of the crystal lattice are possible along the c-axis. Thus the crystal can be grown in the +c orientation (0001) referred to as the III (either Ga, In or Al)-polar and the −c orientation (000-1) referred to as the N-polar [15]. This orientation also determines the direction of the spontaneous polarization of the crystal, thus referring to its polarity orientation. The corresponding polarity determines the possible surface reconstructions available for the c-planes, or simply the number of dangling bonds available on the surface [16]. In the III-polar case, a negative polarization charge at the surface induces a positively charged depletion region near the surface, while for the N-polar surface a positive polarization charge induces a negatively charge region. This negative charge region arises from accumulation of free carrier or a change in the occupation of the surface states. Polarization differences across interfaces between III-nitride layers determine the induce charge at that interface. Such an example is the polarization discontinuity found between the AlGaN/GaN interfaces where electron accumulation forms a 2 dimensional electron gas with resulting high electron mobility. This is the basis for the formation of the conductive channel in the field effect transistors previously described and as such they are typically described as high electron mobility transistors (HEMT).

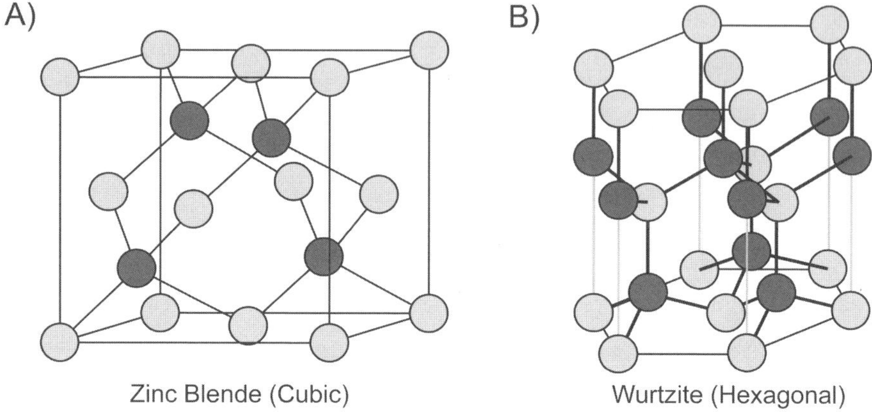

Figure 1. Example diagrams of both Zinc Blende and Wurtzite structures seen in III-nitride compounds.

5.2. Doping of Nitrides

Doping of III-nitride materials has become more common practice in recent years and can be incorporated into the MOCVD growth processes [10]. The addition of dopants to these films allows for tunable conductivity characteristics and selection of the charge carrier type that will be dominant in the film [17]. When defining doping charge carriers, n-type refers to electrons dominating, and p-type being hole dominated conduction. Dopants of a higher valence than the base material, in this case group IV, are added to generate n-type behavior. In contrast, for hole dominated films a lower valence dopant is added, such as a group II material. Depending on the charge carrier type, either donor or acceptor states will be created within the bandgap of a semiconducting material [18]. However, effective doping is only possible if the activation energy of the charge carriers is low enough to provide free electrons or holes with sufficient amounts at room temperature. Here, the activation energy is defined as the energy that is needed to detach a carrier from its dopant. Since this energy is typically provided by the temperature of the material (thermal activation) the activation energy should be $< kT$ which is around 27 meV at room temperature. Furthermore, compensation of dopants due to incorporation of unwanted charged point defects needs to be controlled.

This highlights the need to find shallow dopant and control compensation in III-Nitrides.

Most common dopants in GaN and its ternary materials are Si for n-type behavior and Mg for p-type. These provide good representation for doping in this material system as both are widely used and well researched for III-nitride materials. Si is a shallow donor in GaN with an activation energy of 12 meV. For Si doped GaN a good controllability has been found and free carrier concentrations ranging from 1×10^{16} cm^{-3} to 5×10^{19} cm^{-3} have been demonstrated. Higher free carrier concentrations can be targeted with Ge if needed. However, p-type doping with Mg is found to be more difficult [18,19]. This can be related to the higher activation energy of the Mg-acceptor of 160 meV, the formation of Mg-H complexes and passivation after growth , and limitations in the doping capabilities due to self-compensation for Mg$>3 \times 10^{19}$ cm^{-3}[18,20]. While the Mg-H complex can be activated following fabrication with an additional annealing step in an H free atmosphere and compensation can be controlled using advanced point defect control mechanisms, the activation energy stays the main limiting factor [21]. Consequently, typical free carrier concentrations of MOCVD grown GaN:Mg are in the range of $3\text{-}6 \times 10^{17}$ cm^{-3} for 3×10^{19} cm^{-3} acceptors. Thus, the conductivity of Mg p-type doping as a whole is relatively low given the above factors when compared to the capabilities of Si n-type doping [17,22].

Doping in the other binary III-nitride systems is based on the results and experience from GaN. N-type behavior in AlN has been shown with Si but doping capabilities are limited due the formation of a DX-center and increase of the activation energy to >200 meV [23]. P-doping using Mg has been tried in AlN as well, but no free holes or p-conductivity has been demonstrated as of today. InN possesses native n-type behavior in many cases due to its surface high electron affinity [23,25]. Some reports suggest this is in part to impurity formation within the InN [26-28]. BN is more accessible to dopants due to its smaller atomic radii as shown in work involving Si and Mg [29-31]. Nuances of doping for each of these binary compounds are beyond the scope of this work and is still a field where new up and coming research is being done. As such for more information on doping effects on stability of biosensors readers are encouraged to investigate further with some of the references provided here.

III-nitride ternary compounds are unique from a doping perspective in that both Group III elements in the ternary compound can be treated as dopants of each other depending on desired characteristics and tuning of the desired final structure. Similar to literature mentioned previously on binary III-nitrides research within this field is broad and as such only n-type AlGaN will be discussed in a briefly as an example. N-type doping of low Al-content AlGaN has been shown using Si and provides a good example of the balancing that must be done with the two III-group elements in the final product. In work done by Collazo et al, AlGaN with varying ratios of Al to Ga are explored relative to their Si doped n-type behavior [32]. The activation energies of Si impurity formation within GaN and AlN are approximately 17 meV and 200 meV respectively [33]. Thus, as Al content increases in AlGaN a drop off in n-type behavior is observed resulting in almost complete removal of the behavior at $Al_{0.8}Ga_{0.2}N$ [32]. This decrease in n-type behavior can be further explained by DX center formation and vacancy-complex compensation within the AlGaN. More information into AlGaN's conductive properties and performance relative to growth processes and doping can be found in the extensive review on the subject in Jones et al [10].

In discussing the environmental stability that doped III-nitride materials possess a definition of shallow and deep dopants needs to be established. Shallow dopants like that of Si donors in GaN have relatively small activation energies which can be filled from thermal energy within the compound [34]. These dopants once activated create available charge carriers for the n/p type behavior designed for the material. Here, the majority of shallow dopants will be active or intended to be active once the III-nitride compound is in a conducting mode. Deep dopants however, require a much greater energy to ionize and in many cases will not provide free charge carriers to aid in conduction [10]. Dopants such as these provide compensating centers that can act as recombination sites for electrons and holes, and are commonly referred to as charge carrier traps [35].

Doped GaN, both n-type and p-type, experiences a carrier removal or relocation when exposed to neutron radiation that varies on the type of charge carrier [36]. The methodology of quantifying radiation damage takes into account the number of carriers that no longer contribute to the overall n-type or p-type behavior. This is typically because the carrier or dopant has been moved to a dislocation site within the lattice. In this study by Pearton et al. carrier removal rates were 20 times higher for p-type doping than in n-type doping, however both methods had removal rates $>10^3$ cm^{-1} under proton radiation doses of 10^{14} cm^2 [36]. Under the same testing neutron and electron radiation were examined with neutron radiation having a weaker removal rate. The rate of removal for neutron radiation was $<10^2$ - 10^0 cm^{-1} however, electron radiation yielded a weaker removal at $<10^1 - 10^{-1}$ cm^{-1} over a wider range of dosages. The generation of electron traps within the GaN layers leads to a Fermi level pinning effect (reported at a conduction band energy of -0.95 eV) which can lead to self-compensating effects for conductivity [37]. In general under radiation dosing n-GaN has been observed to become semi-insulating as carriers are displaced. The problem presents itself with Al$_x$Ga$_{1-x}$N heterojunction layers at various levels of Al doping. The self-compensating effects of radiation damage manifested itself here as an overall increase in resistivity of the GaN layers in AlGaN/GaN heterostructures [38]. This increase in resistivity resulted in degradation of the 2DEG in both AlGaN/GaN interfaces and AlN/GaN interfaces, and is largely attributed to the increase in defect density causing a drop in mobility. Cited values for the drop in conductance were as high as 50% for elevated levels of electron dosing [38]. Further work by Polyakov et al. into neutron effects on the same series of AlGaN/GaN and AlN/GaN (additionally InAlN/GaN) showed cause for concern for stability of 2DEG layers under such conditions [36]. It was observed that with increasing Al content in AlGaN/GaN heterostructures (up to Al$_{0.5}$Ga$_{0.5}$N) an increase in trap concentration was observed which in turn lead to decreases in the carrier mobility and conductance of the device [39]. Similar effects were observed in proton irradiated samples of AlGaN/GaN as well [40]. Thus it can be said that radiation effects are of significant concern for GaN based devices and heterostructures if the intended end use requires reliable performance under prolonged and elevated radiation exposure. In context of electrical stability these effects stand out and may present problems over time, but there are ways to mitigate these effects by lowering initial dislocation densities in GaN based materials [37]. There is also evidence for recovery and relaxation of radiation induced carrier migration over time after following a strong radiation exposure [41].

Despite these findings, III-nitrides when compared to other materials have shown evidence of significantly higher radiation hardness.[42] In terms of overall resilience both AlGaN and GaN shown radiation hardness values less than but comparable to diamond [42,43]. AlN has been further investigated for creating resistant layers from UV up to gamma radiation [44-46]. In this work done by Tittmann et al. AlN showed high resilience when exposed to 5.8×10^{18} neutrons cm^{-2} and 26.8 MGy gamma radiation, this was quantified by the lack of change in AlN's piezoelectric properties [45]. InN has not been as widely studied for radiation hardness but has shown interesting responses in relation to charge carrier removal. N-type InN demonstrates an opposite effect to GaN and other III-nitride compounds in that it shows an increase in carrier concentration once exposed to high levels of radiation. This can be explained by the high electron accumulation that occurs in InN and the innate n-type behavior that comes from higher defect densities [47]. Highly crystalline InN nanowires exhibit minimal electron accumulation, thus defects brought on by radiation damage increase this accumulation and undermining deliberate processing to remove electron accumulation along the surface [48]. It is worth noting that heat treating of radiation damaged III-nitrides allows for some recovery of conductive properties. In this same study by Emtsev et al. annealing of InN and GaN allowed for a recovery of properties to values similar to previous measurements done before the radiation treatments [47]. Annealing GaN required temperatures greater than 600°C before substantial electron mobility and charge carrier concentrations recovered. InN showed opposite effects as a reduction in mobility was observed at temperatures between 400° and 500°C. Once temperatures exceeded 500°C electron mobility began to return to that of pre-radiation treatment levels [47].

5.3. Water Stability (resistance to etching or degradation)

Solution based measurements for biosensors require an understanding of a material's behavior when exposed to a variety of liquid environments. Leeching of material into solution or defect exacerbation are examples of mechanisms that can degrade surfaces of biosensing devices. Fortunately, III-nitride materials have been shown to possess a high tolerance for wet etching conditions and have high chemical stability [49,50]. Two types of etching are discussed here in the context of III-nitrides; electrochemical etching and conventional etching. The use of chemical etchants such as hot acids is an example of conventional etching and are used in cleaning and processing of III-nitride materials for biosensors [51]. Electro-chemical etching relies on potential differences between the

III-nitride and the charged species interacting with the surface, examples include anodic etching or electrodeless etching. The primary difference for these two mechanisms is the necessity of charge carriers for electrochemical etching and the absence of them in chemical etching [52]. As such electrochemical processes are more likely to take place in physiological environments encountered by biosensors where a highly diverse and dense concentration of electrolyte species could be found.

The high bond energy of III-nitrides is what allows for this resistance to most etching methods. Bond energies of III-nitrides range between 2.88 eV, 2.2 eV, and 1.93 eV for AlN, GaN, and InN respectively [53]. A benefit of this tolerance for chemical etching relative to biological applications is the ability to remove surface organic compounds and oxide layers without damaging or heavily altering the III-nitride material beneath [54]. HCl etching on GaN for removal of surface contaminants and preparing the surface for functionalization is such an example [55]. Surface stability of GaN under exposure to etchants is affected by the quality of the grown film and its corresponding dislocation density. In the fabrication of III-nitride materials substrates such as Si or sapphire are commonly used and can produce a substantial lattice mismatch at the interface of the film. This interface produces threading dislocations with a density determined by the growth parameters typically in the range of 10^7-10^{11} cm^{-2} for GaN [49,56]. Wet etching can still be realized on III-nitrides both to show dislocations and to remove layers of the surface despite an overall resistance to common etchants. AlN and GaN have been well documented as being susceptible to H_3PO_4, KOH-NaOH, and HF/H_2O_2 [56-59]. Some examples of GaN and AlN etching rates in common chemical etchants can be seen further in Zhuang et al [49].

Stability in aqueous conditions is observable by measuring the concentration of the dissolved group III element or compound into the surrounding solution. GaN for example exhibits very low concentrations in pure DI water at reported values of 3.68 ± 0.54 ppb after 7 days [60]. This demonstrates a high degree of aqueous stability in GaN for possible biosensor applications. Aluminum nitride exhibits similar behavior when bulk samples are placed into solutions. Bulk samples themselves are virtually insoluble in water, however AlN powders exhibit corrosion and hydrolysis in the presence of aqueous environments [61,62]. This suggests a correlation between exposed surface area to solution and stability of the AlN material. Work has shown that AlN powders can be passivated to their aqueous surroundings to prolong lifetime in solution. Research on this is still dependent on a phosphoric or phosphonic acid complex or etch to encapsulate the surface due to AlN showing high resistance to traditional etchants even as a powder [63].

BN also demonstrates highly stable surface properties under exposure to aqueous environments. Oxidation of boron nitride is generally seen as a high temperature process taking place at upwards of 1500°C, with minimal oxidation taking place at room temperature or within feasible temperatures for a biological setting [64]. When under aqueous conditions the primary reaction scheme for BN degradation via hydrolysis is the formation of metaboric acid (HBO_2) and orthoboric acid (H_3BO_3) [65]. However, even these processes were shown to not become substantial until temperatures greater than 700°C, which is far beyond the bounds of what a biological sensor material would encounter. In recent work BN was shown to be a strong candidate for ultra-hydrophobic surfaces due to its chemical inertness and lack of reactivity in water at < 100°C [66]. The growth morphologies given in this study show that even in vertically grown nanosheet conformations water reactivity was negligible. Nanosheet surface morphologies have a much higher amount of exposed surface area than epitaxially grown BN, further illustrating BN's aqueous stability and lack of reactivity. BN's high aqueous stability allows it to be used in applications of perpetual water contact, like those proposed for the use of nano-porous structures of BN in water filtration. Here it was demonstrated that due to BN high resistance to oxidation it is an excellent candidate for use in removal of oils and solvents from water systems[67]. Although it is worth noting that BN stability is dependent on crystal structure. At higher temperatures (1200°C) a hexagonal orientation of BN is more stable while closer to ambient temperatures cubic orientations are preferred [68]. This temperature dependence of BN and other III-nitride is not a focus of this work as operational stability for GaN and III-nitride sensors is seen well past temperatures one would find under biological conditions.

While not the dominant III-nitride of interest in aqueous environments, multiple studies on InN in aqueous environments have exhibited similar characteristics to that of other III-nitrides. InN hydrolysis is detectable by the presence of NH_3 in solution or by surface presence of $In(OH)_3$, both of which are the reaction products of hydrolysis on InN [69]. These are difficult to observe except through XPS as the hydrolysis process does not produce a noticeable amount of material loss from the surface. The effect of hydrolysis in aqueous solutions seems to be negligible as research into InN as an anion sensing platform deemed the material an excellent candidate for such a role and did not observe any degradation effects of the InN surfaces while exposed to various ionic solutions [70,71]. Etching of InN surfaces is expectedly difficult and similar to AlN and GaN. Acidic solutions typically do not have any effect on InN, but basic solutions such as KOH and NaOH are capable of etching the surfaces [72]. However, etching of the surface is a relatively slow process with etching rates from this study ranging from ~5 Å/min at ambient temperature to ~400 Å/min at elevated (60°C) temperatures. In general InN is very stable and not easily etched by any common etchant with the exception of highly basic solutions, and even such at a markedly slow rate.

5.4. Toxicity

Toxicity can be seen as the amount of a material that is needed initiate damage or harm instead of a helpful or neutral response within an organism or cellular system [73]. In contrast, biocompatibility relates the interfacing of tissues and cellular structures with a given biomaterial. In general concentrations of III-nitride materials are very low in natural settings as they are not naturally occurring compounds. III-nitrides are largely chemically inert under biological conditions and in water solutions. The possible concern then from III-nitrides in the environment comes from the leeching of the III group element (Ga, Al, In, B) into surrounding environments. Industrial pollution would be an example of a point source for this type of contamination. Al is well known to be toxic and having detrimental effects to biological systems, which is one reason why many biosensors are focused on Ga-based systems [74]. Ga has yet to show any toxic effects on biological systems and exists in nature at very low concentrations within materials with daily contact. Toxicity of some III group materials such as InN are not well known as they are not naturally occurring in large quantities.

One study examined the accumulation of Ga in 1300 food samples and revealed that Ga concentrations in most (84% of those tested) common foods was beneath the limit of quantification of 0.002 mg kg^{-1} and 83% of foods tested concentrations were beneath the limit of detection of the instrument at 0.001 mg kg^{-1} [75]. The only substantial findings from this study relative to Ga were an increased level of Ga in fats and oil products with an average level of over 0.005 mg kg^{-1} [75]. Shellfish and chocolate were the only other substantial sources at 0.008 and 0.010 mg kg^{-1}, respectively. Indium dosing for humans is also not considered substantial from foods that occur naturally, and is instead linked almost entirely to industry related pollutants and water contamination [76]. In this study done by Ayadi et al. indium was given to mice via injection to gain some insight into effects on mammalian tissues from indium [76]. Results from this work showed highly toxic effects from indium nitrate given at dosages of 28 mg kg^{-1}. Monitoring prolactin concentrations from mouse mammary glands showed a drop to 23% that of levels seen in control mice. This coupled with decreasing body weight and food intake provides evidence that indium is indeed toxic when leeched into biological systems. Similar studies have been done on male rats and found similar results that indium deposits accumulating in cellular lysosomes lead to necrosis of the tissue and loss of weight in the rats [77,78]. Indium's toxic effects have been known for some time despite newer studies highlighting the issues of ingesting large quantities. Work dating back to the early 1970s has shown that indium once dissolved into solution has the same nephrotoxic effects as Al and less so Ga [79]. This study examined both ionic indium and hydrated indium oxide. For ionic indium it was found that almost all damage occurred in the kidneys with some accumulation in the liver when tested in mice. This accumulation was seen as 17.83% of the administered dose per gram persisting in the kidneys after three days for the LD$_{50}$ of 12.5 mg/kg. This increased to 33.14% for an LD$_{100}$ level of dosage at 16.5 mg/kg indium. In both instances necrosis was observed in localized areas near injection sites for both liver and kidney after three days. When not injected directly into organs the accumulation led to the loss of cell structure in both the cytoplasm and nuclei and was observed at LD$_{50}$ levels. Epithelial swelling was also observed. In more recent studies indium containing compounds have been examined for pulmonary effects on both humans and animals as technology has shifted to producing indium compounds such as InN, ITO, and InP. Results in this case were similar to the injected indium studies. These again indicated that the intake of fine powders from indium processing led to accumulation of indium within tissues [80]. Here indium compounds were shown to cause severe damage to pulmonary tissues and possibly demonstrate carcinogenic effects as well depending on the degree of In accumulation within cellular structures. Based off of findings such as these, standards in countries like Japan have instituted a 3 µg/L concentration of indium in biological serum as the upper limit for safe dosage for humans [81].

GaN has been shown to be nontoxic and is thus one of the more promising III-nitrides. This lack of toxicity in GaN extends to some of its other compounds such as gallium nitrate which has been shown to be useful for cancer treatments [82]. In this study done by Kelsen, a continuous supply of gallium nitrate supplied intravenously to patients only showed toxic effects when doses reached 700 mg/M^2 [82]. The units used in such toxicity measurements are commonly expressed in terms of body weight or body surface area (BSA) as used in Kelsen et al, an explanation of these units is provided here [83]. Under such levels nephrotoxic effects were seen as kidneys of the patients were inundated with gallium nitrate. Otherwise the study showed that the kidney remove gallium nitrate from the body dropping gallium levels in the body from 1.9 µg/ml during infusion periods to 0.4 µg/ml several days after.[82] This work done on blood plasma concentrations is dated back to the 1980s, but more recent work has shown similar results and still shows gallium nitrate suitable for metal-based cancer therapies [84]. This work is consistent with the previous findings that gallium nitrate has no strong toxicity to bone marrow and when given through manageable infusions show no overall toxic effects in the body. This is not to say that gallium within the body is inert on its own. It has been shown that gallium at certain concentrations will begin to replace iron in cellular processes, more notably by binding to transferrin proteins which are typically used for iron transport [85]. Gallium nitrate and the subsequent release of Ga into cellular structures have mostly been examined for treatment of lymphoma cells (both Hodgkin's and non-Hodgkin's). One reason for this is the level of radioactive Ga-67 provides a means to measure the metabolic processes that are characteristic of lymphoma cells again by replacing iron in cellular activity [86]. For a more in-depth look into Ga compounds pharmacokinetic behaviors and applications readers are directed toward a review on subject by Chitambar et al [73].

AlN is largely inert in both biological fluids and most etchants, the exception to this rule however is when AlN comes in contact with fluids while in a dispersive powder form. In these circumstances the increased surface area was enough to overcome the inherent chemical inertness of the AlN and allow for some dissolution into the environment. From a toxicity perspective the concentration of AlN is relevant as Al accumulation in kidneys and lysosomes of cells is toxic. One study investigated the degree of dissolution of AlN when in contact with various biological solutions [87]. Here simulated blood serum, gastric juice, and physiological solution were tested relative to the size of AlN particulate. The findings indicate that the finer the powder the greater the degree at which it dissolves into solution up to 95% in gastric and intestinal fluids for nanopowders. Fiber crystals showed only ~10% dissolution in ionic and gastric solutions and almost no dissolution within blood. This is consistent with previously mentioned work that larger AlN substrates and surfaces are markedly more stable in solution.

5.5. Biocompatibility

Biological compatibility refers to the ability of the III-nitride to exist in a biological system without disrupting regular cell activity or harming the host organism. In a broad sense materials used in biosensing applications must be conducive to cell growth by either not causing cell death (apoptosis) or being a surface that cells are willing to grow upon [88]. Within the scope of III-nitrides as biosensing platforms cells are either in contact with the native III-nitride surface or a modified surface layer such as a SAM [89]. Studies quantify this biocompatibility through cell density counts and cell population over time. Methods of doing this involve sterilized surfaces populated with cell lines (PC12 or HeLa for example) which are then monitored over time for removal of cells into solution or apoptosis. This provides information as to how well cells adhere to the surfaces, cell growth relative to surface topology, and cell damage incurred by the III-nitride surface. Examples of cell growth studies can be found in Figure 2 studying both topography and surface functionalization. GaN provides an example of the III-nitrides that has demonstrated a high degree of biocompatibility both as a stand-alone surface and as one modified with biological molecules [60,88]. GaN surfaces did not significantly affect or impede cell growth, and once etched showed a better cell adhesion than etched silicon. Further work has been done on examining bare GaN surfaces with varied topographies and surface adherence of cells [90]. In general it has been shown that cells prefer rougher surfaces to smoother ones for adhesive ability [91]. Mimicking surface roughness of tissue is possible using adlayers of polymer molecules, but similar effects have been shown using bare surfaces put through both etching and polishing. One study in particular found that the effects of polishing and etching GaN were sufficient to create a surface roughness suitable for cell growth and prorogation. This same study additionally tested RF assisted growth of GaN nanowires along the surface and found similar results.[92] Here an increase in cell adhesion on unmodified GaN was observed when surfaces were roughened via polishing. Subsequent increases were again observed with etching, and then finally sharply increasing when nanowires were introduced. The cell phenotypes were dependent upon the surface of the GaN following these treatments [92]. Cells grew with significantly more branching when placed upon polished surfaces (high degree of parallel surface features) than the nanowire arrangement. These findings are in agreement with research that examined modified-nanotextured GaN surfaces with similar values of high cell adhesion capabilities for GaN [93]. Both of these studies conclude that GaN is a suitable surface for biosensing uses and can be fabricated to encourage cell adhesion further. Surface modification was used in both of these cases to create a more favorable surface for cell adhesion.

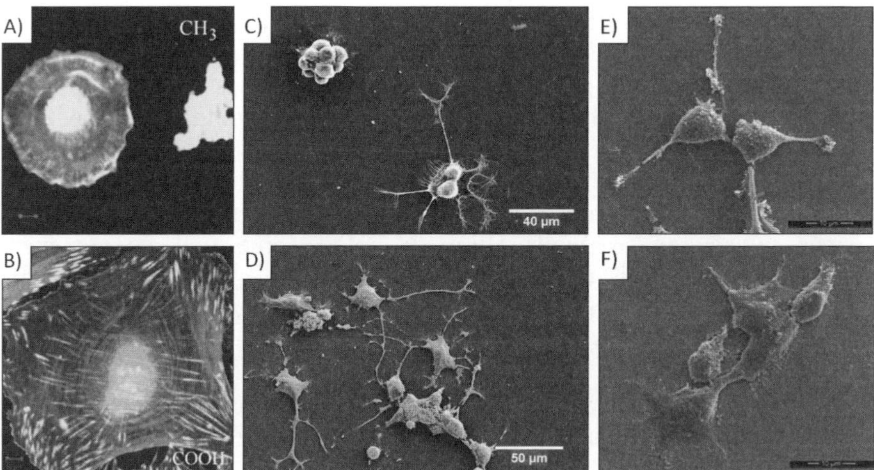

Figure 2. Cell growth on various GaN surfaces; A) and B) fibroblast growth on GaN using SAM modified surfaces, adapted with permissions from Faucheux et al. [89], C) and D) neuron growth on GaN using modified surface topographies, adapted with permission from Bain et al. [92], E) and F) neuron cell growth using IKVAV binding proteins on GaN, adapted with permissions from Jewett et al [60].

The inherent biocompatibility of other III-nitride materials is not as well understood or as studied as GaN. What has been done however, does not yield as promising of results as modified GaN surfaces. Research into AlN provided poor results which align with literature regarding aluminum nitride's toxic effects on different organ systems in mammals [87]. Examination of AlN surfaces for cellular level studies along the AlN surface showed axonal degeneration and damage to glial cells within a 10 day time frame [94]. AlN is remarkably stable in biological media, but leeching of AlN occurs into solution and is dependent on the surface topography.[87] Although damage may not be wide spread this would directly affect the stability of any cellular matrix that is being brought into contact with an AlN device or sensor. Thus AlN seems a poor choice for biological sensing apparatuses.

A similar effect of AlN can be observed in $Al_xGa_{1-x}N$ systems for biocompatible interfacing, although to a lessened and more controllable degree. Examination of this effect revealed a relation between cell death rate and increasing percent Al content within the $Al_xGa_{1-x}N$ matrix [95]. In this work Al content was measured from 0% up to 35% and throughout all examined concentrations of Al within the $Al_xGa_{1-x}N$ matrix there was evidence of cellular activity [95]. Growth rates of cells did suffered with increasing Al, indicating that some of the toxic effects of AlN appear to carry over to AlGaN. The conclusion to be drawn from this is that AlGaN (at least documented up to Al content of 35%) does not directly mirror pure AlN for its lack of biocompatibility. Instead, previous work by Podolska et al. highlighted AlGaN as a promising surface for biological sensor interfacing as well, although noted some cell counts were suboptimal [96]. This also helps support the use of AlGaN/GaN type heterostructures for biosensing devices where cell interfacing or interactions may be a possibility.

InN and InGaN have been shown in literature to have a similar behavior to AlN/AlGaN. Regarding their cellular interfacing, some evidence suggests that the ternary InGaN may behave in similar fashion to AlGaN as described [76]. While InN has shown some promise as a biosensing platform fewer experiments have explored InN behavior with cellular structures. InN has been tested in biological media for the stability of InN surfaces and surface components, but less focus on directly binding cells for tissue interfaces. InN has similar behavior to AlN when placed in a biological system in that increasing In content in mammalian tissue can lead to cellular damage and kidney malfunction [76]. Thus interfacing directly with InN may be unfavorable for cellular systems. One study of InGaN however, revealed that theses surfaces may provide a bio-compatible intermediate between InN and GaN [97]. Here, lymphocytes were grown on flexible LED devices including Si, AlInGaPAs, and InGaN. This may prove similar to AlGaN that when In is the lesser component of a GaN ternary system the toxic effects can be mediated. In turn allowing for a biologically compatible surface to be made from an In based III-nitride compound. Until this is done AlGaN/GaN and GaN systems will likely remain the III-nitride of choice for biocompatible material systems.

5.6. Chemical Functionalization

The surfaces of III-nitride materials have available bonding sites depending on the polarity of the surface from its growth procedure [3]. These dangling bonds are what give III-nitride materials their broad range of functionality for various analytes or chemical acceptors. Here we will cover some III-nitride compounds and selected functionalization methods and schemes.

5.6.1. Aluminum Nitride

The applications of AlN are more limited in biosensor applications due to concerns of aluminum toxicity in biological environments. However, many chemical modification methods have been explored to create a selectively sensitive surface upon AlN. One such surface modification with broader application can be seen with a benzenethiosulfonate (BTS) crosslinker capped organosilane coating [98]. This work highlights the ability of organosilane adlayers to act as a platform for immobilization of relevant compounds such as biotinthiol. These organosilane surface coatings form what is known as a self-assembled monolayer (SAM) and have been well documented as surface modifiers in other semiconductor materials as well [99]. Chemically modifying AlN and other surfaces for such organosilane SAMs typically require etchants such as H_2SO_4 or piranha to remove surface contaminants and organics [89]. Pentafluorophenyl ester and β-propiolactone were among several other chemical silanization methods studied alongside BTS, all of which serve to immobilize a specific functional group onto the surface [100]. Silanization of these surfaces has also been shown using dip coating techniques and subsequent heat treatments for adlayers of polysilazane to generate ceramic coatings atop AlN [101]. These techniques demonstrate the versatility of these surfaces under more precise surface bond modifications and using higher throughput processing techniques such as the dip coat. For processing of AlN with silanization the surfaces are often terminated with hydroxyl groups at available bonding sites. Exposure of surfaces after etching or during etching with an H_2O_2 solution allows for hydroxylation to take place, thus making uniform SAM possible along AlN surfaces [1]. This could eventually open up other chemical means of functionalization by immobilizing non SAM molecules or compounds to these hydroxyl group sites.

5.6.2. Gallium Nitride

With regards to biosensing and functional surface regions, GaN has been more widely researched and explored than AlN. As mentioned in the discussion of AlN, organosilanes and SAMs play an important role in GaN chemical functionalization as well. The use of monolayers like aminopropyltriethoxysilane (APTES) or thioglycolic acid (TGA) have been shown to successfully immobilize surface biomolecules for sensing schemes [102-104]. A schematic of some of the different chemical functionalization methods discussed here can be found in Figure 3.

Further work has been done into exploring chemical termination methods without the use of a silane adlayer for GaN based sensors. Etching techniques on GaN surfaces can be done to terminate a surface with hydroxyl groups, but with additional steps aid in incorporating another molecule onto these hydroxyl terminated binding groups [105]. This has been shown in work done by our laboratory using a phosphoric acid and phosphonic acid mix to create ethephon binding sites for future biomolecule attachment [15,55]. Additional surface functionalization work via organophosphonic acids have been shown to be an alternative to common silanization methodology [106]. Phosphonic acid functionalization methods do not rely on surface hydroxide group concentration, and can bind to both oxide groups and hydrogen terminated sites [107]. Thus functionalization of GaN surfaces using phosphonic acids as a vehicle for other biologically relevant groups is a viable option for future surface modification schemes. As mentioned previously, phosphonic and phosphoric acids are capable of etching III-nitrides where other more common etchants cannot. This slower etching process creates new binding sites for other phosphonic acid derivatives to occupy as the etching takes place [105].

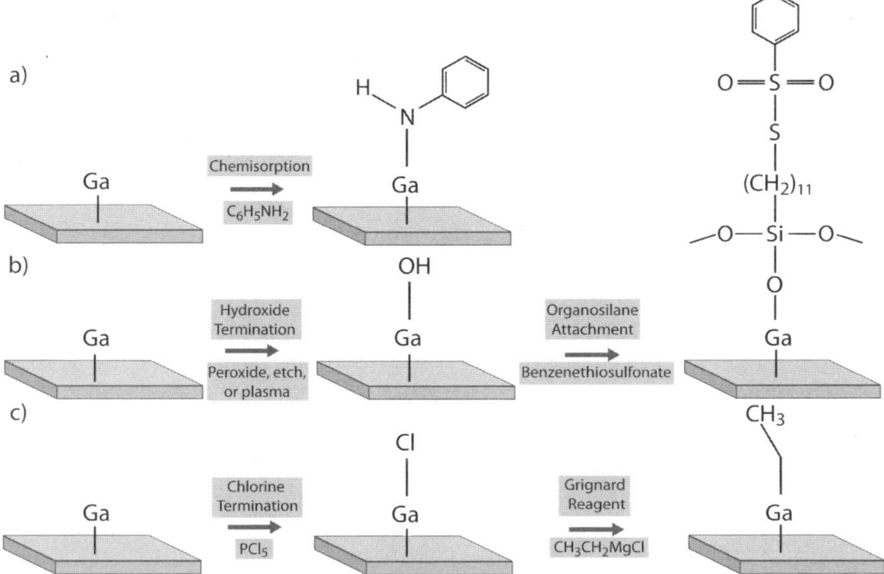

Figure 3. a) Basic chemisorption functionalization with analine following ion bombardment of GaN surfaces under UHV conditions [108]. b) Example of organosilane surface attachment following hydroxide termination to the surface, using the forementioned TUBTS molecule [98]. c) Example of a Grignard reagent attachment to GaN using a chlorine terminated surface [109].

Other methods of binding 1-alkene groups to GaN surfaces have been explored to generate similar layered structures to those seen by silanization and covalent attachment with phosphonic groups. Photochemical assisted binding of alkenes has been performed using ultraviolet (UV) light at a wavelength of 254 nm [110]. Surfaces in this work were hydrogen terminated via inductively coupled hydrogen plasma and then exposed to 254 nm UV once placed in a solution of an alkene-amine molecule (10-trifluoro-acetamide-1-decene, abbreviated TFAAD) [110]. This generates a highly stable surface but lacks the adaptability of phosphonic groups while adding a plasma treatment step [106]. Additionally, UV treatments on biological molecules can lead to degradation of the newly bound molecules or functional groups on the surface. Thermal methods of 1-alkene binding have thus been explored to provide other options for this type of surface chemical functionalization. It was found that heat treating surfaces coated in the same TFAAD solution at 180°C was successful in binding the alkene group of the decene molecule to both hydroxyl groups and oxide groups on GaN surfaces [111]. This shows the ability to attach 1-alkenes to GaN when UV degradation is a concern, and vice versa if thermal degradation is an issue the UV method operates on the same chemical mechanism. Aside from photopatterning and thermal treatments another method of 1-alkene surface termination to GaN has been done using olefin metathesis. Here a first generation Grubbs catalyst (ruthenium chloride complex) was used following alkene surface termination to create an olefin cross-metathesis reaction to bind a 7-bromo-1-heptene marker to the GaN surface [109]. Resultant surfaces proved resistant to oxidation and stable enough to move to attachment of a peptide monolayer to the GaN surfaces [112]. Here a 6-heptenoic acid was bound to the alkene terminated surfaces using the same olefin metathesis method. Peptide was successfully added to these alkene surfaces thus showing this scheme as a functional platform for sensing with a larger variety of peptides or biomolecules [112].

Another method of wet chemical functionalization using the Grignard reaction has been demonstrated for GaN surfaces. Grignard reagents in a broader sense are alkyl halides introduced to magnesium in solution to generate a RMgX molecule, with R representing the alkyl group and X representing the halide [113]. This method is typically used for generating carbon-carbon bonding in organic molecules. However, here it was presented as a means to cap GaP, GaAs, and GaN surfaces for superior surface stability in solution [114]. The termination of the surface in alkyl chains prevented the degradation of GaN surface via oxidation and was able to maintain the electrical properties of the GaN film. Few experiments have been done on this methodology in regard to other crystalline III-nitride surfaces and their functionalization for biosensing or otherwise. Other work has been done using the Grignard reagents as a step within a larger chemical functionalization method, but it was not presented as the primary focus of the study itself [109]. Thus, the full breadth of applications of Grignard reagents on III-nitrides is promising as more focus is given to these materials for tailorable biofunctional surfaces. Additionally there are research opportunities in examining the Grignard reactions on other III-nitride surfaces to explore their functionalization capabilities as well.

Earlier modification techniques utilized vacuum approaches to GaN functionalization and it is worthy of discussion as these processes are still viable options. Chemical functionalization of GaN under such conditions can be seen in the chemisorption of aniline and halogenated anilines under pressures of approximately 5×10^{-11} Torr [108,115]. Aniline and aniline compounds were introduced to GaN surfaces following ion bombardment to remove surface contaminants and vacate surface dangling bonds. This method is successful for depositing high purity compounds onto the surface, but is not suitable for compounds and biomolecules mentioned in the previous chemical functionalization methods for GaN. Thus, research into high vacuum processing for GaN biosensors has not been fully explored, and instead focus has shifted towards wet chemical methods.

The departure from ultra-high vacuum processing for functionalization follows the trend for GaN in that functionalization methods are become progressively more accessible to simpler bench top chemistry procedures. Current methodology stated here has removed much of the time and resource intensive processing and replaced it with schemes and techniques which allow for a wider range of surface chemistries.

5.6.3. Aluminum Gallium Nitride (AlGaN)

As one of the most commonly used tertiary allows from the III-nitride set of materials, AlGaN is used heavily in biosensing research. However, the functionalization of AlGaN is not well explored as it is most commonly used in combination with GaN to form a heterojunction type device. Under these circumstances surface functionalization is similar to GaN in its response to surface bond termination schemes.

5.7. Sensing Devices

Selectivity can be defined as a biosensor's ability to ignore outside interference when taking a measurement or registering an electrochemical interaction between surface bound receptors and analytes present in solution [116]. This interference can be noise or other processes that cause some minor or major surface charge changes to the biosensor. Specificity in similar fashion refers to the actual ability of the biosensor to single out target compounds, biomolecules or ions within solution [117]. One could view this as analyte bycatch, or unwanted entities occupying receptor sites and thus blocking possible interactions that would contribute to the biosensor's total sensitivity. These two things contribute to the overall sensitivity of the biosensor which represents the net output or signal change that a sensor can detect. Higher sensitivity means that a device can detect lower concentrations of compounds in solution with higher accuracy. This also determines the strength of the signal that can be obtained. Research in the field has progressed towards a wider range of surface modifications and improved III-nitride biosensor fabrication techniques enabling modern biosensors to be tailored to a variety of testing modalities.

When operating a given biosensing modality the response of the material or device relative to the environmental factors can be described as the sensing mechanism. In the design and use of III-nitrides for sensing applications there are several different schemes which have been well established in literature [118]. Most research into III-nitride biosensors is done through microfabricated devices that are capable of being produced in large quantities. Sensors such as FETs and HEMTs rely on conductance and charging effects to sense environmental stimuli [15]. Others such as SAWS monitor physical motion of molecules and material and are a type of microelectromechanical system (MEMS). Both of these types of sensors act as transducers for the external stimuli and the electrical signal given off by the sensor and do so by taking advantage of the piezoelectric properties of III-nitride materials. SAWS measure shifting piezoelectric properties for the III-nitride to generate an output. FETs operate as a function of the interfacial conductance effects between their heterostructure layers; examples include AlGaN and GaN or InGaN and GaN in terms of III-nitrides. The 2DEG discussed earlier is an example of this induced channels at the hetero-interface between AlGaN and GaN [119]. Along with the interface layer between the ternary and binary III-nitride two other interfaces are also present; the environment with the surface layer (in this case AlGaN in the AlGaN/GaN system) and the interface of the lower layer with the substrate itself (in this case GaN). A schematic of this device structure can be seen in Figure 4 [120]. Sensing along the surface interface relates the chemistry from the outer layer of AlGaN or GaN to the interior 2DEG and is what determines the sensing capabilities for a biosensing device. When considering both the piezoelectric and spontaneous polarization of the AlGaN and GaN they are often lumped together as a net polarization effect and are both characteristic properties of their material [121]. These different net polarization effects in AlGaN to GaN depend on the aluminum content in the AlGaN. The piezoelectric component of the polarization is affected by strain induced by replacing Ga with Al and causes a mismatch at the lattice interface with bulk GaN [122]. When these properties are viewed as macroscopic polarization effects for each material they can be effectively seen as fixed surface charges [120]. Since the surfaces of the GaN bulk layer are in contact with either the substrate or the AlGaN layer these will ideally remain unchanged when put into operation. AlGaN however has an exposed surface to the environment and any internal electric field caused by polarization is now subject to environmental interactions with the AlGaN surface. Conductance and electrical behaviors will change within the 2DEG present along the AlGaN/GaN interface relative to the effects that surface charging will have on the internal electric field. Thus, monitoring the current flow or potential in AlGaN/GaN FETs becomes a means to quantify environmental effects on the sensor itself. Additionally the interface structure of the AlGaN/GaN devices is highly stable and allows for a high transconductance at low potentials and often without the need for an external electrode [123]. These surfaces are typically functionalized or capped with a receptor specific to a given environment or to an analyte intended to be measured [124]. The focus here will be put upon FET type setups and gas sensors relative to III-nitrides. SAWs will not be covered in this text; readers are encouraged to start with reviews and papers such as these [118c,118d].

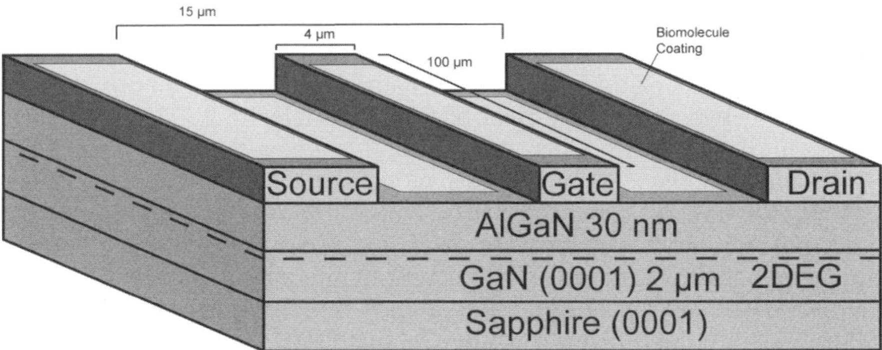

Figure 4. Schematic of biomolecule functionalized AlGaN/GaN FET. Adapted with the permission from Rohrbaugh et al. [54].

One of the more simplistic detection schemes for these AlGaN/GaN and GaN devices is surface ion concentrations. Studies focusing on ion sensing applications rely on either capping these surfaces with an oxide layer, or capping the available GaN surface bonds with ionic elements or compounds. Chloride ion attachment along these surfaces has been performed using HCl etching of oxide layers on GaN and AlGaN [125]. In this study oxide layers were used as a medium for either Cl or OH group attachment atop the surface for pH studies. Oxide layers had to be fabricated due to the etching resistance of both AlGaN and GaN. The findings showed that GaN and AlGaN behave differently as pH sensors depending on their surface oxide layer [125]. Here, GaN has a higher propensity for Cl at lower pH levels than AlGaN, but both behave similarly under high pH conditions. This was in response to work showing that AlGaN/GaN heterostructures show stronger dependence on negative ion concentration in solution than to pH fluctuations [126]. Both of these studies showed OH and Cl affinity of the surfaces and point to these simpler immobilizations as applicable ion sensors, but suggest another modification is needed for pH sensing applications.

Other recent efforts have shown that capping GaN based ion sensors in Ga_xO_y layers using peroxide instead of chemical etching also provides a similar effect to these two previous methods and highlighting the anion affinity for the FET surface [118c]. Ion sensing without altering GaN surfaces has been shown as a potentiometric sensor for various potassium and sodium salts [127]. This functioned by use of the Helmholtz layer properties of each test solution, which provided potential and impedance data relating to the GaN natural affinity for anions and the potential between the GaN electrode and a Pt electrode. While an early work in the field of GaN sensing applications this highlighted the importance of the dangling bonds along the GaN surface [127]. The stability of GaN was useful for these initial ion tests as it was noted that GaN potentiometric sensors were reversible due to the lack of ion binding to the actual film surfaces. The inherent dipole effects present along the surface of the (0001) configuration of GaN are what allow this to happen. More recently, modification of the GaN and AlGaN surfaces has led to detection schemes for heavy metal ions in water sources. An example can be seen in mercury detection due to the concern of its neurotoxic behavior. AlGaN/GaN heterostructures have been used as a fast detection method for mercury when coupled with an Au metallized gate and thioglycolic acid surface layer [128]. Time dependent I_D measurements showed response times as short as 5 s for concentrations of 10^{-5} M Hg^{2+}. When using these FET devices in the field a quick response time combined with a simple solution exposure is advantageous to off-site bench top lab testing for similar confirmations of mercury contaminants. Using thioglycolic acid as a receptor additionally allowed for a very high selectivity for Hg over other control ions; a behavior that was reexamined in a follow up study for even lower concentrations of Hg^{2+} ions in solution [129]. Other examples of AlGaN/GaN heterostructure devices providing a malleable ion sensing platform can be seen in the detection of dissolved nitrate. AlGaN/GaN structures were coated in a 2-nitrophenyloctyl ether and high molecular weight polyvinyl chloride solution to generate a nitrate ionophore membrane [123]. This sensing mechanism relied on the concentration of nitrate adsorbed into the polymer adlayer which in turn affected surface charge and conductivity of the AlGaN/GaN device. Similar to the Hg^{2+} detection scheme the response time is rapid with AlGaN devices (in this case less than 60 s for each KNO_3 solution addition). Both of these schemes take advantage of the high transconductance of the AlGaN/GaN heterostructure near zero exterior potential. This allows for a very high degree of sensitivity when conducting ion sensing and is seen at levels of 10^{-8} M for the Hg^{2+} and $\sim 10^{-6}$ M for NO_3^- [123,129].

InN has its own set of ionic sensing schemes as both an additive layer and a base layer for which surface coatings can be applied. Metallization of AlGaN/GaN gate areas have already been shown using films such as Au, but InN has also been shown as a successful coating as well. InN as mentioned has a very high affinity for electrons and thus negative ions, which provides the motivation for the study. A thin layer of ~ 10 nm in this study was enough to accumulate enough Cl^- ions to increase piezoelectric effects in the electron gas layer of the AlGaN/GaN heterostructure [130]. Previous experiments of similar nature showed AlGaN/GaN FETs response was ~ 20 s and was capable of detecting concentrations down to 100 nM. Further evidence was shown that cation interactions are negligible on AlGaN/GaN surfaces under such sensing schemes [130]. Ion sensing FETs have also been implemented using InN, and in particular a surface modified InN/AlN stack has been shown to provide a stable base for *in vivo* calcium ion detection. In this work a (3-aminopropyl) trimethoxysilane layer was covalently bound to InN surfaces and combined with a phospho-tyrosine end group to provide a phosphate end group for Ca^{2+} ions to bind [131]. Results indicated a stable current measurement for each increasing order of magnitude in Ca^{2+} ion concentration (from 10^{-6} M up to 10^{-1} M) with variance of current at only 1.11%. These experiments yielded logarithmic relations to OH^- and Cl^- in various salt solutions similar to that seen on the bare GaN ion exposed study [70,127]. There was a polarity and growth dependence on the InN sensors similar to the GaN in that InN films of (0001) orientation was needed. However the surface polarity of InN does not play a large role in anion accumulation due to all wurtzite orientations of InN having similar electron accumulation effects which is generally undesirable [132]. Finally, although AlGaN is typically paired with GaN for heterostructure conformations of III-nitrides InN has also been experimented with under similar circumstances. InN/GaN has been shown to be a stable platform for pH sensing especially once a surface In_2O_3 layer has been used to passivate the surface [133]. Under this study unpassivated pH sensors had an average sensitivity of 34.75 mV/pH. However, once passivated showed an increase in sensitivity to an average of 52.04 mV/pH. This use of surface oxide layers in FETs for pH sensing is similar to prior work using Ga_xO_y oxide compounds as ionophoric surface layers [118c]. InN has also been paired with AlN to create a heterostructure base for a Pt electrode in gas sensing applications, and of particular interest for acetone sensing of patients with diabetes [134,135]. Situations such as these are typically seen using a thin film of InN of approximately 5-10 nm to negate any effects of bulk InN interfering with the characteristically high sheet densities ($\sim$1.57 x 10^{13} cm^{-2}) that occur in the first 5 nm of InN films [136].

5.8. Analytes/Receptors

Biosensing devices depend upon an analyte and receptor pairing to provide a scheme for a specific sensing environmental. This pairing can be a simplistic surface response to an ion in solution, or more complicated and involve antibody binding or DNA sequences. Looking at the number of available and documented schemes, it is clear that GaN is the most common of the III-nitride as a platform for biosensor research. While we will discuss AlN, InN, and BN and their respective schemes current literature is more focused on GaN and thus will be the most represented here.

GaN has been established to have a surface that is both biocompatible and modifiable [60]. Biomolecule schemes, like peptides, are a method of this modification that does well in bridging molecules and compounds in solution and the GaN or AlGaN surface. Peptides have two terminal ends and can thus be made to have one functional end adhere to GaN or a III-nitride surface and the other left free in solution. Examples of this can be seen the IKVAV peptide sequence which is derived from laminin and commonly used to support neurite growth [137]. IKVAV peptide was also studied and shown to improve cell adhesion and stability atop GaN surfaces.[109]

Other modes of analyte/receptor pairings have been done through functionalized nanoparticles laid atop AlGaN/GaN heterostructures. Research done here found that gold nanoparticles covered in different functional groups (thiol, carboxyl, amine, methyl) could generate changes in the electrical characteristics of the AlGaN/GaN devices [138]. This shows that gold nanoparticles can operate on similar principles and act as a medium between the AlGaN surface and alkanethiols on the particles. Additionally, the scheme shown here lends credence to idea that physisorption alone is capable of generating recognition properties to the biosensor surface [138].

InN has seen relatively little with regards to actual surface modification to create a biosensor, especially compared to GaN or AlGaN. However, research has shown InN is capable of similar interactions with analytes or biomolecular compounds in solution. One example of this is a physisorbed peptide and phage system that examined 11 different peptides expressed by bacteriophages [139]. Here InN was chosen due to its high surface electron affinity as a possibility for a film that readily accepts peptides without extensive modification or chemical processes. The results of this showed one peptide sequence in particular (YPLLPESPTDAN) had remarkably high affinity for bacteriophages out of solution [139].

Table 1

Material	Type of receptor /Surface Treatment	Receptor	Analyte	Reference
GaN	Oxide adlayer	Ga_xO_y	pH	Chen [118c]
	Bare Surface	Surface Ga atoms	Anionic solution activity (KOH)	Chaniotakis [127]
AlN	Polyimide-Protein layer	DNA, antibody, phage selective proteins	Bacteriophage	Auner [118d]
	SAM, SAM/AuNP	GFP Antibodies	GFP	Chiu [90]
InN	Bare Surface	n-polar InN	pH	Lu [70]
	Peptide	YPLLPESPTDAN	Bacteriophage	Estephan [139]
	Peptide	STLMTTTYHSVS	Bacteriophage	Estephan [139]
	Peptide	QGAHYEYSRTEL	Bacteriophage	Estephan [139]
	Peptide	IPGDAGKGLHMT	Bacteriophage	Estephan [139]
	Peptide	YDTTSSPPRLTR	Bacteriophage	Estephan [139]
	Peptide	GMIKAAHERPLR	Bacteriophage	Estephan [139]

	Peptide	VLSNSLPTAIST	Bacteriophage	Estephan [139]
	Peptide	GMATEATVHELA	Bacteriophage	Estephan [139]
	Peptide	LIRGLLIGFGRN	Bacteriophage	Estephan [139]
	Peptide	NWSAEKAKLYDS	Bacteriophage	Estephan [139]
	Peptide	QNSHRVALENNT	Bacteriophage	Estephan [139]
	SAM	O-phospho-I-tyrosine	Dissolved calcium	Kao [131]
	Metal adlayer	Pt Catalyst	Hydrogen	Chang [134]
	Metal adlayer	Pt Catalyst	Exhaled Acetone	Kao [135]
AlGaN/GaN	Bare Surface	Surface atoms	Ionic compounds	Podolska [126]
	Nanoparticle	Functional Group Terminated Au NPs	Src Kinase	Chu [140]
	Nanoparticle	Functionalized AuNP	DNA, proteins, metabolites	Makowski [138]
	ZnO nanorods	Glucose oxidase	Glucose	Pearton [4], Review

Table 1 *Continued*

	Biomolecule Coating	Immobilized ssDNA	Phosphorescent dye	Fahrenkopf[117]
	Polymer Membrane	PVC with TDA Bromide and 2-nitrophenyloctyl ether	Nitrate Ions	Myers [123]
	Metal adlayer	InN Gate Coating	Chloride ions	Chu [130]
	SAM	Vitellogenin Antibody	Vitellogenin	Chu [102]
	SAM	Thioglycolic Acid	Dissolved mercury	Chen [129]
	SAM	Thioglycolic Acid	Dissolved mercury	Wang [128]
	SAM	Immobilized Penicillinase	Penicillin	Baur [103]
	SAM	Prostate Specific Antibody	Prostate Specific Antigen	Li [118a]0
	SAM	Anti-MIG IgG	Recombinant human MIG	Tulip [141]
InN/GaN	Oxide adlayer	In$_2$O$_3$ Passivation	pH	Lee [133]
BN	Functionalized Multiwalled BNNT	Biotin Fluorescent Ag NPs	pH	Huang [142], Golberg [143]

Table 1 *Continued*

5.9. III-nitride FET Biosensor Stability

More recently work has been done to bring AlGaN/GaN forward as a reliable and reproducible device that could be used for aqueous biosensing systems. Our lab has done work to determine the stability and reliability of various AlGaN/GaN schemes within possible operational conditions. Preliminary studies done by Makowski et al. demonstrated the wide range of methods available to create a functional biosensing platform out of AlGaN/GaN devices [109,112]. Beyond this, multiple studies have explored the actual application of these devices with functionalized surfaces [102,144]. However, as the technology matures research has begun to progress towards determining the viability of III-nitride sensors for sensing and less on proof of concept type scenarios. The various elements that make up biosensor stability are described and diagrammed in Figure 5.

GaN based surfaces offer a stable substrate when placed into aqueous environments, but the stability and reliability of a device becomes more complex as the GaN surface is tailored to specific scenarios. Questions arise as to the persistence of surface functionalization under constant use or the effectiveness of the biosensor over time once put into an operating environment. This determines whether or not the III-nitride devices would be better suited for longer term exposure to solution or if there is a currently a limitation to shorter time frames for effective detection. Storage of III-nitride devices is an example of a situation where a device may not be in operation but would be required to sit idle for an extended period of time. All of these points and concerns come back to the stability of the sensing interface that is created on the surface of the device itself.

There are numerous ways of generating a possible biosensor with a III-nitride device due to the flexibility of the surface to different modifications, see Table 1. Surface terminated ionic groups, peptide layers, and antibody systems are all possibilities. One could look at the degree of coverage as a first challenge in assessing the stability and long term utility of an FET biosensor. In a general sense, the higher the concentration of receptor material or sites present on a biosensor surface the more sensitive and the better the response will be. This however is not always the case due to excessive functionalization on a surface. This can be thought of as an overcrowding of the surface with receptor sites such that it can become overly difficult for an analyte to bind to a surface [117]. Receptor molecules may block one another from fully interacting with the relevant compound within the solution. This phenomenon is dependent on what is known as the Debye screening length (or Debye length) of a molecule. The Debye length represents a molecule or compound's effective distance within a solution to have an electrostatic interaction with another compound [145]. One can think of this as the effective range around a receptor site that allows it to gather a response from the analyte in solution. When this range becomes overly crowded with receptors the bonding to the surface can become unstable and may become restricted entirely. This occurs when not enough electrostatic interactions occur with the surface to affect device conductance [146]. Thus for any reliable sensing applications a series of assays would likely be required to find optimal surface densities and concentrations of receptor molecules for proper detection and sensing. An example of this can be seen in the work of Rebriiev and Starodub who demonstrated several different criteria which can affect device response, one of which being concentration of the receptor enzyme in the FET coating [147]. Here a polymer mixture was dosed with urease for detection of urea within solution, and enzyme concentration was observed to best perform at a value of just 3% enzyme within the polymer adlayer.

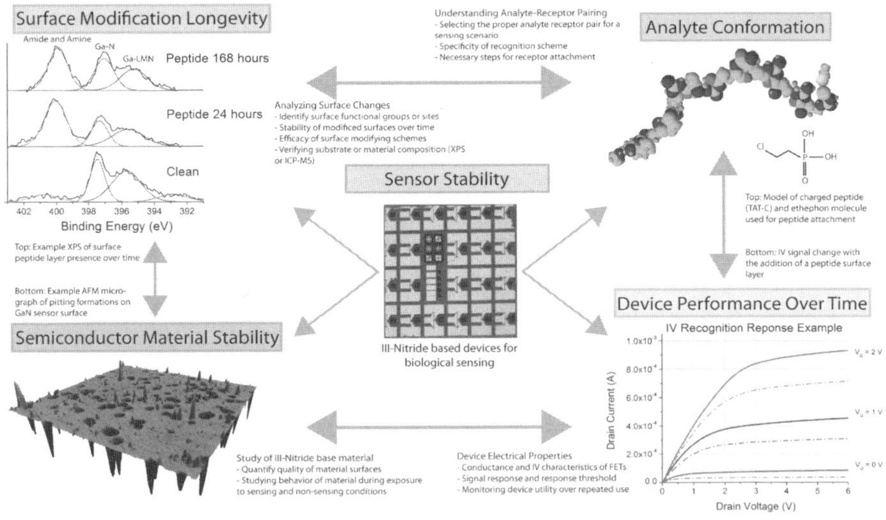

Figure 5. Testing methods and objectives of the different properties that make up the stability of a biological sensor.

Longevity of surface bound receptors for FETs is the next challenge that must be overcome for devices to offer stable performance over a wider range of times. It is likely devices will be placed into storage or may not see operation for some time after fabrication. These elements need to be accounted for when assessing the usable lifetime for any given FET sensor scheme or type. Much work has already been done on the subject and has shown stability of such III-nitride devices past a month of storage time [147]. In almost every case found there is a corresponding decrease in sensitivity after a given number of uses or in some cases without any operational use. Other instances documented in this study showed that storage of fabricated devices did not seem to affect the performance of the FETs [147]. In work done by this author's laboratory the strength of stability of surface bound peptide groups was examined in similar fashion. Initial work tested if a set complex chemical functionalization steps were truly necessary to create recognition on a surface and if a recognition event could be achieved via an incubation of peptide to an AlGaN/GaN FET surface [54]. For this case a solution of a recognition sequence (SVSVGMKPSPRP) was allowed to incubate and dry to the FET surface and then subsequently soaked for various amounts of time to determine the stability of this attachment. Results showed that drain current of the FETs increased an average of 96.43% across ~300 FETs tested. This demonstrated that incubation can successfully be used to produce recognition behavior, however longevity was a concern. Once placed in solution surface peptides appeared to dissociate back into solution within 24 hrs of soaking in water [54]. Thus revealing that the forces between incubated biomolecules and the surface are relatively weak, and some form of chemical functionalization would likely be required.

Silane chemistry and polymer adlayers have been used in other instances for longer term stability of surface receptor sites [144]. However, many of these use materials other than III-nitrides as a base for their sensing applications. As AlGaN/GaN in particular has come out as a stronger candidate for biosensors, more work has been done on these silanization and polymer embedding methods [118a,141,148]. However, research has yet to meet the number of long term stability assessments on AlGaN/GaN over other III-nitride biosensing platforms. In response, a second set of AlGaN/GaN peptide binding studies done by our laboratory was performed to increase the attachment of incubated peptide to device surfaces. Additionally it provides a possible channel to further investigate the longer term stability of the AlGaN/GaN devices which is currently under represented in literature [144]. Previous work done by our lab has shown a novel chemical functionalization technique to covalently bond

functional groups to GaN based surfaces [105,149]. Phosphonic acid and ethephon attachment to the AlGaN/GaN device surfaces was used from these prior experiments in a one-step etch process following cleaning. This created a base to attach peptide (in this case TAT-C) and create covalently bound sites across the breadth of the FETs. Results showed a positive presence of surface bound peptide groups within solution up to 7 days of exposure.[150] Analysis of XPS data did not give statistically significant evidence to a reduction in surface peptide levels over the duration of this test. Thus with regard to stability of a AlGaN/GaN prepared biosensors it has been demonstrated that storage in solution appears to be viable up to a week. This helps in verifying AlGaN/GaN as a platform for short term disposable type biological sensors.

There are further opportunities for research in the specificity of FET operation and response relative to other unwanted compounds or particulate present in solution. Most recognition schemes account for unwanted molecules in solution by design of the receptor choice. Antibody and enzyme immobilized on surfaces are by nature highly selective to their paired analyte, and a strong example of this can be seen in work by Ren et al [102]. This study and others commonly use serum or albumin based solutions to test for the reliability of FETs in detecting specific biomolecules when placed in a general biologically representative fluids. Few studies examine the effect of particulate on functionalized surfaces though. This may not be as strong a concern for sensors that will be resting in solution, but in the scenario that fluids are passing over there is the possibility of accumulation of unwanted particles on the surface. We examine this in a previous work where an excess of unmodified GaN powder was laid atop a recognition peptide functionalized AlGaN/GaN FET to determine if additive layers of sediment or neutral compounds would prove detrimental to device operation [54]. Results of this demonstrated the robustness of the surface groups to unwanted large objects in solution. Surfaces in this study were not chemically modified and used incubated peptides, and despite this weak surface adherence, powders did not show any effect on device performance. These initial findings suggest that AlGaN/GaN FETs and other III-nitride FETs in general will likely behave well within realistic biological settings where complex molecules and larger particles may build up or come in contact with the surfaces.

5.10. Future Directions

Growth techniques have advanced for more consistent films, functionalization techniques are better understood, their reactions within biological systems is being studied, and doping for improved properties is advancing as well. However, this research is largely centered on our basic understanding of the materials and their operation. Understanding these systems is crucial to moving forward to industrial applications, but many of these applications have yet to come to fruition. III-nitride applications in fields such as LEDs and optoelectronics are already in markets and being used in industrial processes. Sensing apparatuses, and especially biosensing applications, are where much of new research into III-nitrides is being focused. As we have covered over the course of this chapter biosensor applications have been shown repeatedly either through SAM modified devices, biological molecule modification, or surface termination. Many avenues exist for devices like FETs to become tailorable platforms for a wide range of end user applications. The remaining research to be done it seems lies within the stability of III-nitride biosensors over time and the reliability in their performance. Under laboratory conditions and short testing cycles these sensors behave well, but more intensive longer assessments need to be done. Questions of repeatability on the same sensor, loss of efficacy of surface bound sites, degrees of coverage and surface concentrations of receptor sites need to be investigated to optimize performance.

In our continued efforts to fully characterize AlGaN/GaN devices for biosensing applications we have shown methods to covalently bond peptide groups to FET surfaces, and that this method provides a stable bond within solution for up to 7 days. Next steps in our research are moving to focus on the specificity and sensitivity over time using this same functionalization method. Inclusions of different peptides and biomolecules will be done to examine the specificity of peptides bound in this manner. Although likely more important is the in depth examination of device sensitivity and response over time to actual analyte compounds. As mentioned previously device schemes involving non III-nitride materials have been widely characterized for this type of behavior and many systems have measured degradation rates from both long term storage and repeated sensing tests [144]. These values do not exist currently for AlGaN/GaN for the wide range of biosensing schemes available to it. We are seeking to change this and better characterize AlGaN/GaN solution response behavior. To do this extensive longer term stability study on the time scale of months is recommended to understand the storage capacity of these devices both in dry and wet conditions. Likewise repeated-use studies need to be done to determine the capabilities of the AlGaN/GaN FETs to undergo multiple cycles of response. If these devices respond well to multiple sensing cycles then assessing their reliability would be the next step. The exploration of InGaN heterojunctions and other III-nitride surfaces for biological sensing applications may be worthwhile, but currently the focus of the field is upon AlGaN/GaN and it becoming industrially viable. Their malleability and stability in sensing environments can provide researchers with a robust platform from which to conduct medical assays, or quantification of environmental compounds. Some of the last remaining work to be done before it becomes commercially viable though is a better understanding of how the III-nitrides and in particular AlGaN/GaN function over time under operable conditions. Stability and reliability of the signals from these devices are paramount in creating not just inexpensive disposable type in-field biosensors, but multi-use and longer term implants into biological systems.

References

1.   Baur, B.; Steinhoff, G.; Hernando, J.; Purrucker, O.; Tanaka, M.; Nickel, B.; Stutzmann, M.; Eickhoff, M., Chemical functionalization of GaN and AlN surfaces. *Applied Physics Letters* **2005,** *87* (26), 3901.

2.   Steinhoff, G.; Purrucker, O.; Tanaka, M.; Stutzmann, M.; Eickhoff, M., AlxGa1–xN—A New Material System for Biosensors. *Advanced Functional Materials* **2003,** *13* (11), 841-846.

3.   Mishra, M.; Krishna, T. C. S.; Aggarwal, N.; Gupta, G., Surface chemistry and electronic structure of nonpolar and polar GaN films. *Applied Surface Science* **2015,** *345* (0), 440-447.

4.   Pearton, S. J.; Ren, F., Gallium Nitride-Based Gas, Chemical and Biomedical Sensors. *IEEE Instrum. Meas. Mag.* **2012,** *15* (1), 16-21.

5.   Ambacher, O., Growth and applications of Group III-nitrides. *Journal of Physics D: Applied Physics* **1998,** *31* (20), 2653.

6.   Levinshtein, M. E., Properties of advanced semiconductor materials GaN, AlN, InN, BN, SiC, SiGe. *Scitech Book News* Jun 2001, 2001.

7. Georgakilas, A. In *InN: The low bandgap III-nitride semiconductor*, Semiconductor Conference, 2008. CAS 2008. International, 13-15 Oct. 2008; 2008; pp 43-52.

8. Sari, E.; Nizamoglu, S.; Ozel, T.; Demir, H. V.; Inal, A.; Ulker, E.; Ozbay, E.; Dikme, Y.; Heuken, M. In *InGaN/GaN based LEDs with electroluminescence in violet, blue, and green tuned by epitaxial growth temperature*, Lasers and Electro-Optics Society, 2006. LEOS 2006. 19th Annual Meeting of the IEEE, Oct. 2006; 2006; pp 34-35.

9. Kuykendall, T.; Ulrich, P.; Aloni, S.; Yang, P., Complete composition tunability of InGaN nanowires using a combinatorial approach. *Nat Mater* **2007,** *6* (12), 951-956.

10. Jones, K. A.; Chow, T. P.; Wraback, M.; Shatalov, M.; Sitar, Z.; Shahedipour, F.; Udwary, K.; Tompa, G. S., AlGaN devices and growth of device structures. *Journal of Materials Science* **2015,** *50* (9), 3267-3307.

11. Kunert, B.; Volz, K.; Koch, J.; Stolz, W., Direct-band-gap Ga(NAsP)-material system pseudomorphically grown on GaP substrate. *Applied Physics Letters* **2006,** *88* (18), 182108.

12. Gan, C. K.; Srolovitz, D. J., First-principles study of wurtzite InN (0001) and (000-1) surfaces. *Physical Review B* **2006,** *74* (11), 115319.

13. Cai, D.; Guo, G.-Y., Tuning linear and nonlinear optical properties of wurtzite GaN by c-axial stress. *Journal of Physics D-Applied Physics* **2009,** *42* (18), 185107.

14. Bernardini, F.; Fiorentini, V.; Vanderbilt, D., Spontaneous polarization and piezoelectric constants of III-V nitrides. *Physical Review B* **1997,** *56* (16), 10024-10027.

15. Kirste, R.; Rohrbaugh, N.; Bryan, I.; Bryan, Z.; Collazo, R.; Ivanisevic, A., Electronic Biosensors Based on III-Nitride Semiconductors. *Annual Review of Analytical Chemistry* **2014,** *8*, 149-169.

16. Fritsch, J.; Sankey, O. F.; Schmidt, K. E.; Page, J. B., Ab initio calculation of the stoichiometry and structure of the (0001) surfaces of GaN and AlN. *Physical Review B* **1998,** *57* (24), 15360-15371.

17. Kivisaari, P.; Oksanen, J.; Tulkki, J., Polarization doping and the efficiency of III-nitride optoelectronic devices. *Applied Physics Letters* **2013,** *103* (21), 211118.

18. Feneberg, M.; Leute, R. A. R.; Neuschl, B.; Thonke, K.; Bickermann, M., High-excitation and high-resolution photoluminescence spectra of bulk AlN. *Physical Review B* **2010,** *82* (7), 075208.

19. Guarneros, C.; Sánchez, V., Magnesium doped GaN grown by MOCVD. *Materials Science and Engineering: B* **2010,** *174* (1–3), 263-265.

20. Monemar, B.; Paskov, P. P.; Pozina, G.; Hemmingsson, C.; Bergman, J. P.; Kawashima, T.; Amano, H.; Akasaki, I.; Paskova, T.; Figge, S.; Hommel, D.; Usui, A., Evidence for Two Mg Related Acceptors in GaN. *Physical Review Letters* **2009,** *102* (23), 235501.

21. Castiglia, A.; Carlin, J.-F.; Grandjean, N., Role of stable and metastable Mg–H complexes in p-type GaN for cw blue laser diodes. *Applied Physics Letters* **2011**, *98* (21), 213505.

22. Köhler, K.; Stephan, T.; Perona, A.; Wiegert, J.; Maier, M.; Kunzer, M.; Wagner, J., Control of the Mg doping profile in III-N light-emitting diodes and its effect on the electroluminescence efficiency. *Journal of Applied Physics* **2005**, *97* (10), 104914.

23. Eiting, C. J.; Grudowski, P. A.; Dupuis, R. D., P- and N-type doping of GaN and AlGaN epitaxial layers grown by metalorganic chemical vapor deposition. *Journal of Elec Materi* **1998**, *27* (4), 206-209.

24. Van de Walle, C. G.; Lyons, J. L.; Janotti, A., Controlling the conductivity of InN. *physica status solidi (a)* **2010**, *207* (5), 1024-1036.

25. Gallinat, C. S.; Koblmüller, G.; Speck, J. S., The role of threading dislocations and unintentionally incorporated impurities on the bulk electron conductivity of In-face InN. *Applied Physics Letters* **2009**, *95* (2), 022103.

26. Stampfl, C.; Van de Walle, C. G.; Vogel, D.; Krüger, P.; Pollmann, J., Native defects and impurities in InN: First-principles studies using the local-density approximation and self-interaction and relaxation-corrected pseudopotentials. *Physical Review B* **2000**, *61* (12), R7846-R7849.

27. Van de Walle, C. G.; Stampfl, C.; Neugebauer, J., Theory of doping and defects in III–V nitrides. *J. Cryst. Growth* **1998**, *189*, 505-510.

28. Wilke, I.; Ascazubi, R.; Lu, H.; Schaff, W. J., Terahertz emission from silicon and magnesium doped indium nitride. *Applied Physics Letters* **2008**, *93* (22), 221113.

29. Ruoxi, W.; Dongju, Z.; Yongjun, L.; Chengbu, L., A theoretical study of silicon-doped boron nitride nanotubes serving as a potential chemical sensor for hydrogen cyanide. *Nanotechnology* **2009**, *20* (50), 505704.

30. Zhao, C.; Xu, Z.; Wang, H.; Wei, J.; Wang, W.; Bai, X.; Wang, E., Carbon-Doped Boron Nitride Nanosheets with Ferromagnetism above Room Temperature. *Advanced Functional Materials* **2014**, *24* (38), 5985-5992.

31. Gubanov, V. A.; Pentaleri, E. A.; Fong, C. Y.; Klein, B. M., Electronic structure of defects and impurities in III-V nitrides. II. Be, Mg, and Si in cubic boron nitride. *Physical Review B* **1997**, *56* (20), 13077-13086.

32. Collazo, R.; Mita, S.; Xie, J.; Rice, A.; Tweedie, J.; Dalmau, R.; Sitar, Z., Progress on n-type doping of AlGaN alloys on AlN single crystal substrates for UV optoelectronic applications. *physica status solidi (c)* **2011**, *8* (7-8), 2031-2033.

33. Zhu, K.; Nakarmi, M. L.; Kim, K. H.; Lin, J. Y.; Jiang, H. X., Silicon doping dependence of highly conductive n-type Al0.7Ga0.3N. *Applied Physics Letters* **2004**, *85* (20), 4669-4671.

34. Davis, R. F.; Roskowski, A. M.; Preble, E. A.; Speck, J. S.; Heying, B.; Freitas, J. A., Jr.; Glaser, E. R.; Carlos, W. E., Gallium nitride materials - progress, status, and potential roadblocks. *Proceedings of the IEEE* **2002**, *90* (6), 993-1005.

35. Polyakov, A. Y.; Lee, I.-H., Deep traps in GaN-based structures as affecting the performance of GaN devices. *Materials Science and Engineering: R: Reports* **2015**, *94*, 1-56.

36. Pearton, S. J.; Deist, R.; Ren, F.; Liu, L.; Polyakov, A. Y.; Kim, J., Review of radiation damage in GaN-based materials and devices. *Journal of Vacuum Science & Technology A* **2013**, *31* (5), 050801.

37. Polyakov, A. Y.; Smirnov, N. B.; Govorkov, A. V.; Markov, A. V.; Yakimov, E. B.; Vergeles, P. S.; Kolin, N. G.; Merkurisov, D. I.; Boiko, V. M.; Lee, I.-H.; Lee, C.-R.; Pearton, S. J., Neutron Radiation Effects in Epitaxially Laterally Overgrown GaN Films. *Journal of Elec Materi* **2007**, *36* (10), 1320-1325.

38. Polyakov, A. Y.; Smirnov, N. B.; Govorkov, A. V.; Markov, A. V.; Pearton, S. J.; Dabiran, A. M.; Wowchak, A. M.; Cui, B.; Osinsky, A. V.; Chow, P. P.; Kolin, N. G.; Boiko, V. M.; Merkurisov, D. I., Electron irradiation of AlGaN/GaN and AlN/GaN heterojunctions. *Applied Physics Letters* **2008**, *93* (15), 152101.

39. Polyakov, A. Y.; Smirnov, N. B.; Govorkov, A. V.; Kozhukhova, E. A.; Pearton, S. J.; Ren, F.; Liu, L.; Johnson, J. W.; Lim, W.; Kolin, N. G.; Veryovkin, S. S.; Ermakov, V. S., Comparison of neutron irradiation effects in AlGaN/AlN/GaN, AlGaN/GaN, and InAlN/GaN heterojunctions. *Journal of Vacuum Science & Technology B* **2012**, *30* (6), 061207.

40. White, B. D.; Bataiev, M.; Goss, S. H.; Hu, X.; Karmarkar, A.; Fleetwood, D. M.; Schrimpf, R. D.; Schaff, W. J.; Brillson, L. J., Electrical, spectral, and chemical properties of 1.8 MeV proton irradiated AlGaN/GaN HEMT structures as a function of proton fluence. *Nuclear Science, IEEE Transactions on* **2003**, *50* (6), 1934-1941.

41. Jha, S.; Jelenković, E. V.; Pejović, M. M.; Ristić, G. S.; Pejović, M.; Tong, K. Y.; Surya, C.; Bello, I.; Zhang, W. J., Stability of submicron AlGaN/GaN HEMT devices irradiated by gamma rays. *Microelectronic Engineering* **2009**, *86* (1), 37-40.

42. Ionascut-Nedelcescu, A.; Carlone, C.; Houdayer, A.; Von Bardeleben, H.; Cantin, J.-L.; Raymond, S., Radiation hardness of gallium nitride. *Nuclear Science, IEEE Transactions on* **2002**, *49* (6), 2733-2738.

43. Gissot, S.; BenMoussa, A.; Giordanengo, B.; Soltani, A.; Saito, T.; Schuhle, U.; Kroth, U.; Gottwald, A. In *Design and radiation hardness of next generation solar UV radiometers*, Radiation Effects Data Workshop (REDW), 2014 IEEE, IEEE: 2014; pp 1-6.

44. Parks, D.; Tittmann, B., Radiation tolerance of piezoelectric bulk single-crystal aluminum nitride. *Ultrasonics, Ferroelectrics, and Frequency Control, IEEE Transactions on* **2014**, *61* (7), 1216-1222.

45. Tittmann, B.; Reinhardt, B.; Parks, D. In *Nuclear radiation tolerance of single crystal Aluminum Nitride ultrasonic transducer*, Ultrasonics Symposium (IUS), 2014 IEEE International, IEEE: 2014; pp 507-510.

46. Wu, J.; Walukiewicz, W.; Yu, K. M.; Shan, W.; Ager, J. W.; Haller, E. E.; Lu, H.; Schaff, W. J.; Metzger, W. K.; Kurtz, S., Superior radiation resistance of In1−xGaxN alloys: Full-solar-spectrum photovoltaic material system. *Journal of Applied Physics* **2003**, *94* (10), 6477-6482.

47. Emtsev, V. V.; Davydov, V. Y.; Haller, E. E.; Klochikhin, A. A.; Kozlovskii, V. V.; Oganesyan, G. A.; Poloskin, D. S.; Shmidt, N. M.; Vekshin, V. A.; Usikov, A. S., Radiation-induced defects in n-type GaN and InN. *Physica B: Condensed Matter* **2001**, *308–310*, 58-61.

48. Zhao, S.; Fathololoumi, S.; Bevan, K. H.; Liu, D. P.; Kibria, M. G.; Li, Q.; Wang, G. T.; Guo, H.; Mi, Z., Tuning the Surface Charge Properties of Epitaxial InN Nanowires. *Nano Letters* **2012**, *12* (6), 2877-2882.

49. Zhuang, D.; Edgar, J. H., Wet etching of GaN, AlN, and SiC: a review. *Materials Science and Engineering: R: Reports* **2005**, *48* (1), 1-46.

50. Adesida, I.; Youtsey, C.; Ping, A. T.; Khan, F.; Romano, L. T.; Bulman, G., Dry and wet etching for group III-nitrides. *MRS Proceedings* **1999**, *537* (G1.4).

51. Reiner, M.; Reiss, M.; Brünig, T.; Knuuttila, L.; Pietschnig, R.; Ostermaier, C., Chemical understanding and utility of H3PO4 etching of group-III- nitrides. *Phys. Status Solidi B* **2015**, *252* (5), 1121-1126.

52. Pearton, S.; Lim, W.; Ren, F.; Norton, D., Wet Chemical Etching of Wide Bandgap Semiconductors- GaN, ZnO and SiC. *ECS Transactions* **2007**, *6* (2), 501-512.

53. Morkoc,, H., *Handbook of Nitride Semiconductors and Devices, GaN-based Optical and Electronic Devices*. Wiley-VCH: Hoboken, NJ, USA, 2009.

54. Rohrbaugh, N.; Bryan, I.; Bryan, Z.; Arellano, C.; Collazo, R.; Ivanisevic, A., AlGaN/GaN field effect transistors functionalized with recognition peptides. *Applied Physics Letters* **2014**, *105* (13).

55. Berg, N. G.; Nolan, M. W.; Paskova, T.; Ivanisevic, A., Surface Characterization of Gallium Nitride Modified with Peptides before and after Exposure to Ionizing Radiation in Solution. *Langmuir* **2014**, *30* (51), 15477-15485.

56. Gao, Y.; Craven, M. D.; Speck, J. S.; Den Baars, S. P.; Hu, E. L., Dislocation and crystallographic dependent photoelectrochemical wet etching of gallium nitride. *Applied Physics Letters* **2004**, *84* (17), 3322-3324.

57. Li, X.; Kim, Y.-W.; Bohn, P. W.; Adesida, I., In-plane bandgap control in porous GaN through electroless wet chemical etching. *Applied Physics Letters* **2002**, *80* (6), 980-982.

58. Ababneh, A.; Kreher, H.; Schmid, U., Etching behaviour of sputter-deposited aluminium nitride thin films in H3PO4 and KOH solutions. *Microsyst Technol* **2008**, *14* (4-5), 567-573.

59. Bickermann, M.; Schmidt, S.; Epelbaum, B. M.; Heimann, P.; Nagata, S.; Winnacker, A., Wet KOH etching of freestanding AlN single crystals. *J. Cryst. Growth* **2007**, *300* (2), 299-307.

60. Jewett, S. A.; Makowski, M. S.; Andrews, B.; Manfra, M. J.; Ivanisevic, A., Gallium nitride is biocompatible and non-toxic before and after functionalization with peptides. *Acta Biomaterialia* **2012,** *8* (2), 728-733.

61. Li, J.; Nakamura, M.; Shirai, T.; Matsumaru, K.; Ishizaki, C.; Ishizaki, K., Mechanism and Kinetics of Aluminum Nitride Powder Degradation in Moist Air. *Journal of the American Ceramic Society* **2006,** *89* (3), 937-943.

62. Fukumoto, S.; Hookabe, T.; Tsubakino, H., Hydrolysis behavior of aluminum nitride in various solutions. *Journal of Materials Science* **2000,** *35* (11), 2743-2748.

63. Olhero, S.; Alves, F.; Ferreira, J., *Last Advances in Aqueous Processing of Aluminium Nitride (AlN)-A Review.* INTECH Open Access Publisher: 2011.

64. Liu, Z.; Gong, Y.; Zhou, W.; Ma, L.; Yu, J.; Idrobo, J. C.; Jung, J.; MacDonald, A. H.; Vajtai, R.; Lou, J.; Ajayan, P. M., Ultrathin high-temperature oxidation-resistant coatings of hexagonal boron nitride. *Nat Commun* **2013,** *4.*

65. Cofer, C. G.; Economy, J., Oxidative and hydrolytic stability of boron nitride — A new approach to improving the oxidation resistance of carbonaceous structures. *Carbon* **1995,** *33* (4), 389-395.

66. Pakdel, A.; Zhi, C.; Bando, Y.; Nakayama, T.; Golberg, D., Boron Nitride Nanosheet Coatings with Controllable Water Repellency. *ACS Nano* **2011,** *5* (8), 6507-6515.

67. Lei, W.; Portehault, D.; Liu, D.; Qin, S.; Chen, Y., Porous boron nitride nanosheets for effective water cleaning. *Nat Commun* **2013,** *4,* 1777.

68. Dai, J.; Wu, X.; Yang, J.; Zeng, X. C., Unusual Metallic Microporous Boron Nitride Networks. *The Journal of Physical Chemistry Letters* **2013,** *4* (20), 3484-3488.

69. Butcher, K. S. A.; Fernandes, A. J.; Chen, P. P.-T.; Wintrebert-Fouquet, M.; Timmers, H.; Shrestha, S. K.; Hirshy, H.; Perks, R. M.; Usher, B. F., The nature of nitrogen related point defects in common forms of InN. *Journal of Applied Physics* **2007,** *101* (12), 123702.

70. Lu, Y.-S.; Huang, C.-C.; Yeh, J. A.; Chen, C.-F.; Gwo, S., InN-based anion selective sensors in aqueous solutions. *Applied Physics Letters* **2007,** *91* (20), 202109.

71. Lindgren, T.; Larsson, M.; Lindquist, S.-E., Photoelectrochemical characterisation of indium nitride and tin nitride in aqueous solution. *Solar Energy Materials and Solar Cells* **2002,** *73* (4), 377-389.

72. Guo, Q. X.; Kato, O.; Yoshida, A., Chemical Etching of Indium Nitride. *Journal of The Electrochemical Society* **1992,** *139* (7), 2008-2009.

73. Chitambar, C. R., Medical applications and toxicities of gallium compounds. *International journal of environmental research and public health* **2010,** *7* (5), 2337-2361.

74. Ahlem, A.; Samira, M.; Ali, E. H.; Pierre, G.; Leila, T., Ultrastructural study of the intracellular behavior of four mineral elements in the lactating mammary gland cells: Study using conventional transmission electron microscopy. *Microscopy Research and Technique* **2008,** *71* (12), 849-855.

75. Millour, S.; Noel, L.; Chekri, R.; Vastel, C.; Kadar, A.; Sirot, V.; Leblanc, J. C.; Guerin, T., Strontium, silver, tin, iron, tellurium, gallium, germanium, barium and vanadium levels in foodstuffs from the Second French Total Diet Study. *J. Food Compos. Anal.* **2012,** *25* (2), 108-129.

76. Ayadi, A.; Maghraoui, S.; Kammoun, S.; Tekaya, L., Effects of the presence of indium on the mammary gland ultrastructure, body weight, food intake and plasmatic prolactin concentration. *Microscopy* **2014,** *63* (5), 383-389.

77. Samira, M.; Ahlem, A.; Aouatef, B. A.; Habib, J. M.; Leila, T., Histological and ultrastructural study of the intracellular behavior of indium in the testicular tissues. *Microscopy Research and Technique* **2011,** *74* (6), 546-550.

78. Maghraoui, S.; Ayadi, A.; Ben Ammar, A.; Jaafoura, M.-H.; El Hili, A.; Galle, P.; Tekaya, L., Microscopy and microanalysis study of the indium (In) behavior in the intestinal mucosa, the liver, the kidney and the testicle. *Journal of Electron Microscopy* **2011,** *60* (2), 183-190.

79. Castronovo, F. P.; Wagner, H. N., Factors Affecting the Toxicity of the Element Indium. *British Journal of Experimental Pathology* **1971,** *52* (5), 543-559.

80. Tanaka, A.; Hirata, M.; Kiyohara, Y.; Nakano, M.; Omae, K.; Shiratani, M.; Koga, K., Review of pulmonary toxicity of indium compounds to animals and humans. *Thin Solid Films* **2010,** *518* (11), 2934-2936.

81. Lim, C. H.; Han, J.-H.; Cho, H.-W.; Kang, M., Studies on the Toxicity and Distribution of Indium Compounds According to Particle Size in Sprague-Dawley Rats. *Toxicological Research* **2014,** *30* (1), 55-63.

82. Kelsen, D. P.; Alcock, N.; Yeh, S.; Brown, J.; Young, C., Pharmacokinetics of gallium nitrate in man. *Cancer* **1980,** *46* (9), 2009-2013.

83. Reagan-Shaw, S.; Nihal, M.; Ahmad, N., Dose translation from animal to human studies revisited. *The FASEB Journal* **2008,** *22* (3), 659-661.

84. Chitambar, C. R.; Purpi, D. P.; Woodliff, J.; Yang, M.; Wereley, J. P., Development of Gallium Compounds for Treatment of Lymphoma: Gallium Maltolate, a Novel Hydroxypyrone Gallium Compound, Induces Apoptosis and Circumvents Lymphoma Cell Resistance to Gallium Nitrate. *Journal of Pharmacology and Experimental Therapeutics* **2007,** *322* (3), 1228-1236.

85. Vallabhajosula, S. R.; Harwig, J. F.; Wolf, W., The mechanism of tumor localization of gallium-67 citrate: Role of transferrin binding and effect of tumor pH. *International Journal of Nuclear Medicine and Biology* **1981,** *8* (4), 363-370.

86. Salloum, E.; Brandt, D. S.; Caride, V. J.; Cornelius, E.; Zelterman, D.; Schubert, W.; Mannino, T.; Cooper, D. L., Gallium scans in the management of patients with Hodgkin's disease: a study of 101 patients. *Journal of Clinical Oncology* **1997,** *15* (2), 518-27.

87. Boshitskaya, N. V.; Lavrenko, V. A.; Bartnitskaya, T. S.; Makarenko, G. N.; Shkurko, G. A.; Danilenko, N. V., Reaction of aluminum nitride powder with biochemical media. *Powder Metall Met Ceram* **2000,** *39* (3-4), 157-162.

88. Yu-Ting, K.; Ta-Jen, Y. In *Evaluating biocompatibility of semiconductive gallium nitride, flat and nano-structured silicon chips by cell viability, adhesion and growth,* Nanoelectronics Conference (INEC), 2010 3rd International, 3-8 Jan. 2010; 2010; pp 811-812.

89. Faucheux, N.; Schweiss, R.; Lützow, K.; Werner, C.; Groth, T., Self-assembled monolayers with different terminating groups as model substrates for cell adhesion studies. *Biomaterials* **2004,** *25* (14), 2721-2730.

90. Bain, L. E.; Ivanisevic, A., Engineering the Cell–Semiconductor Interface: A Materials Modification Approach using II-VI and III-V Semiconductor Materials. *Small* **2015,** *11* (7), 768-780.

91. Miller, D. C.; Thapa, A.; Haberstroh, K. M.; Webster, T. J., Endothelial and vascular smooth muscle cell function on poly(lactic-co-glycolic acid) with nano-structured surface features. *Biomaterials* **2004,** *25* (1), 53-61.

92. Bain, L. E.; Collazo, R.; Hsu, S.-h.; Latham, N. P.; Manfra, M. J.; Ivanisevic, A., Surface topography and chemistry shape cellular behavior on wide band-gap semiconductors. *Acta Biomaterialia* **2014,** *10* (6), 2455-2462.

93. Li, J.; Han, Q.; Wang, X.; Yang, R.; Wang, C., Enhanced cell growth on nanotextured GaN surface treated by UV illumination and fibronectin adsorption. *Colloids and Surfaces B: Biointerfaces* **2014,** *123,* 293-301.

94. McAllister, J. P.; Li, J.; Deren, K.; Finlayson, P. G.; Jaboro, C.; Auner, G. A.; Baird, R.; Lagman, A.; Iezzi, R.; Abrams, G. W. In *Chronic in vivo biocompatibility testing of materials used for visual prostheses,* Investigative Ophthalmology & Visual Science, 2004; pp 4214-4214.

95. Podolska, A.; Tham, S.; Hart, R. D.; Seeber, R. M.; Kocan, M.; Kocan, M.; Mishra, U. K.; Pfleger, K. D. G.; Parish, G.; Nener, B. D., Biocompatibility of semiconducting AlGaN/GaN material with living cells. *Sensors and Actuators B: Chemical* **2012,** *169,* 401-406.

96. Podolska, A.; Seeber, R. M.; Kocan, M.; Pfleger, K. D. G.; Parish, G.; Nener, B. D. In *Cell growth and attachment to AlGaN surfaces for biosensor applications,* Nanoscience and Nanotechnology (ICONN), 2010 International Conference on, 22-26 Feb. 2010; 2010; pp 264-267.

97. Park, G.; Chung, H. J.; Kim, K.; Lim, S. A.; Kim, J.; Kim, Y. S.; Liu, Y.; Yeo, W. H.; Kim, R. H.; Kim, S. S., Immunologic and tissue biocompatibility of Flexible/Stretchable electronics and optoelectronics. *Advanced healthcare materials* **2014,** *3* (4), 515-525.

98. Chan, E.; Jackson, N.; Mathewson, A.; Galvin, P.; Alamin Dow, A. B.; Kherani, N. P.; Blaszykowski, C.; Thompson, M., Surface modification of piezoelectric

aluminum nitride with functionalizable organosilane adlayers. *Applied Surface Science* **2013**, *282* (0), 709-713.

99. Chiu, C.-S.; Lee, H.-M.; Gwo, S., Site-Selective Biofunctionalization of Aluminum Nitride Surfaces Using Patterned Organosilane Self-Assembled Monolayers. *Langmuir* **2010**, *26* (4), 2969-2974.

100. Blaszykowski, C.; Sheikh, S.; Benvenuto, P.; Thompson, M., New Functionalizable Alkyltrichlorosilane Surface Modifiers for Biosensor and Biomedical Applications. *Langmuir* **2012**, *28* (5), 2318-2322.

101. Chiu, H. T.; Sukachonmakul, T.; Kuo, M. T.; Wang, Y. H.; Wattanakul, K., Surface modification of aluminum nitride by polysilazane and its polymer-derived amorphous silicon oxycarbide ceramic for the enhancement of thermal conductivity in silicone rubber composite. *Applied Surface Science* **2014**, *292* (0), 928-936.

102. Chu, B. H.; Chang, C. Y.; Kroll, K.; Denslow, N.; Wang, Y.-L.; Pearton, S. J.; Dabiran, A. M.; Wowchak, A. M.; Cui, B.; Chow, P. P.; Ren, F., Detection of an endocrine disrupter biomarker, vitellogenin, in largemouth bass serum using AlGaN/GaN high electron mobility transistors. *Applied Physics Letters* **2010**, *96* (1), 013701.

103. Baur, B.; Howgate, J.; von Ribbeck, H. G.; Gawlina, Y.; Bandalo, V.; Steinhoff, G.; Stutzmann, M.; Eickhoff, M., Catalytic activity of enzymes immobilized on AlGaN/GaN solution gate field-effect transistors. *Applied Physics Letters* **2006**, *89* (18), 3901

104. Wang, Y. L.; Chu, B. H.; Chang, C. Y.; Lo, C. F.; Pearton, S. J.; Dabiran, A.; Chow, P. P.; Ren, F., Long-term stability study of botulinum toxin detection with AlGaN/GaN high electron mobility transistor based sensors. *Sens. Actuator B-Chem.* **2010**, *146* (1), 349-352.

105. Wilkins, S. J.; Greenough, M.; Arellano, C.; Paskova, T.; Ivanisevic, A., In Situ Chemical Functionalization of Gallium Nitride with Phosphonic Acid Derivatives during Etching. *Langmuir* **2014**, *30* (8), 2038-2046.

106. Kim, H.; Colavita, P. E.; Paoprasert, P.; Gopalan, P.; Kuech, T. F.; Hamers, R. J., Grafting of molecular layers to oxidized gallium nitride surfaces via phosphonic acid linkages. *Surface Science* **2008**, *602* (14), 2382-2388.

107. Hanson, E. L.; Schwartz, J.; Nickel, B.; Koch, N.; Danisman, M. F., Bonding self-assembled, compact organophosphonate monolayers to the native oxide surface of silicon. *J. Am. Chem. Soc.* **2003**, *125* (51), 16074-16080.

108. Bermudez, V. M., Functionalizing the GaN(0001)-(1×1) surface I. The chemisorption of aniline. *Surface Science* **2002**, *499* (2–3), 109-123.

109. Makowski, M. S.; Zemlyanov, D. Y.; Ivanisevic, A., Olefin metathesis reaction on GaN (0001) surfaces. *Applied Surface Science* **2011**, *257* (10), 4625-4632.

110. Kim, H.; Colavita, P. E.; Metz, K. M.; Nichols, B. M.; Sun, B.; Uhlrich, J.; Wang, X.; Kuech, T. F.; Hamers, R. J., Photochemical Functionalization of Gallium Nitride

Thin Films with Molecular and Biomolecular Layers. *Langmuir* **2006**, *22* (19), 8121-8126.

111. Schwarz, S. U.; Cimalla, V.; Eichapfel, G.; Himmerlich, M.; Krischok, S.; Ambacher, O., Thermal Functionalization of GaN Surfaces with 1-Alkenes. *Langmuir* **2013**, *29* (21), 6296-6301.

112. Makowski, M. S.; Zemlyanov, D. Y.; Lindsey, J. A.; Bernhard, J. C.; Hagen, E. M.; Chan, B. K.; Petersohn, A. A.; Medow, M. R.; Wendel, L. E.; Chen, D.; Canter, J. M.; Ivanisevic, A., Covalent attachment of a peptide to the surface of gallium nitride. *Surface Science* **2011**, *605* (15–16), 1466-1475.

113. Mukherjee, J.; Peczonczyk, S.; Maldonado, S., Wet Chemical Functionalization of III−V Semiconductor Surfaces: Alkylation of Gallium Phosphide Using a Grignard Reaction Sequence. *Langmuir* **2010**, *26* (13), 10890-10896.

114. Peczonczyk, S. L.; Mukherjee, J.; Carim, A. I.; Maldonado, S., Wet Chemical Functionalization of III–V Semiconductor Surfaces: Alkylation of Gallium Arsenide and Gallium Nitride by a Grignard Reaction Sequence. *Langmuir* **2012,** *28* (10), 4672-4682.

115. Bermudez, V. M., Adsorption and photodissociation of 4-haloanilines on GaN(0001). *Surface Science* **2002,** *519* (3), 173-184.

116. Xue, H.; Shen, Z.; Li, C., Improved selectivity and stability of glucose biosensor based on in situ electropolymerized polyaniline–polyacrylonitrile composite film. *Biosensors and Bioelectronics* **2005,** *20* (11), 2330-2334.

117. Fahrenkopf, N. M.; Shahedipour-Sandvik, F.; Tokranova, N.; Bergkvist, M.; Cady, N. C., Direct attachment of DNA to semiconducting surfaces for biosensor applications. *Journal of Biotechnology* **2010,** *150* (3), 312-314.

118. (a) Jia-dong, L.; Jun-jie, C.; Bin, M.; Xiao-wei, W.; Jie, X.; Jin-cheng, Z.; Zhi-qiang, Z.; Dong-min, W., Detection of prostate-specific antigen with biomolecule-gated AlGaN/GaN high electron mobility transistors. *Journal of Micromechanics and Microengineering* **2014,** *24* (7), 075023; (b) Pearton, S. J.; Kang, B. S.; Kim, S. K.; Ren, F.; Gila, B. P.; Abernathy, C. R.; Lin, J. S.; Chu, S. N. G., GaN-based diodes and transistors for chemical, gas, biological and pressure sensing. *J. Phys.-Condes. Matter* **2004,** *16* (29), R961-R994; (c) Chen, C.-C.; Chen, H.-I.; Liu, H.-Y.; Chou, P.-C.; Liou, J.-K.; Liu, W.-C., On a GaN-based ion sensitive field-effect transistor (ISFET) with a hydrogen peroxide surface treatment. *Sensors and Actuators B: Chemical* **2015,** *209,* 658-663; (d) Auner, G.; Shreve, G.; Ying, H.; Newaz, G.; Hughes, C.; Xu, J. Z., Dual mode acoustic wave biosensors microarrays. In *Bioengineered and Bioinspired Systems*, RodriguezVazquez, A.; Abbott, D.; Carmona, R., Eds. Spie-Int Soc Optical Engineering: Bellingham, 2003; Vol. 5119, pp 129-139; (e) Lange, K.; Rapp, B. E.; Rapp, M., Surface acoustic wave biosensors: a review. *Anal. Bioanal. Chem.* **2008,** *391* (5), 1509-1519.

119. Ambacher, O.; Foutz, B.; Smart, J.; Shealy, J. R.; Weimann, N. G.; Chu, K.; Murphy, M.; Sierakowski, A. J.; Schaff, W. J.; Eastman, L. F.; Dimitrov, R.; Mitchell, A.;

Stutzmann, M., Two dimensional electron gases induced by spontaneous and piezoelectric polarization in undoped and doped AlGaN/GaN heterostructures. *Journal of Applied Physics* **2000,** *87* (1), 334-344.

120. Stutzmann, M.; Steinhoff, G.; Eickhoff, M.; Ambacher, O.; Nebel, C. E.; Schalwig, J.; Neuberger, R.; Müller, G., GaN-based heterostructures for sensor applications. *Diamond and Related Materials* **2002,** *11* (3–6), 886-891.

121. Ibbetson, J. P.; Fini, P. T.; Ness, K. D.; DenBaars, S. P.; Speck, J. S.; Mishra, U. K., Polarization effects, surface states, and the source of electrons in AlGaN/GaN heterostructure field effect transistors. *Applied Physics Letters* **2000,** *77* (2), 250-252.

122. Lenka, T. R.; Panda, A. K., Characteristics study of 2DEG transport properties of AlGaN/GaN and AlGaAs/GaAs-based HEMT. *Semiconductors* **2011,** *45* (5), 650-656.

123. Myers, M.; Khir, F. L. M.; Podolska, A.; Umana-Membreno, G. A.; Nener, B.; Baker, M.; Parish, G., Nitrate ion detection using AlGaN/GaN heterostructure-based devices without a reference electrode. *Sensors and Actuators B: Chemical* **2013,** *181,* 301-305.

124. Xuejin, W.; Schuette, M. L.; Gupta, S. K.; Nicholson, T. R.; Lee, S. C.; Lu, W., Improved Sensitivity of AlGaN/GaN Field Effect Transistor Biosensors by Optimized Surface Functionalization. *Sensors Journal, IEEE* **2011,** *11* (8), 1726-1735.

125. Khir, F. L. M.; Myers, M.; Podolska, A.; Sanders, T. M.; Baker, M. V.; Nener, B. D.; Parish, G., Synchrotron-based XPS studies of AlGaN and GaN surface chemistry and its relationship to ion sensor behaviour. *Applied Surface Science* **2014,** *314,* 850-857.

126. Podolska, A.; Kocan, M.; Cabezas, A. M. G.; Wilson, T. D.; Umana-Membreno, G. A.; Nener, B. D.; Parish, G.; Keller, S.; Mishra, U. K., Ion versus pH sensitivity of ungated AlGaN/GaN heterostructure-based devices. *Applied Physics Letters* **2010,** *97* (1), 012108.

127. Chaniotakis, N. A.; Alifragis, Y.; Konstantinidis, G.; Georgakilas, A., Gallium nitride-based potentiometric anion sensor. *Anal. Chem.* **2004,** *76* (18), 5552-5556.

128. Wang, H.-T.; Kang, B.; Chancellor, T.; Lele, T.; Tseng, Y.; Ren, F.; Pearton, S.; Johnson, W.; Rajagopal, P.; Roberts, J., Fast electrical detection of Hg (II) ions with AlGaN/GaN high electron mobility transistors. *Applied Physics Letters* **2007,** *91* (4), 42114-42114.

129. Chen, K. H.; Wang, H. W.; Kang, B. S.; Chang, C. Y.; Wang, Y. L.; Lele, T. P.; Ren, F.; Pearton, S. J.; Dabiran, A.; Osinsky, A.; Chow, P. P., Low Hg(II) ion concentration electrical detection with AlGaN/GaN high electron mobility transistors. *Sensors and Actuators B: Chemical* **2008,** *134* (2), 386-389.

130. Chu, B.-H.; Lin, H.-W.; Gwo, S.; Wang, Y.-L.; Pearton, S. J.; Johnson, J. W.; Rajagopal, P.; Roberts, J. C.; Piner, E. L.; Linthicuni, K. J.; Ren, F., Chloride ion

detection by InN gated AlGaN/GaN high electron mobility transistors. *Journal of Vacuum Science Technology B* **2010,** *28* (1), L5-L8.

131. Kao, K.-W.; Su, Y.-W.; Lu, Y.-S.; Gwo, S.; Yeh, J. A., Calcium ion detection using miniaturized InN-based ion sensitive field effect transistors. *International Journal of Automation and Smart Technology* **2012,** *2* (1), 49-54.

132. King, P. D. C.; Veal, T. D.; McConville, C. F.; Fuchs, F.; Furthmüller, J.; Bechstedt, F.; Schley, P.; Goldhahn, R.; Schörmann, J.; As, D. J.; Lischka, K.; Muto, D.; Naoi, H.; Nanishi, Y.; Lu, H.; Schaff, W. J., Universality of electron accumulation at wurtzite c- and a-plane and zinc-blende InN surfaces. *Applied Physics Letters* **2007,** *91* (9), 092101.

133. Lee, C.-T.; Chiu, Y.-S.; Wang, X.-Q., Performance enhancement mechanisms of passivated InN/GaN-heterostructured ion-selective field-effect-transistor pH sensors. *Sensors and Actuators B: Chemical* **2013,** *181*, 810-815.

134. Chang, Y. H.; Chang, K. K.; Gwo, S.; Yeh, J. A., Highly Sensitive Hydrogen Detection Using a Pt-Catalyzed InN Epilayer. *Appl. Phys. Express* **2010,** *3* (11), 114101.

135. Kao, K.-W.; Hsu, M.-C.; Chang, Y.-H.; Gwo, S.; Yeh, J. A., A Sub-ppm Acetone Gas Sensor for Diabetes Detection Using 10 nm Thick Ultrathin InN FETs. *Sensors (Basel, Switzerland)* **2012,** *12* (6), 7157-7168.

136. Lu, H.; Schaff, W. J.; Eastman, L. F.; Stutz, C. E., Surface charge accumulation of InN films grown by molecular-beam epitaxy. *Applied Physics Letters* **2003,** *82* (11), 1736-1738.

137. Yamada, M.; Kadoya, Y.; Kasai, S.; Kato, K.; Mochizuki, M.; Nishi, N.; Watanabe, N.; Kleinman, H. K.; Yamada, Y.; Nomizu, M., Ile-Lys-Val-Ala-Val (IKVAV)-containing laminin α1 chain peptides form amyloid-like fibrils. *FEBS Letters* **2002,** *530* (1–3), 48-52.

138. Makowski, M. S.; Kim, S.; Gaillard, M.; Janes, D.; Manfra, M. J.; Bryan, I.; Sitar, Z.; Arellano, C.; Xie, J.; Collazo, R.; Ivanisevic, A., Physisorption of functionalized gold nanoparticles on AlGaN/GaN high electron mobility transistors for sensing applications. *Applied Physics Letters* **2013,** *102* (7), 074102.

139. Estephan, E.; Saab, M.-B.; Martin, M.; Larroque, C.; Cuisinier, F. J. G.; Briot, O.; Ruffenach, S.; Moret, M.; Gergely, C., Phages recognizing the Indium Nitride semiconductor surface via their peptides. *Journal of Peptide Science* **2011,** *17* (2), 143-147.

140. Chu, B. H.; Chang, C.; Wang, Y.; Pearton, S.; Ren, F., Surface Immobilizations of AlGaN/GaN High Electron Mobility Transistor Based Sensors. *ECS Transactions* **2010,** *33* (13), 3-22.

141. Tulip, F. S.; Eteshola, E.; Desai, S.; Mostafa, S.; Roopa, S.; Evans, B.; Islam, S. K., Direct Label-Free Electrical Immunodetection of Transplant Rejection Protein

Biomarker in Physiological Buffer Using Floating Gate AlGaN/GaN High Electron Mobility Transistors. *NanoBioscience, IEEE Transactions on* **2014,** *13* (2), 138-145.

142. Huang, Q.; Bando, Y.; Zhao, L.; Zhi, C. Y.; Golberg, D., pH sensor based on boron nitride nanotubes. *Nanotechnology* **2009,** *20* (41), 415501.

143. Golberg, D.; Bando, Y.; Huang, Y.; Terao, T.; Mitome, M.; Tang, C.; Zhi, C., Boron Nitride Nanotubes and Nanosheets. *ACS Nano* **2010,** *4* (6), 2979-2993.

144. Makowski, M. S.; Ivanisevic, A., Molecular Analysis of Blood with Micro-/Nanoscale Field-Effect-Transistor Biosensors. *Small* **2011,** *7* (14), 1863-1875.

145. Stern, E.; Wagner, R.; Sigworth, F. J.; Breaker, R.; Fahmy, T. M.; Reed, M. A., Importance of the Debye Screening Length on Nanowire Field Effect Transistor Sensors. *Nano Letters* **2007,** *7* (11), 3405-3409.

146. Kim, J. P.; Lee, B. Y.; Lee, J.; Hong, S.; Sim, S. J., Enhancement of sensitivity and specificity by surface modification of carbon nanotubes in diagnosis of prostate cancer based on carbon nanotube field effect transistors. *Biosensors and Bioelectronics* **2009,** *24* (11), 3372-3378.

147. Rebriiev, A. V.; Starodub, N. F., Enzymatic biosensor based on the ISFET and photopolymeric membrane for the determinaion of urea. *Electroanalysis* **2004,** *16* (22), 1891-1895.

148. Wen, X.; Wang, S.; Wang, Y.; Lee, L.; Lu, W., AlGaN/GaN heterostructure field transistor for label-free detection of DNA hybridization. *Chin. Sci. Bull.* **2013,** *58* (21), 2601-2605.

149. Makowski, M. S. Surface functionalization of gallium nitride for biosensor platform development. Purdue University, ProQuest Dissertations, 2013.

150. Rohrbaugh, N.; Bryan, I.; Bryan, Z.; Collazo, R.; Ivanisevic, A., Long-term stability assessment of AlGaN/GaN field effect transistors modified with peptides: Device characteristics vs. surface properties. *AIP Advances* **2015,** *5* (9), 097102.

151. Walukiewicz, W., Narrow band gap group III-Nitride alloys. *Physica E: Low-dimensional Systems and Nanostructures* **2004,** *20* (3-4), 300-307.

Chapter 6

GaN-Based Hydrogen Sensors

Kwang Hyeon Baik*, Soohwan Jang[†]

*School of Materials Science and Engineering, Hongik University, Jochiwon, Sejong 339-701, Republic of Korea
[†] Department of Chemical Engineering, Dankook University, Yongin, 16890, Republic of Korea

6.1. Introduction

Interest in hydrogen as an alternative energy source and a viable commercial carrier has grown significantly in recent years. It is an emission-free fuel that when reacted with oxygen can produce heat combustion of 141.9 kJ/g. By comparison, only 47.0 and 45.0 kJ/g is yielded on average from gasoline and diesel, respectively. Electrochemical cells and combustion in internal engines are used to transport hydrogen-generated energy to power vehicles and electric devices. Ironically, most commercial hydrogen is produced through steam reforming from hydrocarbons which originate from fossil fuels. In addition, this process creates unwanted carbon dioxide and carbon monoxide bi-products [1-2]. In response to these concerns, researchers have examined more efficient and economical means of producing hydrogen. These include radiolysis and water splitting through photo-electrochemical, photocatalytic and photo-biological processes.

Commonly, hydrogen is consumed in petrochemical plants for hydride-alkylation, hydride-sulfurization, and hydrocracking. It is also used as a coolant in power stations since it has the highest specific thermal conductivity of all gases. It is further relied upon as an energy carrier in fuel-cell vehicles, and in the propulsion of spacecraft. Hydrogen gas is colorless, odorless, extremely reactive with oxygen, and has low ignition energy [3]. Unfortunately, when it leaks from a pressurized container, its temperature rises quickly due to its negative Joule-Thomson coefficient. This in turn can induce spontaneous flammable ignition, posing great risk to humans and commercial equipment. Thus, hydrogen gas detection is a critical issue for hydrogen-related industrial processes and product use. For hydrogen sensors, this entails adequately detecting flammability, as well as proper monitoring of the feed stream for the fuel-cell vehicle to ensure an efficient energy conversion. In general, market-driven sensors require a 0.1-10% detection range, less than 1 s response time, -30-80°C operational temperature, an accuracy reading within 5% of full scale, good reliability, and low cost. For fuel-cell vehicle applications, these requirements are even stricter. Therefore, a GaN-based material system is very suitable for constructing hydrogen sensors under these conditions. Specifically, the wide bandgap of GaN can enable a device to operate at high temperatures and in harsher radiative environments. In addition, the mechanical and chemical robustness of GaN can ensure greater product reliability and durability [4–8]. Many types of GaN based hydrogen sensing devices have been developed, including Schottky diodes, metal oxide semiconductor (MOS) diodes, GaN nanowires and AlGaN/GaN high electron mobility transistors (HEMTs). Many of these employ either platinum or palladium film as a catalytic active layer to achieve fast and sensitive detection of hydrogen [9-15].

Figure 1 shows the number of published papers on GaN based hydrogen sensors each year since 2000. Of particular interest is the AlGaN/GaN HEMT structure with a two dimensional electron gas channel (2-DEG) induced by piezoelectric and spontaneous polarization between the AlGaN and GaN layers. This structure shows highly sensitive current changes to surface charges created by a catalytic reaction of target gas on the active layer, which are amplified through the HEMTs [6, 16]. Typically, field effect transistors with higher drain current exhibit better sensitivity to gas detection. With a 30 % Al concentration in the AlGaN layer, 5~10 times higher channel sheet electron densities are obtained compared to GaAs or InP HEMTs [17, 18].

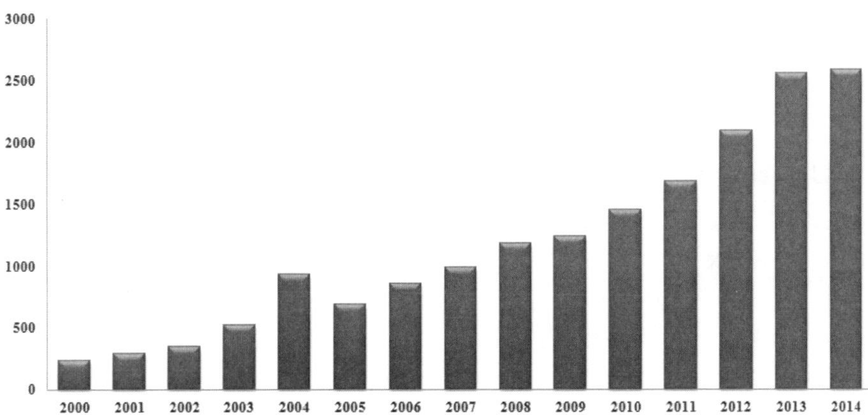

Figure 1. The number of published papers on GaN-based hydrogen sensors per year since 2000.

In this chapter, the effects of crystal polarity and plane on hydrogen sensing characteristics of GaN based diodes are examined. Also, AlGaN/GaN HEMT based devices with dramatically increased active sites on the platinum nanostructures, along with GaN based diodes bearing large etched surface areas, are investigated as means to improving the sensitivity of hydrogen sensors.

6.2. GaN Schottky Diodes on Various Crystal Planes

The basic structure of a GaN-based Schottky diode hydrogen sensor consists of GaN, an Ohmic contact, a Schottky contact, and a passivation layer, as shown in Figure 3. The Schottky contact is made of platinum or palladium where catalytic decomposition of the hydrogen molecules occurs. Once hydrogen gas is exposed to the device, the hydrogen molecules are dissociated into atoms on the catalytic Pt or Pd film. Next, they diffuse into the GaN surface and form dipoles, thus reducing the Schottky barrier height, as illustrated in Figure 3 [19-21]. In effect, the current of the Schottky diode increases in relation to the concentration of hydrogen gas present.

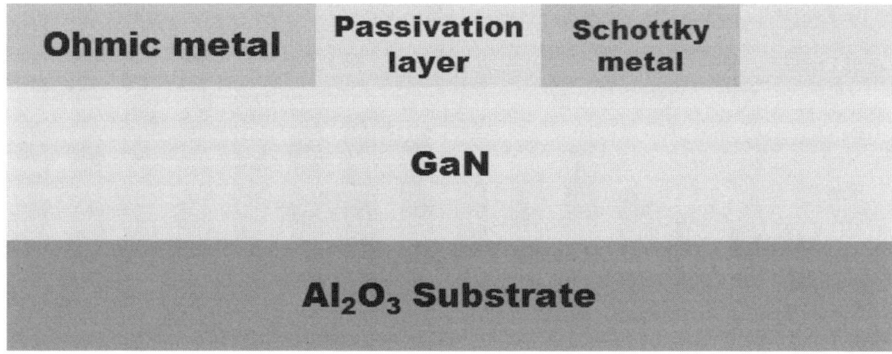

Figure 2. Schematic diagram of GaN-based Schottky diode.

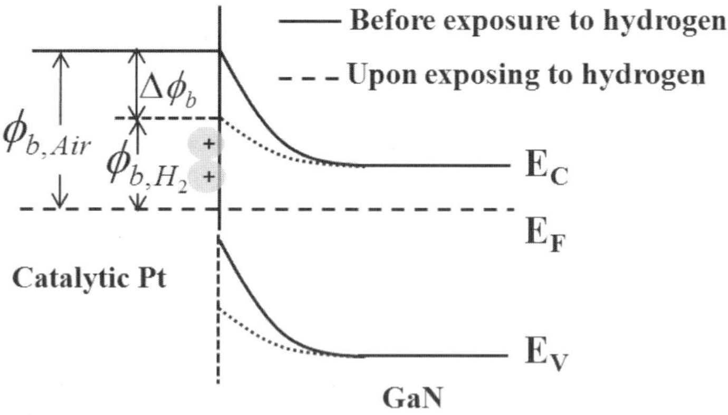

Figure 3. Schematic band diagram of GaN-based Pt Schottky diode for hydrogen exposure.

The Wurtzite structure of GaN has a polar c-plane (0001), a nonpolar *m*-plane ($1\bar{1}00$) and an *a*-plane ($11\bar{2}0$), as well as semipolar planes including ($11\bar{2}2$), ($10\bar{1}\bar{1}$), ($20\bar{2}1$), and ($20\bar{2}\bar{1}$). Nonpolar and semi-polar GaN crystal planes have been intensively researched to eliminate or reduce the quantum-confined Stark effect induced by the spontaneous and piezoelectric polarization fields, which can significantly limit the output performance of a conventional *c*-plane InGaN/GaN quantum well light-emitting diode (LED) under high current injections [22-26]. The crystal plane and polarity of GaN are also of great interest since each crystal plane has its own atomic arrangement that reacts differently to hydrogen [27-32]. Pt Schottky diodes fabricated on a polar c-plane Ga-face (c-GaN(Ga)) and N-face (c-GaN(N)), a nonpolar a-plane (a-GaN), and a semipolar($11\bar{2}2$)GaN (s-GaN) are therefore examined and compared below for their hydrogen sensing characteristics.

In this experiment, fabrication of the diodes began with Ohmic contacts consisting of Ti/Al/Ni/Au deposited by e-beam evaporation and patterned by a lift-off method. These contacts were subsequently annealed at 850°C for 60 s under N_2 ambient in an RTA (rapid thermal annealing) system. The surface was encapsulated with 2000 Å of plasma enhanced chemical vapor deposited SiN_x at 300°C. Windows in the SiN_x were opened using a buffered oxide etchant. 100 Å thick Pt film was deposited by e-beam evaporation for Schottky contacts. The Ti/Au final metal pads were formed for interconnection contacts. The current-voltage (I-V) characteristics of the Schottky diodes were measured at 25°C in a gas test chamber in ambient of N_2 or 4% hydrogen in nitrogen using an Agilent 4156C parameter analyzer.

As shown in Figure 4, all of the diodes show an increase in current in the sweeping bias range when exposed to hydrogen. This occurred due to a reduction in the Schottky barrier height from the positively charged hydrogen ion dissociated at the catalytic Pt layer. The c-GaN(N), a-GaN, and s-GaN sensors whose surface atomic configurations include nitrogen have a much larger current change than the c-GaN(Ga) diode. Notably, the c-GaN(N) Schottky diode shows a shift from rectifying to Ohmic behavior after H_2 exposure, whereas the other three types of diodes remain rectified.

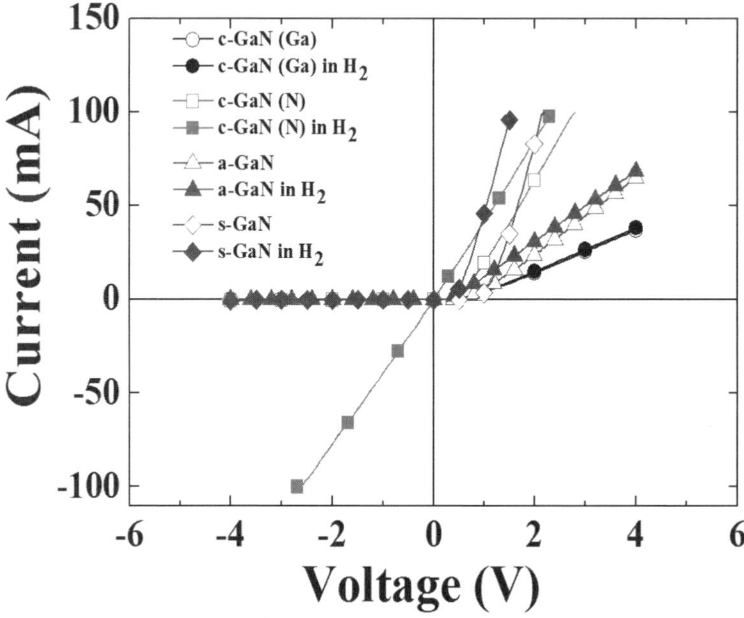

Figure 4. Current-voltage (I-V) characteristics of c-GaN(Ga), c-GaN(N), a-GaN, and s-GaN Schottky diodes before and after exposure to 4% H_2 in N_2 at 25°C.

Figure 5 shows the current responses of the Schottky diodes under examination with different GaN crystal planes exposed to repeated 4% H_2 in N_2 gas injections into the test chamber. A fixed bias voltage of 0.75 V was applied to the c-GaN(N), a-GaN, and s-GaN diodes, while 1 V was used for the c-GaN(Ga) diode. The c-GaN(Ga) and s-GaN diodes exhibited stable repeatability in current change and the ability to cycle this current in response to repeated introductions of hydrogen into the ambient. After switching out of the hydrogen–containing ambient into nitrogen, the diode current decayed exponentially back to its initial value. By contrast, the c-GaN(N) and a-GaN devices did not recover their original currents during the measurement time frame. Density functional theory suggests that hydrogen has a much higher affinity for nitrogen on a GaN surface than gallium. Thus, both the crystal plane and the polarity of the surface play vital roles in determining hydrogen sensing characteristics [30-33]. The strong affinity of nitrogen to hydrogen may also lead to a slower recovery in c-GaN(N) and a-GaN diodes. They exhibited much shorter rise time than c-GaN(Ga) diode.

The relative current change to hydrogen exposure as a percentage can be defined as $(I_H - I_N)/I_N \times 100\%$, where I_N and I_H denote diode currents measured under nitrogen and 4% hydrogen in nitrogen, respectively. As presented in Figure 6, significant differences in relative current changes among all the devices with different surface atomic configurations are observed. The maximum relative current change percentages for the c-GaN(N), a-GaN, and s-GaN diodes are of the order of 10^6, but only 80 for the c-GaN(Ga) diode. The nitrogen atoms on the GaN surface have stronger affinities for hydrogen ions than the gallium [27-29]. The estimated nitrogen densities on the surfaces of c-GaN(N) and a-GaN are 1.1×10^{15} and 7.0×10^{14} cm^{-2} respectively.

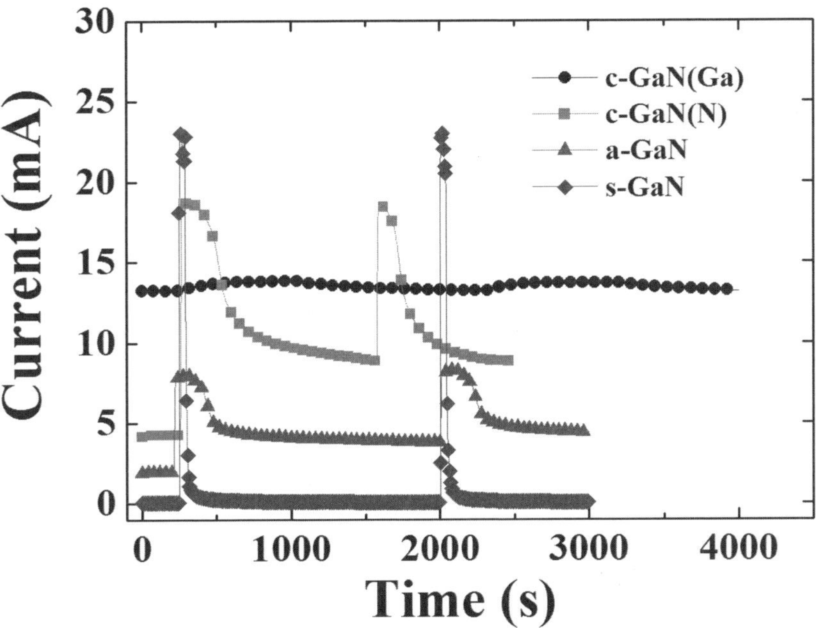

Figure 5. Time dependence of current changes in the c-GaN(Ga), c-GaN(N), a-GaN, and s-GaN sensors as a function of cycling the ambient from N$_2$ to 4% H$_2$ in N$_2$.

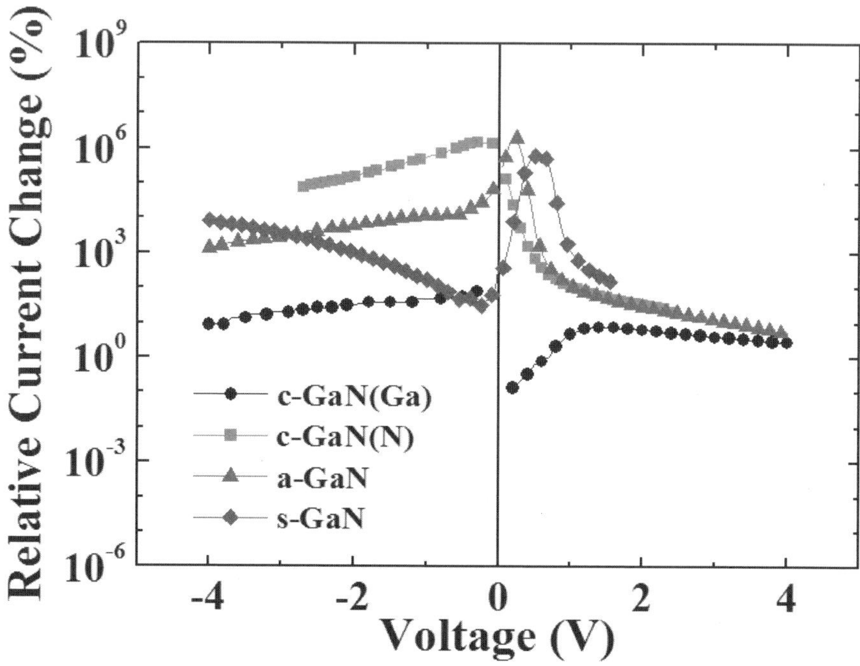

Figure 6. Relative current change percent as a function of voltage for c-GaN(Ga), c-GaN(N), a-GaN, and s-GaN diodes.

The s-GaN diode also displays a significant response to hydrogen, though its surface atomic configuration is known to have Ga polarity [34, 35]. A schematic diagram of the atomic cross-sectionsof c-GaN(Ga), c-GaN(N), a-GaN, and s-GaN is illustrated in Figure 7. It is important to note that the surface of s-GaN is terminated by Ga atoms, but neighboring nitrogen atoms with dangling bonds are exposed beneath the surface. The calculated nitrogen density of s-GaN is 6.0×10^{14} cm^{-2}. Hence, the diffused hydrogen ions are prone to be bound to uncovered nitrogen atoms next to the surface gallium. This results in a reduction of Schottky barrier height and a large increase in current within the diode. The densities of Ga and N atoms on the c-GaN(Ga), c-GaN(N), a-GaN, and s-GaN surfaces and the maximum relative current changes of the diode sensors is summarized in Table 1.

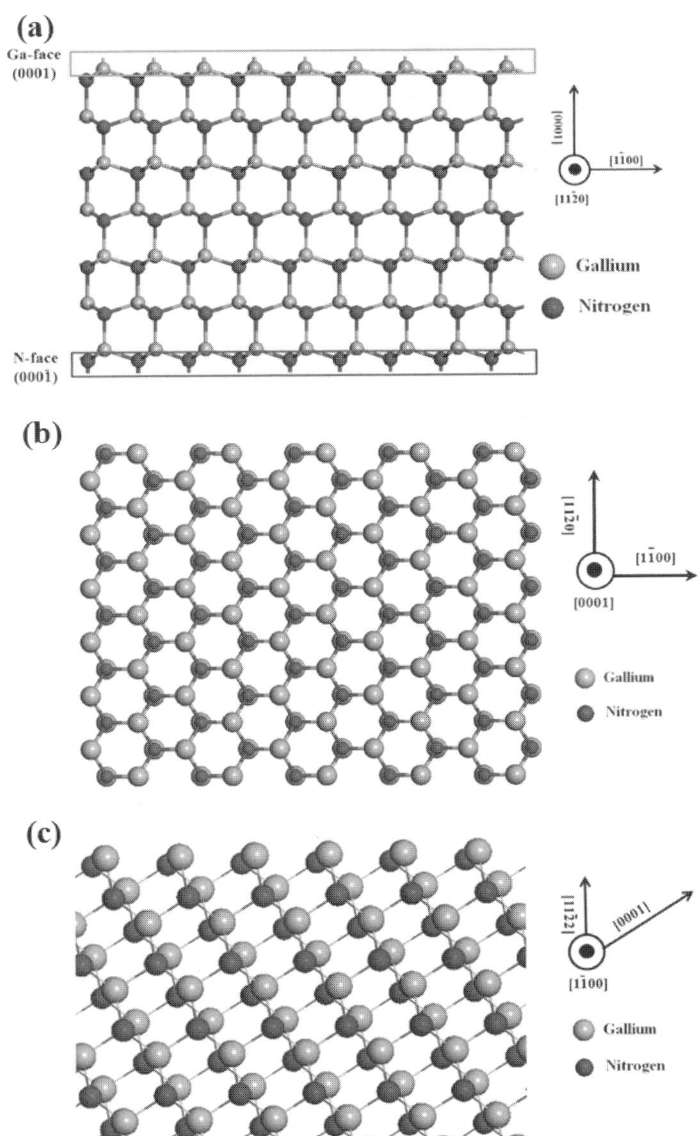

Figure 7. Schematic diagram of cross-section of (a) c-GaN(N), (b) a-GaN, and (c) s-GaN.

Table 0.1. Densities of Ga and N atoms on the c-GaN(Ga), c-GaN(N), a-GaN, and s-GaN surfaces and the maximum relative current changes of the diode sensors.

Crystal Planes	Ga ($\times 10^{15}$ cm^{-2})	N ($\times 10^{15}$ cm^{-2})	Max. Relative Current Change (%)
c-GaN(Ga)	1.10	-	~80
c-GaN(N)	-	1.1	
a-GaN	0.70	0.70	~10^6
s-GaN	0.60	0.60	

6.3. AlGaN/GaN HEMT Hydrogen Sensors with Catalytic Pt Nanostructure

One of the most advantageous GaN material systems is the AlGaN/GaN HEMT structure, as shown in Figure 8. A two dimensional electron gas (2DEG) channel with high carrier density and mobility exists at the interface beneath the AlGaN layer due to piezoelectric and spontaneous polarization without extrinsic doping. This 2DEG channel is very sensitive to changes in AlGaN surface charge. By functionalizing the gate region of HEMT with platinum which is catalytically active to the decomposition reaction of hydrogen, AlGaN/GaN HEMT can be used as a hydrogen sensor. When the HEMT is exposed to hydrogen gas, hydrogen molecules are adsorbed on the active sites of the platinum before being decomposed into hydrogen ions. Then, the dissociated hydrogen ions diffuse into the AlGaN interface to form effective positive surface charges, thereby enhancing 2DEG channel conductance and drain current. [6, 19, 36]. In effect, the drain current response to hydrogen is amplified through a functionalized gate electrode.

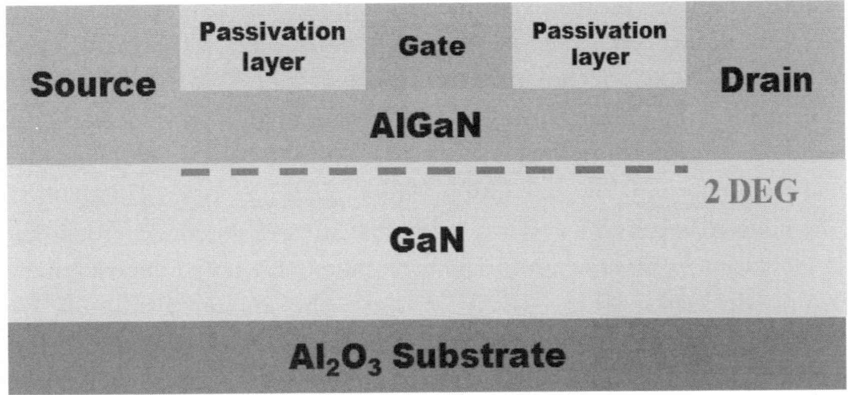

Figure 8. Schematic diagram of AlGaN/GaN HEMT structure.

To further improve the sensitivity of AlGaN/GaN HEMT based hydrogen sensors, a nano-structured platinum layer with a large surface to volume ratio can be employed [16, 37]. Platinum is a well-known transition metal catalyst in many applications including oil cracking and hydrogenation reactions [38]. In these instances, Pt nano-particles with a large surface area can improve catalytic reactivity, as well as overcome high cost and limited supply concerns in commercial usage. In this particular case, various shapes of Pt nanostructures such as particles, nano-wires, tubes, and dendrites were evident [39-42]. Among the many synthesizing techniques available, the solution phase method is one of the most promising methods. This entails using platinum nanostructures made by reducing the platinum ions from a platinum precursor enclosed in an organic surfactant. The approach is simple, environmentally friendly, and readily scalable on a commercial level. Moreover, when the active gate region of an AlGaN/GaN sensor is functionalized with platinum nanostructures that contain a larger surface area offering more active sites for hydrogen molecules to be adsorbed, the drain current response can be dramatically improved.

A platinum nano-network is a random connection of nano-rods in a nanostructure. In this experiment, a sample was synthesized using a simple solution phase method at room temperature [43]. 10 mL of 20 mM K_2PtCl_4 aqueous solution was mixed with 10 ml of 40 mM cetyltrimethylammonium bromide (CTAB) in chloroform. After a few minutes of vigorous stirring, 10 mL of 300 mM $NaBH_4$ was added to reduce the platinum complex into platinum nano-networks. The diameter of the nano-networks was controlled by the stirring speed. As the stirring rate increased from 300 rpm to 1500 rpm, the diameter of the each nano-network decreased from 10 to 2 nm. The mixture solution was centrifuged to remove any supernatants and surfactants, and the platinum nano-networks were extensively washed with acetone, ethanol and de-ionized water several times to completely remove any surfactants. Synthesized platinum nano-networks were then spin-coated on the gate region of the AlGaN/GaN HEMT selectively defined by conventional lithography. The current–voltage characteristics of the AlGaN/GaN HEMT sensors exposed to hydrogen and dry air were measured at room temperature in a gas test chamber using an Agilent 4155C semiconductor parameter analyzer.

Platinum nano-networks with diameters of 2-3 nm were synthesized from a simple and low-cost solution growth method as shown in Figure 9. These synthesized nano-networks were applied to a semiconductor surface through spin coating. The density and distribution of the networks on the surface were controlled by the number of spin coating cycles. Also, it was possible for the platinum nano-networks to be deposited on the gate region of the AlGaN/GaN HEMT selectively using a conventional and facile lift-off technique. Once the nano-networks were attached to the solid surface, they would not be detached during the semiconductor fabrication processes afterwards which included the cleaning and lift-off stages.

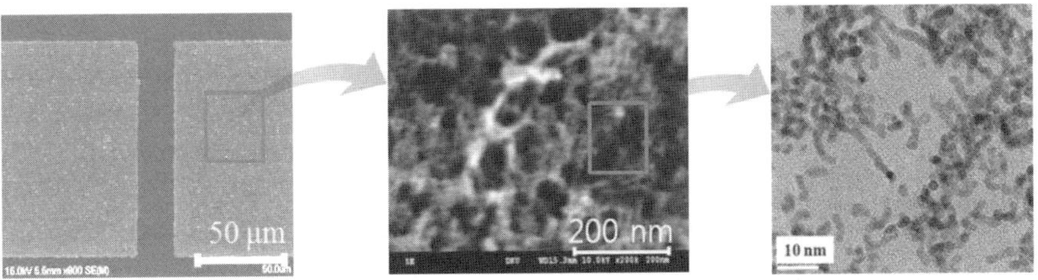

Figure 9. SEM and TEM images of synthesized platinum nano-networks.

Figure 10 shows the source-drain current-voltage (I_{DS}-V_{DS}) characteristics of the HEMT with platinum nano-networks and conventional film in both air and 500 ppm H_2 ambient conditions. It is important to note that the hydrogen induced drain current increases of the nano-network sensor were significantly larger than that of conventional platinum film HEMT. After exposure to 500 ppm hydrogen, the drain current increase at V_{DS} of 6 V and V_{GS} of 0V was 4.5 mA for the nano-network HEMT, while only 0.7 mA for the Pt film. In effect, when hydrogen molecules are introduced to the HEMT, they reach the platinum nano-network surface and are decomposed into hydrogen ions. Then, the dissociated hydrogen ions diffuse into the AlGaN interface to form effective positive charges, thereby improving the 2DEG channel and increasing the drain current. Platinum nano-networks grown by simple solution synthesis can have a huge surface area of 53m^2/g [43]. Subsequently, a larger number of hydrogen molecules can be absorbed which react with the abundant catalytic sites on the platinum thin film. These in turn induce a greater current response to the hydrogen gas. It is interesting to note that the drain currents of nano-network HEMT drop consistently as the gate voltage decreases from 2 to -4 V in increments of -1 V. This distinctive modulation of drain current for the platinum nano-network gate indicates the physical and electrical connection of nano-wires, as well as effective adhesion of the nano-networks to the AlGaN surface.

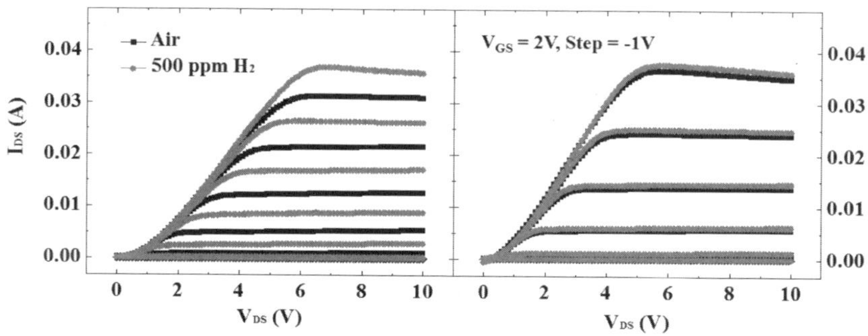

Figure 10. I$_{DS}$-V$_{DS}$ characteristics of platinum nano-network (left) and platinum film (right) gated HEMTs for air and 500 ppm H$_2$ exposures.

Figure 11 shows drain current– gate voltage (I$_{DS}$-V$_{GS}$) plots, and Figure 12 illustrates the transconductances for platinumnanonetwork and film HEMTs exposed to 500 ppm hydrogen. In Figure 11, the nano-network HEMT shows good pinch-off and drain current modulation. This confirms that all of the connected platinum nano-networks covered the entire gate area. Also, stable drain currents without any hysteresis for repeated gate voltage sweep were detected. In addition, a significant threshold voltage shift to a negative bias upon hydrogen exposure was observed in the nanonetwork sensor. This result can be attributed to the enhancement of the 2DEG channel. With a larger number of dissociated hydrogen atoms on the platinum nano-network having more active sites, transconductance may increase further, as shown in Fig. 1.12 [16, 37].

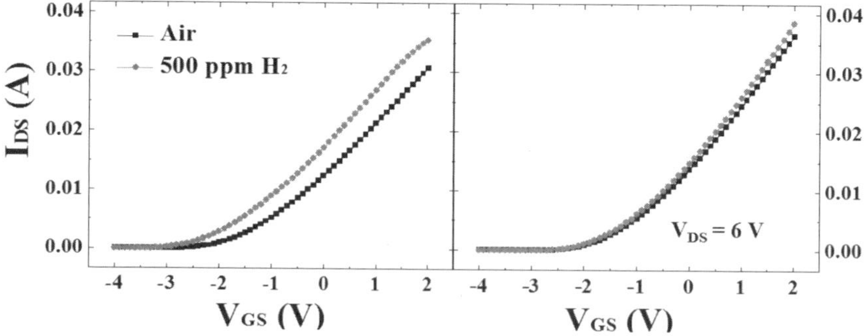

Figure 11. I$_{DS}$-V$_{GS}$ characteristics of platinum nano-network (left) and platinum film (right) gated HEMTs for air and 500 ppm H$_2$ exposures.

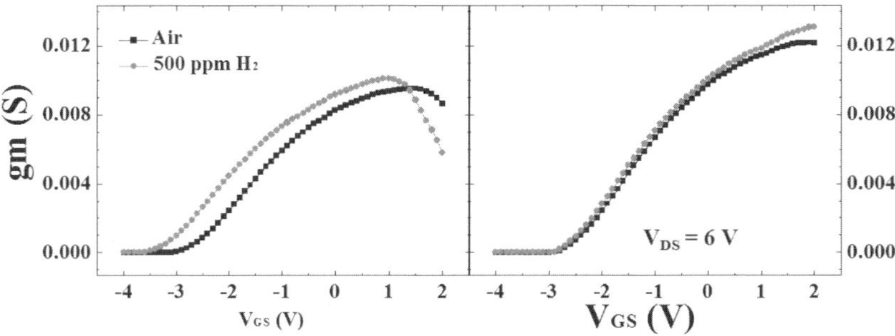

Figure 12. Extrinsic transconductances of platinum nano-network (left) and platinum film (right) gated HEMTs for air and 500 ppm H_2 exposures.

Gate leakage currents for the gate voltage sweep (I_{GS}-V_{GS}) of the platinum nano-network and film gated HEMTs measured under air and 500 ppm H_2 are shown in Fig.1.13. In this measurement condition, the devices operated as Schottky diode mode for hydrogen sensing [37]. An effective reduction of the Schottky barrier height was achieved for the platinum nano-network gate by 0.11 eV. By contrast, the platinum film showed a reduction value of only 0.05 eV. This greater reduction in Schottky barrier height can be attributed to the larger surface area of the platinum nano-network where more hydrogen molecules are adsorbed. As shown in Figure 13, the nano-network Schottky diode shows a much stronger increase in gate current.

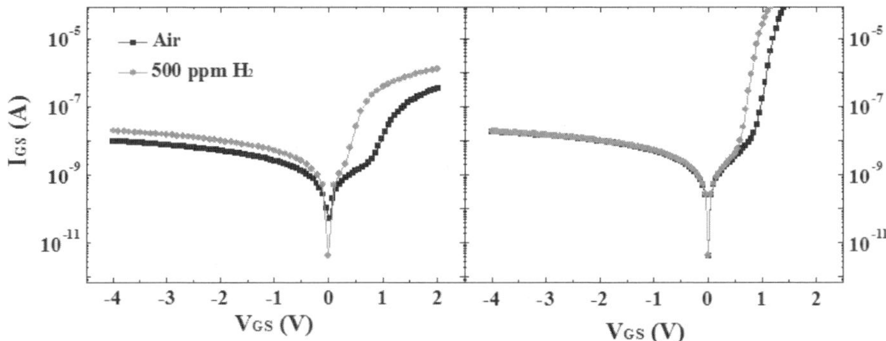

Figure 13. I_{DS}-V_{GS} characteristics of platinum nano-network (left) and platinum film (right) gated HEMTs for air and 500 ppm H_2 exposures.

Relative current changes of platinum nanonetwork and film sensors according to gate bias voltages at V_{DS} of 6 V for 500 ppm hydrogen gas exposure are shown in Figure 14. The maximum percentage of current change for the platinum nanonetwork sensor is 3.3×10^6 % at V_{GS} of -3.3 V, while just 2.5×10^2 % at V_{GS} of -2.9 V for the Pt film. The AlGaN/GaNHEMT with platinum nano-networks has a notably improved current response to hydrogen due to the larger active surface area of the nanostruture. The nanonetwork sensor also shows stable repeatability and rapid responsiveness to cyclical hydrogen exposures, as shown in Figure 15. A series of 500 ppm hydrogen gas injections were put into the device biased at V_{DS} of 6 V ($V_{GS} = 0$ V) in the gas test chamber. The results show a distinctively large current increase and a very short response time of less than 5 s.

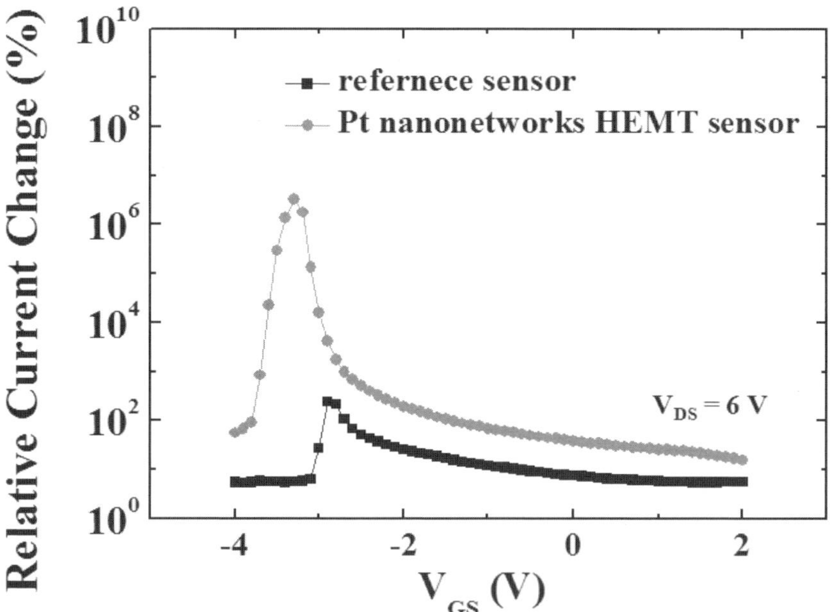

Figure 14. Relative current change of platinum nano-network and platinum film gated HEMTs measured at V_{DS} of 6 V under air and 500 ppm H_2 ambient.

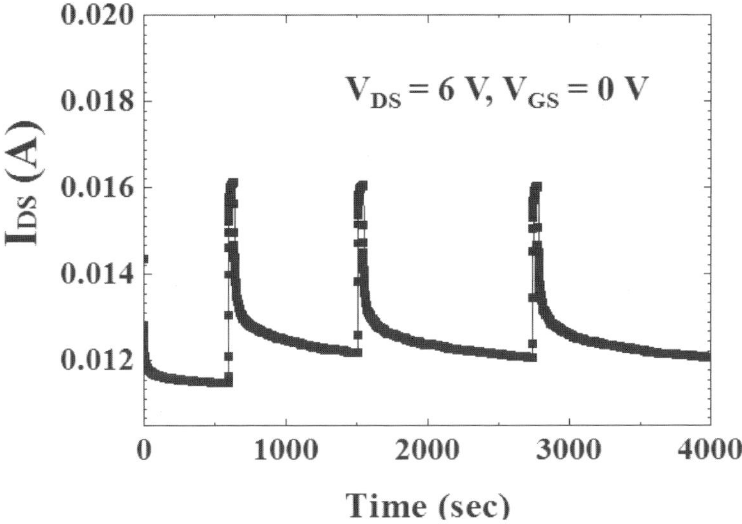

Figure 15. Time dependence of I_{DS} changes of platinum nano-network gated HEMT for cyclic 500 ppm H_2 exposure.

6.4. GaN-based Diode Sensor with Etched Large Surface Area

The hydrogen sensitivity of GaN-based hydrogen sensors can be improved not only by functionalizing the active gate area containing nanostructures and oxide films, but also by incorporating surface etching on the active contact area using wet chemical solutions. Wet etching with ultraviolet illuminations, also known as photo-electrochemical (PEC) wet etching, is widely accepted in thin-film type GaN LEDs. Its purpose is to enhance extraction efficiency from simple and damage-free etching. Interestingly enough, the surface morphologies of nonpolar and semi-polar GaN films can be dramatically modified due to their surface atomic arrangements and chemical ion reaction mechanisms. The crystallographic wet etching of nonpolar and semi-polar GaN films could be well controlled using molten KOH solutions as well as H_3PO_4 [44-46].

213

Figure 16 presents SEM images of the etched surface morphology of the Schottky contact area on nonpolar (a) and semi-polar (c) GaN films after 7 min of PEC wet etching using the KOH solution. Striated trigonal prisms with submicron widths are observed on the top surface of the nonpolar a- GaN films. In addition, inclined trigonal unit cells are exposed on the semi-polar s-GaN films. In both cases, the specific crystallographic planes of {100} and (0001) were similarly exposed, indicating the planes were chemically stable due to atomic bond configurations and a smaller density of the atoms. By using a rough surface, a larger number of absorption sites could be created which in turn helped improve overall hydrogen detection sensitivity. Thus, as more hydrogen atoms are induced on the surface, changes in the Schottky barrier height can be achieved.

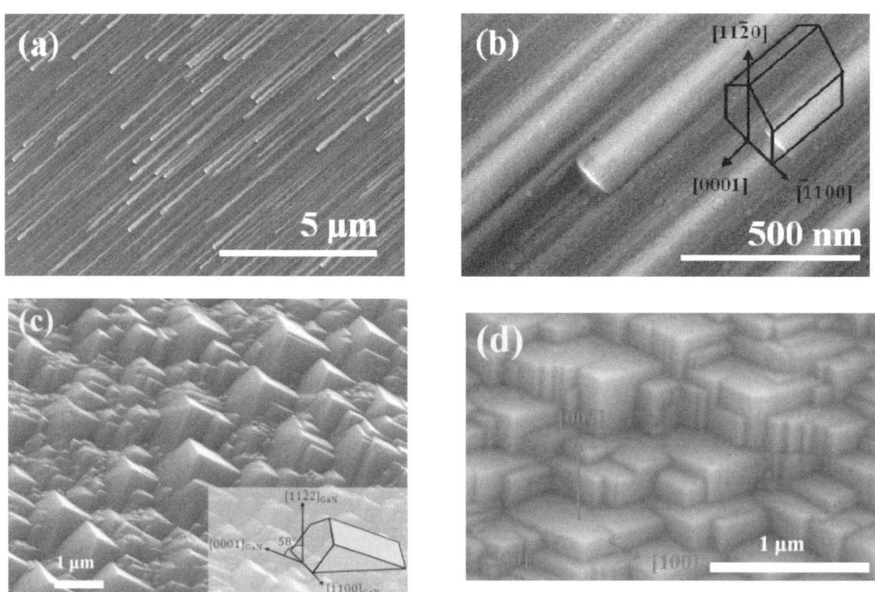

Figure 16. The SEM images of etched surface for (a), (b) nonpolar a-GaN and (c), (d) semipolar s-GaN.

Figure 17 shows the current-voltage (I-V) characteristics for the Pt Schottky diodes on nonpolar a-GaN and semipolar s-GaN films with and without wet etching before and after exposure to 4% H_2 in N_2 at room temperature. Both Pt Schottky diodes showed remarkable current changes when exposed to hydrogen in the sweeping bias range. Clearly, the surface-etched Pt Schottky diodes exhibited a better response to H_2 than the non-etched ones. The benefit of this approach is that the surface polarity and atomic configurations are advantageous when responding to H_2 molecules, due to a much higher affinity of hydrogen to nitrogen in the GaN surface. Specifically, a large density of exposed neighboring nitrogen atoms with dangling bonds exists both on the nonpolar GaN surface and beneath the semipolar GaN surface. This, in turn, can effectively enhance variations in the Schottky barrier height.

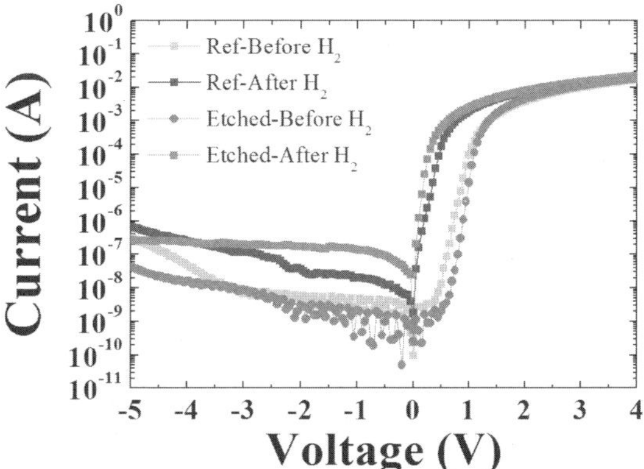

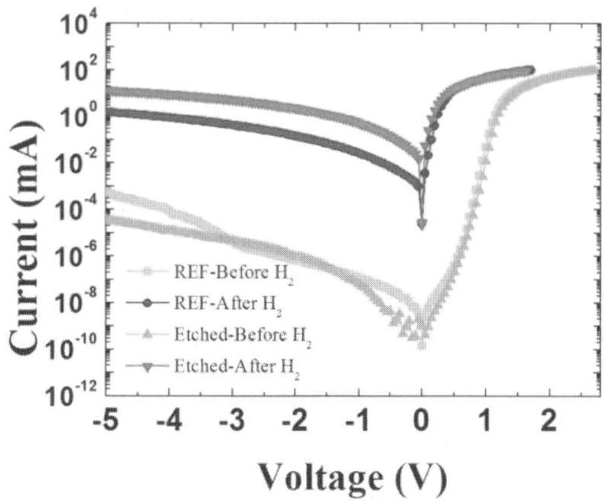

Figure 17. The current-voltage (I-V) characteristics for the Pt Schottky diodes on nonpolar a-GaN (top) and semi-polar s-GaN (bottom) films with wet etching before and after exposure to 4% H_2 in N_2.

Figure 18 shows the relative current change to hydrogen exposure as a percentage value for the Pt Schottky diodes with and without wet etching. The etched Schottky diodes showed a larger current change over the sweeping bias voltages than the non-etched one, especially one or two orders of magnitude higher in the reverse bias range. It also confirms that the etched Schottky diodes exhibited a larger current response when exposed to an H_2-containing ambient brought on by a decrease in effective barrier height. The maximum hydrogen sensitivity of the etched diode is 6-7 times higher than that of the non-etched one. Evidently, combining a rough surface morphology with wet etching is highly beneficial in enhancing the hydrogen detection sensitivity for s-GaN Schottky diodes. The hydrogen gas sensors showed stable and reproducible current changes, as well as the ability to cycle this current in response to repeated introductions of hydrogen into the ambient. The rapid response and sharp recovery of the Schottky diodes were observed within 5 sec. These results suggest that hydrogen detection sensitivity can be improved by using surface etching with KOH solutions for Pt Schottky diodes fabricated on nonpolar a-GaN and semi-polar s-GaN films. This rough surface is expected to improve hydrogen detection sensitivity due to the presence of more available adsorption sites, resulting in effective variations of the Schottky barrier height.

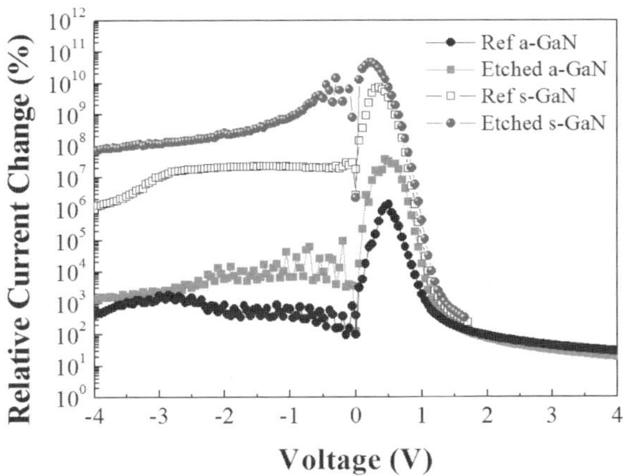

Figure 18. The relative current change to hydrogen exposure as a percentage value for the Pt Schottky diodes with and without wet etching for nonpolar a-GaN and semi-polar s-GaN Schottky diodes.

References

1. R. J. Press, K. S. V. Santhanam, M. J. Miri, A. V. Bailey, and G. A. Takacs, "Introduction to hydrogen Technology", Wiley-Interscience (2008).
2. P. Häussinger, R. Lohmüller, A. M. Watson, "Hydrogen, 1. Properties and Occurrence, Ullmann's Encyclopedia of Industrial Chemistry", John Wiley & Sons (2011)
3. G.R. Astbury and S.J. Hawksworth, Int. J. Hydrogen Energy, 32, 2178 (2007)
4. V. Dobrokhotov, D.N. McIlroy, M.G. Norton, A. Abuzir, W.J. Yeh, L. Stevenson, R. Pouy, J. Bochenek, M. Cartwright, L. Wang, J. Dawson, M. Beaux, and C. Berven, J. Appl. Phys., 99, 104302 (2006)
5. L. Voss, B.P. Gila, S.J. Pearton, H. Wang, and F. Ren, J. Vac. Sci. Technol. B 23, 2373 (2005)
6. H. Wang, T.J. Anderson, F. Ren, C. Li, Z. Low, J. Lin, B.P. Gila, S.J. Pearton, A. Osinsky, and A. Dabiran, Appl. Phys. Lett. 89, 242111 (2006)

7. H. Hasegawa, and M. Akazawa, J. Vac. Sci. Technol. B 25, 1495 (2007)

8. T.H. Tsai, J.R. Huang, K.W. Lin, C.W. Hung, W.C. Hsu, H.I. Chen, and W.C. Liu, Electrochem. Solid. St. 10, J158 (2007)

9. S. J. Pearton, B. S. Kang, S. Kim, F. Ren, B. P. Gila, C. R. Abernathy, J. Lin and S. N. G. Chu, J. Phys. Condens. Matter 16, R961-R994 (2004)

10. C. Chang, G.C. Chi, W. Wang, L. Chen, K. H. Chen, F. Ren and S. J. Pearton, J. Electron. Mater. 35, 738 (2006)

11. Y.-L. Wang, F. Ren, U. Zhang, Q. Sun, C. D. Yerino, T. S. Ko, Y. S. Cho, I. H. Lee, J. Han, and S. J. Pearton, Appl. Phys. Lett. 94, 212108 (2009)

12. W. Lim, J. S. Wright, B. P. Gila, J. L. Johnson, A. Ural, T. Anderson, F. Ren, and S. J. Pearton, Appl. Phys. Lett. 93, 072110 (2008)

13. H.-T. Wang, T. J. Anderson, B. S. Kang, F. Ren, C. Li, Z.-N. Low, J. Lin, B. P. Gila, S. J. Pearton, A. Osinsky and A. Dabiran, Appl. Phys. Lett. 90, 252109 (2007)

14. H.T. Wang, B.S. Kang, F. Ren, R.C. Fitch, J. Gillespie, N. Moser, G. Jessen, R. Dettmer, B.P. Gila, C.R. Abernathy and S.J. Pearton, Appl. Phys. Lett. 87, 172105 (2005)

15. B.S. Kang, R. Mehandru, S. Kim, F. Ren, R. Fitch, J. Gillespie, N. Moser, G. Jessen, T. Jenkins, R. Dettmer, D. Via, A. Crespo, B.P. Gila, C. R. Abernathy and S. J. Pearton, Appl. Phys. Lett. 84, 4635-4637 (2004)

16. H. Kim, and S. Jang, Curr. Appl. Phys. 13, 1746 (2013)

17. B. Luo, J. W. Johnson, J. Kim, R. M. Mehandru, F. Ren, B. P. Gila, A. H. Onstine, C. R. Abernathy, S. J. Pearton, A. G. Baca, R. D. Briggs, R. J. Shul, C. Monier and J. Han, Appl. Phys. Lett. 80, 1661 (2002)

18. H. Morkoç, R. Cingolani and B. Gil, Solid-State Electron. 43, 1909 (1999)

19. M. Eickhoff, J. Schalwig, G. Steinhoff, O. Weidmann, L. Gorgens, R. Neuberger, M. Hermann, B. Baur, G. Muller, O. Ambacher, and M. Stutzmann, Phys. Status Solidi C, 6, 1908 (2003)

20. J. Song and W. Lu, IEEE Sensors J. 8, 903 (2008)

21. Y.L. W, F. Ren, W. Lim, S.J Pearton, K. Baik, S. Hwang, Y. Seo and S. Jang, Curr. Appl. Phys. 10, 1029 (2010)

22. T. Detchprohm, M. Zhu, Y. Li, Y. Xia, C. Wetzel, E.A. Preble, L. Liu, T. Paskova, and D. Hanser, Appl. Phys. Lett. 92, 241109 (2008)

23. S.-M. Hwang, Y. G. Seo, K. H. Baik, I.-S. Cho, J. H. Baek, S. Jung, T. G. Kim, and M. Cho, Appl. Phys. Lett. 95, 071101 (2009)

24. A. Chakraborty, B.A. Haskell, S. Keller, J.S. Speck, S.P. DenBarrs, S. Nakamura, and U. K. Mishra, Jpn. J. Appl. Phys. 44, L173 (2005).

25. M. Funato, M. Ueda, Y. Kawakami, Y. Narukawa, T. Kosugi, M. Takahashi, and T. Mukai, Jpn. J. Appl. Phys. 45, L659 (2006).

26. B. Neubert, T. Wunderer, P. Bruckner, F. Scholz, M. Feneberg, F. Lipski, M. Schirra, and K. Thonke, J. Cryst. Growth, 298, 706 (2007).

27. U. Starke, S. Sloboshanin, F.S. Tautz, A. Seubert, and J.A. Schaefer, Phys. Status Solidi A, 177, 5 (2000)

28. J.E. Northrup, and J. Neugebauer, Appl. Phys. Lett. 85, 3429 (2004)

29. M. Mayumi, F. Satoh, Y. Kumagai, K. Takemoto, and A. Koukitu, Phys. Status Solidi B 228, 537 (2001)

30. K. Baik, H. Kim, S. Lee, E. Lim, S.J. Pearton, F. Ren and S. Jang, Appl. Phys. Lett. 104, 072103 (2014)

31. H. Kim, K. Baik, F. Ren, S.J. Pearton, and S. Jang, ECS Trans. 61, 353 (2014)

32. S. Jang, P. Son, J. Kim, S. Lee, and K. Baik, Sens. Actuat. B 222, 43 (2016)

33. B.S. Kang, F. Ren, B.P. Gila, C.R. Abernathy and S.J. Pearton, Appl. Phys. Lett. 84, 1123 (2004)

34. M. J. Kappers, J. L. Hollander, C. Mc Aleese, C. F. Johnson, R. F. Broom, J. S. Barnard, M. E. Vickers, and C. J. Humphreys, J. Cryst. Growth, 300, 155 (2007)

35. M. Frentrup, S. Ploch, M. Pristovsek, and M. Kneissl, Phys. Status Solidi B 248, 583 (2011)

36. C.-S. Hsu, H.-I. Chen, C.-F. Chang, T.-Y. Chen, C.-C. Huang, P.-C. Chou, and W.-C. Liu, Sens. Actuat. B 165, 19 (2012)

37. H. Kim, W. Lim, J.-H. Lee, S.J. Pearton, F. Ren, and S. Jang, Sens. Actuators B 164, 64 (2012)

38. A. Rouxoux, J. Schulz, and H. Patin, Chem. Rev. 102, 3757 (2002)

39. Z. Peng, and H. Yang, Nano Today, 4, 143 (2009)

40. Y.J. Song, S.R. Challa, C.J. Medforth, Y. Qiu, R.K. Watt, D. Pena, J.E. Miller, F. van Swol, and J.A. Shelnutt, Chem. Commun. 9, 1044 (2004)

41. Y.J. Song, Y.B. Jiang, H.R. Wang, D.A. Pena, Y. Qiu, J.E. Miller, and J.A. Shelnutt, Nanotechnology, 17, 1300 (2006)

42. Y. J. Song, Y. Yang, C. J. Medforth, E. Pereira, A. K. Singh, H. F. Xu, Y. B. Jiang, C. J. Brinker, F. V. Swol, and J. A. Shelnutt, J. Am. Chem. Soc. 126, 635 (2004)

43. Y. Song, R. M. Garcia, R. M. Dorin, H. Wang, Y. Qiu, E. N. Coker, W. A. Steen, J. E. Miller, and J. A. Shelnutt, Nano Lett. 7, 3650 (2007)

44. K. H. Baik, H.-Y. Song, S.-M. Hwang, Y. Jung, J. Ahn and J. Kim, J. Electrochem. Soc. 158 D196 (2011)

45. Y. Jung, J. Ahn, K. H. Baik, D. Kim, S. J. Pearton, F. Ren and J. Kim, J. Electrochem. Soc. 159, H117 (2012)

46. Y. Jung, K. H. Baik, M.A. Mastro, J. Hite, C.R. Eddy Jr., and J. Kim, Phys. Chem. Chem. Phys. 16, 15780 (2015)

Chapter 7

Graphene-based Chemical Sensors

Geonyeop Lee, Younghun Jung and Jihyun Kim

*Department of Chemical and Biological Engineering, Korea University
Anam-dong, Sungbuk-gu, Seoul 136-713, Korea*

7.1. Introduction

Carbon nanomaterials, such as carbon nanofibers, carbon black, fullerenes, carbon nanotubes, graphene, and reduced graphene oxide, have attracted a lot of research interest in recent years for applications in (opto) electronics, sensors, composites, displays and energy devices. Among them, particular interest in the use of graphene as sensors has been fueled by its excellent properties, such as high carrier mobility [1-4], high surface area [5], gas adsorption properties [6], and charge transfer interaction with molecules [7,8]. In addition, graphene possesses a large specific surface area (2630 $m^2 \cdot g^{-1}$), which indicates that, theoretically, all atoms of graphene can contribute to its detection capacity [9]. The crystal lattice of graphene allows a low level of Johnson noise ($1/f$), which, in turn, makes the electrical conductivity responsive to small changes in carrier concentrations. These properties have made graphene a promising candidate for next generation sensors with ultrahigh sensitivity and responsivity [10,11]. Due to its excellent mechanical and electrical properties under a mechanical strain, graphene sensors can be integrated in various areas including flexible and wearable devices [12]. In this chapter, we will focus on graphene as a sensing material.

7.2. Synthesis of Graphene

Synthesis techniques for graphene sheets have undergone significant progress since, the very first time, Novoselov and co-workers [1] mechanically exfoliated single-layer graphene (SLG) from highly ordered pyrolytic graphite (HOPG) using the scotch tape method, as shown in Figure. 1.

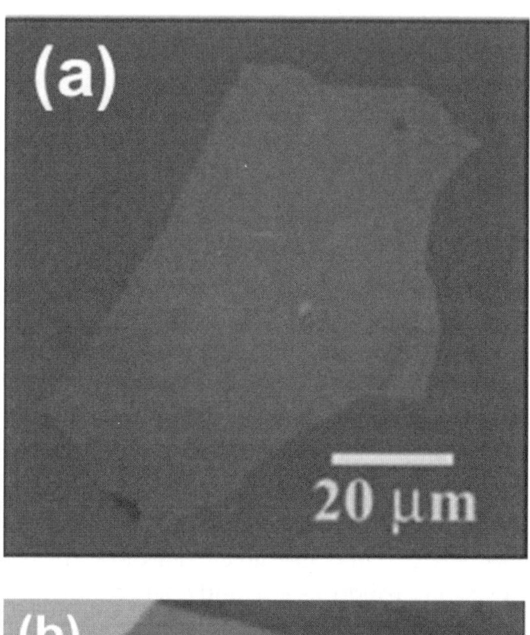

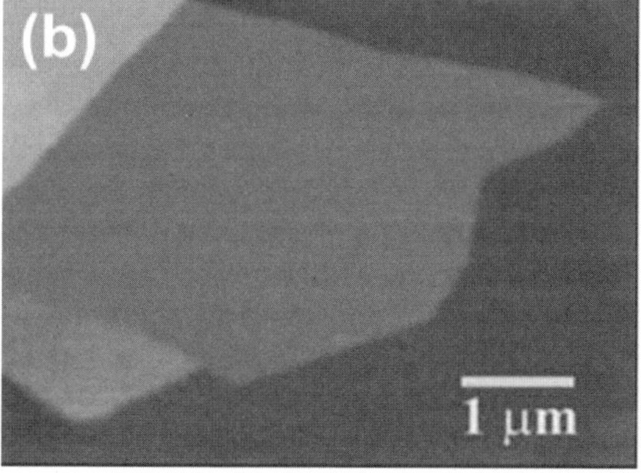

Figure 1. (a) Photograph of a multilayer graphene flake on an oxidized Si wafer. (b) Atomic force microscope image of single-layer graphene (brown-red region; 0.8 nm height) [1]. Reprinted with permission from K. S. Novoselov,A. K. Geim,S. V. Morozov,D. Jiang,Y. Zhang,S. V. Dubonos,I. V. Grigorieva,A. A. Firsov, Science **306** (5696), 666-669 (2004).

Although the mechanical exfoliation method can produce high-quality graphene sheets using simple and inexpensive equipment, this method is not suited for large-scale applications due to its low productivity. To overcome this issue, chemical reduction of graphene oxide (GO) is used for mass production of graphene sheets [13-17]. GO can be prepared according to the Hummers' Method, which involves the oxidation of graphite using strong oxidizing agents such as $KMnO_4$ and $NaNO_3$ in H_2SO_4/H_3PO_4. Single layer graphene oxide can be obtained by ultrasonication in dimethylformamide (DMF)/water solution. Reduced graphene oxide (rGO) of high quality, similar to that of graphene, can be prepared via chemical, thermal, or electrochemical reduction of GO [18]. Hydrazine monohydrate is one of the strong reducing agents and has weak reactivity towards water [19]. Hydroxylamine [20], glucose [21], phenylhydrazine [22], hydroquinone [23], pyrrole [24], and ascorbic acid [25] have also been used as reducing agents. Moreover, spray coating techniques can be used to deposit patterned rGO films with a wide range of thicknesses (mono-layer to micrometer) over large areas of various substrates [26].

The most promising large-scale synthesis method for high quality graphene is chemical vapor deposition (CVD) on transition metal substrates such as Cu [27], Pd [28], Ru [29], Ir [30], and Ni [31]. When growing graphene on a metal substrate using the CVD technique, wrinkles associated with differences in the thermal expansion coefficient between the metal and graphene are commonly formed along the metal grain boundaries, as shown in Figure 2 (a). Figure 2 (b–d) shows that large-area high quality mono-layer graphene can be synthesized via CVD technique. The formation of graphene on a metal substrate by CVD is generally a two-step process. The first step involves carbon source saturation of the transition metal, which is achieved by exposure of the metal to a hydrocarbon gas (methane, ethylene, acetylene, and benzene) at high temperatures. During the cooling step, carbon layers precipitate on the surface of the metal as carbon solubility in the transition metal decreases. Carbon solubility determines the number of graphene layers deposited on the metal substrate. Therefore, it is challenging to grow mono-layer graphene on the Ni film because it is difficult to control the amount of dissolved carbon atoms in the metal due to high carbon solubility of Ni (~1 at.%). However, the Cu film has a low carbon solubility of ~0.03 at.%; therefore, graphitization occurs only on the Cu surface. Mono-layer graphene grown on the Cu surface interrupts the dissolution of carbon atoms into the Cu surface. The graphitization process ceases when graphene completely covers the Cu film. In addition, a radio frequency plasma-enhanced CVD process has been developed to reduce the energy consumption and prevent the formation of amorphous carbon [32,33]. Graphene layers formed via CVD methods can be easily transferred onto arbitrary substrates using simple wet-based transfer processes.

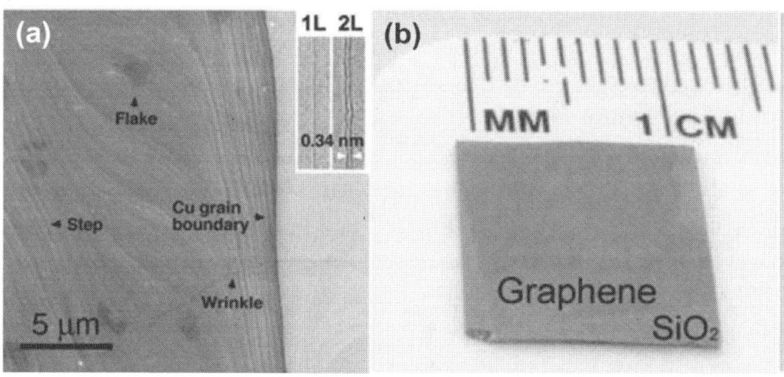

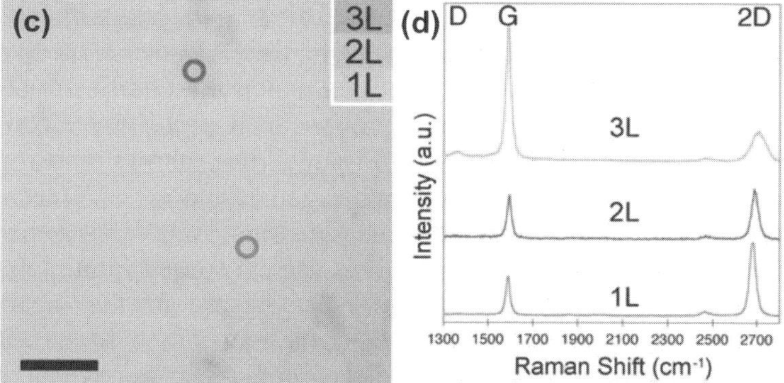

Figure 2. (a) SEM image showing a Cu grain boundary and steps, two- and three-layer graphene flakes, and graphene wrinkles. Inset in (a) shows TEM images of folded graphene edges. (b) Graphene films transferred onto a SiO₂/Si substrate. (c) Optical microscope image of graphene transferred on SiO₂/Si substrate. (d) Raman spectra at the marked spots with corresponding colored circles or arrows showing the presence of one, two, and three layers of graphene [27]. Reprinted with permission from X. Li, W. Cai, J. An, S. Kim, J. Nah, D. Yang, R. Piner, A. Velamakanni, I. Jung, E. Tutuc, S. K. Banerjee, L. Colombo, and R. S. Ruoff, Science **324** (5932), 1312-1314 (2009).

In addition, the CVD method has been employed on transition metal (Cu, Ni)-deposited SiO_2/Si substrates to grow high-quality and wafer-scale graphene layers on dielectric substrates without the need of the graphene transfer process [34,35]. Graphene layers were formed on the top and bottom sides of the metal film by diffusion of carbon atoms through the metal grain boundaries to the interface between the metal films and underlying insulating substrates. Graphene was then directly obtained on a SiO_2/Si substrate after the removal of the transition metal film.

An ion implantation method involving the high-temperature annealing of Ni thin films containing the implanted carbon atoms can be implemented to grow wafer-scale graphene layers on the semiconductor substrate [36-38]. The ion implantation technique allows for the accurate determination of the number of graphene layers by controlling the dose of carbon ions. In addition, the high-temperature annealing of SiC in an ultra-high-vacuum chamber was used to manage the direct growth of graphene [39-42]. In this process, graphene layers are formed by the carbon atoms that remain after the decomposition of silicon atoms from the SiC surface. Epitaxial graphene layers thermally grown on SiC substrates can be patterned using standard lithography methods, making them compatible with the present semiconductor industry. The morphology of graphene on SiC is highly influenced by the underlying SiC crystal structure. Heer *et al.* demonstrated the self-organized growth of graphene on a templated SiC substrate with controlled facets prepared by photolithography and fluorine-based reactive ion etching [43]. These techniques of the direct growth of graphene layers eliminate the need for transfer steps, which can damage the graphene.

7.3. Sensing Mechanism of Graphene-based Gas Sensors

The sensing mechanism of graphene is based mainly on the charge transfer between graphene and adsorbed gas molecules. The latter, which can act as chemical dopants, change local carrier concentrations in graphene and thus vary its electrical resistivity [10,11,44]. This can also be explained by the transfer characteristics of graphene field-effect transistor (FETs). When graphene is exposed to various gases, its charge neutrality point

(CNP) is readily shifted depending on chemical doping caused by the adsorption of gas molecules. Therefore, the gas molecules that change the chemical doping influence the electrical conductance of graphene. Electron acceptors (e.g. NO_2, H_2O, and I_2) increase the hole concentrations in graphene when they are adsorbed on graphene (i.e. p-doping). As a result, CNP is positively shifted, which can increase the electrical conductance. Figure 3 shows the ambipolar characteristics of graphene FETs and the shift of CNP when it was exposed to NO_2 gas molecules. The conductance of graphene at $V_g = 0$ increased with increasing chemical doping. On the other hand, electron donors (e.g. CO, ethanol, and NH_3) induce the electron concentrations in graphene and negatively shift CNP (n-doping) [11,44,45].

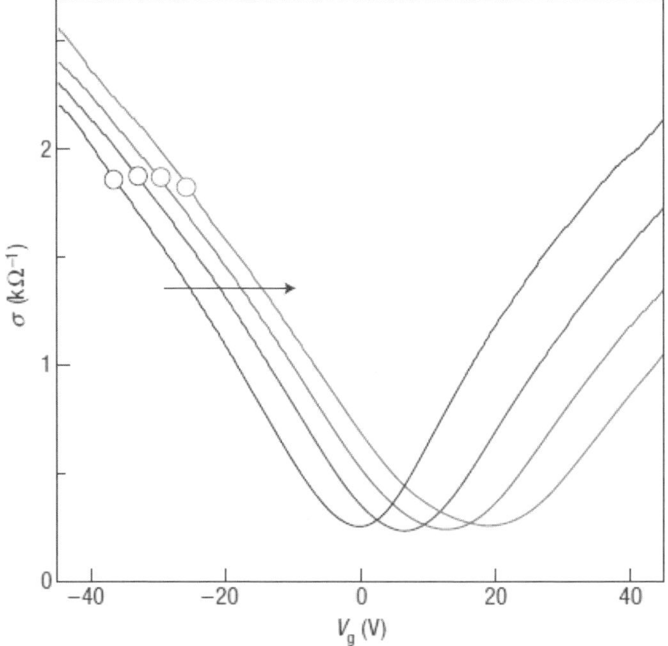

Figure 3. The influence of p-type doping on the shift of CNP. CNP was shifted from zero (black curve) to ~ +20 V (red curve) with increasing exposure of graphene to NO_2 [11]. Reprinted with permission from F. Schedin, A. K. Geim, S. V. Morozov, E. W. Hill, P. Blake, M. I. Katsnelson, and K. S. Novoselov, Nat. Mater 6 (9), 652-655 (2007).

7.4. Pristine Graphene-based Gas Sensors

Since the sensing properties of graphene were reported by Schedin *et al.* in 2007, graphene-based chemical sensors have been widely studied. Schedin *et al.* fabricated graphene-based gas sensors using mechanically exfoliated graphene and demonstrated the sensing properties of graphene towards NO_2, NH_3, H_2O, and CO, as illustrated in Figure 4[11]. In this study, the researchers exposed graphene sensors to various gases down to a concentration of 1 ppm. The gas molecules were adsorbed and then desorbed by thermal annealing at 150 °C. The authors observed step-like changes in resistivity and distinguished electron-donating molecules from electron-withdrawing molecules.

After this pioneering work, numerous studies on the sensing performance of graphene have been reported. Leenaerts *et al.* calculated the adsorption energy and charge transfer of each gas molecule including H_2O, NH_3, CO, NO, and NO_2 using density functional theory (DFT) calculations, which theoretically explained the responsivity and charge transfer of graphene upon exposure to each molecule [45]. Dan *et al.* studied the intrinsic response of graphene by removing contamination on the graphene surface. They performed an annealing process under a reducing (H_2/Ar) atmosphere to clean the surface of graphene. They found that the contamination layer acted as an adsorbent layer, which enhanced the responsivity of the sensors. After completing the cleaning, the intrinsic response of the sensors was surprisingly small upon exposure to ammonia vapor [46]. Furthermore, Ko *et al.* reported the responsivity of graphene sensors to NO_2 under ambient atmospheric conditions. The response in the current-voltage characteristics was immediate, but the recovery process was very slow. The desorption of the molecules from the graphene surface can be enhanced by illumination with high energy photons and thermal annealing under vacuum [47].

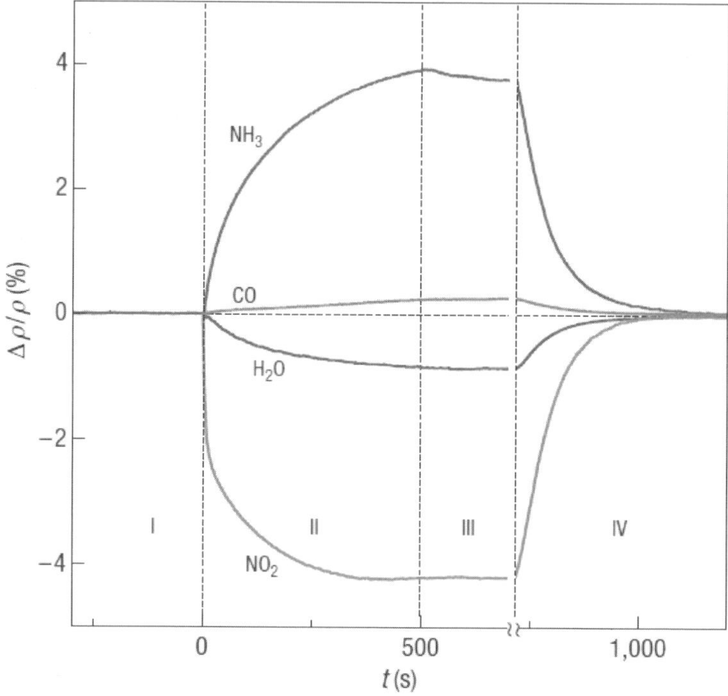

Figure 4. Sensitivity of graphene toward various gases [11]. Reprinted with permission from F. Schedin, A. K. Geim, S. V. Morozov, E. W. Hill, P. Blake, M. I. Katsnelson, and K. S. Novoselov, Nat. Mater 6 (9), 652-655 (2007).

Different methods for producing graphene have been reported, including chemical exfoliation of HOPG, epitaxial growth, and CVD. Among these techniques, CVD-grown graphene has been widely studied for fabricating gas sensors owing to its productivity and high quality. The sensitivity of CVD-grown graphene towards O_2 was reported by Chen *et al.* O_2 molecules acted as p-type dopants and the detection limit was 1.25% volume [48]. Rumyantsev *et al.* found that the behaviors of noise frequency and relative resistance changes are different for different gases. Therefore, they proposed that selective gas sensing could be possible by using these two characteristics [49]. Gautam *et al.* also investigated the response of CVD-grown graphene to NH_3, CH_4, and H_2 [50]. Flexible graphene-based gas sensors using polyethylene terephthalate and paper substrates were reported for potential applications in wearable and foldable sensors. In these studies, the graphene gas sensors showed high sensitivity under conditions of strain [51,52].

7.5. Defective Graphene-based Gas Sensors

Defects in graphene have received great attention since they are inevitable during the synthesis and subsequent handling (or fabrication) process. They are formed during the transfer processes as well as from the polymethyl methacrylate (PMMA) residues [53]. The intrinsic characteristics of graphene, such as its electrical, magnetic, and mechanical properties, can be affected by the presence of defects [54-57]. Here, we will focus on the vacancy and sp^3 defects in graphene. Figure 5 shows two types of defect regions in graphene.

The properties of graphene can typically be characterized by Raman spectroscopy. In Raman spectra (Fig. 6(a)), peak G (centered approximately at 1580 cm^{-1}) and peak G' (~2680 cm^{-1}) are known to be related to strong symmetry in the sp^2 bonding of graphene. Peaks D (~1340 cm^{-1}) and D' (~1620 cm^{-1}) are associated with disorders in graphene. The ratios of peaks D to G and D to D' imply the defect density of graphene [58]. Ferrari and Robertson classified the evolution of defect stages in carbon materials as crystalline, nanocrystalline, and amorphous states [59]. Furthermore, Eckmann *et al.* classified the evolution of defect states as sp^3- and vacancy-type dominant regions, based on the ratio of each Raman peak (Fig. 6 (b)) [55,60].

Another issue related to defects in graphene is the precise control of defect density because, when optimized, defect engineering can enhance the charge transfer between the graphene and adsorbed molecules. In previous studies, defects were experimentally introduced into the graphene using Ar-ion bombardment, electron-beam irradiation, and UV ozone treatment [61-64]. Cançado *et al.* calculated the crystal size via Raman spectral analysis and scanning tunneling microscopy (STM) according to the increase in the density of defects introduced by Ar-ion bombardment [61]. Defects induced by electron beam irradiation were studied based on the experimental results and molecular dynamic simulations [62]. Oxygen radicals produced by UV illumination creates defective sites in graphene, which were controlled by the duration of UV exposure. The change in defect density after UV treatment was measured using atomic force microscopy (AFM) [63]. Furthermore, oxygen plasma was applied to introduce defects in graphene using a conventional reactive ion etching (RIE) system. The direct bombardment of ions and plasma can accelerate the formation of defective sites in graphene. To avoid this, the configuration of graphene in the RIE chamber can be modified; the graphene was loaded face-down to avoid direct ion bombardment and to make it react only with oxygen radicals that diffuse between the posts. The etching rate was controlled by adjusting the power, oxygen flow, and exposure time [57,64].

Structural modification (or defect engineering) of graphene significantly influences the electrical response of graphene-based sensors. The reactivity and charge transfer in graphene generally increase with increasing defect density [44,65-70]. Zhang *et al.* studied the gas sensing properties of graphene using DFT calculations and reported that defective graphene with vacancies can have higher adsorption energy and exhibit greater charge transfer than pristine graphene [44]. Defects in graphene were introduced by plasma or ion irradiation (Fig. 7(a)) and that gas sensors made from it exhibited improved sensing properties toward NO_2 gas [65,70]. UV-ozone treatment has also been used to fabricate defects in graphene, and the sensing performance was examined under various conditions of ozone exposure time, NO_2 concentration, and temperature (Figs. 7(b) and 7(c)) [67]. Kumar *et al.* reported that external defects formed in the substrate affect the sensing performance of graphene sensors [68]. Paul *et al.* fabricated a graphene nanomesh using a combination of the techniques of nanosphere lithography and RIE. The graphene nanomesh structure provided a lot of defective sites and showed higher sensitivity toward NO_2 and NH_3 than pristine graphene-based sensors [69].

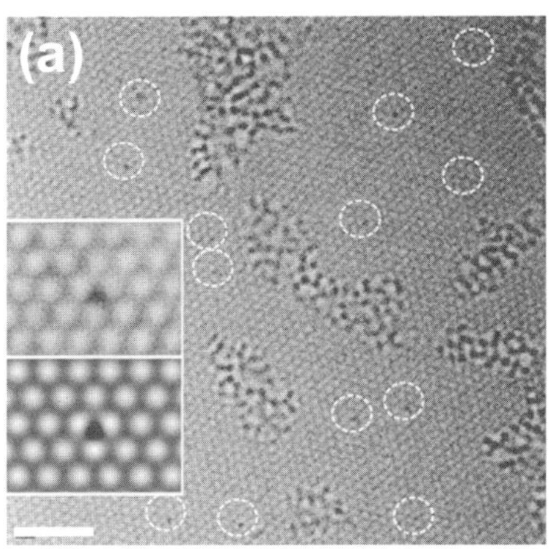

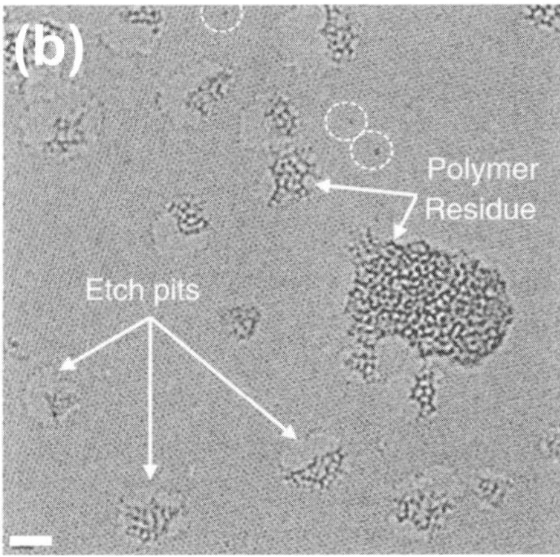

Figure 5. Defects in graphene. AC-HRTEM images of (a) sp^3-type dominant (b) vacancy (nanocavity) dominant region. The black dots circled with dashed lines in (a) and (b) indicate oxygen adatoms [57]. Reprinted with permission from A. Zandiatashbar, G.-H. Lee, S. J. An, S. Lee, N. Mathew, M. Terrones, T. Hayashi, C. R. Picu, J. Hone, and N. Koratkar, Nat Commun 5 (2014).

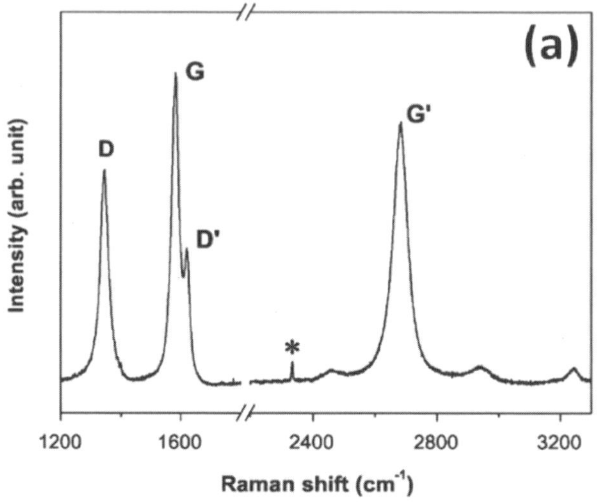

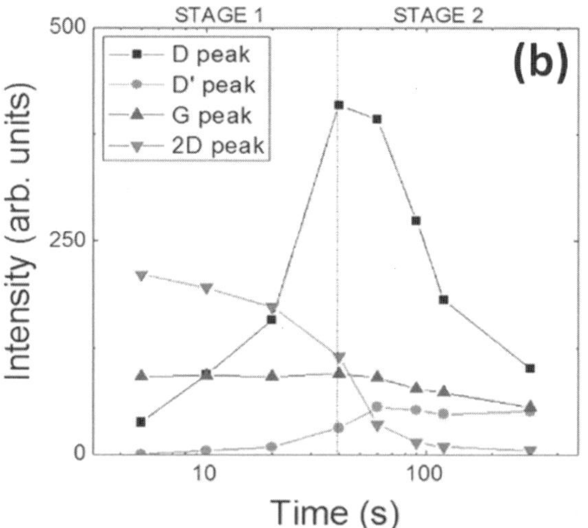

Figure 6. (a) Raman spectra of defective graphene [71]. Reprinted with permission from M. A. Pimenta, G. Dresselhaus, M. S. Dresselhaus, L. G. Cancado, A. Jorio, and R. Saito, Physical Chemistry Chemical Physics 9 (11), 1276-1290 (2007). (b) Defect states of oxidized graphene under increasing plasma exposure [55]. Reprinted with permission from A. Eckmann, A. Felten, A. Mishchenko, L. Britnell, R. Krupke, K. S. Novoselov, and C. Casiraghi, Nano Letters 12 (8), 3925-3930 (2012).

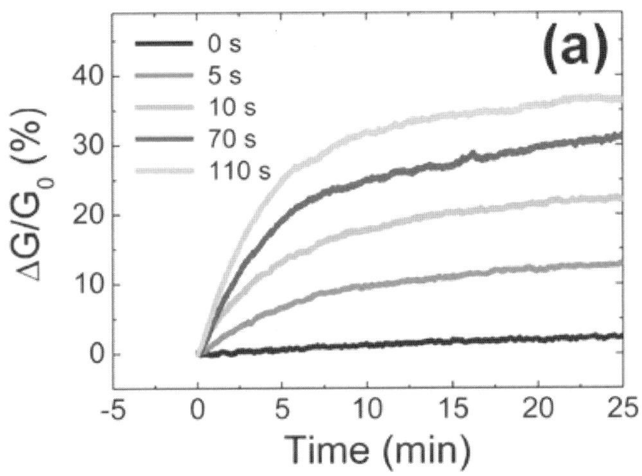
0 s
5 s
10 s
70 s
110 s
(a)
40
30
20
10
0
-5   0   5   10   15   20   25
Time (min)
$\Delta G/G_0$ (%)

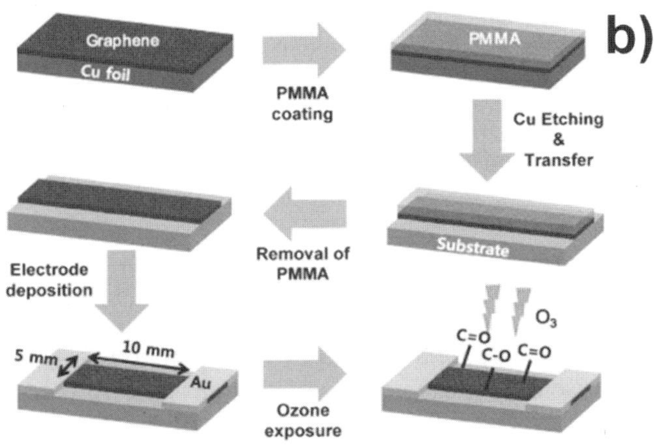
Graphene
Cu foil
PMMA
b)
PMMA coating
Cu Etching & Transfer
Substrate
Removal of PMMA
Electrode deposition
5 mm
10 mm
Au
Ozone exposure
O_3
C=O C-O C=O

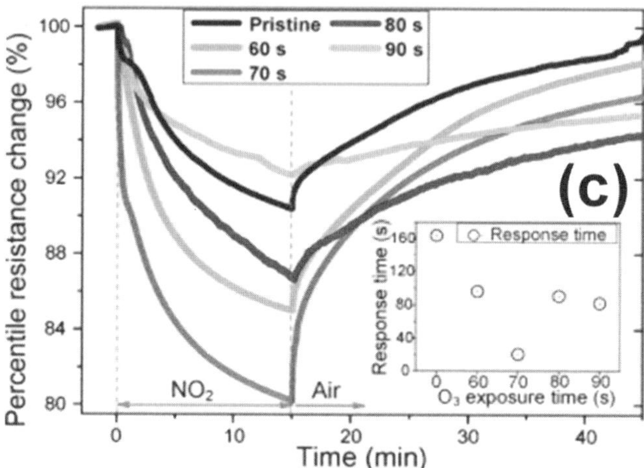

Figure 7. (a) The increased sensitivity of graphene according to time of exposure to Ar-ion plasma [65]. Reprinted with permission from K. Kim, H. Kang, C. Y. Lee, and W. S. Yun, Journal of Vacuum Science & Technology B 31 (3), 030602 (2013). (b) Schematic of the ozone treatment of graphene [67] (c) Percentile resistance change (%) of defective graphene with increasing time of ozone treatment [67]. Reprinted with permission from M. G. Chung, D. H. Kim, H. M. Lee, T. Kim, J. H. Choi, D. k. Seo, J.-B. Yoo, S.-H. Hong, T. J. Kang, and Y. H. Kim, Sensors and Actuators B: Chemical 166–167, 172-176 (2012).

7.6. Configurations for Enhancing the Sensitivity

Graphene chem-resistor sensors find a wide range of applications owing to their direct measurement capabilities and simple fabrication processes. In this type of sensors, the detection of analytes is performed by measuring the change in electrical resistance caused by the attachment and detachment of gas molecules. Field-effect transistor (FET) structures have great potential for use in graphene-based sensors [10,72,73]. Kumar *et al.* utilized FET structures in graphene sensors, which enhanced their sensitivity; they swept the back-gate voltage from −20 V to +20 V and observed that the sensitivity of sensors operating at V_g = +20 V was approximately 10 times that of sensors operated at V_g = −20 V. The recovery time varied depending on the type of chemical dopant. In the case of 1,2 dichlorobenzene (DCB), which is considered as an electron acceptor, recovery time gradually increased as V_g varied from −20 V to +20 V. On the contrary, recovery time decreased when the sensors were exposed to dimethyl methylphosphonate (DMMP), an electron donor [68].

The influence of a supporting substrate on enhancing the sensitivity of graphene sensors has also been studied. Mica substrate, a known p-type doping material to graphene, was used to enhance the sensitivity towards n-type dopants. The effects of p-doping with mica on graphene were observed by measuring the source-drain current versus gate voltage characteristics. When exposed to NH_3, the CNP of graphene on mica was largely shifted and it showed higher sensitivity than graphene on SiO_2 [74]. Suspended graphene can also be used to improve the sensitivity of graphene sensors because of a very low 1/f noise [75]. In addition, suspended structure can provide larger surface (both top and bottom surfaces) than non-suspended structure if the chemical can reach the backside of the graphene.

7.7. Graphene-based Sensors Decorated with Metal Particles

Pristine graphene is chemically inert owing to the nature of its sp^2 bonded carbon structure. Therefore, the incorporation of metal or metal oxides into graphene is an effective method for enhancing the sensitivity or selectivity. Metals such as palladium, platinum, and gold work as catalysts to dissociate gas molecules. The unique properties of graphene, such as high surface to volume ratio, high carrier mobility, and low noise level, make metal-functionalized graphene a promising candidate for chemical sensors [10,72,73,76].

Theoretical studies have been performed to understand the sensing mechanism of metal-functionalized graphene. Zhou *et al.* reported that graphene embedded with various transition metals exhibited enhanced chemical reactivity; this finding is supported by DFT calculations. DFT studies on graphene decorated with various metals have helped to reveal the mechanism of atomic and molecular gas adsorption [77].

Chung *et al.* fabricated hydrogen sensors using graphene decorated with palladium nanoparticles. They deposited palladium nanoparticles onto graphene using a thermal evaporation method and showed that sensing performance depends on the size of nanoparticles and hydrogen concentration [78]. Chu *et al.* reported that platinum-coated graphene grown on SiC exhibits higher sensitivity toward hydrogen gas than its pristine graphene counterpart. The sensing behavior of graphene decorated with gold nanoparticles has also been studied [79]. Gautam *et al.* deposited gold nanoparticles by reducing $HAuCl_4 \cdot 3H_2O$ solution onto CVD-grown graphene. NH_3 gas molecules were dissociated at the surface of Au nanoparticles, resulting in more electrons being transferred from Au nanoparticles back to the graphene. This reduced the p-type tendencies of the graphene and enhanced NH_3 sensing performance [80].

7.8. Prospects of Graphene-based Gas Sensors

Graphene has been considered as a strong candidate for solid-state chemical sensors because of its excellent properties. Its unique two-dimensional structure and remarkable electrical properties, such as high electrical conductivity and low level of noise, allow graphene to have a high response to adsorbed gas molecules. However, the commercialization of graphene-based chemical sensors poses several challenges. Firstly, more studies on the selectivity of graphene-based sensors are required. The majority of reported sensors could not distinguish specific gas molecules when exposed to a mixture of gases. Several studies claimed that selectivity could be acquired based on the sensor's sensitivity towards each gas, but it still does not work in an environment where various gases exist simultaneously. Secondly, repeatability and durability should be improved. The degradation of graphene owing to aging effects is a challenge for the commercial implementation. Finally, recovery time is relatively slow compared to other solid-state chemical sensors. Successful commercialization of graphene-based gas sensors requires a full under-standing of the sensing mechanisms and properties of graphene.

7.9. Graphene-based Sensors in Aqueous Environments

Graphene FET sensors with a solution-gate structure have been widely used to detect various species in aqueous solution because pH, metal ions, and bio-species are measured in aqueous environments. Its sensing mechanism can be explained by a combination of doping and electrostatic effects. The gate electrode, commonly composed of Ag/AgCl, is immersed in the electrolyte and the source-drain electrodes are protected by dielectric materials that prevent current leakage through the electrolyte.

A gate potential is applied through the electric double layer capacitance formed at the channel-solution interfaces. Normally, the solution-gate FET structure is known to be over two orders of magnitude more sensitive than the typical back-gate FET structures [81-85]. Ohno *et al.* studied the sensing properties of electrolyte-gate graphene FETs for pH and protein adsorption. They observed a linear increase in the Dirac point with increasing pH when non-functionalized single-layer graphene was used as the channel. Furthermore, they noticed an increase in the conductance upon exposure to a protein at concentrations of several hundred picomolar. Solution-gated epitaxial graphene was also used as a pH sensor and showed similar results with increasing pH [83]. Maehashi *et al.* fabricated selective ion sensors using ionophore-modified graphene. They observed a shift in the Dirac point, depending on the concentration of K^+ or Na^+ ions, on graphene coated with valinomycin, which acts as an ionophore. Researchers showed that the Dirac point was negatively shifted upon increasing the concentration of K^+ ions, while no changes were observed upon increasing the concentration of Na^+ ions [85]. Zhang *et al.* arrayed ion sensitive sensors (based on layer-by-layer self-assembled graphene) with various ionophores. Each sensor showed good selectivity in solutions of NaCl, KCL, $CaCl_2$, and HCl. This study also reported fast response times of around 15 s for 1 mM of different ions [84].

Nucleic acid graphene sensors for the detection of selected nucleic acid sequences related to human diseases can offer high selectivity and sensitivity along with low fabrication cost. Nanopore structures have great potential in nucleic acid detection because they are label- and amplification-free and provide a single-molecule approach [86]. In addition, graphene is suitable for the fabrication of biosensors due to its biocompatibility [87-89]. Drndic *et al.* first created the nanopore arrays in suspended graphene films using a focused electron beam technique [90]. Using this method, they fabricated a nanopore on CVD-grown few-layer graphene (3-15 layers) and measured electrical properties when DNA sequences translocated through a graphene nanopore [91].

Golovchenko *et al.* developed a graphene-based trans-electrode membrane using mono-layer graphene, as shown in Fig. 8 (a) [92]. The electron beam drilling method was used to create a nanopore on the graphene sheet (Inset of Fig. 8 (a)). The membrane separates two ionic solutions in contact with Ag/AgCl electrodes. The conductance of the graphene membrane containing the nanopore was increased by orders of magnitude comparable to that of an as-prepared membrane sample (Fig. 8 (b)). Bashir *et al.* reported a stacked graphene-Al_2O_3 nanopore sensor for detecting a protein-DNA complex that demonstrated high sensitivity, mechanical robustness, and low electrical noise [93]. Graphene has demonstrated great successes in biological sensing due to its unique properties including bio-compatibility, large surface area, and impermeability. Graphene-based biosensors exhibit excellent performance in terms of sensitivity, selectivity, and resolution. However, more studies are necessary to overcome the issues of reproducibility and selectivity.

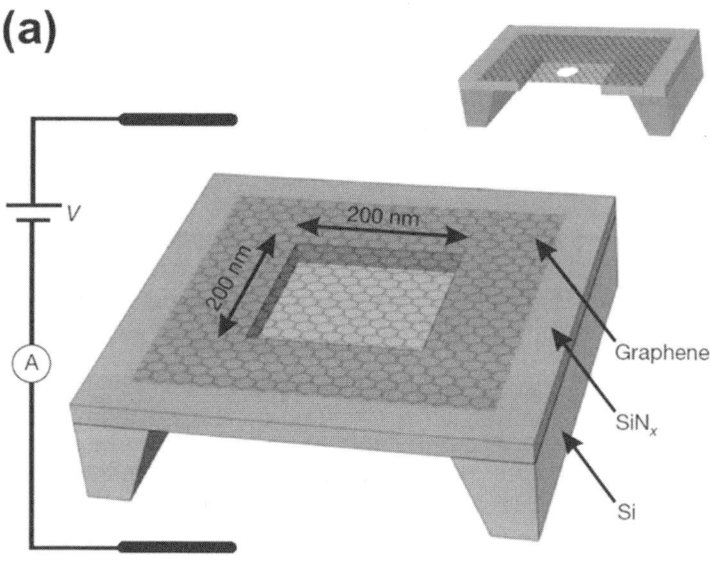

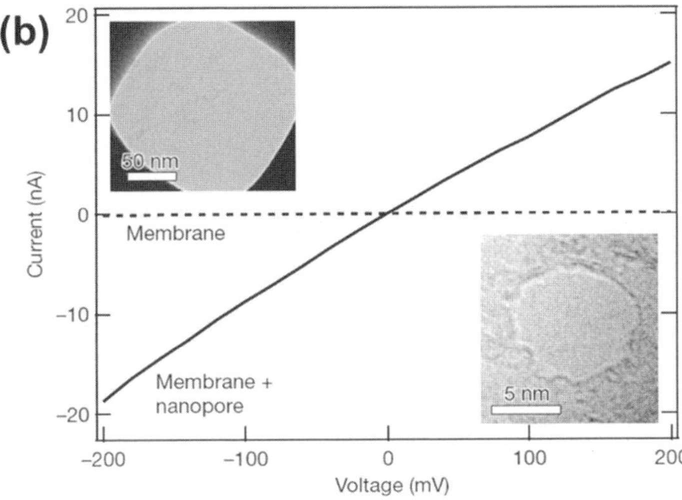

Figure 8. (a) Schematic diagram of a graphene membrane mounted on an SiN$_x$/Si frame with Ag/AgCl electrodes and two ionic solutions (not shown). (b) I-V characteristics for an as-grown graphene membrane (dashed line) and a membrane with an 8-nm pore (solid line) [92]. Reprinted with permission from S. Garaj, W. Hubbard, A. Reina, J. Kong, D. Branton, and J. A. Golovchenko, Nature 467 (7312), 190-193 (2010).

References

1. K. S. Novoselov, A. K. Geim, S. V. Morozov, D. Jiang, Y. Zhang, S. V. Dubonos, I. V. Grigorieva, and A. A. Firsov, Science **306** (5696), 666-669 (2004).
2. K. S. Novoselov, A. K. Geim, S. V. Morozov, D. Jiang, M. I. Katsnelson, I. V. Grigorieva, S. V. Dubonos, and A. A. Firsov, Nature **438** (7065), 197-200 (2005).
3. J.-H. Chen, C. Jang, S. Xiao, M. Ishigami, and M. S. Fuhrer, Nat Nano **3** (4), 206-209 (2008).
4. C. N. R. Rao, A. K. Sood, K. S. Subrahmanyam, and A. Govindaraj, Angewandte Chemie International Edition **48** (42), 7752-7777 (2009).
5. H. K. Chae, D. Y. Siberio-Perez, J. Kim, Y. Go, M. Eddaoudi, A. J. Matzger, M. O'Keeffe, and O. M. Yaghi, Nature **427** (6974), 523-527 (2004).
6. C. N. R. Rao, A. K. Sood, R. Voggu, and K. S. Subrahmanyam, The Journal of Physical Chemistry Letters **1** (2), 572-580 (2010).
7. B. Das, R. Voggu, C. S. Rout, and C. N. R. Rao, Chemical Communications (41), 5155-5157 (2008).

8. C. N. R. Rao and R. Voggu, Materials Today **13** (9), 34-40 (2010).
9. G. Srinivas, Y. Zhu, R. Piner, N. Skipper, M. Ellerby, and R. Ruoff, Carbon **48** (3), 630-635 (2010).
10. F. Yavari and N. Koratkar, The Journal of Physical Chemistry Letters **3** (13), 1746-1753 (2012).
11. F. Schedin, A. K. Geim, S. V. Morozov, E. W. Hill, P. Blake, M. I. Katsnelson, and K. S. Novoselov, Nat Mater **6** (9), 652-655 (2007).
12. A. K. Geim and K. S. Novoselov, Nat Mater **6** (3), 183-191 (2007).
13. W. S. Hummers and R. E. Offeman, Journal of the American Chemical Society **80** (6), 1339-1339 (1958).
14. G. Eda, G. Fanchini, and M. Chhowalla, Nat Nano **3** (5), 270-274 (2008).
15. S. Park, J. An, I. Jung, R. D. Piner, S. J. An, X. Li, A. Velamakanni, and R. S. Ruoff, Nano Letters **9** (4), 1593-1597 (2009).
16. V. C. Tung, M. J. Allen, Y. Yang, and R. B. Kaner, Nat Nano **4** (1), 25-29 (2009).
17. D. C. Marcano, D. V. Kosynkin, J. M. Berlin, A. Sinitskii, Z. Sun, A. Slesarev, L. B. Alemany, W. Lu, and J. M. Tour, ACS Nano **4** (8), 4806-4814 (2010).
18. D. R. Dreyer, S. Park, C. W. Bielawski, and R. S. Ruoff, Chemical Society Reviews **39** (1), 228-240 (2010).
19. S. Stankovich, D. A. Dikin, R. D. Piner, K. A. Kohlhaas, A. Kleinhammes, Y. Jia, Y. Wu, S. T. Nguyen, and R. S. Ruoff, Carbon **45** (7), 1558-1565 (2007).
20. X. Zhou, J. Zhang, H. Wu, H. Yang, J. Zhang, and S. Guo, The Journal of Physical Chemistry C **115** (24), 11957-11961 (2011).
21. C. Zhu, S. Guo, Y. Fang, and S. Dong, ACS Nano **4** (4), 2429-2437 (2010).
22. V. H. Pham, T. V. Cuong, T.-D. Nguyen-Phan, H. D. Pham, E. J. Kim, S. H. Hur, E. W. Shin, S. Kim, and J. S. Chung, Chemical Communications **46** (24), 4375-4377 (2010).
23. G. Wang, J. Yang, J. Park, X. Gou, B. Wang, H. Liu, and J. Yao, The Journal of Physical Chemistry C **112** (22), 8192-8195 (2008).
24. C. A. Amarnath, C. E. Hong, N. H. Kim, B.-C. Ku, T. Kuila, and J. H. Lee, Carbon **49** (11), 3497-3502 (2011).
25. J. Zhang, H. Yang, G. Shen, P. Cheng, J. Zhang, and S. Guo, Chemical Communications **46** (7), 1112-1114 (2010).
26. M. Zhou, Y. Wang, Y. Zhai, J. Zhai, W. Ren, F. Wang, and S. Dong, Chemistry — A European Journal **15** (25), 6116-6120 (2009).
27. X. Li, W. Cai, J. An, S. Kim, J. Nah, D. Yang, R. Piner, A. Velamakanni, I. Jung, E. Tutuc, S. K. Banerjee, L. Colombo, and R. S. Ruoff, Science **324** (5932), 1312-1314 (2009).
28. S.-Y. Kwon, C. V. Ciobanu, V. Petrova, V. B. Shenoy, J. Bareño, V. Gambin, I. Petrov, and S. Kodambaka, Nano Letters **9** (12), 3985-3990 (2009).
29. P. W. Sutter, J.-I. Flege, and E. A. Sutter, Nat Mater **7** (5), 406-411 (2008).
30. J. Coraux, A. T. N'Diaye, C. Busse, and T. Michely, Nano Letters **8** (2), 565-570 (2008).
31. K. S. Kim, Y. Zhao, H. Jang, S. Y. Lee, J. M. Kim, K. S. Kim, J.-H. Ahn, P. Kim, J.-Y. Choi, and B. H. Hong, Nature **457** (7230), 706-710 (2009).

32. J. J. Wang, M. Y. Zhu, R. A. Outlaw, X. Zhao, D. M. Manos, B. C. Holloway, and V. P. Mammana, Applied Physics Letters **85** (7), 1265-1267 (2004).

33. A. Dato, V. Radmilovic, Z. Lee, J. Phillips, and M. Frenklach, Nano Letters **8** (7), 2012-2016 (2008).

34. Z. Peng, Z. Yan, Z. Sun, and J. M. Tour, ACS Nano **5** (10), 8241-8247 (2011).

35. C.-Y. Su, A.-Y. Lu, C.-Y. Wu, Y.-T. Li, K.-K. Liu, W. Zhang, S.-Y. Lin, Z.-Y. Juang, Y.-L. Zhong, F.-R. Chen, and L.-J. Li, Nano Letters **11** (9), 3612-3616 (2011).

36. B. Laurent, H. Zhanbing, L. Chang Seok, M. Jean-Luc, C. Costel Sorin, G.-L. Anne-Françoise, L. Young Hee, and P. Didier, Nanotechnology **22** (8), 085601 (2011).

37. G. Gutierrez, F. Le Normand, D. Muller, F. Aweke, C. Speisser, F. Antoni, Y. Le Gall, C. S. Lee, and C. S. Cojocaru, Carbon **66**, 1-10 (2014).

38. J. Kim, G. Lee, and J. Kim, Applied Physics Letters **107** (3), 033104 (2015).

39. C. Berger, Z. Song, X. Li, X. Wu, N. Brown, C. Naud, D. Mayou, T. Li, J. Hass, A. N. Marchenkov, E. H. Conrad, P. N. First, and W. A. de Heer, Science **312** (5777), 1191-1196 (2006).

40. N. Ferralis, R. Maboudian, and C. Carraro, Physical Review Letters **101** (15), 156801 (2008).

41. J. Hass, W. A. d. Heer, and E. H. Conrad, Journal of Physics: Condensed Matter **20** (32), 323202 (2008).

42. H. Huang, W. Chen, S. Chen, and A. T. S. Wee, ACS Nano **2** (12), 2513-2518 (2008).

43. SprinkleM, RuanM, HuY, HankinsonJ, M. Rubio Roy, ZhangB, WuX, BergerC, and W. A. de Heer, Nat Nano **5** (10), 727-731 (2010).

44. Y.-H. Zhang, Y.-B. Chen, Z. Kai-Ge, L. Cai-Hong, Z. Jing, Z. Hao-Li, and P. Yong, Nanotechnology **20** (18), 185504.

45. O. Leenaerts, B. Partoens, and F. M. Peeters, Physical Review B **77** (12), 125416 (2008).

46. Y. Dan, Y. Lu, N. J. Kybert, Z. Luo, and A. T. C. Johnson, Nano Letters **9** (4), 1472-1475 (2009).

47. G. Ko, H. Y. Kim, J. Ahn, Y. M. Park, K. Y. Lee, and J. Kim, Current Applied Physics **10** (4), 1002-1004 (2010).

48. C. W. Chen, S. C. Hung, M. D. Yang, C. W. Yeh, C. H. Wu, G. C. Chi, F. Ren, and S. J. Pearton, Applied Physics Letters **99** (24), 243502 (2011).

49. S. Rumyantsev, G. Liu, M. S. Shur, R. A. Potyrailo, and A. A. Balandin, Nano Letters **12** (5), 2294-2298 (2012).

50. M. Gautam and A. H. Jayatissa, Materials Science and Engineering: C **31** (7), 1405-1411 (2011).

51. C. Lee, J. Ahn, K. B. Lee, D. Kim, and J. Kim, Thin Solid Films **520** (16), 5459-5462 (2012).

52. G. Yang, C. Lee, J. Kim, F. Ren, and S. J. Pearton, Physical Chemistry Chemical Physics **15** (6), 1798-1801 (2013).

53.   A. Pirkle, J. Chan, A. Venugopal, D. Hinojos, C. W. Magnuson, S. McDonnell, L. Colombo, E. M. Vogel, R. S. Ruoff, and R. M. Wallace, Applied Physics Letters **99** (12), 122108 (2011).

54.   F. Banhart, J. Kotakoski, and A. V. Krasheninnikov, ACS Nano **5** (1), 26-41 (2011).

55.   A. Eckmann, A. Felten, A. Mishchenko, L. Britnell, R. Krupke, K. S. Novoselov, and C. Casiraghi, Nano Letters **12** (8), 3925-3930 (2012).

56.   O. V. Yazyev and L. Helm, Physical Review B **75** (12), 125408 (2007).

57.   A. Zandiatashbar, G.-H. Lee, S. J. An, S. Lee, N. Mathew, M. Terrones, T. Hayashi, C. R. Picu, J. Hone, and N. Koratkar, Nat Commun **5** (2014).

58.   H. Raza, *Graphene Nanoelectronics*. (Springer, 2012).

59.   A. C. Ferrari and J. Robertson, Physical Review B **64** (7), 075414 (2001).

60.   A. Eckmann, A. Felten, I. Verzhbitskiy, R. Davey, and C. Casiraghi, Physical Review B **88** (3), 035426 (2013).

61.   L. G. Cançado, A. Jorio, E. H. M. Ferreira, F. Stavale, C. A. Achete, R. B. Capaz, M. V. O. Moutinho, A. Lombardo, T. S. Kulmala, and A. C. Ferrari, Nano Letters **11** (8), 3190-3196 (2011).

62.   M. Yamamoto, Y. Asayama, M. Yasuda, H. Kawata, and Y. Hirai, Journal of Vacuum Science & Technology B **32** (6), 06FK01 (2014).

63.   H. Tao, J. Moser, F. Alzina, Q. Wang, and C. M. Sotomayor-Torres, The Journal of Physical Chemistry C **115** (37), 18257-18260 (2011).

64.   G. Lee, J. Kim, K. Kim, and J. W. Han, Journal of Vacuum Science & Technology A **33** (6), 060602 (2015).

65.   K. Kim, H. Kang, C. Y. Lee, and W. S. Yun, Journal of Vacuum Science & Technology B **31** (3), 030602 (2013).

66.   Y.-B. Zhou, Z.-M. Liao, Y.-F. Wang, G. S. Duesberg, J. Xu, Q. Fu, X.-S. Wu, and D.-P. Yu, The Journal of Chemical Physics **133** (23), 234703 (2010).

67.   M. G. Chung, D. H. Kim, H. M. Lee, T. Kim, J. H. Choi, D. k. Seo, J.-B. Yoo, S.-H. Hong, T. J. Kang, and Y. H. Kim, Sensors and Actuators B: Chemical **166–167**, 172-176 (2012).

68.   B. Kumar, K. Min, M. Bashirzadeh, A. B. Farimani, M. H. Bae, D. Estrada, Y. D. Kim, P. Yasaei, Y. D. Park, E. Pop, N. R. Aluru, and A. Salehi-Khojin, Nano Letters **13** (5), 1962-1968 (2013).

69.   R. K. Paul, S. Badhulika, N. M. Saucedo, and A. Mulchandani, Analytical Chemistry **84** (19), 8171-8178 (2012).

70.   Y. Hajati, T. Blom, S. H. M. Jafri, S. Haldar, S. Bhandary, M. Z. Shoushtari, O. Eriksson, B. Sanyal, and K. Leifer, Nanotechnology **23** (50), 505501 (2012).

71.   M. A. Pimenta, G. Dresselhaus, M. S. Dresselhaus, L. G. Cancado, A. Jorio, and R. Saito, Physical Chemistry Chemical Physics **9** (11), 1276-1290 (2007).

72.   T. Gan and S. Hu, Microchim Acta **175** (1-2), 1-19 (2011).

73.   M. Pumera, A. Ambrosi, A. Bonanni, E. L. K. Chng, and H. L. Poh, TrAC Trends in Analytical Chemistry **29** (9), 954-965 (2010).

74.   Z. Ben Aziza, Q. Zhang, and D. Baillargeat, Applied Physics Letters **105** (25), 254102 (2014).

75. Z. Cheng, Q. Li, Z. Li, Q. Zhou, and Y. Fang, Nano Letters **10** (5), 1864-1868 (2010).

76. S. Wu, Q. He, C. Tan, Y. Wang, and H. Zhang, Small **9** (8), 1160-1172 (2013).

77. Z. Miao, L. Yun-Hao, C. Yong-Qing, Z. Chun, and F. Yuan-Ping, Nanotechnology **22** (38), 385502 (2011).

78. M. G. Chung, D.-H. Kim, D. K. Seo, T. Kim, H. U. Im, H. M. Lee, J.-B. Yoo, S.-H. Hong, T. J. Kang, and Y. H. Kim, Sensors and Actuators B: Chemical **169**, 387-392 (2012).

79. B. H. Chu, J. Nicolosi, C. F. Lo, W. Strupinski, S. J. Pearton, and F. Ren, Electrochemical and Solid-State Letters **14** (7), K43-K45 (2011).

80. M. Gautam and A. H. Jayatissa, Solid-State Electronics **78**, 159-165 (2012).

81. W. Yuan and G. Shi, Journal of Materials Chemistry A **1** (35), 10078-10091 (2013).

82. P. K. Ang, W. Chen, A. T. S. Wee, and K. P. Loh, Journal of the American Chemical Society **130** (44), 14392-14393 (2008).

83. Y. Ohno, K. Maehashi, Y. Yamashiro, and K. Matsumoto, Nano Letters **9** (9), 3318-3322 (2009).

84. B. Zhang and T. Cui, Sensors and Actuators A: Physical **177**, 110-114 (2012).

85. K. Maehashi, Y. Sofue, S. Okamoto, Y. Ohno, K. Inoue, and K. Matsumoto, Sensors and Actuators B: Chemical **187**, 45-49 (2013).

86. B. M. Venkatesan and R. Bashir, Nat Nano **6** (10), 615-624 (2011).

87. H. Chen, M. B. Müller, K. J. Gilmore, G. G. Wallace, and D. Li, Advanced Materials **20** (18), 3557-3561 (2008).

88. N. Mohanty and V. Berry, Nano Letters **8** (12), 4469-4476 (2008).

89. O. Akhavan and E. Ghaderi, The Journal of Physical Chemistry C **113** (47), 20214-20220 (2009).

90. M. D. Fischbein and M. Drndić, Applied Physics Letters **93** (11), 113107 (2008).

91. C. A. Merchant, K. Healy, M. Wanunu, V. Ray, N. Peterman, J. Bartel, M. D. Fischbein, K. Venta, Z. Luo, A. T. C. Johnson, and M. Drndić, Nano Letters **10** (8), 2915-2921 (2010).

92. S. Garaj, W. Hubbard, A. Reina, J. Kong, D. Branton, and J. A. Golovchenko, Nature **467** (7312), 190-193 (2010).

93. B. M. Venkatesan, D. Estrada, S. Banerjee, X. Jin, V. E. Dorgan, M.-H. Bae, N. R. Aluru, E. Pop, and R. Bashir, ACS Nano **6** (1), 441-450 (2012).

Chapter 8

Electronic Micro-Sensors for Metabolite Detection Based on Conductivity Change of Polyaniline

Yu-Lin Wang, Kuan-Chung Fang, Chia-Ho Chu, Jung-Ying Fang, Indu Sarangadharan

Institute of Nanoengineering and Microsystems,
National Tsing Hua University
Hsin Chu, Taiwan 30013

8.1. Introduction

Electronic micro-sensors are attractive due to their small size, low weight, low cost, high sensitivity and electrical output signals, which are suitable for portable devices or point-of-care testing (POCT). Conducting polymers, such as polyaniline, has been widely used as electron transfer materials in the electrodes of electrochemical sensors. There are some unique advantages when using conducting polymer for biosensors, including low cost, simple protein immobilization, adaptive process, adjustable conductivity, and good biocompatibility. Immobilization of proteins, such as enzymes, in the polymer is usually done by encapsulating proteins physically with the porous structure of the polymer film, and the electrostatic interaction between the protein and the functional groups of the polymer. The physical immobilization of protein in polymer is very strong and much simpler compared to chemical protein immobilization. The force between proteins and the polymer is very strong and steady. Once proteins are trapped in the pores of polymer, it is very difficult to wash the proteins away. This can be explained as the protein is trapped in 3-dimensional space, which is much more efficient than 2-dimensional immobilization (surface immobilization). That is why for many solid materials, which can only be done with surface immobilization, has to do chemical immobilization with covalent bonding, because physical immobilization on surface is

245

relatively weak and un-stable. However, the chemical protein immobilization is not convenient, and may have risk to change the property of proteins. Protein encapsulated in 3-dimensional space brings another advantage, that is, the charge transfer efficiency between the protein (enzyme) and the polymer is better than that of 2-dimensional surface because of the larger contact area between protein and the polymer in 3-dimensional space. More contact area gives higher probability for electrons to transfer.

Due to the above-mentioned reasons, enzymes are very suitable to be used with conducting polymer, because many enzyme reactions are either oxidative or reductive reactions. Enzymes may lose or gain electrons when they catalyze biological reactions. In electrochemical sensors, enzymes are immobilized in conducting polymer and catalyze the oxidation or reduction of metabolites. The oxidative or reductive current level can represent the concentration of the target metabolic molecule. Improvement of the charge transfer rate between the metabolic molecule and the electrode is obviously critical for the sensitivity of the conventionally voltammetric electrochemical sensor. Voltammetry is widely used for electrochemical sensors. The signals are based on the highest oxidative or reductive current, which means that the "rate" of electron transfer dominates the amplitude of the signals. On the other hand, the signal of the sensor may also be limited to the rate in this methodology. If enzyme reactions are spontaneous, external bias will not be needed to activate the reactions. For these spontaneous reactions, the conductivity of the polymer will be changed because external bias is not provided to the polymer. The reactions stop when the conducting polymer is not bale to provide more changes to the enzymes. In this different methodology, the conducting polymer becomes the transducer, giving the signals as the conductivity change of itself. It turns out that the signals can be accumulated because every reaction between the target molecules and the enzymes can be accumulated and does not reverse, without external bias. The sensitivity of the sensor can be amplified by collecting signals with "accumulation", instead of the "rate" of charge transfer. Ultra-sensitivity of the sensor can be realized with the new method.

Polyaniline has good chemical and electrical stability. Polyaniline has three different forms, including fully reduced form (leucoemeraldine), conducting form (emeraldine salt), and fully oxidized form (pernigraniline), as shown in Fig. 2. Only the emeraldine salt is conductive. Hydrogen proton delocalizes the electron and turns the emeradine base become conductive. Hydrogen proton is doped to increase the conductivity with acid. Not only protonation, but also sultonation can make the emeraldine base become conductive. Propane sultone is used frequently to dope polyaniline. Sultonated emeraldine salt is electrically steady in neutral pH.

In this chapter, PANI was spin-coated on Au electrodes on a Si_3N_4/Si substrate, modified with propane sultone (N-Alkylated PANI), and then immobilized with HRP for hydrogen peroxide detection. Cholesterol oxidase is immobilized in dialysis membrane placed on the hydrogen peroxide sensor for cholesterol detection. Finally, the conductive polyaniline is used to detect hydroxyl radicals. Polyaniline is a good radical scavenger. The conductivity change due to the interaction between the radicals and the polymer can be used to monitor the concentration of radicals in solutions.

8.2. Ultra-Sensitivity Hydrogen Peroxide Sensors

8.2.1. Detection of Hydrogen Peroxide

Hydrogen peroxide is an important signaling molecule for many biological reactions (Giorgio *et al.*, 2007; D'Autréaux *et al.*, 2007). Hydrogen peroxide is also one of the reactive oxygen species (ROS), which is usually monitored for oxidative stress. Oxidative stress is currently an important research topic and is believed to be relevant to tumors, cancers, parkinsons disease, and aging (Burns *et al.*, 2012). However, the concentration of hydrogen peroxide in biological systems is quite low. The intramitochondrial H_2O_2 concentration is estimated as 4.8 nM only (Cadenas *et al.*, 2000). Hydrogen peroxide is also a by-product in many enzymatic reactions, such as glucose oxidase, lactate oxidase, cholesterol oxidase, alcohol oxidase, urate oxidase, aldehyde oxidase, and oxalate oxidase, which are implemented with hydrogen peroxide sensor to detect glucose, lactic acid, cholesterol, ethanol, urea, formaldehyde, and oxalate, respectively. There is an increasing demand on the development of cheap and disposable biosensors to detect hydrogen peroxide for characterizing these metabolic biomarkers in point-of-care health monitoring. Therefore, it is required to develop highly sensitive H_2O_2 sensor for these applications. Conventional approaches, including fluorimetry (Albers *et al.*, 2006), chemiluminescence (Yang *et al.*, 2009), spectrometry (Matsubara *et al.*, 1992), high performance liquid phase (HPLC) chromatography (Stainberg, 2013), and electrochemistry (Chen *et al.*, 2012), were used to detect H_2O_2. However, they are only used in laboratory. Instead, electronic biosensors, including the electrochemical and resistive sensors, are much cheap, small and thus suitable for disposable sensors, thanks to the mature microfabrication techniques. It is demanding to develop cheap and disposable electronic microsensors with high sensitivity, which can detect hydrogen peroxide in nM or even sub-nM. However, many H_2O_2 sensors can only detect μM of H_2O_2, and few can detect nM or sub-nM of H_2O_2 (Chen *et al.*, 2013; Ahammad, 2013). HRP-immobilized PANI (Bartlett *et al.*, 1996; Gorton *et al.*, 1992; Bartlett *et al.*, 1998; Solanki *et al.*, 2011; Sheng *et al.*, 2011) and organic thin film transistors (Royer *et al.*, 2012) were used to detect H_2O_2. PANI provides an efficient surface for the direct electrochemical reduction of HRP. The detection technology in which the oxidation of HRP results from the reduction of H_2O_2, hence leading to the oxidation of PANI, then resulting in the decrease of PANI conductivity, has been reported in literature (Bartlett *et al.*, 1996; Gorton *et al.*, 1992; Bartlett *et al.*, 1998). Doping PANI to ensure the high conductivity is often conducted with hydrogen proton by adding acids. However, in most biological environments, pH value is usually near neutral, which causes low conductivity of PANI. Electrosynthesized N-Alkylated PANI has shown good conductivity at neutral pH and been fabricated as a microelectrochemical enzyme transistor, which shows detection limit less than 1 ppm (25 μM) of H_2O_2 (Raffa *et al.*, 1995). Electrochemical sensors usually work in chronometry or amperometry for the detection of H_2O_2 (Raffa *et al.*, 1995). However, in this prepared sensors, the conductance change of PANI depends on (i) the spontaneous reaction of the oxidation of PANI via HRP-catalyzed enzymatic reaction without any external bias, and (ii) the full consumption of hydrogen peroxide, to characterize the concentrations of H_2O_2 in buffer solution.

8.2.2. Fabrication of Hydrogen Peroxide Sensors

Polyaniline emeraldine base 0.3 g was dissolved in 5 ml of dimethyl sulfoxide (DMSO) with stirring for 6 hours. The polyaniline solution was then mixed with the same volume of 0.5 M sulfuric acid for 24 hours. The sulfuric acid can increase the conductivity and the stability of the PANI thin film (Chen and Hwang, 1995). After that, 1.5 μl of the PANI solution was dropped using a micro pipette and spin-coated, followed by baking at 60°C for 30 minutes in air. The microchip consists of two metal electrodes made by 200 Å Ti and 1000 Å Au deposited with an e-beam evaporator on a Si_3N_4/Si substrate. The length and the width of the Au electrodes are 500 μm and 100 μm, respectively. The separation between the two metal electrodes is 10 μm. The PANI/Au interface was confirmed to be ohmic by measuring the current-voltage characteristics of the device. Propanesultone was dropped on the PANI film and allowed to wait for 8 hours. The device was then washed with DI water to remove the excess of propanesultone. The HRP enzyme was prepared in 120 units cm^{-3} in a citrate phosphate buffer solution (pH=5.5), with 25 mM of 1,4-diaminobenzene. The prepared device was then placed in the HRP enzyme solution for 20 minutes. The sensor was then washed with DI water. Figure 1(a) presents the schematic of the H_2O_2 sensor, and the HRP-catalyzed oxidation of PANI by hydrogen peroxide. Figure 1(b) shows the photograph of a packaged sensor.

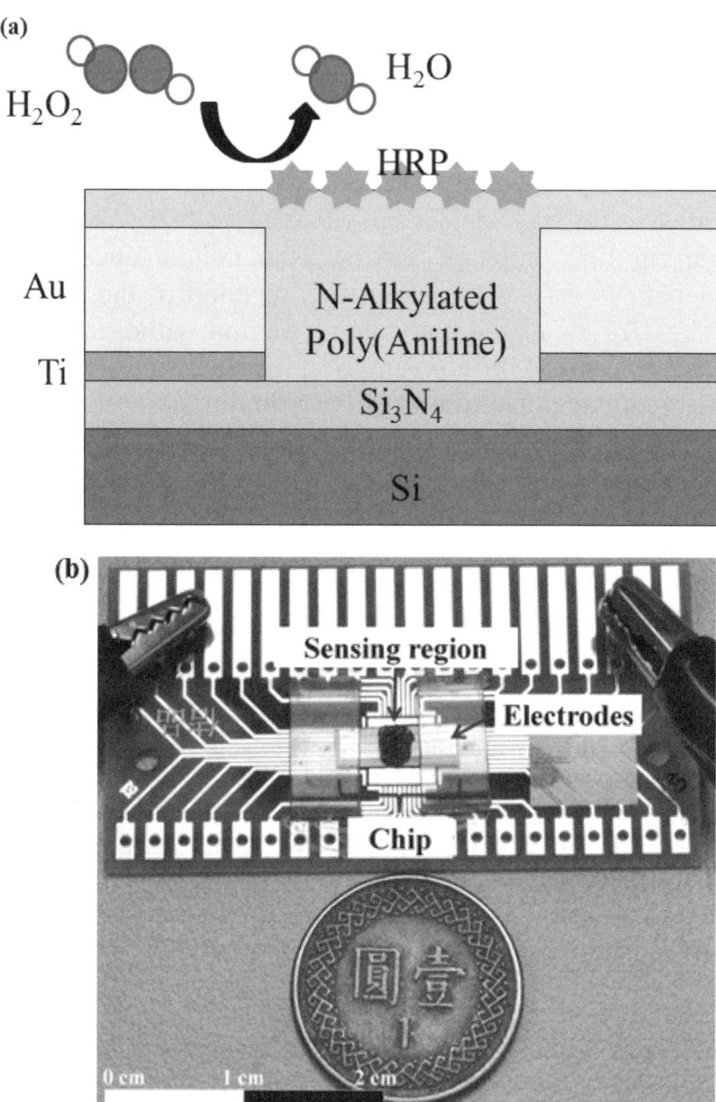

Figure 1. (a) Schematic of the hydrogen peroxide sensor and the enzymatic reactions (b) plan-view photograph of a packaged hydrogen peroxide sensor.

8.2.3. Detection Range of the Hydrogen Peroxide Sensors

Figure 2(a), 2(b), 2(c), and 2(d) show the real time detection of 10 µl of 0.1 nM, 1 nM, 10 nM, and 100 nM H_2O_2, respectively, at a constant bias of 100 mV with fresh sensors. The currents of the sensors were measured every minute during the detection. When 10 µl of 0.1 nM H_2O_2 was dropped onto the surface of the PANI, there was no significant current change, as shown in Figure 2(a). In sharp contrast, when 10 µl of 1 nM H_2O_2 was dropped on a fresh sensor, a sharp current change was observed in one minute as the system reached a steady state, as shown in Figure 2(b). The current change keeps increasing as the concentration of H_2O_2 increases until 1 µM. Beyond 1 µM, the current change would not increase significantly, as shown in Figure 3.

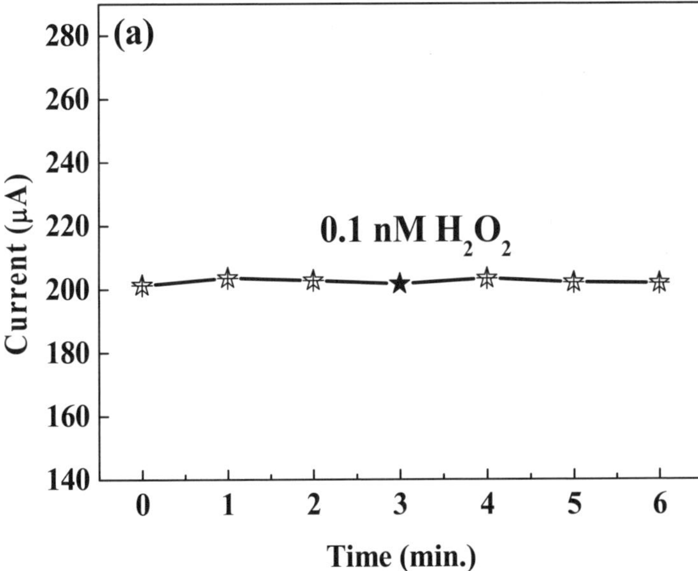

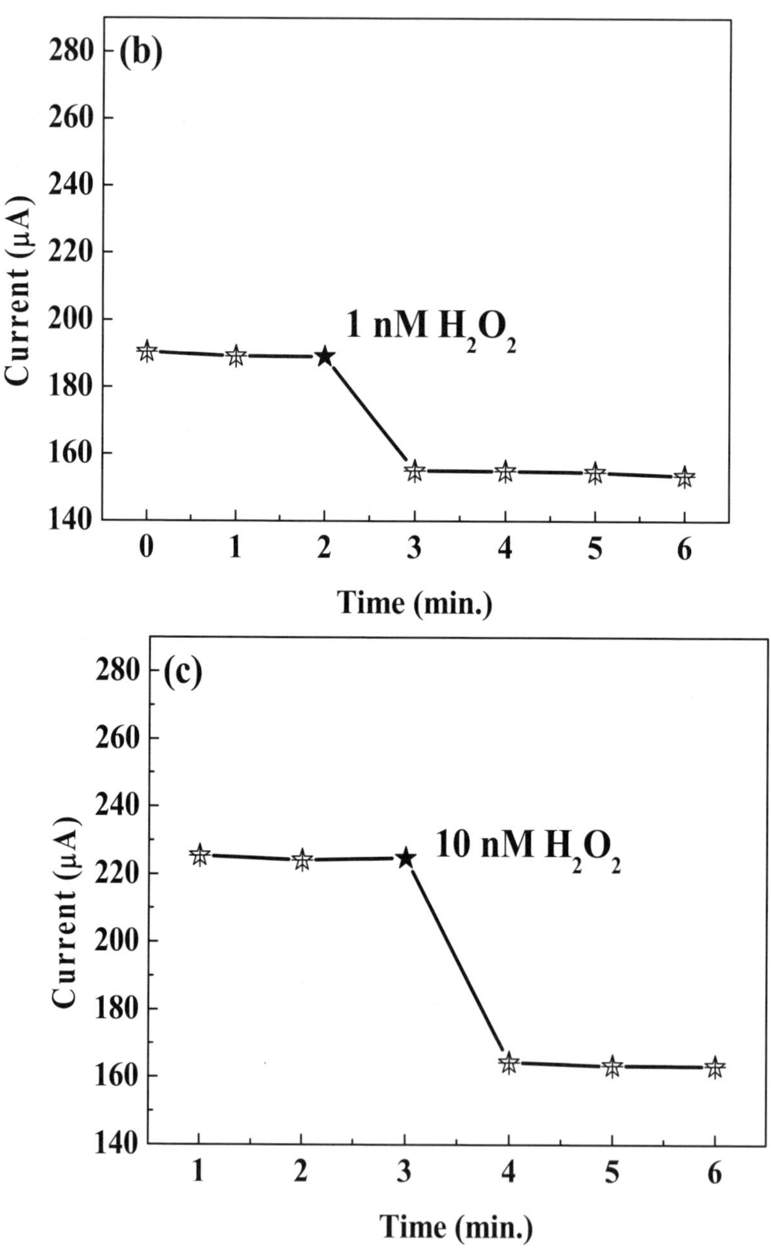

(b)
1 nM H$_2$O$_2$
Current (μA)
Time (min.)
(c)
10 nM H$_2$O$_2$
Current (μA)
Time (min.)

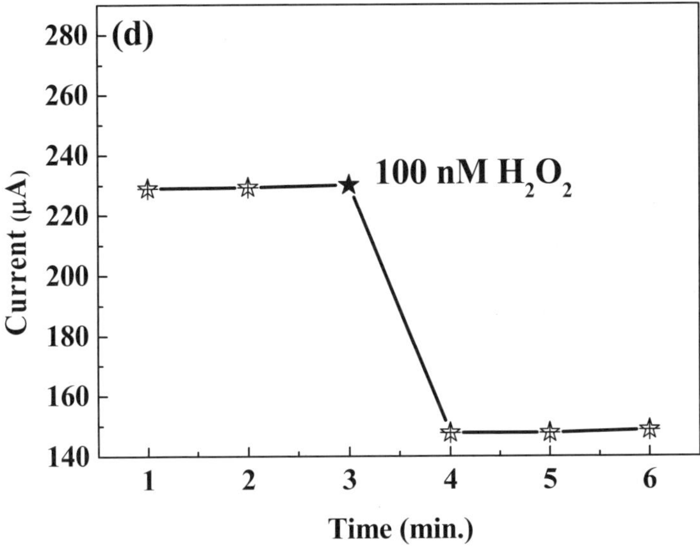

Figure 2. Real time detection of 10 µl of (a) 0.1 nM, (b) 1 nM, (c) 10 nM, (d) 100 nM H_2O_2, respectively, at constant bias of 100 mV with fresh sensors.

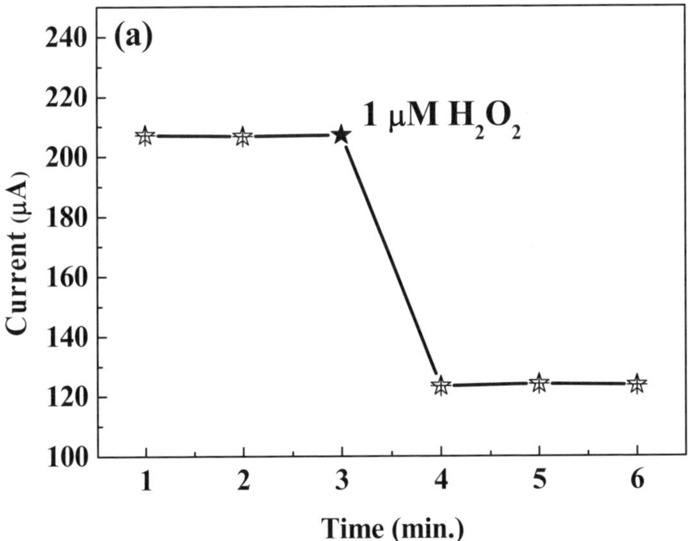

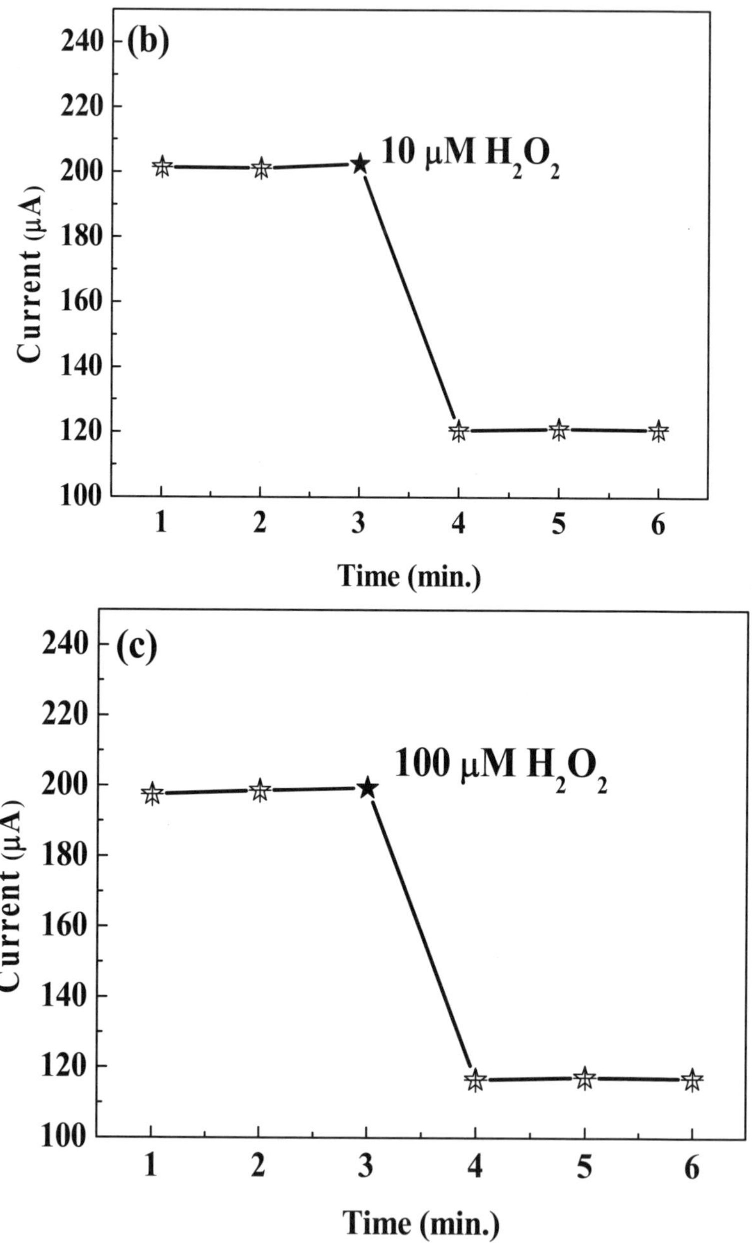
(b)
10 μM H₂O₂
Current (μA)
Time (min.)
(c)
100 μM H₂O₂
Current (μA)
Time (min.)

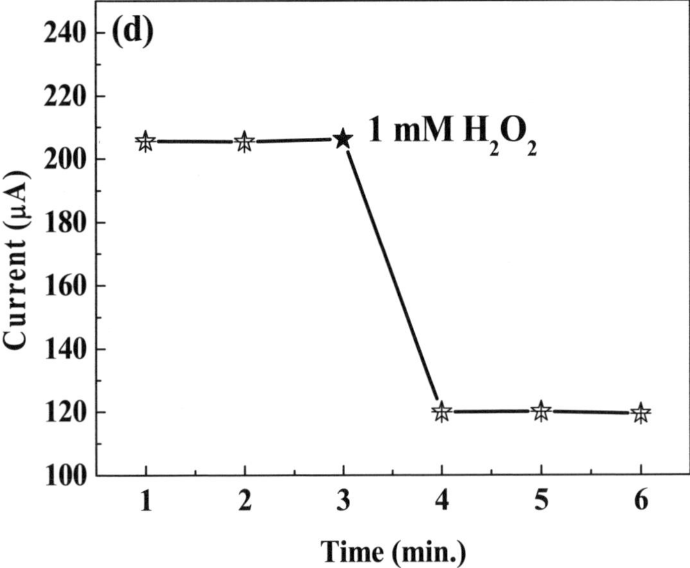

Figure 3. Real time detection of 10 µl of (a) 1 µM, (b) 10 µM, (c) 100 µM, and (d) 1 mM H_2O_2, respectively, at constant bias of 100 mV with fresh sensors.

To find out the detection limit, 10 µl of 0.5 nM, 0.6 nM, 0.7 nM, 0.8 nM, and 0.9 nM H_2O_2 were further tested with fresh sensors. The results of the real-time measurements for the sub-nM H_2O_2 are shown in Figure 4.

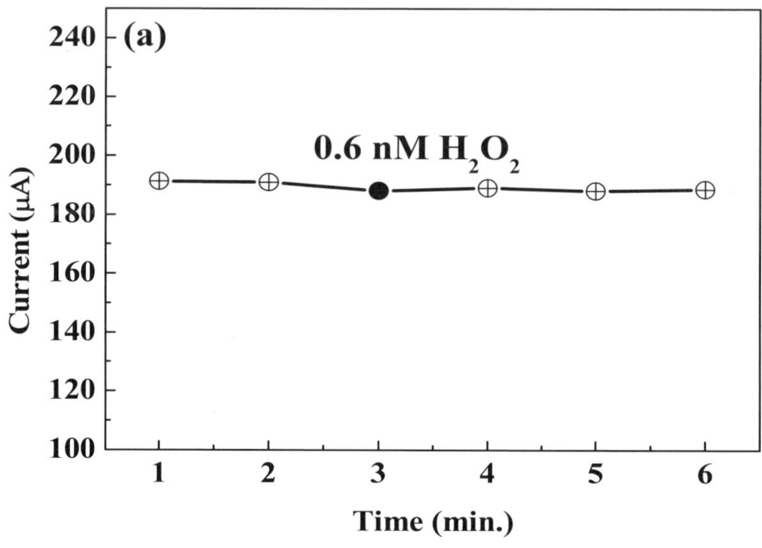
(a)
0.6 nM H$_2$O$_2$
Current (μA)
Time (min.)

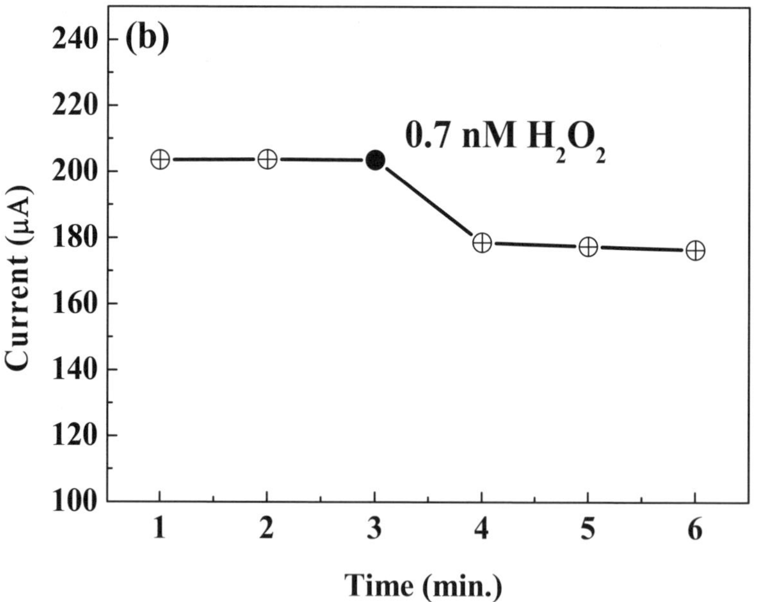
(b)
0.7 nM H$_2$O$_2$
Current (μA)
Time (min.)

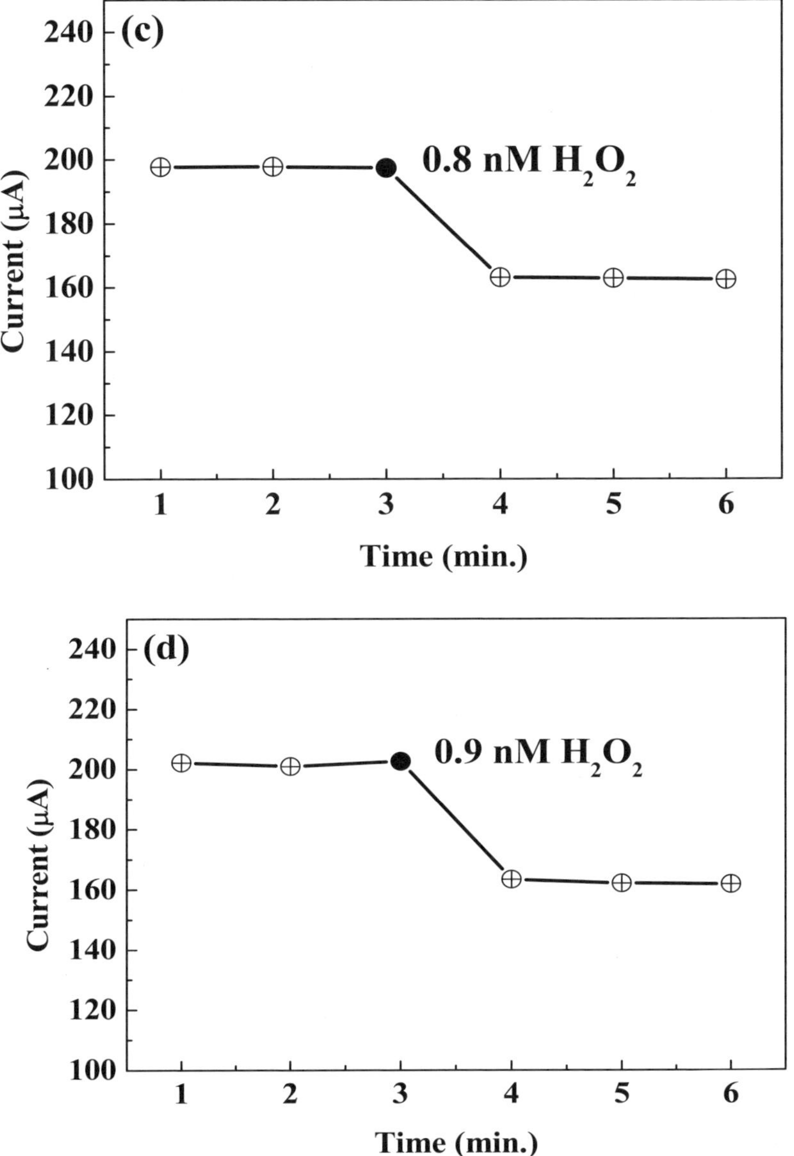

Figure 4. Real time measurement of 10 μl of (a) 0.6 nM, (b) 0.7 nM, (c) 0.8 nM, and (d) 0.9 nM H_2O_2, respectively, at 0.1 V.

The average percentage of conductance changes and the error bars (standard deviation) from 5 measurements on fresh sensors for H_2O_2 concentration ranging from 0.1 nM to 1 nM and from 0.1 nM to 1 mM were shown in Figure 5(a) and 5(b), respectively. The result shows the detection limit of this sensor is 0.7 nM H_2O_2, which is one of the lowest concentrations that have ever been reported (Chen *et al.*, 2013; Ahammad, 2013). From Figure 5(a) and 5(b), the observed detectable concentration of H_2O_2 for this sensor is from 0.7 nM to 1 μM. The sensor gradually saturates when hydrogen peroxide higher than 1 μM. Real-time measurements were also tested with sensors without HRP immobilized under different concentrations of hydrogen peroxide at pH 5.5 either with or without 1,4-diaminobenzene, as shown in Figure 5(c) and 5(d), respectively. The results show no significant current change from 1 nM to 1 mM H_2O_2 for both tests, which demonstrate that hydrogen peroxide would not react with PANI if HRP is not present. The oxidation of PANI is only occurring when HRP is present, thus indicating that the enzymatic reaction is necessitated in the process and the affinity constant between HRP and H_2O_2 would govern the sensitivity of the sensor.

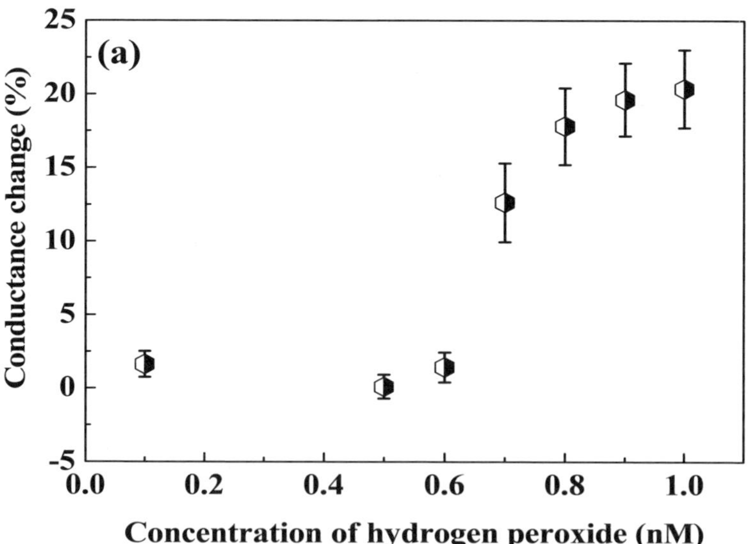

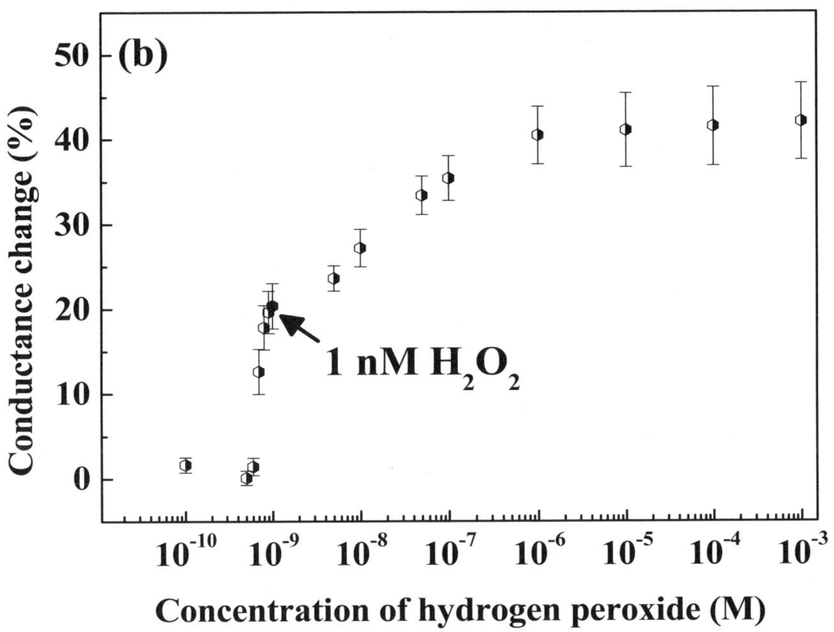
(b)
50
40
30
20
10
0
Conductance change (%)
1 nM H$_2$O$_2$
10^{-10} 10^{-9} 10^{-8} 10^{-7} 10^{-6} 10^{-5} 10^{-4} 10^{-3}
Concentration of hydrogen peroxide (M)

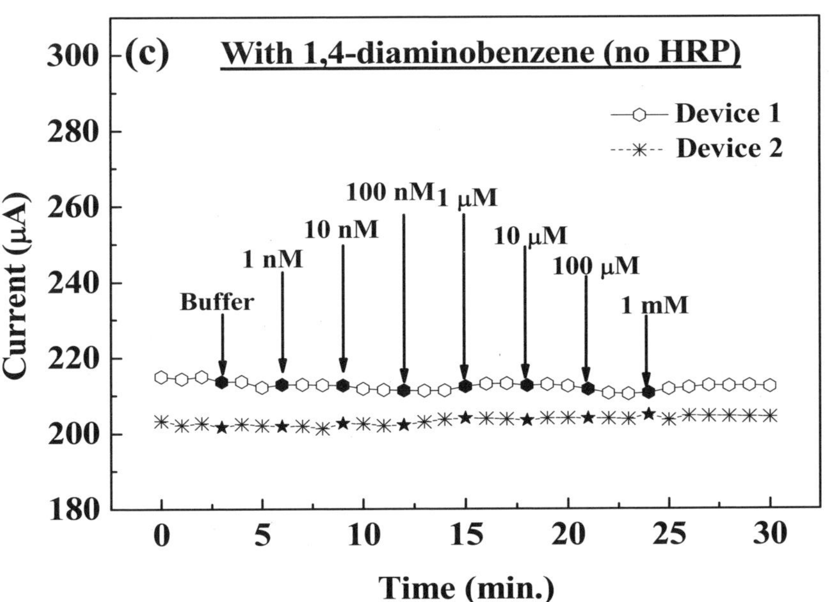
300
280
260
240
220
200
180
Current (µA)
(c) With 1,4-diaminobenzene (no HRP)
Device 1
Device 2
Buffer
1 nM
10 nM
100 nM
1 µM
10 µM
100 µM
1 mM
0 5 10 15 20 25 30
Time (min.)

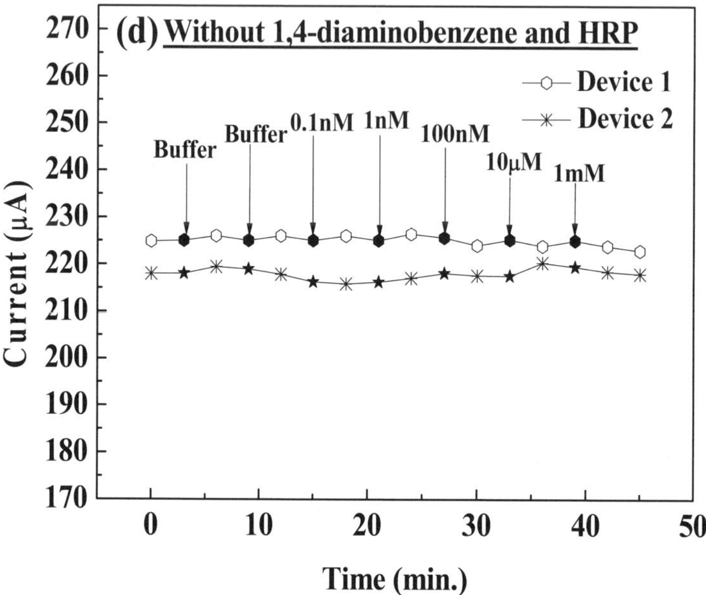

Figure 5. The average percentage of conductance changes with error bars (standard deviation) versus H_2O_2 concentration from (a) 0.1 nM, to 1 nM, and (b) the overall concentration from 0.1nM to 1mM. The results were measured from 5 fresh sensors for each concentration. The current of sultonated PANI without HRP immobilization are tested with different target concentrations of hydrogen peroxide at pH 5.5 (c) with or (d) without 1,4-diaminobenzene, respectively.

8.2.4. Reactivity of the Hydrogen Peroxide Sensors

If the H_2O_2 solution is removed from the sensor after testing, ideally, the sensor should be capable of reuse because the PANI still has good conductivity, indicating the considerable carrier concentration remaining in the polymer available for further oxidation of PANI and reduction of H_2O_2 and HRP. However, the HRP has to be active or re-activated for reuse by capturing electrons from an electron donor, the PANI, after being oxidized by H_2O_2. Therefore, it is proposed that the conjugated system of the PANI can efficiently allow electrons, which are far away from the oxidized HRP, to be transferred to the oxidized HRP and thus allow the oxidized HRP to be reduced. Therefore, a sensor was reused to detect H_2O_2 for three times, as shown in Figure 6(a), to confirm its reusability. The process steps, <1>, <2>, <3>, <4>, <5>, and <6>, depicted in Figure 6(a) are corresponding to the same notations shown in Figure 7(a) and Figure 7(b), for 1 nM and 5 μM H_2O_2, respectively. Figure 7(a) shows the detection of 1 nM hydrogen peroxide with a fresh sensor for the 1st, the 2nd, and the 3rd measurements. Figure 7(b) shows the detection of 5 μM hydrogen peroxide with another fresh sensor for the 1st, the 2nd, and the 3rd measurements. The result shows that the reused sensor has a reduced current level, but still can sense hydrogen peroxide with concentration as low as 1 nM. Practically, if the sensor needs to be reused, the calibration curve may need to be collected for the 2nd or the 3rd measurement, respectively. However, the cost of this microsensor is not high, so it can be used just once and disposed. The calibration curves shown in Figure 5(a) and 5(b) are only prepared based on the first measurement on fresh sensors.

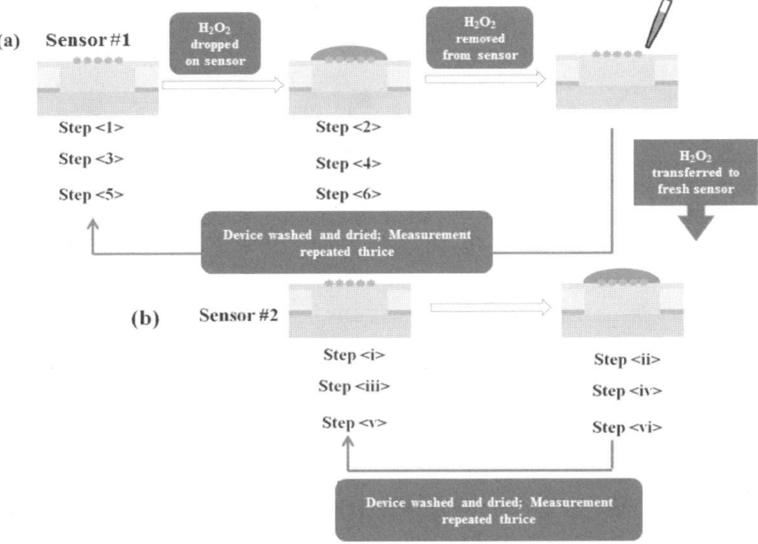

Figure 6. (a) The measurement procedures for sensor reusability tests. Step <1> and <2> represent the current measured before and after the H_2O_2 solution was dropped on the device, respectively. Step <3> and <4> perform the second measurement. Step <5> and <6> are the third one. (b) Procedures for remaining H_2O_2 detection. Step <i> and <ii> represent the current measured before and after the H_2O_2 solution, which is transferred from the first sensor after measurement, dropped on the second sensor to detect remaining H_2O_2. Step <iii> and <iv> perform the second test. Step <v> and <vi> are the third one.

The process steps, <1>,<2>,<3>,<4>,<5>, and <6>, are corresponding to the same notations shown in Figure 7(a) and (b). Steps <i>, <ii>, <iii>, <iv>, <iii> and <iv> are also corresponding to the same notations in Figure 7(c) and (d). Similar reusability of the sensor for 10 µl of 100 nM and 1 µM H_2O_2 detection was also conducted, as shown in Figure 8.

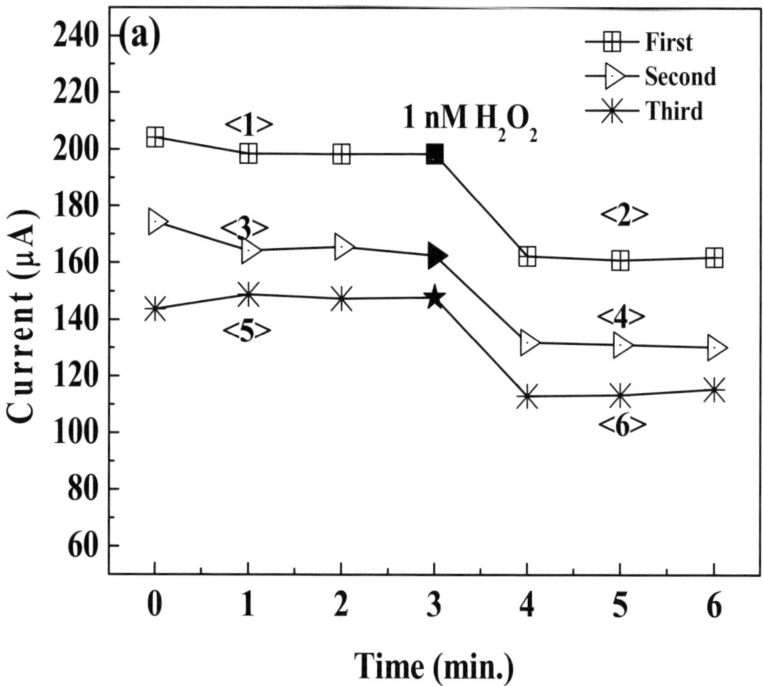

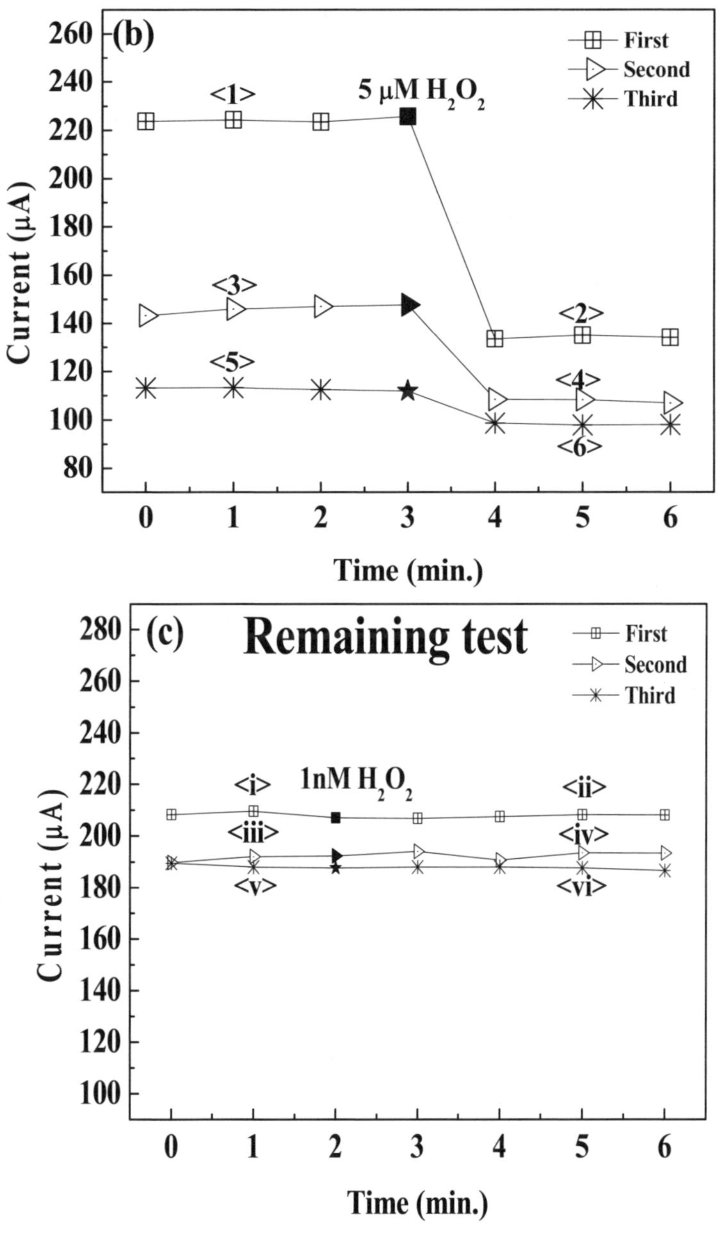
(b)
First
Second
Third
5 μM H₂O₂
<1>
<3>
<5>
<2>
<4>
<6>
Current (μA)
Time (min.)
(c)
Remaining test
First
Second
Third
1nM H₂O₂
<i>
<ii>
<iii>
<iv>
<v>
<vi>
Current (μA)
Time (min.)

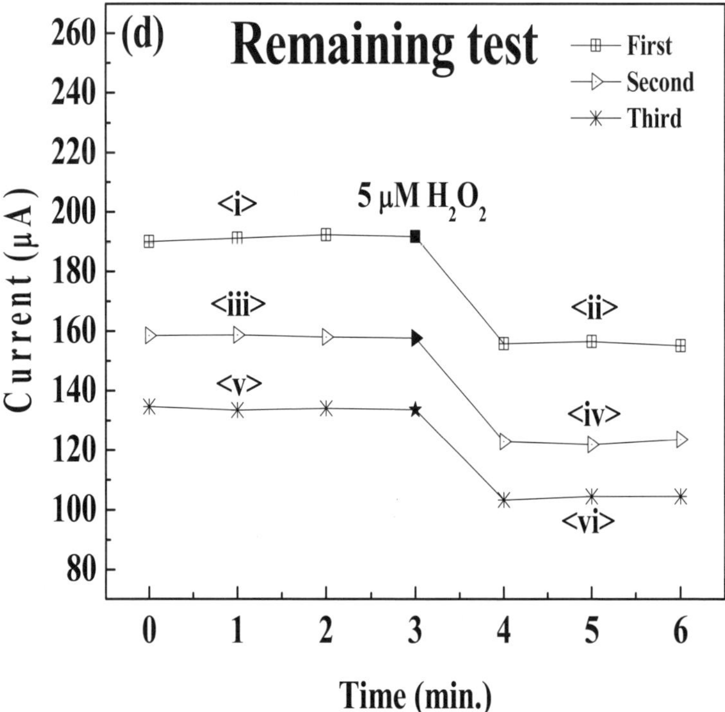

Figure 7. Real-time detection 10 µl of (a) 1 nM, and (b) 5 µM of H₂O₂, respectively, for 3 repeated measurements on each sensor. (c) and (d) represent the measurements of H₂O₂ remaining in solutions transferred from the first sensor after measurements, corresponding to (a), and (b), respectively. The notations shown here represent the corresponding steps in Figure 6.

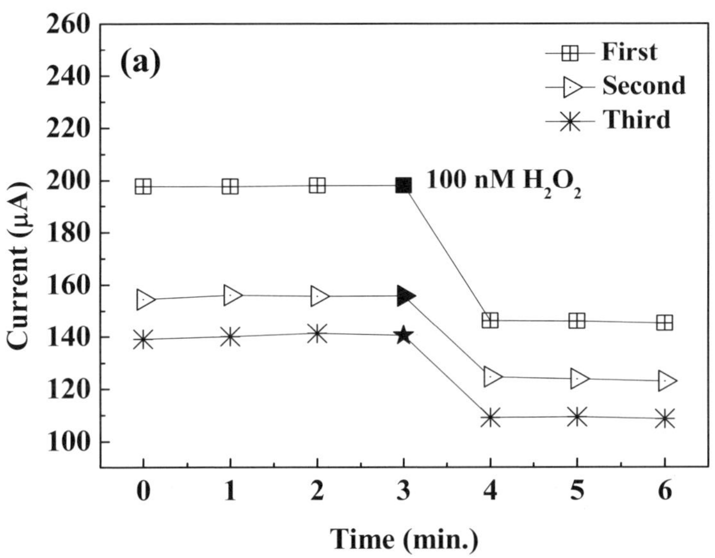
(a)
First
Second
Third
100 nM H_2O_2
Current (μA)
Time (min.)

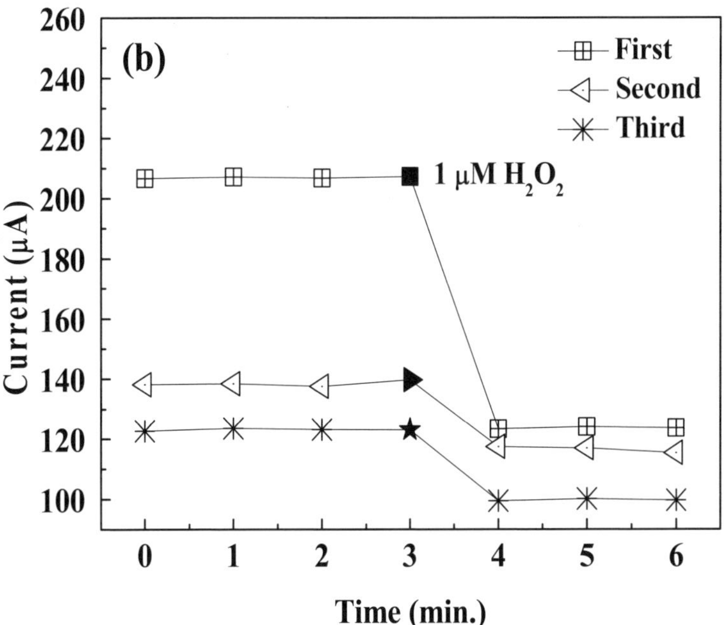
(b)
First
Second
Third
1 μM H_2O_2
Current (μA)
Time (min.)

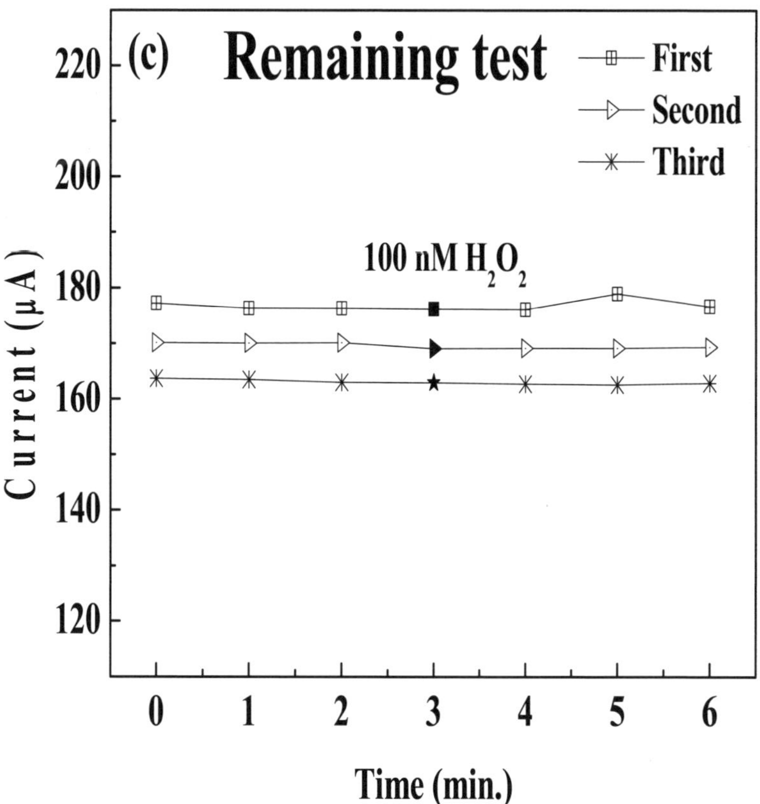

(c)
Remaining test
First
Second
Third
100 nM H₂O₂
Current (µA)
220
200
180
160
140
120
0
1
2
3
4
5
6
Time (min.)

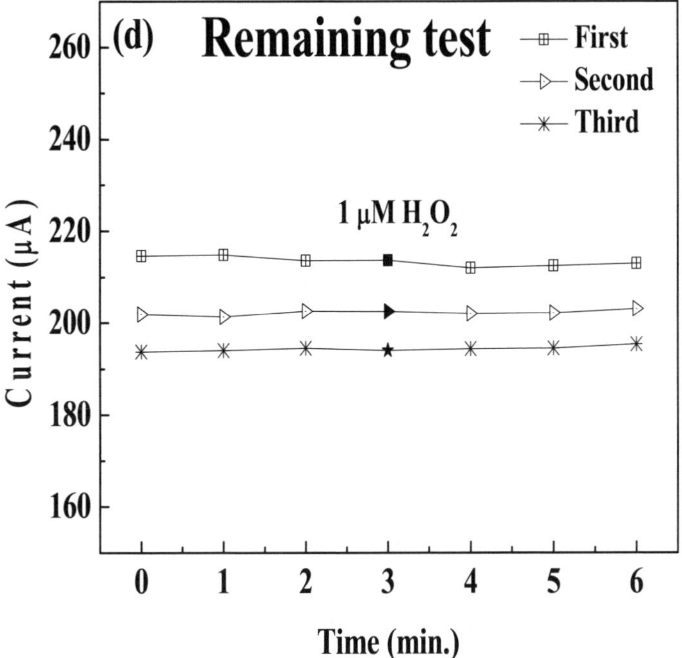

Figure 8. Real-time detection of 10 μl of (a) 100 nM, and (b) 1 μM H₂O₂, respectively, for 3 repeated measurements on each sensor. (c) and (d) represent the measurements of H₂O₂ remaining in solutions transferred from the first sensor after measurements, corresponding to (a), and (b), respectively.

8.2.5. Sensing Mechanisms

Ideally, before the sensor shows saturation at higher concentration of H₂O₂ (<1 μM), the total amount of H₂O₂ molecules should be fully consumed by HRP. To evaluate whether all H₂O₂ molecules were consumed after the test, we re-tested the remaining 10 μl of H₂O₂ solution, which was transferred from the initial H₂O₂ sensor used in Figure 6(a) after the test, with another fresh sensor, as depicted in Figure 6(b). The process steps, <i>, <ii>, <iii>, <iv>, <v>, and <vi>, depicted in Figure 6(b) are corresponding to the same notations shown in Figure 7(c) and Figure 7(d), for 1 nM and 5 μM H₂O₂, respectively. Figure 7(c) shows the 1st, the 2nd, and the 3rd re-tests of the remaining H_2O_2 solutions that were initially tested in 1 nM hydrogen peroxide, as shown in Figure 7(a). There is no current change observed in Figure 7(c), showing that all H_2O_2 were consumed after each measurement of 1 nM hydrogen peroxide in Figure 7(a). Similar experiments were conducted for higher H_2O_2 concentrations. Hydrogen peroxide was not detected in the remaining tests for 100 nM and 1 μM H_2O_2, but was found in the same test in 5 μM H_2O_2. Figure 7(d) shows the measurements of H_2O_2 remaining after the initial detection of 5 μM H_2O_2, as shown in Figure 7(b). There is no H_2O_2 remaining after each initial detection for all concentrations except the 5 μM H_2O_2. The results of the remaining tests for the 100 nM and 1 μM H_2O_2 are shown in Figure 8(c) and (d). The results demonstrate that in the detectable concentration range, H_2O_2 was fully consumed and the signal of the sensor was determined by the total amount of H_2O_2 in solution. This may arise an interesting question, that is, whether the detection limit can be pushed even lower by increasing the total amount of the H_2O_2 but with a lower concentration, such as 0.1 nM? Or it is only dependent on the concentration of H_2O_2. To prove the abovementioned hypotheses, 100 μl of 0.1 nM H_2O_2, which has the same amount of H_2O_2 as in 10 μl of 1 nM H_2O_2 solution was tested. Figure 9(a) shows the real-time detection of 100 μl of 0.1 nM H_2O_2 with a fresh sensor. There is only a very tiny current change in a very long reaction time (over 30 minutes), compared to the result from the same amount of H_2O_2 in 10 μl of 1 nM H_2O_2 solution shown in Figure 2(b). This reveals that only when the H_2O_2 concentration is higher than a critical value, the sensor can sense H_2O_2 and the magnitude of the signal is relevant to the total amount of H_2O_2. It is suggested that H_2O_2 has to bind with HRP to form a complex before it is reduced. Therefore, the affinity (or the association constant) between H_2O_2 and the HRP is critical. When H_2O_2 concentration is lower than the dissociation constant of the H_2O_2-HRP complex, the signal of the sensor is dominated by the concentration, not the total amount of H_2O_2. If the H_2O_2 concentration is much lower than the dissociation constant (reciprocal of the association constant) of the H_2O_2-HRP complex, the reaction rate will be very slow, leading to no significant change or just tiny signal. The detection limit is also resulted from the slow diffusion of hydrogen peroxide, leading to a slow and small signal response, as shown in Figure 9(a). Therefore, we conclude that in our method, in a short period, the detection limit of our sensor is determined by the dissociation constant of the H_2O_2-HRP complex, and the slow diffusion of hydrogen peroxide in a very low concentration level, but not the total amount of H_2O_2 in solution.

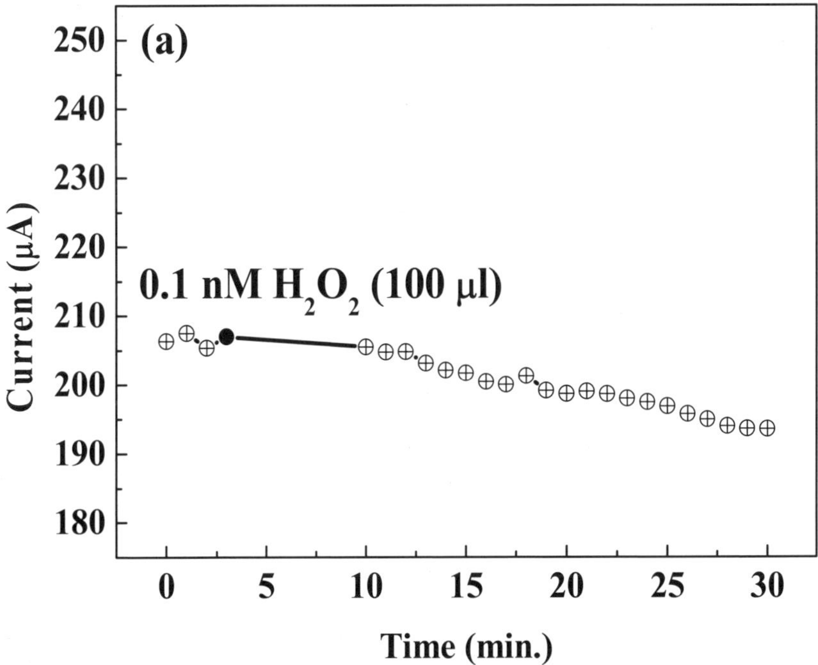

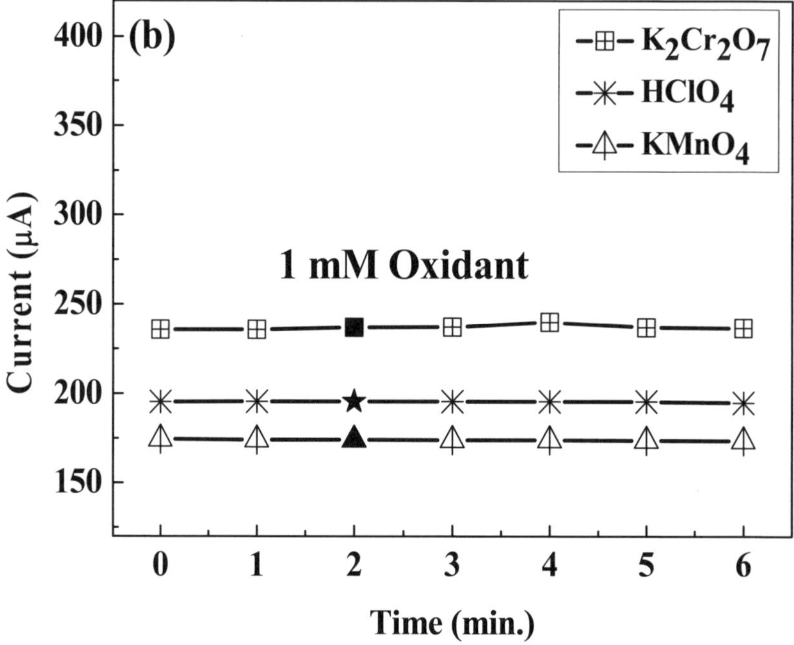

Figure 9. (a) Real-time detection of 100 µl of 0.1 nM H_2O_2 at 0.1 V (b) Commonly seen oxidants tested with hydrogen peroxide sensors.

As we compare this sensor with other hydrogen peroxide sensors, which are mostly electrochemical sensors, how can we explain the high sensitivity of this sensor? In electrochemistry, amperometric detection is frequently used to directly measure the current generated by the analytes at the electrode. This method has limited detection of concentration around 10^{-6} M, due to interferences or nonfaradaic process. The detection limit of potentiometric sensors is similar to amperometric devices. Stripping voltammetry has been reaching lower detection limit, by accumulating analytes near the surface of the electrodes in a known period of time and subsequently detecting them. The signal is then integrated over the period of time, therefore improving the detection limit (Bartlett *et al.*, 1998). In our method, beyond the detection limit, most hydrogen molecules are consumed within one minute and fully contribute into the oxidation of PANI, resulting in the low detection limit. The essential idea to reach high sensitivity for this sensor and the stripping voltammetry is somewhat similar. However, this sensor can really achieve fully consumption of hydrogen peroxide and high efficiency of charge transfer from PANI to HRP. These features explain the high sensitivity of this sensor.

The sensor shows saturation when the H_2O_2 concentration is higher than 1 μM, shown is Figure 5(b). There is H_2O_2 remaining after the initial detection of 10 μl of 5 μM H_2O_2 solution which is confirmed with another fresh sensor, as shown in Figure 7(d). Because this sensor still has good conductivity and sensitivity after two tests of 10 μl of 5 μM, as shown in Figure 7(d), we can exclude the possibility that the PANI may lose its activity as an electron donor after the first measurement. Therefore, the saturation of the sensor at high H_2O_2 concentration, is not resulted from lower carrier concentration in PANI, or from lower efficiency of electron transfer from PANI to the oxidized HRP, but may be resulted from the inhibition of the HRP enzyme activity by excessive H_2O_2, which were reported previously (Arnao *et al.*, 1990; Vlasits, 2010). However, further investigation will be needed to explore the complex mechanism in detail to confirm this point of view.

8.2.6. Stability and Robustness

Back to Figure 1(a), the signal of the sensor is generated due to the oxidation of PANI by hydrogen peroxide via the HRP-catalyzed enzymatic reaction. This mechanism may arise a question, that is, "Will other oxidants also oxidize the PANI, causing interference during the hydrogen peroxide detection?" Figure 9(b) shows the real-time tests of commonly seen strong oxidants, including $KMnO_4$, $K_2Cr_2O_7$, and $HClO_4$ with fresh hydrogen peroxide sensors. There is no any current change caused by these strong oxidants. This result demonstrates that this hydrogen peroxide sensor is very robust. Previously, we have shown that hydrogen peroxide cannot oxidize PANI if HRP is not present. In fact, hydrogen peroxide is a strong oxidant, which has a standard reduction potential of 1.77V, when reduced into water (Sanli 2013). The above-mentioned strong oxidants have smaller reduction potentials compared to that of hydrogen peroxide. Therefore, we think that only when oxidants have even higher potentials, they may have chances to oxidize PANI spontaneously. For example, the hydroxyl radical, HO, has a reduction potential of 2.3V (Buettner 1993) when reduced into water. If free radicals of reactive oxygen species can oxidize PANI, we can use sensor arrays, including sensors with or without HRP immobilized PANI, to distinguish the signal generated by hydrogen peroxide from that by free radicals. In next section we will show how hydroxyl radical change the conductivity of PANI.

8.3. Cholesterol Sensors

8.3.1. Detection of Cholesterol

Hydrogen peroxide is a by-product in many enzyme-catalytic reactions, such as glucose oxidase, lactate oxidase, cholesterol oxidase, alcohol oxidase, urate oxidase, aldehyde oxidase, and oxalate oxidase, which are implemented to detect glucose, lactic acid, cholesterol, ethanol, urea, formaldehyde, and oxalate, respectively. These biomolecules are significant markers in many biologically metabolic reactions. For example, cholesterol level may indicate clinical disorders such as heart disease, hypertension, arteriosclerosis, and cerebral thrombosis etc. (Chetna *et al.*, 2011). Monitoring cholesterol level in blood can evaluate the potential risk of cardiovascular disease and prevent deaths from the disease. The demand on cheap and disposable electronic cholesterol sensors for the point-of-care health monitoring is increasing. In the previous section, we have shown an ultra-sensitive hydrogen peroxide microsensor using horseradish peroxidase (HRP)-immobilized polyaniline (PANI) to detect H_2O_2. In cholesterol detection, cholesterol oxidase (ChOx) is used to react with cholesterol and produce hydrogen peroxide. The produced hydrogen peroxide is then detected by the hydrogen peroxide sensor described previously. The reactions are described as following.

$$Cholesterol + O_2 \xrightarrow{\quad ChOx \quad} Cholestenone + H_2O_2 \qquad (1)$$

$$H_2O_2 + PANI_{(red)} \xrightarrow{\quad HRP \quad} H_2O + PANI_{(ox)} \qquad (2)$$

where the $PANI_{(red)}$ and $PANI_{(ox)}$ represent the reduced and oxidized PANI thin film, respectively.

Both the generation of hydrogen peroxide in reaction (1) and the oxidation of PANI in the reaction (2) are all spontaneous. Therefore it is not necessary to apply any external bias to trigger any oxidation or reduction reactions, which can avoid any interference from un-necessary oxidation or reduction resulted from external bias. The signal measurement for this sensor is easy and straightforward. To accommodate these two reactions, as mentioned as in the previous section, PANI was first spin-coated on Au electrodes on a Si_3N_4/Si substrate, followed by being modified with propane sultone (N-Alkylated PANI) to ensure its high conductivity in neutral solution, and then immobilized with HRP. A ChOx-immobilized dialysis membrane was then placed aside the HRP-immobilized PANI for cholesterol detection, as shown in Figure 10(a). During the detection of cholesterol, the solution is able to cover both the dialysis membrane and the HRP-immobilized PANI. The developed cholesterol microsensor has the advantages such as low cost, small size, fast response and ease of operation, which make it a perfect candidate for the routine personal healthcare monitoring.

(a)

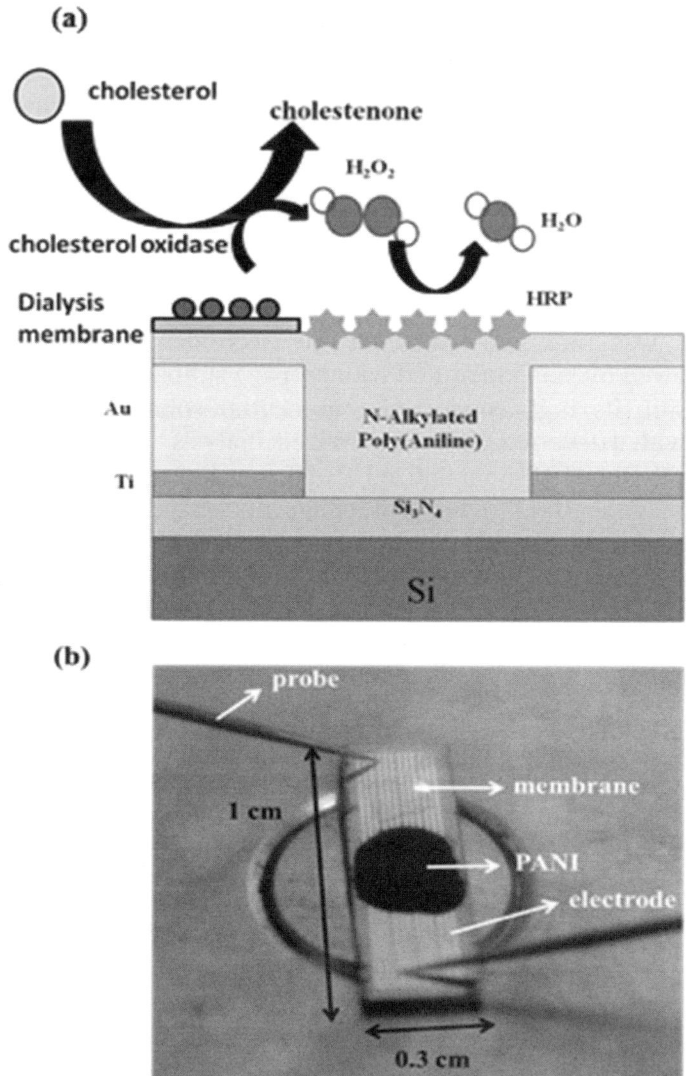

(b)

Figure. 10. (a) The schematic of the cholesterol sensor and the enzymatic reactions. (b) The top view of the photography of a cholesterol sensor.

8.3.2. Real Time Detection

Cholesterol in physiological levels, ranging from 100mg/dl to 400 mg/dl, prepared in citrate PBS buffer solutions (pH = 7.0) were directly dropped onto fresh sensors and the currents were measured at 0.1 V every minute. Figure 11(a), 11(b), 11(c), 11(d), 11(e), and 11(f) show the real time detection of 5 μl of 100 mg/dl, 150 mg/dl, 200 mg/dl, 250 mg/dl, 300 mg/dl, and 400 mg/dl water-soluble cholesterol, respectively. Before the cholesterol solution was added on the sensor, the current was measured and showed very good stability. When the cholesterol solution was added, within only 1 min., the current quickly decreased and stabilized, showing the fast response of the cholesterol sensor. The fast response of the cholesterol sensor is attributed to the high reaction rates of the two enzyme reactions. In addition, because the cholesterol solution covered on both the ChOx-immobilized dialysis membrane and the HRP-immobilized PANI, the hydrogen peroxide generated from the region of the dialysis membrane needs to diffuse to the region of the HRP-immobilized PANI. Thanks to the small volume of the cholesterol sample solution (5 μl) used in the detection, the diffusion of hydrogen peroxide did not take too long. The response time for all the concentrations of cholesterol that were tested for each measurement with fresh sensors, is less than 1 min., which is acceptable for point-of-care testing.

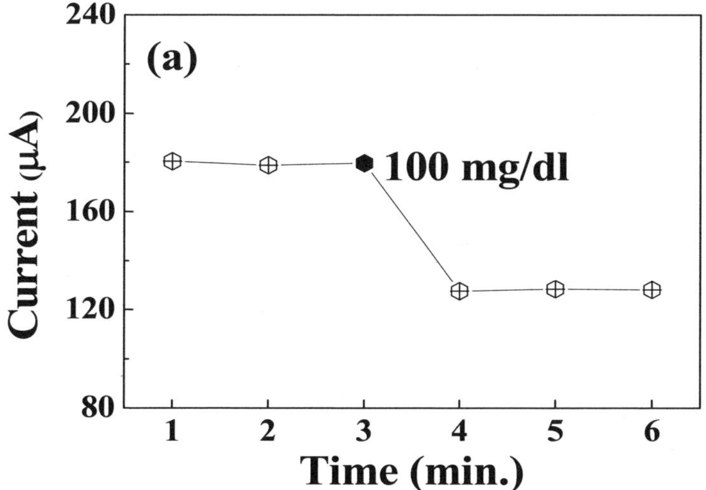

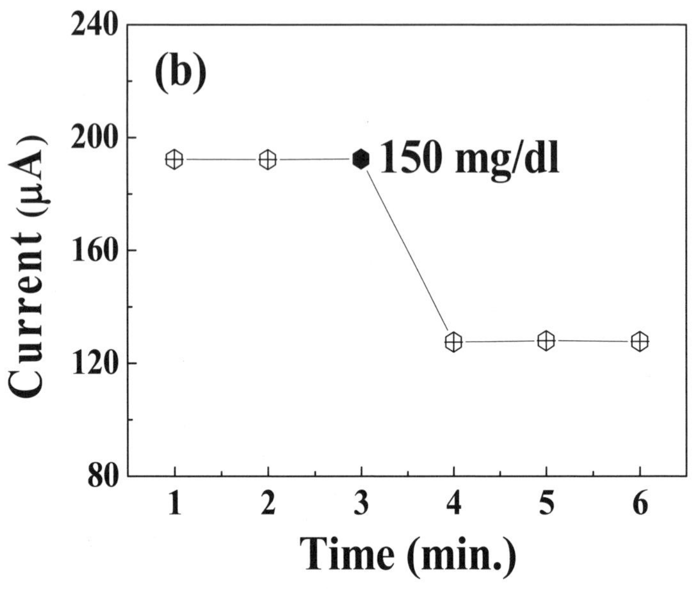

(b)
150 mg/dl

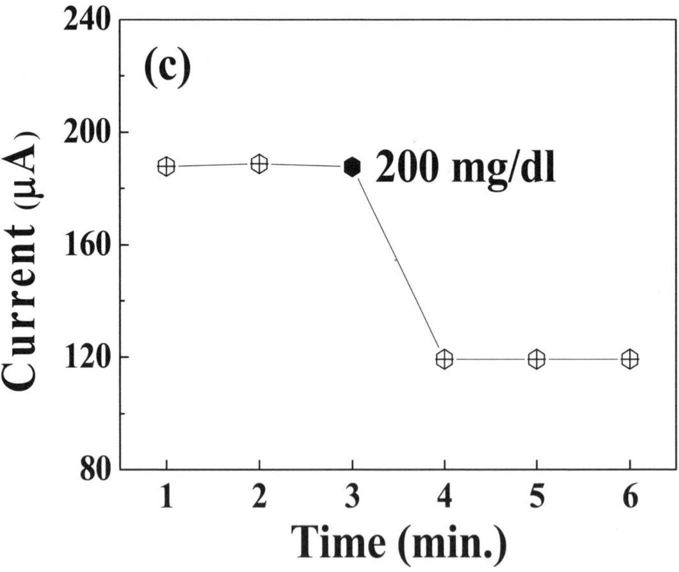

(c)
200 mg/dl

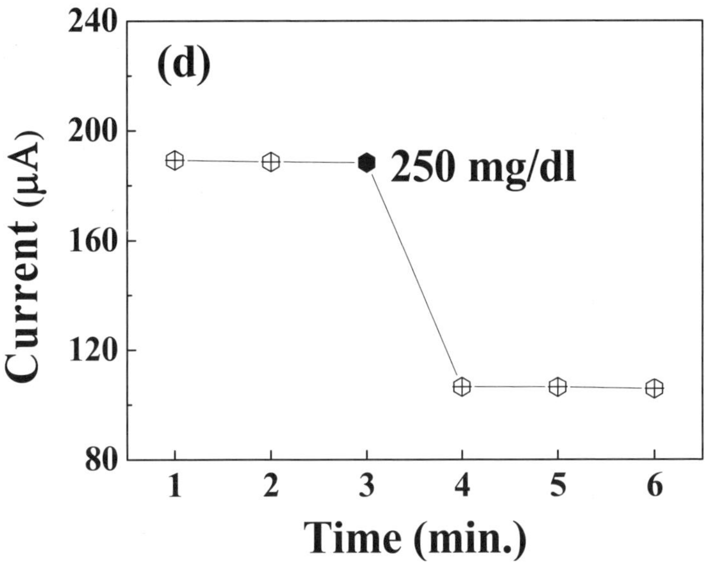

(d)
250 mg/dl
Current (μA)
Time (min.)

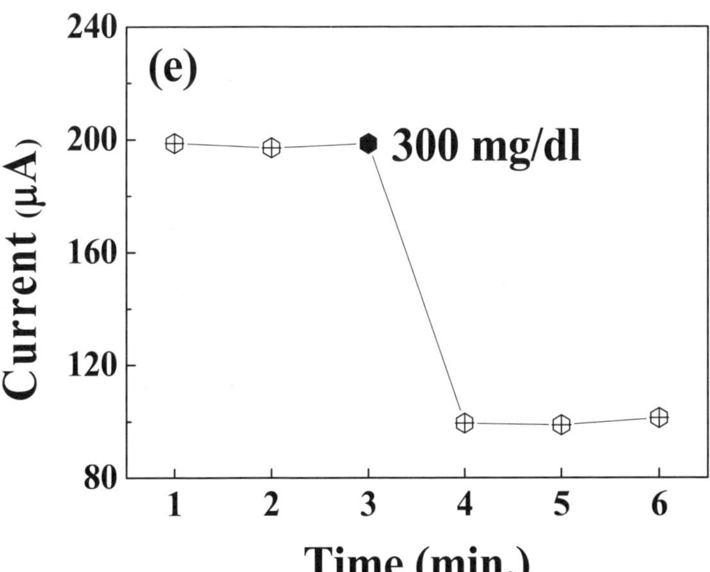

(e)
300 mg/dl
Current (μA)
Time (min.)

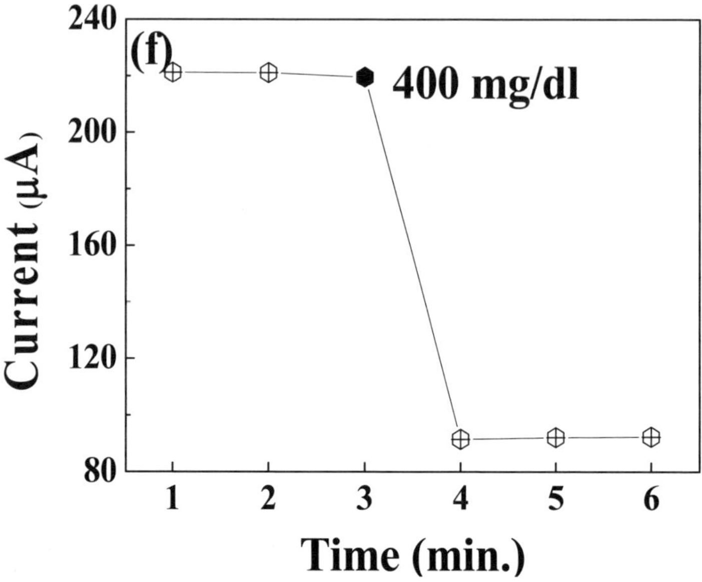

Figure 11. Real time detection of 5 μl of (a) 100 mg/dl, (b) 150 mg/dl, (c) 200 mg/dl, (d) 250 mg/dl, (e) 300 mg/dl, and (f) 400 mg/dl cholesterol, at constant bias of 100 mV with fresh sensors.

The conductivity change of the microsensors shows a very good linear dependence on the cholesterol concentration, as shown in Figure 12(a), indicating a very good linearity of sensor response. A real-time measurement was also tested with sensors without ChOx immobilized under several consecutive drops of 5ul of 400 mg/dl cholesterol, as shown in Figure 12(b). The results show no significant current change for every drop of the 5ul of cholesterol solution on the sensor without ChOx. The oxidation of PANI is only occurring when ChOx and HRP are both present, therefore indicating that the enzymatic reactions are necessitated in the whole sensing process.

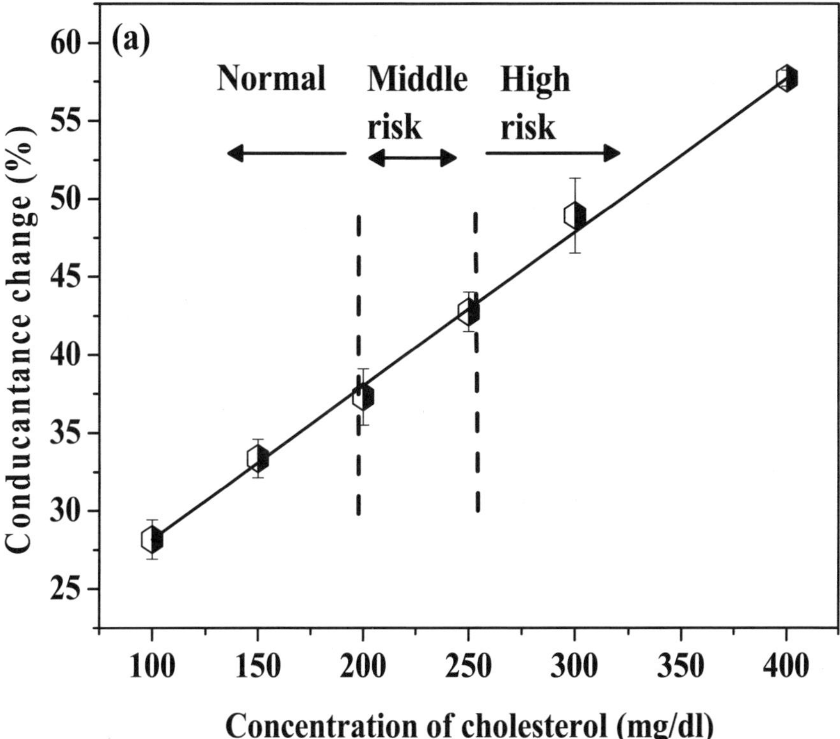
(a)
Normal
Middle
risk
High
risk
60
55
50
45
40
35
30
25
Conducantance change (%)
100
150
200
250
300
350
400
Concentration of cholesterol (mg/dl)

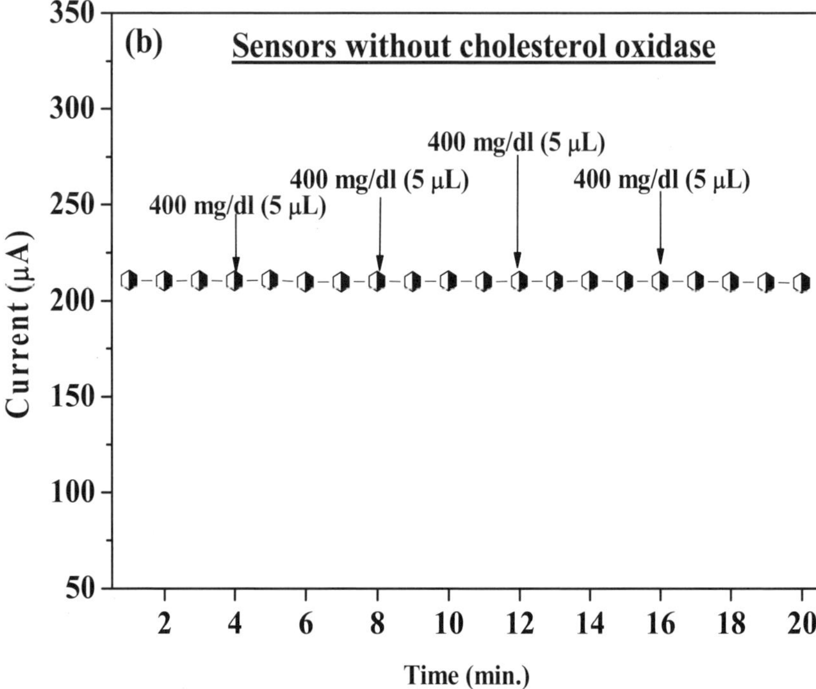

Figure. 12. (a) The average percentage of conductance changes versus cholesterol concentration. (b) The current of the cholesterol sensor without cholesterol oxidase immobilized on the dialysis membrane was tested with 400 mg/dl cholesterol in PBS solution at pH 7.0 for four times.

Interference studies in presence of other analytes are conducted. Figure 13(a) shows that the typical cholesterol sensor was dropped with 5 µl of 10 mM glucose twice and no any current change was observed, until 5 µl of 300 mg/dL cholesterol was added on the sensor. The percentage of the conductance change of the sensor lies in the value for 100 mg/dL cholesterol, which is exactly the final diluted cholesterol concentration at equilibrium. However, the response time for this test is 3~4 minutes, which is much longer than that of the direct test of cholesterol solution. The longer response time is due to the slow diffusion of the cholesterol before it reaches equilibrium. To verify this idea, a similar test was conducted by replacing the glucose solution with buffer solution in the same volume, as shown in Figure 13(b). The response time shown in Figure 13(b) for buffer solution is also 3~4 minutes, as it is in Figure 13(a) for glucose solution. This demonstrates that the longer response time is really resulted from the mass transport of the cholesterol. The interference was also studied by direct measurement of 100mg/dL cholesterol solution, followed by dropping 10mM glucose solution, as shown in Figure 13(c). Although the sensor has still high conductivity after reacting with cholesterol solution, the following glucose solution cannot generate any significant conductance change of the PANI. The percentage of the conductance change in Figure 13(a), 13(b) and 13(c) are 26.74%, 25.29%, and 27.06%, respectively, which are consistent to the average value (28.17%) caused by 100mg/dL cholesterol as shown in Figure 12(a). These interference tests have shown that the cholesterol sensors have very high specificity and low interference.

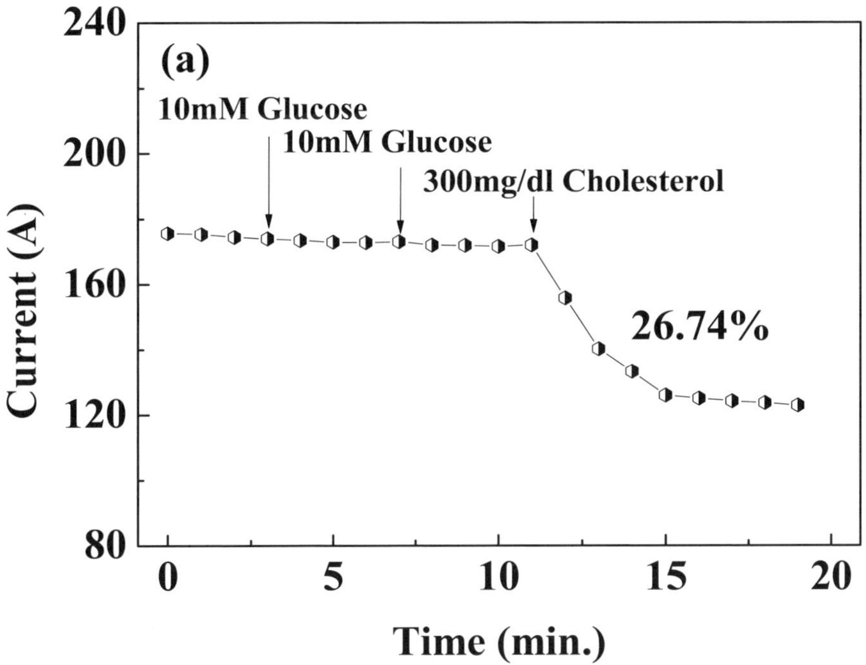

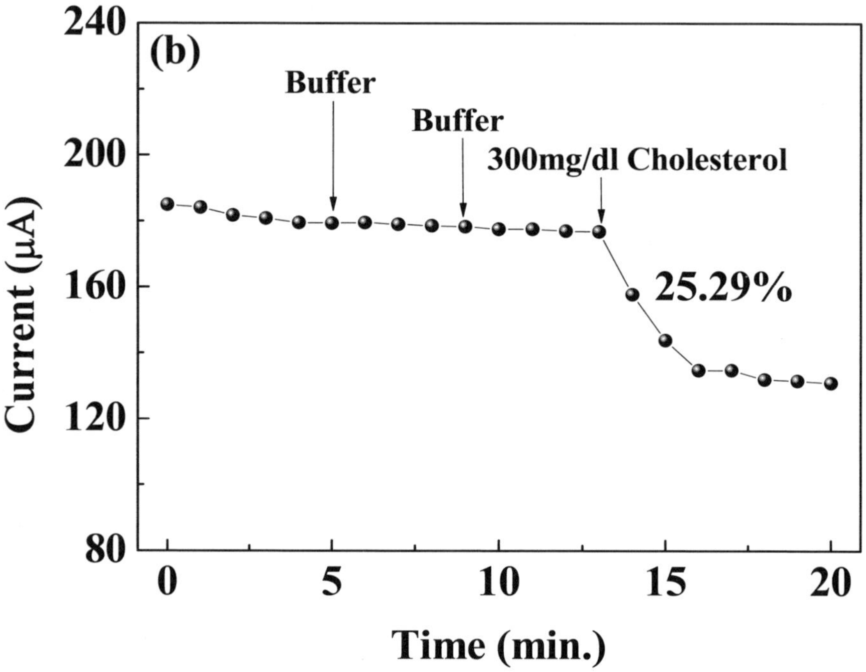
(b)
Buffer
Buffer
300mg/dl Cholesterol
25.29%
240
200
160
120
80
Current (μA)
0
5
10
15
20
Time (min.)

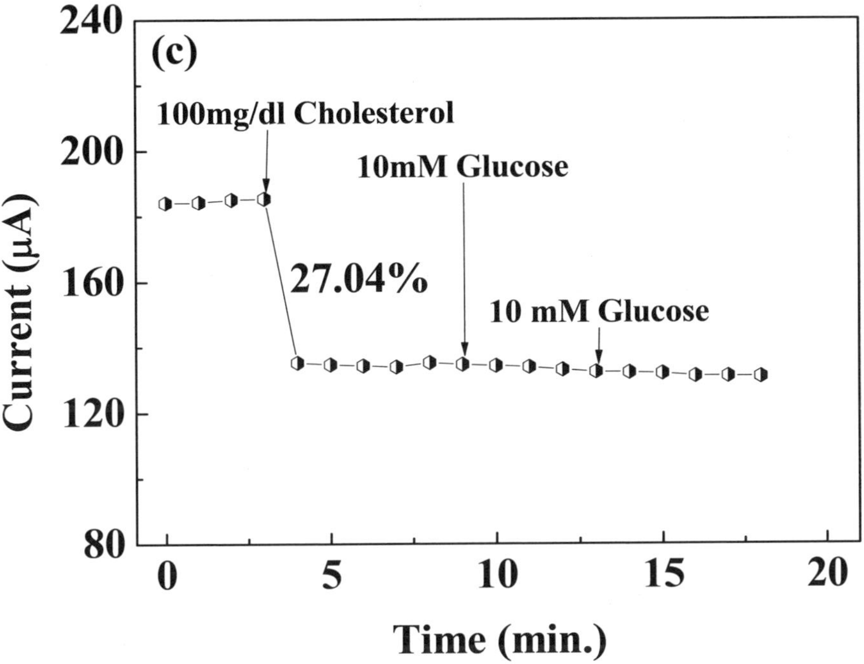

Figure. 13. Interference tests for cholesterol sensors at constant bias of 100 mV. (a) The current of a fresh sensor tested with 5 µl of 10 mM glucose solution twice followed by 5 µl of 300 mg/dl cholesterol solution. (b) The current of a fresh sensor tested with 5 µl of 10 mM glucose solution twice followed by 5 µl of 300 mg/dl cholesterol solution. (c) The current of a fresh sensor directly tested with 5 µl of 100 mg/dl cholesterol solution followed by the glucose solution.

8.4. Hydroxyl Radical Sensors

8.4.1. Detection of Hydroxyl Radicals

Oxidative stress has become an important topic because it has been shown to be relevant to tumors, cancers, Parkinson's disease, and aging (Burns et al. 2012). Oxidative stress is induced by reactive oxygen species (ROS), including hydrogen peroxide H_2O_2, hydroxyl radical OH^{-1}, superoxide anion O_2^{-2}, and singlet oxygen $^1O_2$. Hydroxyl radical was reported to be the most reactive free radical (Yapici et al. 2011;

Zhang   et al. 2014). Hydroxyl radical was found to be able to damage DNA (Malins et al. 1996), RNA (Jacobs et al. 2010), proteins (Gieseg et al. 1993) and lipids (Tien et al. 1982), leading to abnormal cell response and cell apoptosis (Ren et al. 2001) in physiological and pathological environment (Li et al. 2010; Liu et al. 2008). Hydroxyl radicals can be generated by Fenton reaction or Haber-Weiss reaction in vivo (Lemire et al. 2013; Zhang et al. 2014). Hydrogen peroxide is much easier to be detected due to its longest life-time. However, detecting radicals, especially for hydroxyl radicals, is challenging due to its very short life-time (~μs) (Burns et al. 2012). Electron spin resonance (ESR) spectroscopy is widely used to detect radicals by characterizing the electron paramagnetic spectrum. However, the life-time of hydroxyl radicals is too short to be detected directly by ESR. Instead, a spin trap molecules, is required to bind with hydroxyl radicals to form a complex, resulting in a longer half-life for ESR (Burns et al. 2012; Cohn et al. 2008). Fluorescence spectroscopy has also frequently been used to quantify hydroxyl radicals by the oxidation of fluorescent probes with hydroxyl radicals. (Cohn et al. 2008; Price et al. 2009; Yapici et al. 2011; Zhuang et al. 2014). Chemiluminescence (CL) is enhanced with the chemiluminogenic probes when reacting with hydroxyl radicals (Page et al. 2010). Ultraviolet−visible (UV−vis) spectroscopy is based on the absorbance change of a probe molecule after reacting with hydroxyl radicals, such as Br⁻ (Westerhoff et al. 2007), Crocin (Bors et al. 1982), Ferrocyanide (Schuler et al. 1981), or Rhodamine B (Yu et al. 2008). High pressure liquid chromatography (HPLC) has also been employed for hydroxyl radical detection (Teismann and Ferger 2000). The above-mentioned techniques are currently the most commonly used methods for detecting hydroxyl radicals. However, these techniques require expensive instruments, complicated process, and degradation of probe molecules. Electronic microsensors are suitable for simple, cost-effective, and highly sensitive hydroxyl radical detection.

PANI was reported as a good radical scavenger (Gizdavic-Nikolaidis et al. 2004). Hydrogen proton-doped PANI was proved to be able to prevent red blood cells from the attack by hydroxyl radicals, which was generated from Fenton reaction in the presence of hydrogen peroxide (Banerjee et al. 2010). Accordingly, it is possible to fabricate electronic hydroxyl radical microsensors by utilizing highly doped PANI and measuring the conductivity change of the doped-PANI after it react with hydroxyl radicals. It is believed that the doped-PANI can be oxidized by hydroxyl radicals and the oxidation of doped-PANI will result in the decreased conductivity, which can be related to the concentration of hydroxyl radicals in solutions.

8.4.2. Sensing Mechanism and Calibration of Hydroxyl Radicals

The hydroxyl radical was generated from Fenton reaction as shown below in reaction 1:

$$Fe^{2+} + H_2O_2 \rightarrow Fe^{3+} + OH^{\cdot} + OH^{-} \qquad k_1 = 63 \ M^{-1}s^{-1} \qquad (1)$$

Doped PANI with high conductivity was spin-coated on substrates as described in previous sections. The PANI/Au interface was confirmed to be ohmic contact by measuring the current-voltage characteristics of the device. Figure 14(a) shows that the process steps for the reaction between PANI and hydroxyl radicals. Figure 14(b) shows the top view photography of the PANI-coated microchip.

(a)

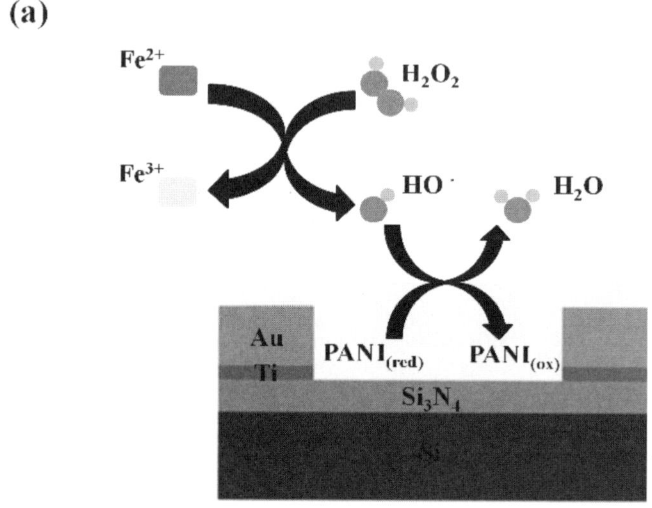

(b)

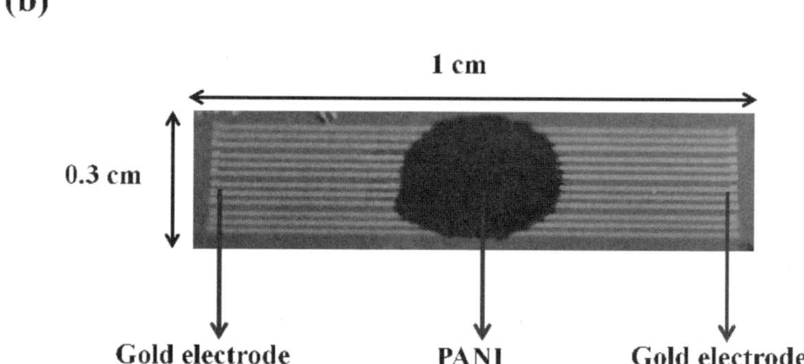

Figure 14. (a) The schematic of the hydroxyl radical sensor and the interaction between hydroxyl radicals and PANI (b) The top view photography of the microsensor.

The real time detection of hydroxyl radicals with fresh sensors by preparing Fenton reagents with the target concentrations of ferrous ion at 1.25 mM and hydrogen peroxide at 1.56 mM, 3.13 mM, 4.69 mM, 6.25 mM, 7.81 mM, and 9.38 mM, were conducted at a constant bias of 100 mV. Figure 15(a), 15(b), 15(c), and 15(d) show the real time detection of hydroxyl radicals for hydrogen concentrations at 1.56 mM, 4.69 mM, 7.81 mM, and 9.38 mM, respectively, at fixed ferrous ion concentration (Fe^{2+}=1.25 mM). It is significantly observed that hydrogen peroxide do not affect the conductivity of the PANI in the absence of ferrous ions. The current changed only when ferrous ion was added into the hydrogen peroxide solution on the sensor. This result showed that the conductivity change of the PANI was attributed to the existence of hydroxyl radicals, which was generated by Fenton reagents. Regarding to the small rate constant of the Fenton reaction as shown in equation 1, the long response time in the real-time measurement of hydroxyl radicals was attributed to the slow generation rate of hydroxyl radicals (Lewis et al. 2009).

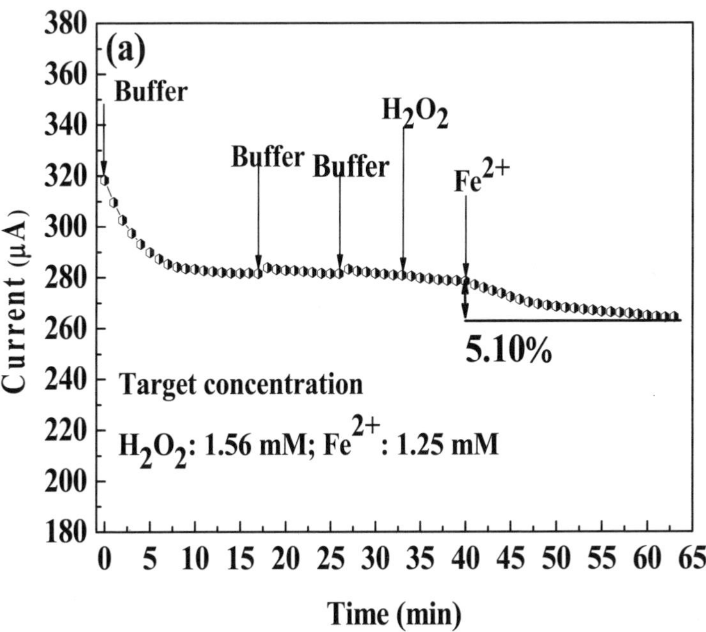

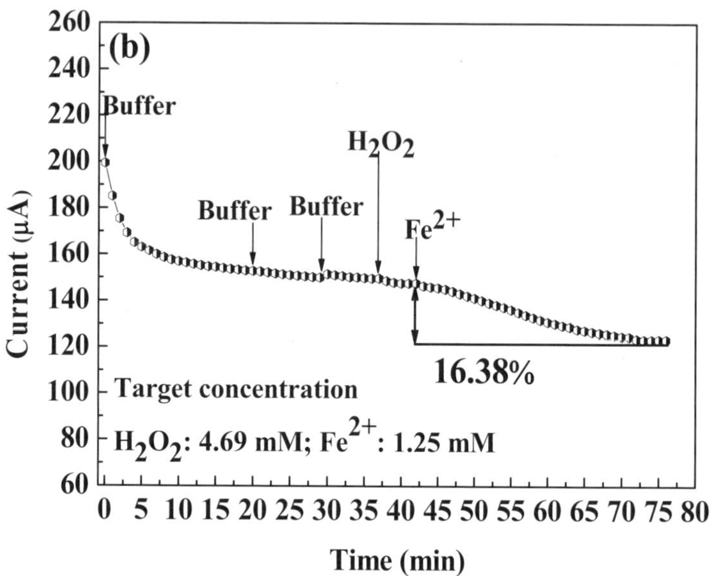

(b)
Current (μA)
Buffer
Buffer
Buffer
H₂O₂
Fe²⁺
16.38%
Target concentration
H₂O₂: 4.69 mM; Fe²⁺: 1.25 mM
Time (min)

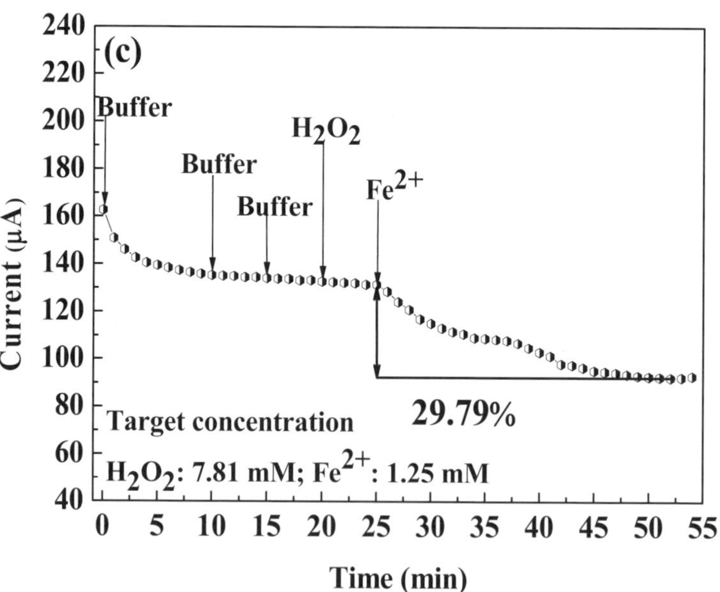

(c)
Current (μA)
Buffer
Buffer
Buffer
H₂O₂
Fe²⁺
29.79%
Target concentration
H₂O₂: 7.81 mM; Fe²⁺: 1.25 mM
Time (min)

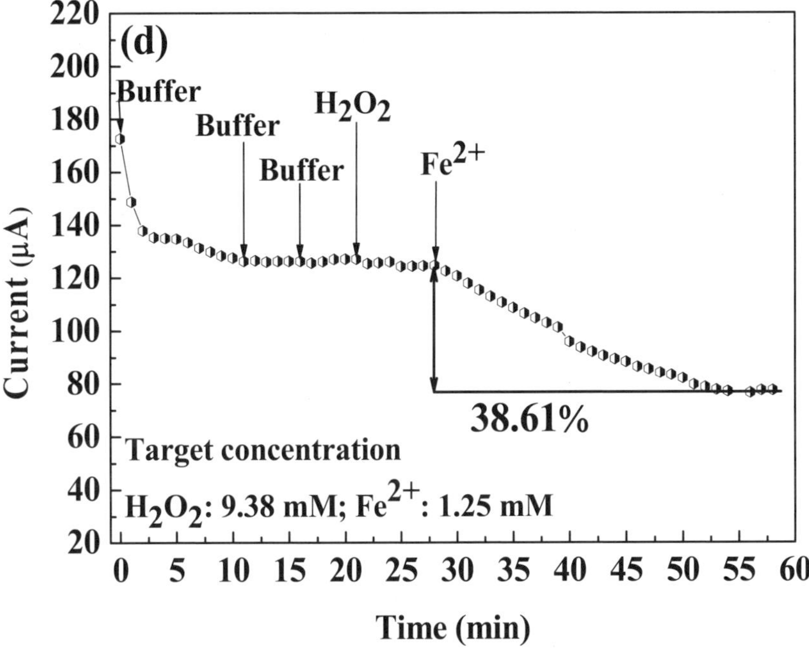

Figure 15. Real time detection of hydroxyl radicals generated by Fenton reaction, using Fenton reagents, including fixed ferrous ion at 1.25 mM and hydrogen peroxide at (a) 1.56 mM, (b) 4.69 mM, (c) 7.81 mM, and (d) 9.38 mM, at constant bias of 100 mV with fresh sensors.

Figure 16 shows the percentage of the conductivity change of PANI versus different hydrogen peroxide concentrations, including 1.56 mM, 3.13 mM, 4.69 mM, 6.25 mM, 7.81 mM, and 9.38 mM, with fixed ferrous ion at 1.25 mM in Fenton reaction. Four fresh sensors were measured for each concentration of hydrogen peroxide. The average and the error bar (the standard deviation) of the percentage of the conductivity change were calculated for each concentration of hydrogen peroxide. Although different sensors may have different initial currents, the percentage of the conductivity change of PANI was shown to be strongly and linearly dependent on the concentration of hydrogen peroxide, in the presence of ferrous ions.

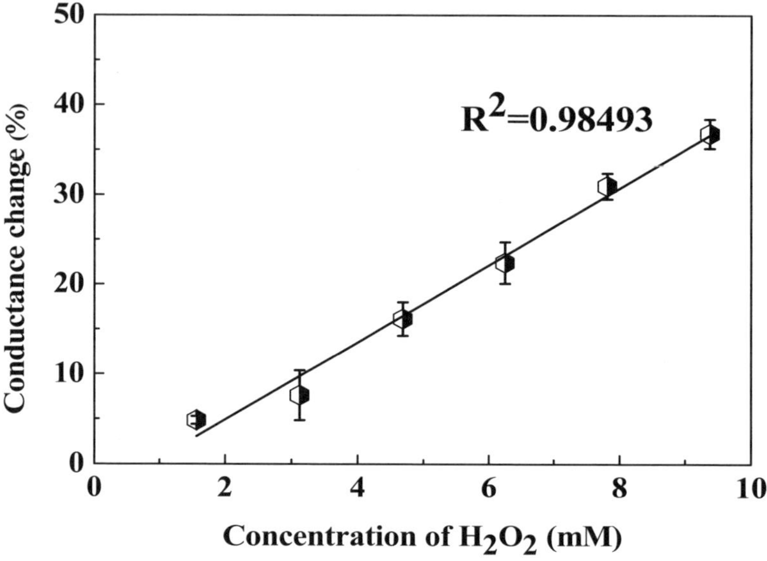

Figure 16. The average percentage of conductivity changes of PANI with error bars versus the concentration of hydrogen peroxide with fixed ferrous ion at 1.25 mM in Fenton reaction.

Hydroxyl radicals, generated in Fenton reacton, may be reacted and consumed by multiple following reactions. Lewis et al. had summarized the multiple reactions following the Fenton reaction (Lewis et al. 2009), as below.

$$OH^{\cdot} + Fe^{2+} \rightarrow OH^{-} + Fe^{3+} \qquad k_2 = 3.2 \times 10^8 \, M^{-1}s^{-1} \qquad (2)$$

$$H_2O_2 + OH^{\cdot} \rightarrow H_2O + HO_2^{\cdot} \qquad k_3 = 3.3 \times 10^7 \, M^{-1}s^{-1} \qquad (3)$$

$$OH^{\cdot} + OH^{\cdot} \rightarrow H_2O_2 \qquad k_4 = 6 \times 10^9 \, M^{-1}s^{-1} \qquad (4)$$

$$OH^{\cdot} + HO_2^{\cdot} \rightarrow H_2O + O_2 \qquad k_5 = 7.1 \times 10^9 \, M^{-1}s^{-1} \qquad (5)$$

From reaction 2 to 5, hydroxyl radicals are quickly reacted and consumed. The rate constants of reaction 2, 3, 4, and 5, are 7, 6, 8, and 8 orders of magnitude higher than that of the Fenton reaction (reaction 1), respectively. The much larger rate constants of reaction 2, 3, 4, and 5, compared with that of reaction 1, explaining the short half-life (~µs) (Burns et al. 2012) and the low concentration of hydroxyl radicals generated by Fenton reaction in the steady state. Therefore, the effective hydroxyl radical concentrations generated in our sample solutions were much lower than the concentration of hydrogen peroxide or the ferrous ion. The concentration of hydrogen peroxide shown in Figure 16 cannot represent the equivalent concentrations of hydroxyl radicals. Thus, the equivalent concentration of hydroxyl radicals needs to be elucidated and calibrated with other techniques. The resolved concentration of hydroxyl radicals by other techniques can then be re-calibrated with the percentage of conductivity change of PANI.

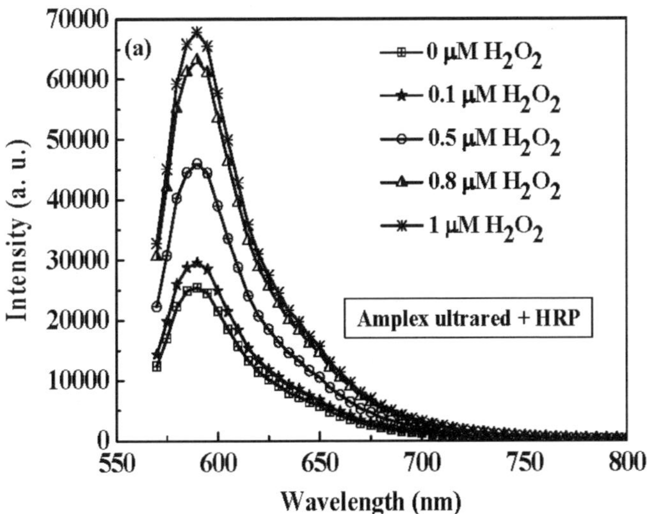

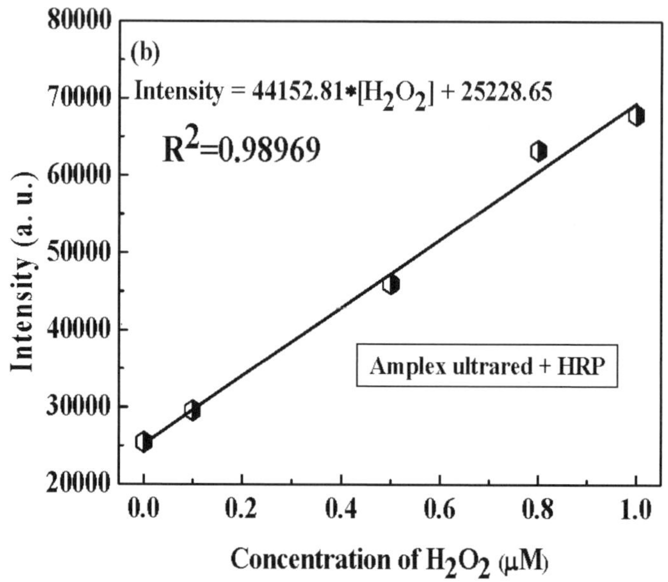

(b)
Intensity = 44152.81*[H_2O_2] + 25228.65
R^2=0.98969
Amplex ultrared + HRP
Intensity (a. u.)
Concentration of H_2O_2 (μM)

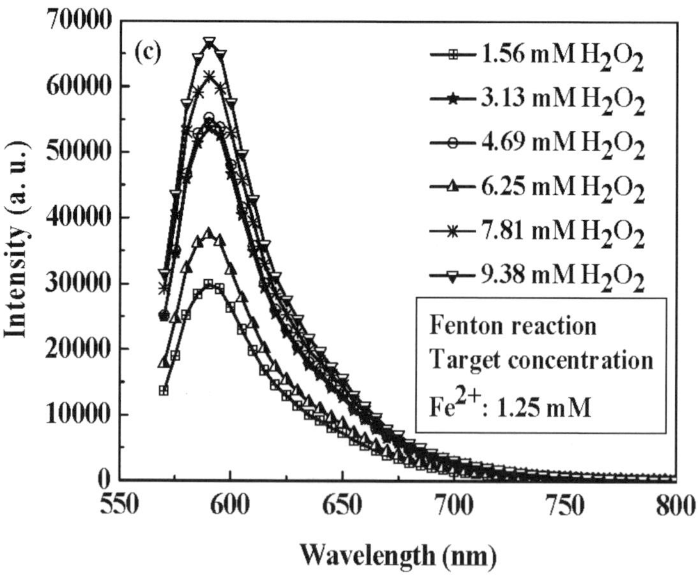

(c)
1.56 mM H_2O_2
3.13 mM H_2O_2
4.69 mM H_2O_2
6.25 mM H_2O_2
7.81 mM H_2O_2
9.38 mM H_2O_2
Fenton reaction
Target concentration
Fe^2+: 1.25 mM
Intensity (a. u.)
Wavelength (nm)

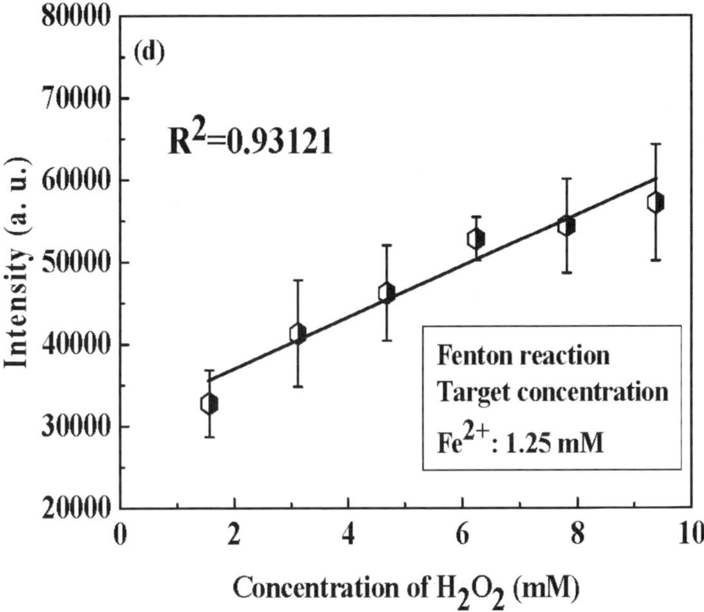

Figure 17. (a) Fluorescence emission at 590 nm from oxidized Amplex ultrared in the presence of HRP and different concentrations of hydrogen peroxide (b) Fluorescence intensity from oxidized Amplex ultrared versus the concentration of hydrogen peroxide in the presence of HRP. (c) Fluorescence emission at 590 nm from oxidized Amplex ultrared in the presence of hydroxyl radicals generated in Fenton reaction, with different concentrations of hydrogen peroxide. (d) Fluorescence emission at 590 nm from oxidized Amplex ultrared in the presence of hydroxyl radicals versus the concentration of hydrogen peroxide.

Amplex ultrared was utilized to be oxidized by hydrogen peroxide via the catalytic effect of HRP. The oxidized dye gives a strong fluorescence emission at 590 nm. The original (un-oxidized) dye is inactive in fluorescence emission, and thus did not interfere the fluorescence emission from the oxidized dye. Since the dye can also be oxidized by hydroxyl radical, there should be fluorescence emission from the oxidized dye, as long as the radical-oxidized dye is the same as the one oxidized by hydrogen peroxide. By comparing the fluorescence intensities of hydrogen peroxide-oxidized and radical-oxidized dyes, the concentration of hydroxyl radicals is resolved, as shown in Figure 17. The concentration of hydroxyl radicals is thereby elucidated in Fenton reaction. Figure 18 shows the concentration of hydroxyl radicals versus the concentration of hydrogen peroxide in Fenton reaction at constant concentration of ferrous ions.

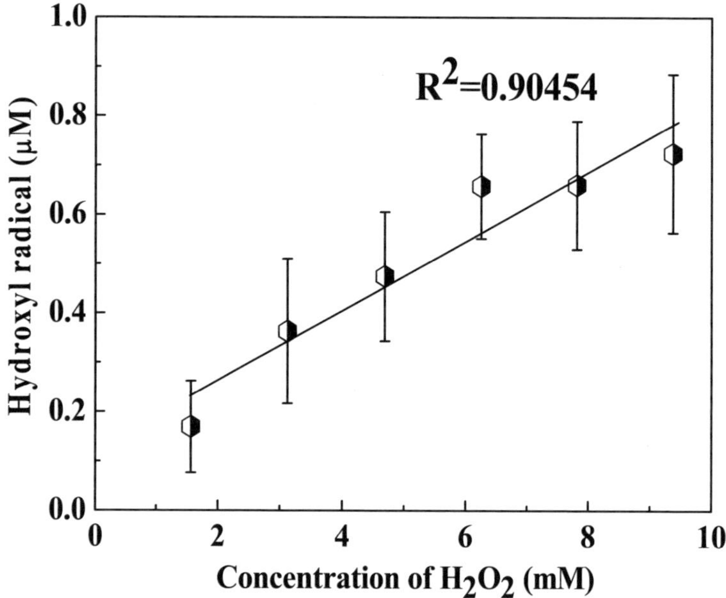

Figure 18. The concentration of hydroxyl radicals versus the concentration of hydrogen peroxide with fixed ferrous ions at 1.25mM in Fenton reaction.

By combining Figure 16 and Figure 18, the dependence of the conductivity change on the concentration of hydroxyl radicals can be established, as shown in Figure 19. In Figure 19, the x-axis and the y-axis are from the line fitted in Figure 18, which represents the concentration of hydroxyl radicals in Fenton reaction, and the y-axis of Figure 16, respectively. The percentage of the conductivity change of

PANI versus the concentration of hydroxyl radicals is shown in Figure 19. The percentage of the conductivity change of PANI shows linear dependence on the concentration of hydroxyl radicals. The detectable range for the hydroxyl radicals is from 0.2 μM to 0.8 μM. The limit of detection is around 0.2 μM. The result shows that this electronic hydroxyl radical microsensor has comparable sensitivity and limit of detection to the frequently used techniques, such as the fluorescence spectroscopy, ESR, or UV-Vis (absorbance) (Burns et al. 2012).

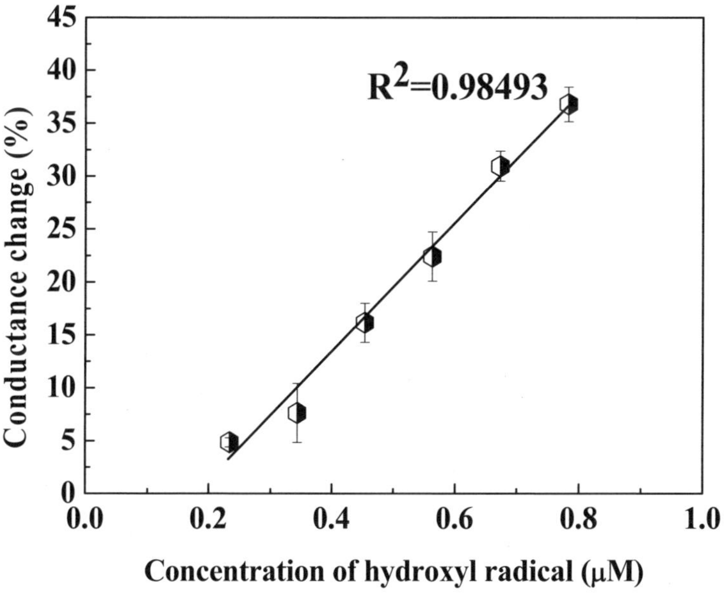

Figure 19. The percentage of conductivity change of PANI versus the concentration of hydroxyl radicals.

8.5. Summary

In this chapter, the conducting polymer, polyaniline, has been fabricated as hydrogen peroxide sensors, cholesterol sensors and hydroxyl radical sensors, based on the conductivity change of polyaniline. Enzymes are easily immobilized in the polymer and charge transfer is efficient. Ultra-sensitive hydrogen peroxide sensors are achieved by accumulation of the conductivity change caused by each reaction between hydrogen peroxide and HRP. Because hydrogen peroxide is the product of many reactions of oxidases, a lot of metabolites can be detected with hydrogen peroxide sensor incorporated with oxidase, as illustrated as HRP and cholesterol oxidase for cholesterol detection. Reaction between hydroxyl radicals and polyaniline simply cause the conductivity change of the polyaniline. The calibration curve for hydroxyl radicals and conductivity change of polyaniline is established. The polyaniline has been shown a promising polymer, which can be used for electronic microsensors for low cost and high sensitivity detection of metabolites.

References

1. Giorgio, M., Trinel, M., Migliaccio, E., Pelicci, P.G., 2007. Nat. Rev. Mol. Cell Biol. 8, 722–728.
2. D'Autréaux, B., Toledano, M.B., 2007. Nat. Rev. Mol. Cell Biol. 8, 813–824.
3. Burns, J.M., Cooper, W.J., Ferry, J.L., King, D.W., DiMento, B.P., McNeill, K., Miller, C.J., Miller, W.L., Peake, B.M., Rusak, S.A.,Rose A.L.,Waite T.D., 2012. Aqua. Scien. 74, 683–734.
4. Cadenas, E., Davies, K.J., 2000. Free. Radical Bio. Med. 29(3-4), 222-30.
5. Albers , A.E.,Okreglak, V.S., Chang, C.J., 2006. J. the Am. Chemi. Soci. 128 (30), 9640–9641.
6. Yang, X., Guo, Y., Mei, Z., 2009. Anal. Biochem. 393, 56–61.
7. Matsubara, C.,Kawamoto, N.,Takamura, K., 1992. Analyst 117, 1781-1784.
8. Steinberg, S.M., 2013. Environ. Monitor. Assess. 185, 3749–3757.
9. Chen, W., Cai, S., Ren, Q.Q., Wen, W., Zhao Y.D., 2012. Anal. 137, 49-58.
10. Chen, S., Yuan, R., Chai, Y., Hu, F., 2013. Microchim. Acta. 180, 15–32.
11. Ahammad, A.J.S., 2013. Biosens. Bioelectron. S9:001.
12. Bartlett, P.N., Birkin, P.R., Palmisano, F., Benedetto, G.D., 1996. J. Chemi. Soci., Faraday Transactions 92, 3123-3130.
13. Gorton, L., Jonsson-Pettersson, G., Csoregi, E., Johansson, K., Dominguezt , E., Marko-Varga, G., 1992. Anal. 117, 1235-1341.
14. Bartlett, P.N., Birkin, P.R., Wang, J.H., 1998. Anal. Chem. 70, 3685-3694.
15. Raffa, D., Leung, K.T., Battaglini, F., 2003. Anal. Chem. 75, 4983-4987.
16. Chen, S.A., Hwang, G.W.,1995. J. the Am. Chemi. Soci. 117, 10055-10062.
17. Arnao, M.B., Acosta, M., del Río, J.A., Varón, R., García-Cánovas, F.,1990. Biochimica et Biophysica Acta 1041(1), 43-47.

18. Vlasits, J., Jakopitsch, C., Bernroitner, M., Zamocky, M., Furtmüller, P.G., Obinger, C., 2010.   Arch. Biochem. Biophys. 500, 74–81.
19. Sanli, A.E., 2013. Int. J. Energy Res. 37, 1488–1497.
20. Buettner, G.R., 1993. Arch. Biochem. Biophys. 300, 535-543.
21. Chetna Dhand, Maumita Das, Monika Datta, and B. D. Malhotra, 2011. Biosens. Bioelectron. 26, 2811.
22. Yapici, N.B., Jockusch, S., Moscatelli, A., Mandalapu, S.R., Itagaki, Y., Bates, D.K., Wiseman, S., Gibson, K.M., Turro, N.J., Bi, L., 2011. Org. Lett. 14, 50-53.
23. Zhang, L., Liang, R.-P., Xiao, S.-J., Bai, J.-M., Zheng, L.-L., Zhan, L., Zhao, X.-J., Qiu, J.-D., Huang, C.-Z., 2014. Talanta. 118, 339-347.
24. Malins, D.C., Polissar, N.L., Gunselman, S.J., 1996. PNAS 93, 2557-2563.
25. Jacobs, A.C., Resendiz, M.J.E., Greenberg, M.M., 2010. J. Am. Chem. Soc. 132, 3668-3669.
26. Gieseg, S.P., Simpson, J.A., Charlton, T.S., Duncan, M.W., Dean, R.T., 1993. Biochemistry 32, 4780-4786.
27. Tien, M., Svingen, B.A., Aust, S.D., 1982. Arch. Biochem. Biophys. 216, 142-151.
28. Ren, J.-G., Xia, H.-L., Just, T., Dai, Y.-R., 2001. FEBS Lett. 488, 123-132.
29. Li, X., Liu, Y., Zhu, A., Luo, Y., Deng, Z., Tian, Y., 2010. Anal. Chem. 82, 6512-6518.
30. Liu, H., Tian, Y., Xia, P., 2008. Langmuir 24, 6359-6366.
31. Lemire, J., Harrison, J., Turner, R., 2013. Nat. Rev. Micro. 11, 371-384.
32. Cohn, C., Simon, S., Schoonen, M., 2008. Part. Fibre Toxicol. 5, 1-9.
33. Price, M., Reiners, J.J., Santiago, A.M., Kessel, D., 2009. Photochem. Photobiol. 85, 1177-1181.
34. Zhuang, M., Ding, C., Zhu, A., Tian, Y., 2014. Anal. Chem. 86, 1829-1836.
35. Page, S.E., Wilke, K.T., Pierre, V.C., 2010. Chem. Commun. 46, 2423-2425.
36. Westerhoff, P., Mezyk, S.P., Cooper, W.J., Minakata, D., 2007. Environ. Sci. Technol. Lett. 41, 4640-4646.
37. Bors, W., Saran, M., Michel, C., 1982. Int. J. Radiat. Biol. 41, 493-501.
38. Schuler, R.H., Hartzell, A.L., Behar, B., 1981. J. Phys. Chem. A 85, 192-199.
39. Yu, F., Xu, D., Lei, R., Li, N., Li, K.a., 2008. J. Agric. Food. Chem .56, 730-735.
40. Teismann, P., Ferger, B., 2000. Brain Res. Protoc. 5, 204-210.
41. Gizdavic-Nikolaidis, M., Travas-Sejdic, J., Kilmartin, P.A., Bowmaker, G.A., Cooney, R.P., 2004. Curr. Appl. Phys. 4, 343-346.
42. Banerjee, S., Saikia, J.P., Kumar, A., Konwar, B., 2010. Nanotechnology 21, 045101.
43. Lewis, S., Lynch, A., Bachas, L., Hampson, S., Ormsbee, L., Bhattacharyya, D., 2009. Environ. Eng. Sci. 26, 849-859.

<h1 style="text-align:center">Chapter 9</h1>

<h2 style="text-align:center">ZnO Nanorod Based Sensors</h2>

Chen-Fong Lo[*], Byung S. Kang[*], Stephen J. Pearton[*],
Ivan I. Kravchenko[†] and Fan Ren[*]

*   University of Florida, Gainesville, FL 32611, USA

†   Center for Nanophase Materials Sciences, Oak Ridge National
Laboratory, Oak Ridge, Tennessee 37830

9.1. Introduction

AlGaN/GaN high electron mobility transistors (HEMTs) have shown promise for bio-sensing applications [1-6], since they include a high electron sheet carrier concentration channel induced by piezoelectric polarization of the strained AlGaN layer and spontaneous polarization [7-9]. There are positive counter charges at the HEMT surface layer induced by the two-dimensional electron gas (2DEG) located at the AlGaN/GaN interface. Any slight changes in the ambient can affect the surface charge of the HEMT, thus changing the electron concentration in the channel at the AlGaN/GaN interface. The typical detection time for the AlGaN/GaN HEMT sensors is in the range of 5-20 seconds [1-6]. Additionally, due to the wide energy bandgap, the AlGaN/GaN material system is extremely chemically stable. The detection is real time and the technology can be easily interfaced with wireless electronics. With proper surface functionalization, the AlGaN/GaN HEMT sensors have been used for different sensing applications, such as gas sensing for detecting hydrogen, carbon monoxide, carbon dioxide, and ammonium, as well as liquid sensing for detecting proteins, pH values of aqueous solutions, lactic acid, breast cancer in saliva, DNA, kidney injury molecules, prostate cancer, glucose, chloride ion, and mercury ions [1-6].

9.2. Sensor Fabrication

9.2.1. HEMT Fabrication

HEMT layer structures were grown on c-plane sapphire or (111) silicon substrate with a molecular beam epitaxy (MBE) or a metal organic chemical vapor systems. The HEMT layer structure included an initial 2 μm thick undoped GaN buffer followed by a 25 nm thick unintentionally doped $Al_{0.25}Ga_{0.75}N$ layer. Sensor fabrication began with the Ohmic contact deposition to form 50 μm $\times$ 50 μm Ohmic contact pads separated by a gap of 20-50 μm with the standard lift-off of e-beam evaporated Ti/Al/Ni/Au-based metallization. The metal contact were subsequently annealed at 850°C for 45 seconds in N_2 ambient. Device isolation was achieved with multiple energies and doses of nitrogen ion implantations using photoresist as the mask to protect active regions of the devices [10]. E-beam deposited Ti/Au (200 Å/1000 Å) was employed as interconnection metal. 250 nm of silicon nitride layer deposited with a plasma-enhanced chemical vapor deposition (PECVD) system was used to passivate the source/drain regions [19]. CF_4/O_2 based plasma etching was employed to open the contact windows on the silicon nitride layer. The gate area of the sensor was functionalized with ZnO nanorods for CO or glucose sensing. Figure 1 illustrates a schematic of HEMT with ZnO nanorods grown on the gate area.

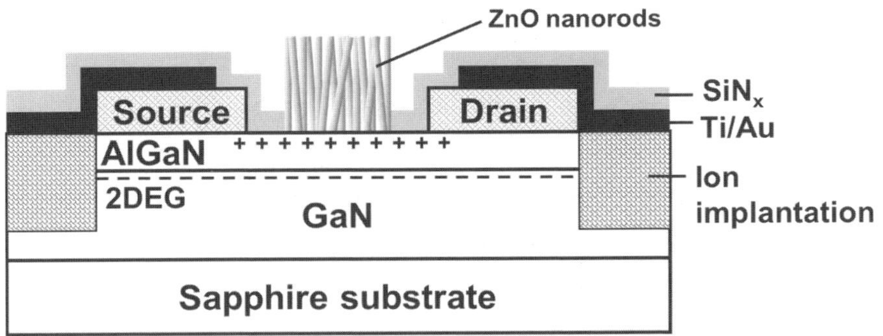

Figure 1. Schematic of AlGaN/GAN HEMT sensor with ZnO nanorods grown on the gate area.

9.2.2. ZnO nanorod growth

The growth ZnO nanorods started with ZnO nano-crystal seed preparation on the gate area ZnO nano-crystal seed solution was prepared by slowly adding 30 mM NaOH solution in methanol to a 10 mM zinc acetate dihydrate solution kept at 60°C. Then, the ZnO nano-crystal seed solution was spun on the HEMT sample. Photoresist was used to pattern the gate area prior to the coating of seed solution, and the resist were removed with acetone after the coating of seed solution.

The nano-crystalline seed coated HEMT sensor chips were then immersed in a beaker containing an aqueous mixture of 20 mM zinc nitrate hexahydrate and 20 mM hexamethylenetetramine, then put the beaker in the oven at ~94°C for 3 hrs [11]. After the nanorod growth, the device was removed from the solution, thoroughly rinsed with de-ionized water to remove any residual salts, and dried with nitrogen gas. Figure 2(a) shows the scanning electron microscope picture of the ZnO nanorods on the gate area of AlGaN/GaN HEMT sensors as illustrated in Figure 2(b) and 2(c).

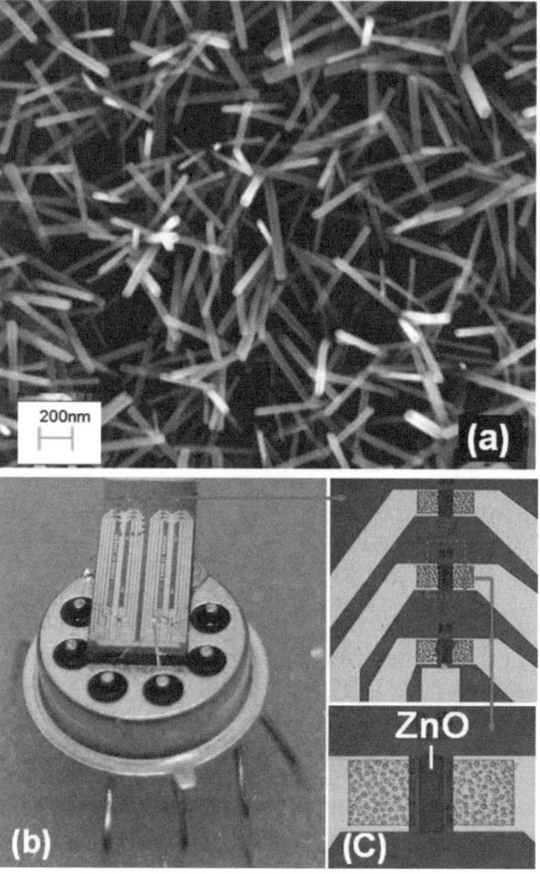

Figure 2(a). Scanning electron microscope picture of the ZnO nanorods on the gate area of the HEMT sensor.   2(b) Microscope picture of the ZnO nanorods grown on the gate area of the HEMT sensor. 2(c) Microscope picture of the AlGan/gaN HMET sensor chip mounted a TC05 carrier with ZnO nanorods grown on the gate area [21]. Reprinted with permission from Chien-Fong Lo, Lu Liu, Byung-Hwan Chu, Fan Ren, Stephen J. Pearton, Sylvain Dore´, Chien-Hsing Hsu, Jihyun Kim, Amir M. Dabiran and Peter P. Chow, J. Vac. Sci. Technol. B30, 010606 (2012).

ZnO nanorods could be selectively grown on specific spots of substrate. As shown in Figure 3, a shape of gator head, the logo of the University of Florida, displays with a bunch of ZnO nanorod dots. Each dot consists of hundreds of ZnO nanorods.

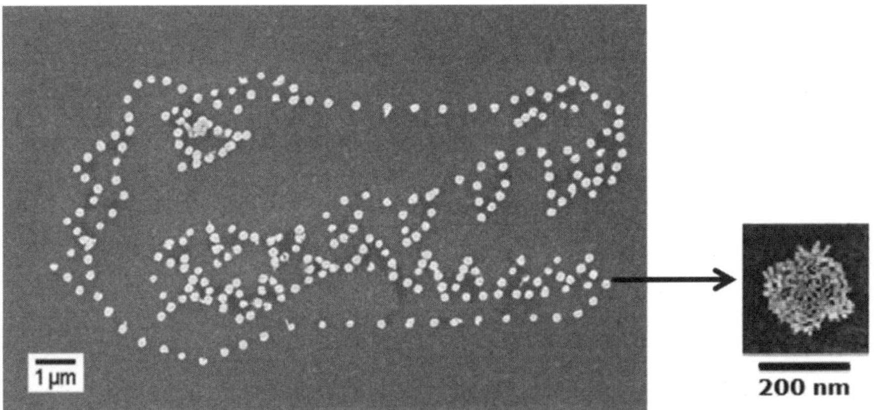

Figure 3. A shape of gator head, the logo of the University of Florida, displayed with a bunch of ZnO nanorod dots; each ZnO nanorod dot consists of hundreds of ZnO nanorods.

9.3. Carbon Monoxide Detection

Carbon monoxide (CO), a colorless, odorless, tasteless, non-irritating but extremely toxic gas, is generally difficult to detect. It has been reported that exposures of 100 ppm or greater concentrations of CO is dangerous to human health [12]. Human brain as well as heart may be further injured with carboxyhemoglobin (COHb) levels exceeding 20% in the blood [13]. Interestingly, after ischemic brain injury, a small amount of CO introduced to patients can be neuroprotective[14]. Inhalation of low CO concentrations provided neuroprotection for the brain of the mice with transient middle cerebral artery occlusions after cerebral ischemia. Therefore it is very critical to modulate the amount of CO inhaled to avoid overdose.

Various metal oxide gas sensors based on the change of material conductances have been widely studied [15-18]. However the performances of these oxide based sensors suffer from very low detection sensitivities and slow response times in the range of several minutes. In order to boost the sensitivity, high temperature operation (>300°C) was essential for these metal oxide based sensors [6-8]. Recently, sensors using AlGaN/GaN high electron mobility transistors (HEMTs) have shown excellent potential for a variety of gas and biochemical sensing applications by functionalizing specific sensing materials, including hydrogen, carbon dioxide, oxygen, mercury ions, breast cancer and glucose [1-6]. The two-dimensional electron gas (2DEG) channel of the AlGaN/GaN HEMT based sensors located 20-30 nm below the surface at the interface between the AlGaN and GaN layers is very sensitive to the surface charges[7-9]. Thus, any small changes of charges in the ambient near the HEMT surface can affect the conductance in 2DEG right away and the sensor response time can be dramatically shortened. The tiny charge variations will be amplified through the transistor operation, thus the detection limits of the AlGaN/GaN HEMTs are significantly improved over the conventional simple conductance measurement based metal oxide sensors. Recently, chemical vapor deposited (CVD) ZnO nano-wires functionalized AlGaN/GaN HEMT sensors have shown a CO detection sensitivity of 400 ppm [19], however, the CO concentration of 400 ppm is too high for the neuroprotective applications. In this work, we have studied the ambient temperature effect ranging from 25°C to 400°C on the CO detection limit of ZnO nano-rod functionalized AlGaN/GaN HEMTs. The ZnO nanorods were prepared with a sol-gel method to lower the cost of the metal oxide growth.

9.3.1. Carbon Monoxide Sensing System

The gas sensing experiments were performed in a tube-furnace that contained electrical feed-throughs connected to either an HP4145 parameter analyzer or a current–voltage (I–V) measurement system [20]. The CO concentration was adjusted by diluting 500 ppm CO-balanced with nitrogen with pure nitrogen, as shown in Figure 4.

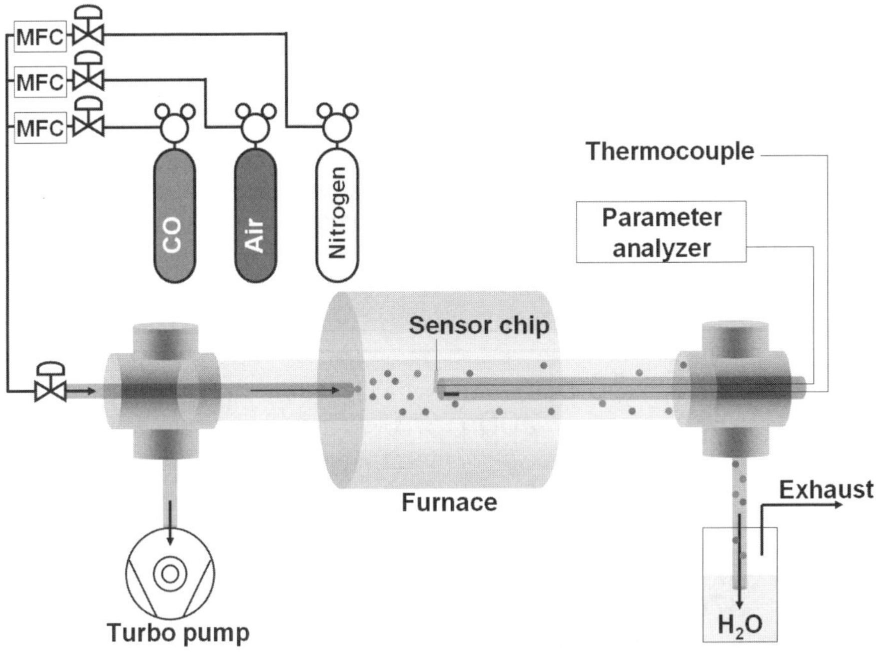

Figure 4. Schematic of the CO sensing system. Reprinted with permission from Chien-Fong Lo, Yuyin Xia, Lu Liu, Stephen J. Pearton, Sylvain Doré, Chien-Hsing Hsu, Amir M. Dabiran, Peter P. Chow, Fan Ren, Sensors and Actuators B176, 708-712 (2013).

9.3.2. Carbon Monoxide Limit of Detection

Figure 5 illustrates normalized percent drain current change, $\Delta I/I$, and time-dependent drain current of the ZnO nanorod-gated AlGaN/GaN HEMTs, as the gas ambient switched from N$_2$ to different CO concentrations in N$_2$ ambient, and then back to N$_2$ at three different ambient temperatures; room temperature, 150, and 250ºC.

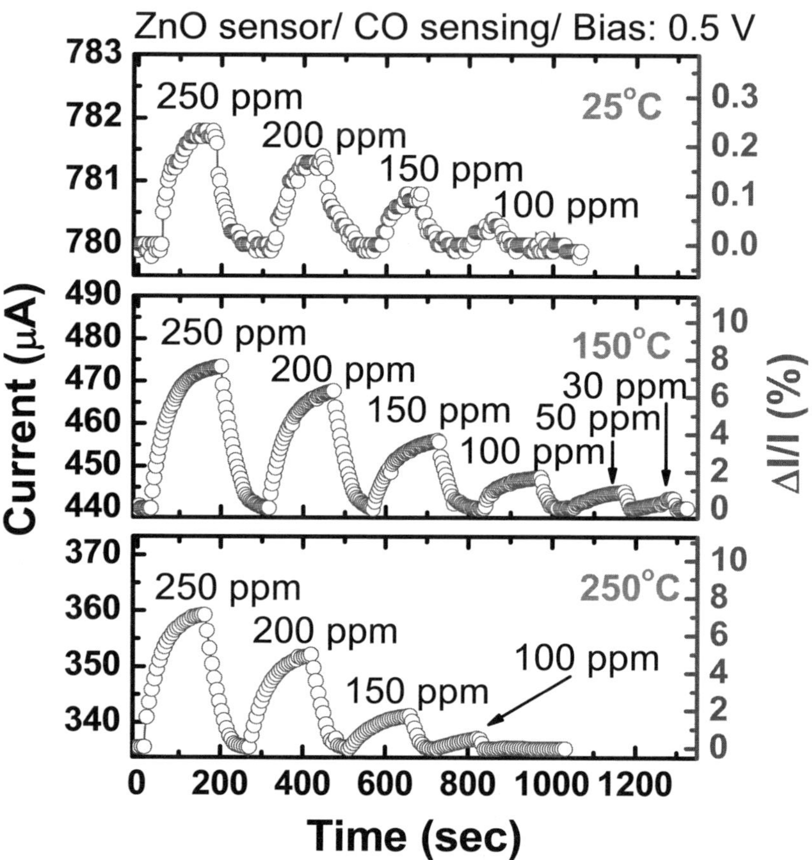

Figure 5. Time-dependent drain current of the ZnO nanorod gated AlGaN/GaN HEMTs and time dependant normalized percent drain current change with different CO concentration at 25ºC, 150ºC, and 250ºC [21]. Reprinted with permission from Chien-Fong Lo, Lu Liu, Byung-Hwan Chu, Fan Ren, Stephen J. Pearton, Sylvain Dore´, Chien-Hsing Hsu, Jihyun Kim, Amir M. Dabiran and Peter P. Chow, J. Vac. Sci. Technol. B30, 010606 (2012).

The drain current of the sensors increased upon exposure to the CO ambient and the drain current changes were proportional to the

CO concentration. Although, the sensing sensitivity was proportional to the bias voltage at the drain electrode, the drain electrode of the sensor was biased at a constant voltage of 0.5 V to minimize the sensor power consumption and maintain a reasonable sensing sensitivity. Interestingly, the ZnO nano-rod functionalized HEMT sensors could sense CO ranging from 250 ppm to 100 ppm due to the larger sensing surface provided by the ZnO nanorods at room temperature. A reference sensor was also fabricated on the same sample without the ZnO nanorods in the gate area. The reference sensor did not show any response to CO ambient. Smilar experiments were also conducted at both 150 and 250°C. As shown in Figure 4 (b) and (c), the HEMT sensor showed sensitivities to CO concentrations ranging from 250 ppm to 30 ppm and from 250 ppm to ~100 ppm at 150 and 250°C, respectively. The CO percent detection sensitivities for 250 ppm CO concentrations measured at 150 and 250°C were around 7.5% and 7%, which were 30 times larger than the detection sensitivity of ~0.23% measured at room temperature. The detection limits were also improved to 30 ppm and 100 ppm at 150°C and 250 °C, respectively, as compared to the detection limit of ~100 ppm at room temperature.

9.3.3. Sensing Mechanism

It is well known that the chemisorbed oxygen ions on the oxide surface react with CO to form CO_2 and release electrons to the oxide surface [22]. There are three different states of oxygen ions, O_2^-, O^-, and O^{2-}, present on the ZnO surface dependent on ZnO surface temperature; O_2^- is the dominant ions below 100°C, O^- ions between 100 and 300°C, and O^{2-} ions above 300°C [22]. Thus, the predominant reaction at room temperature is between CO and O_2^- ions; two molecules of CO react with one O_2^- ion to produce two molecules of CO_2 and one electron.

$$2CO + O_2^- \rightleftharpoons 2CO_2 + e^- \tag{1}$$

At 150°C, the main reaction involves CO and O^- ions; one CO molecule reacts with one O^- ion to produce one CO_2 molecule and one electron.

$$CO + O^- \rightleftharpoons CO_2 + e^- \tag{2}$$

Above 300°C, one molecule of CO reacts with one O^{2-} ion to yield one CO_2 molecule and two electrons.

$$CO + O^{-2} \rightleftharpoons CO_2 + 2e^- \tag{3}$$

Those electrons generated from the interactions between CO molecules and oxygen ions make the oxide surface more negatively charged, and the negative charges on the oxide surface induced additional positive charges on the AlGaN surface and extra electrons in the 2DEG channel. By functionalized ZnO nano-rods on the gate area of the HEMTs, the sensor has a larger surface area as compared to ZnO thin film, and more electrons would be produced during the CO sensing to achieve better CO detection sensitivity. A schematic of the sensing mechanism of the ZnO nanorod-gated AlGaN/GaN HEMTs is shown in Figure 6.

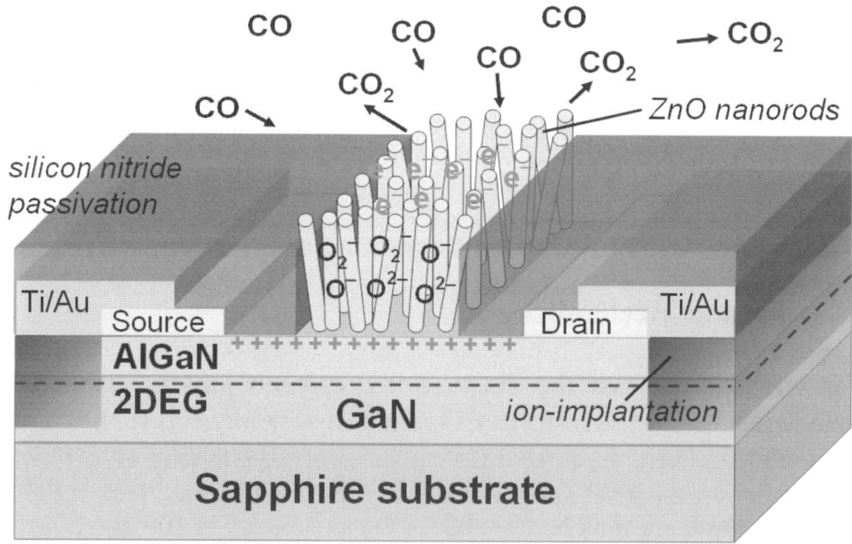

Figure 6. Schematic of ZnO nanorod-gated AlGaN/GaN HEMT sensing mechanism [21]. Reprinted with permission from Chien-Fong Lo, Lu Liu, Byung-Hwan Chu, Fan Ren, Stephen J. Pearton, Sylvain Dore´, Chien-Hsing Hsu, Jihyun Kim, Amir M. Dabiran and Peter P. Chow, J. Vac. Sci. Technol. B30, 010606 (2012).

Figure 7 shows the drain currents of the ZnO nanorod functionalized AlGaN HEMT sensors as a function of the temperature ranging from 25°C to 400°C during the exposures of 250 ppm CO or pure nitrogen. The detection sensitivity defined as the drain current differences between the 250 ppm CO and pure nitrogen increased from 25°C to ~200°C due to additional electrons produced during the CO exposure at the elevated temperature as previous discussed. However the difference of drain current slightly decreased once the ambient temperature passed ~200°C. This could be due to the reduction of oxygen molecules absorbed on the oxide surface at higher temperature and less electrons generated through the oxygen and CO reaction [24-26].

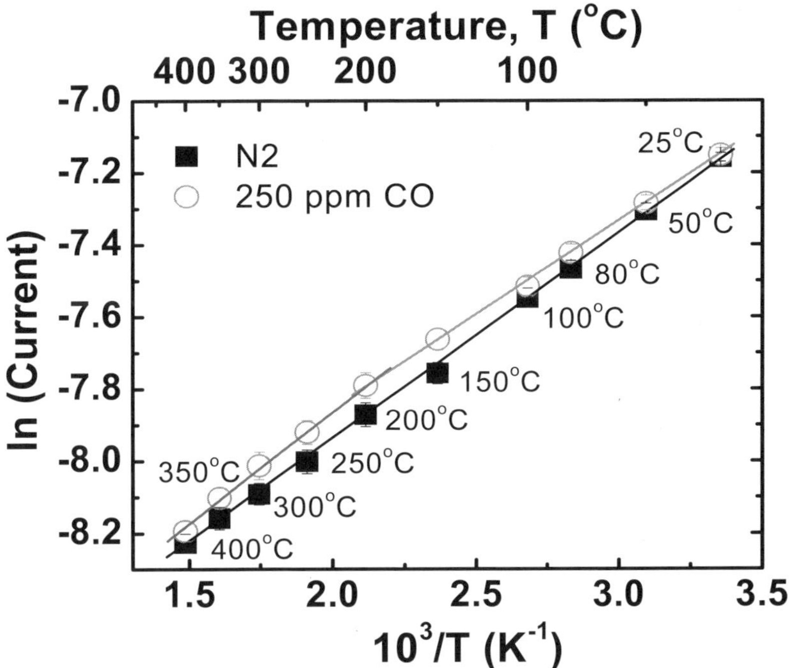

Figure 7. Drain currents of the ZnO nanorod functionalized AlGaN HEMT sensors as a function of ambient temperature ranging from 25°C to 400°C during the exposures of 250 ppm CO or pure nitrogen [21]. Reprinted with permission from Chien-Fong Lo, Lu Liu, Byung-Hwan Chu, Fan Ren, Stephen J. Pearton, Sylvain Dore´, Chien-Hsing Hsu, Jihyun Kim, Amir M. Dabiran and Peter P. Chow, J. Vac. Sci. Technol. B30, 010606 (2012).

9.3.4. Sensing repeatability

Figure 8 illustrated sensing repeatability of the ZnO nanorod functionalized AlGaN/GaN HEMT sensor measured at 150°C. The drain current responded to the CO gas repeatedly by switching the CO concentration back and forth from 0 to 50 ppm balanced with nitrogen. The rise and fall time of the CO sensing were less than 100 and 40 seconds, respectively, for 50 ppm CO detection. These delay were dominated by the dimension of the gas chamber used to sensing experiment; the diameter of the gas chamber and the distance between the gas injector head and the sample were 5.4 and 10 cm, respectively. It takes time to introduce CO to the sensor and purge out the CO from the chamber. Thus, the response times could be further decreased with a smaller diameter gas chamber and a shorter distance between the gas injector head and the sensor. ZnO nanorod-gated AlGaN HEMT sensors have demonstrated a tremendous potential to be employed for detecting low CO concentration with fast response times.

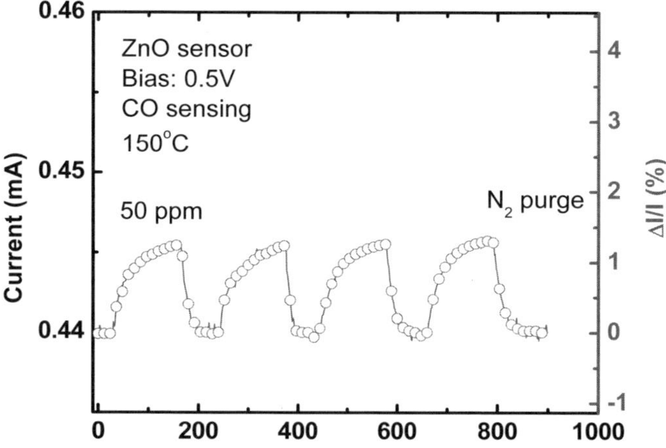

Figure 8. ZnO nanorod functionalized AlGaN/GaN HEMT sensing repeatability measured at 150°C [21]. Reprinted with permission from Chien-Fong Lo, Lu Liu, Byung-Hwan Chu, Fan Ren, Stephen J. Pearton, Sylvain Dore′, Chien-Hsing Hsu, Jihyun Kim, Amir M. Dabiran and Peter P. Chow, J. Vac. Sci. Technol. B30, 010606 (2012).

9.4. Glucose Detection

Diabetes is a leading cause of death and disability worldwide affecting more than 135 million people. Projected number of diabetics will rise to 300 million by year 2025. Majority of current consumer tests for glucose need to draw blood for testing, and cause discomfort and inconvenience. Therefore, it is highly desirable to develop a non-invasive, low-cost, portable and wireless-capable device for glucose sensing. The glucose level in exhaled breath condensate (EBC) was reported to correlate well with the blood glucose level [13]. Besides blood, EBC is the most easy-accessed body fluids for testing glucose. The non-invasive EBC based technology allows point-of-need testing with quick and accurate results and has the potential for many new other opportunities such as in the medical, home, law enforcement and industrial diagnostic markets [1-12]. While it is a very exciting prospect for the approach detecting glucose in EBC, glucose level in EBC can be roughly two orders of magnitude lower than the glucose level in the blood. Hence, the accuracy, sensitivity, and robustness of the EBC based sensors need to be much better than that of the conventional blood-based glucose monitors.

9.4.1. Functionalize ZnO Nanorods with Glucose Oxidase Enzyme (GO$_X$) and Glucose Detection

The fabrication of the ZnO nanorod functionalized AlGaN/GaN HEMT was described in section 9.2.1. After fabricating the device, 5 µl of ~100 U/mg GO$_x$ solution was introduced to the gate area of the HEMT using a micro-liter pipette, and the GOx solution was prepared by dissolving GOx in 10 mM phosphate buffer saline (PBS). Then, the functinalized sensor chip was kept at 4°C for 48 hours for GO$_x$ immobilization on the ZnO nanorod arrays followed by rinsing the sensor chip with Di water to remove the un-immobilized GO$_x$.

9.4.2. Glucose Sensing Sensitivity

Prior to the glucose detection, the functionalized HEMT chip was kept in the incubator at around 37°C for 30 minutes to make the enzyme active. Then, the target glucose was applied to the device through a micro-pipette

(2-20μl). The current-voltage characteristics of the HEMT chip were recorded using an Agilent 4156C parameter analyzer. The real time glucose detection in PBS buffer solution was illustrated as the drain current change of the HEMT sensor chip with a constant bias of 250 mV on the drain electrode, as shown in Figure 9. There was no drain current change detected by exposing the sensor with PBS buffer solution at around 200 sec, demonstrating the specificity and stability of the sensor chip. By sharp contrast, rapid current changes within 1-2 seconds were observed when sensor chip was exposed to target glucoses. The response of drain current change is linear proportional to glucose concentration from 0.5 nM to 14.5 μM with a limit of detection at 0.5 nM. The ZnO nanorod matrix provides a micro-environment for immobilizing negatively charged GO_x, which interact with glucose and produce electrons. The negative charges of the electrons increase the positive charges in the AlGaN layer. More positive charges would increase higher drain current of the HEMT.

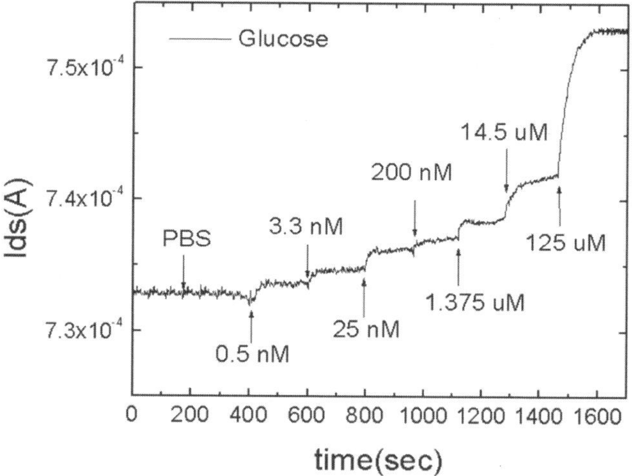

Figure 9. Drain current versus time with successive exposure of glucose from 500 pM to 125 μM in 10 mM phosphate buffer saline [22]. Reprinted with permission from B. S. Kang, H. T. Wang, and F. Ren, S. J. Pearton, T. E. Morey and D. M. Dennis, J. W. Johnson, P. Rajagopal, J. C. Roberts, E. L. Piner, and K. J. Linthicum, Appl. Phys. Lett., 91, 25106 (2007).

9.4.3. Detection Mechanism

For glucose detection, the AlGaN/GaN gate area was functionalized with oxidase enzyme attached to ZnO nanorods. The glucose oxidase enzyme (GO_X) is commonly used in biosensors to detect levels of glucose for diabetics. By keeping track of the number of electrons passed through the enzyme, the concentration of glucose can be measured. Due to the importance and difficulty of glucose immobilization, numerous studies have been focused on the techniques of immobilization of glucose with carbon nanotubes, ZnO nano-materials, and gold particles [28, 29]. ZnO-based nano-materials are especially interesting due to their non-toxic properties, low cost of fabrication and favorable electrostatic interaction between the ZnO and GO_x.

The glucose sensor was achieved with a selective-area ZnO nanorod growth on the gate area of the glucose sensing HEMT. By incorporating the nanorods on the HEMT gate sensing area, the total sensing area and sensitivity could be increased significantly. The ZnO nanorod matrix provided a micro-environment for immobilizing negatively charged GOx while retaining its bioactivity, and passed charges produced during the GOx and glucose interaction to the AlGaN/GaN HEMT. Figure 10 shows the drain current change in different concentrations of glucose dissolved in the PBS buffer solution. No current change could be seen with the addition of buffer solution, showing the specificity and stability of the device. By sharp contrast, the current change showed a rapid response of less than 1-2 seconds when target glucose was added to the surface. When the glucose sensor was used in a pH controlled environment, the drain current stayed fairly constant [12-14]. We have previously shown that HEMT sensors functionalized with Sc_2O_3 thin films has been demonstrated to have excellent response to the pH value of the solution [15-17] and these can be integrated on the same chip as the glucose sensors change in different concentrations of glucose dissolved in the PBS buffer solution.

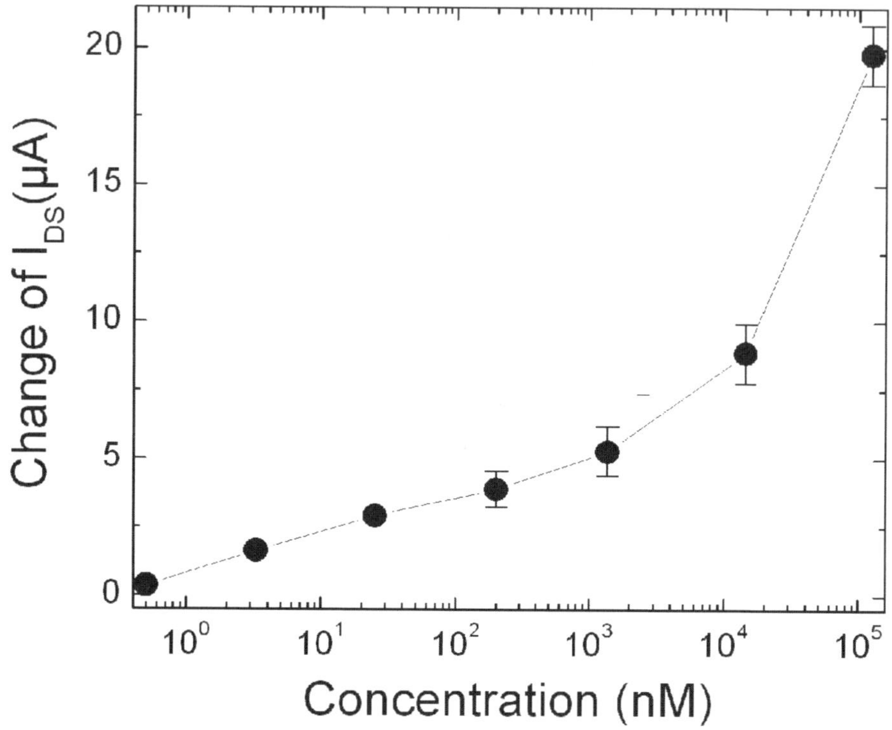

Figure 10. Change of drain current as a function of glucose concentrations from 500 pM to 125 μM in 10 mM phosphate buffer saline with a pH value of 7.4 [22]. Reprinted with permission from B. S. Kang, H. T. Wang, and F. Ren, S. J. Pearton, T. E. Morey and D. M. Dennis, J. W. Johnson, P. Rajagopal, J. C. Roberts, E. L. Piner, and K. J. Linthicum, Appl. Phys. Lett., 91, 25106 (2007).

9.4.4. Exhale Breath Condensate Collector Design

Presently, Au nano-particle, ZnO nanorod and nanocomb, or carbon nanotube material immobilized with GO_x has been used to detect glucose electrochemically, which requires to place a reference electrode in the glucose containing solution [23-25]. Thus the minimal volume of glucose sample can not be less than 1 milli-liter. However, each tidal breath roughly contains around 3 μl of the EBC, it will take more than 20 mins to collect 1 milli-liter of EBC. It is high desire to detect the glucose in BEC with only one tidal breath. Thus, these approaches are not practical for detecting glucose in EBC. On the other hand, the gate sensing area of the HEMT sensor is 4×50 μm^2. There is no reference electrode required for the HEMT based sensor, the amount of sample only needs to cover the gate area of the HEMT sensor. The volume of a half sphere with a diameter of 50 μm is around 3×10^{-11} liter. This volume is orders less than the 3 μl of EBC in each of tidal breath, therefore the EBC in one tidal breath would be more than enough to determine the glucose concentration in the EBC.

Besides the issue of the sample size, the sensing sensitivity of HEMT based sensor is superior to electrochemical detecting sensors, which measure the current density by applying a fixed potential applied between nano-materials and the reference electrode. The typical detection limit of these sensors is around 0.5-70 μM, which would be able to detect the glucose concentration in blood but not in EBC. Although AlGaN/GaN HEMT based sensor used the same ZnO nanorods and GO_x immobilization on the gate of the HEMT. The glucose detection signal was amplified through the HEMT. Although the number of electrons generated between glucose and immobilized GO_x were identical for both approach, a much lower detection limit of 0.5 nM was achieved for the HEMT based sensor due to this transistor amplification effect.

To take the advantage of quick response (around 1-2 sec) of the HEMT sensor, a real-time EBC collector is required to effectively condense the vapor in the tidal breath. Each tidal breath roughly contains around 3 μl of the EBC. To condense 3 μl of water vapor from each tidal breath, only around ~ 7 J of energy are needed to be removed. Such small energy can be simply taken out with a thermal electric module, a Peltier device, by mounting HEMT sensor directly on the top of the Peltier unit. The Peltier unit can be cooled to precise temperatures with a short response time around 1 second by applying known voltages and currents to the unit. For example, by applying bias of 0.7 V at 0.2 A to the Peltier unit employed in this study, the temperature of hotter and colder plater plate of the Peltier unit became 21 and 7°C, respectively, and the sensor took less than 2 sec reaching thermal equilibrium with the Peltier unit. This allows the EBC in the exhaled breath to instantaneously condense on the gate region of the HEMT sensor.

9.5. Conclusion

ZnO nanorod-gated AlGaN/GaN HEMT sensors have been demonstrated to be very effectively to detect low concentration of carbon monoxide. By increasing the ambient temperature, the limit of detection and sensing sensitivity significantly were improved. The detection capability of low carbon monoxide concentration and the behavior of rapid sensing response show ZnO nanorod-gated AlGaN/GaN HEMT sensors to be an excellent candidate of carbon monoxide sensing for clinical or industrial applications.

By immobilizing ZnO nanorod grow on the gate area of the AlGaN/Gan HEMT with glucose oxidase enzyme, AlGaN/GaN HEMT based sensor could detect glucose with a limit of detection of 0.5 nM. This electronic detection of biomolecules is a significant step towards a compact sensor chip, which can be integrated with a commercial available hand-held wireless transmitter to realize a portable, fast response and high sensitivity glucose detection.

References

1. X. Yu, Z. N. Low, J. Lin, T. J. Anderson, H. T.Wang, F. Ren, Y. L.Wang, C. Y. Chang, X. Yua, C. Li, S. J. Pearton, C. H. Hsu, A. Osinsky, A. Dabiran, P. Chow, C. Balaban, J. Painter, Sensors & Actuators, **B135,** 188 (2008).
2. H. T. Wang, B. S. Kang, T. F. Chancellor, Jr., T. P. Lele, Y. Tseng, F. Ren, S. J. Pearton, A. Dabiran, A. Osinsky, and P. P. Chow, Electrochem. & Solid-State Lett., **10,** J150-J153 (2007).
3. C. Y. Chang, B. S. Kang, H. T. Wang, F. Ren, Y. L. Wang, S. J. Pearton, D. M. Dennis, J. W. Johnson, P. Rajagopal, J. C. Roberts, E. L. Piner, and K. J. Linthicum, Appl. Phys. Lett., **92,** 232102 (2008).
4. K. H. Chen, B. S. Kang, H. T. Wang, T. P. Lele, F. Ren, Y. L. Wang, C. Y. Chang, S. J. Pearton, D. M. Dennis, J. W. Johnson, P. Rajagopal, J. C. Roberts, E. L. Piner, and K. J. Linthicum, Appl. Phys. Lett., **92,** 192103(2008).
5. Yu-Lin Wang, C. Y. Chang, Wantae Lim, S. J. Pearton, D. P. Norton, B. H. Chu, C. F. Lo, F. Ren, J. W. Johnson, P. Rajagopal, J. C. Roberts, E. L. Piner, and K. J. Linthicum, J. Vac. Sci. Technol. **B 28,** 376 (2010).
6. B. S. Kang, H. T. Wang, F. Ren, and S. J. Pearton, J. Appl. Phys., **104,** 031101 (2008).
7. A. P. Zhang, L. B. Rowland, E. B. Kaminsky, V. Tilak, J. C. Grande, J. Teetsov, A. Vertiatchikh, and L. F. Eastman, J. Electron. Mater., **32,** 388 (2003).

8. O. Ambacher, M. Eickhoff, G. Steinhoff, M. Hermann, L. Gorgens, V. Werss, B. Baur, M. Stutzmann, R. Neuterger, J. Schalwig, G. Muller, V. Tilak, B. Green, B. Schafft, L. F. Eastman, F. Bernadini, and V. Fiorienbini, Proc. ECS **02-14**, 27 (2002).

9. S. C. Hung, C. W. Chen, C. Y. Shieh, G. C. Chi, R. Fan, and S. J. Pearton, Appl. Phys. Lett., **98**, 223504 (2011).

10. C. F. Lo, T. S. Kang, L. Liu, C. Y. Chang, S. J. Pearton, I. I. Kravchenko, O. Laboutin, J. W. Johnson, and F. Ren, Appl. Phys. Lett., **97**, 262116 (2010).

11. B. S. Kang, S. J. Pearton, and F. Ren, Appl. Phys. Lett., **90**, 083104 (2008).

12. L. D. Prockop and R. I., Chichkova, J. Neurology. Sci., **262**, 122 (2007).

13. M. C. Dolan, T. L. Haltom, G. H. Barrows, C. S. Short, and K. M. Ferriell, Annals of Emergency Medicine, **16**, 782 (1987).

14. E. Zeynalov and S. Doré, Neurotox Res., **15**, 133 (2009).

15. S. Roy, H.Saha, C. K. Sarkar, International J. Smart Sensnig & Intelligent systems, **3**, 605 (2010).

16. M. Radecka, K. Zakrzewska, M. Rekas, Sensors & Actuators, **B47**, pp. 194–204 (1998).

17. D. Kohl, Sensors & Actuators, **18**, pp. 71-113 (1989).

18. W. Schmid, N. Barsan, U. Weimar, Sensors & Actuators, **B103**, 362 (2004).

19. S. C. Hung, C. W. Chen, C. Y. Shieh, G. C. Chi, R. Fan, and S. J. Pearton, Appl. Phys. Lett., **98**, 223504 (2011).

20. J. Kim, B. P. Gila, C. R. Abernathy, G. Y. Chung, F. Ren, and S. J. Pearton, Solid-State Electronics, **47**, 1487–1490 (2003).

21. C. F. Lo, L. Liu, B. H. Chu, F. Ren, S. J. Pearton, S. Doré, C. H. Hsu, J. Kim, A. Dabiran, and P. P. Chow, J. Vac. Sci. Technol. B, **30**, 010606 (2012).

22. B. S. Kang, H. T. Wang, F. Ren, S. J. Pearton, T. E. Morey, D. M. Dennis, J. W. Johnson, P. Rajagopal, J. C. Roberts, E. L. Piner, and K. J.Linthicum, Appl. Phys. Lett. 91, 252103 (2007).

23. S. Park, H. Boo, T. D. Chung, Anal. Chi. Act., 556, 46 (2006).

24. S. K. Jung, Y. R. Chae, J. M. Yoon, B. W. Cho, K. G. Ryu, J. Microbiol. Biotechnol. 15(2), 234 (2005).

25. P. Pandey, S. P. Singh, S. K. Arya, V. Gupta, M. Datta, S. Singh, B. D. Malhotra, Langmuir, 23, 3333 (2007).

Chapter 10

Scalable Nanomanufacturing of Broadband Antireflection Coatings on Semiconductors

Pratik Kothary, Sin-Yen Leo, Danielle Liu, Peng Jiang

Department of Chemical Engineering
University of Florida, Gainesville, FL 32611, USA

10.1. Introduction

The microelectronics revolution sparked by the invention of transistors and the miniaturization of integrated electronic circuits has affected almost every aspect of our daily lives. Semiconductors than can control the flow of electrons lie at the heart of microelectronics. In an effort toward further high-density integration and system performance, scientists are now turning to light as the information carrier [1]. The travel speed of light in a dielectric material is greater than that of an electron in a metallic wire. The bandwidth of dielectric materials is about 3 to 4 orders of magnitude larger than that of metals [2]. Moreover, light particles (photons) are not as strongly interacting as electrons, helping to reduce energy losses. Unfortunately, when integrating optical functions into existing microelectronic devices, a key technical challenge surfaces. Semiconductor usually has a much higher refractive index than other dielectric materials [3]. For instance, silicon, which is the most widely used semiconductor in microelectronics, has a refractive index of ~3.4. This high refractive index leads to very large reflection of incident light (> 30%) at a single silicon-air interface if an AR coating is not applied on the silicon surface. Therefore, AR coatings are widely used in a spectrum of optical and optoelectronic devices, such as monitors, car dashboards, optical lenses, photodiodes, and solar cells [4-12]. Traditional quarter-wavelength AR coatings have been extensively developed to suppress unwanted optical reflection from different substrates [13-19]. They are designed to enable destructive interference between the reflected light from the surrounding medium/AR coating interface and the AR coating/substrate interface to eliminate reflection and maximize transmission [20]. Based on the Snell's law, optical reflection can be efficiently suppressed if the refractive index of the coating equals to the geometric mean of the refractive indices of the two media at the interface and the thickness of the coating is a quarter of the wavelength of light [21]. For instance, vacuum-deposited silicon nitride (SiN_x) AR coatings with a refractive index of ~1.9 (close to the geometric mean of the refractive indices of silicon and air) and thickness of ~150 nm are widely used on crystalline silicon solar cells to suppress optical reflection around 600 nm wavelength, where the energy intensity is maximized in the solar spectrum[22].

Unfortunately, traditional quarter-wavelength AR coatings have several major drawbacks that greatly limit the performance of the final optoelectronic devices. Single-layer dielectric AR coatings are narrowband and they only target a specific wavelength (e.g., ~600 nm wavelength for solar cells).[8] In addition, the incident light needs to be close to the normal incidence for destructive interference to occur [23]. Broadband dielectric multilayer AR coatings with alternating layers of high and low refractive index materials have therefore been developed to suppress optical reflection over a wide range of wavelengths and incident angles [24-27]. However, multilayer AR coatings, which are typically fabricated by consecutive deposition of different dielectrics in high vacuum through plasma-enhanced chemical vapor deposition (PECVD) [13-19], electron-beam evaporation [28-30], and sputtering[31,32], suffer from high cost, complex fabrication processes, and limited material selection for different substrates. Moreover, the vacuum-deposited dielectric layers could have very different coefficients of thermal expansion than the semiconductor substrates, leading to poor durability and thermal stability when operated at high temperature [33].

To overcome the aforementioned key technical barriers of traditional dielectric AR coatings, scientists are turning to the mother-nature to find technological inspiration [34-37]. Through millions of years of natural selection and evolution, nanostructured architectures with unique functions are abundant in biological systems [38]. One preeminent example can be found on the cornea of some nocturnal moths. They have a thin layer of non-close-packed hexagonal arrays of nanostructures with sub-300 nm height and spacing, named nipple arrays (Figure 1) [10,34,39]. These subwavelength-structured nipple arrays induce a gradual transition of refractive index from the surface of the eye to the air, and can therefore greatly reduce glare from the moths' eyes, preventing predators from pinpointing their locations during nights. This gradual transition in refractive index results in reducing reflection over a broad range of wavelengths and angles of incidence [40-46].[40-46] Therefore, subwavelength-structured moth-eye AR coatings are broadband [45,47,48].

To fabricate periodic nanostructured moth-eye AR coatings, many top-down nanomanufacturing technologies, such as photolithography[49], electron-beam lithography[50],nanoimprint lithography[48,51], and interference lithography[34,52-54], have been explored. However, attaining high-throughput and large-area fabrication continues to be a major challenge with these top-down approaches. By contrast, bottom-up self-assembly and subsequent templating nanofabrication provides a much simpler, faster, and inexpensive alternative to top-down nanolithography [51,55-78]. Some exemplary bottom-up technologies for fabricating subwavelength moth-eye AR coatings include templating processes using self-assembled colloidal particles and by using anodized aluminum oxide (AAO) films [42,43,79-82]. Unfortunately, currently available bottom-up techniques are mostly not compatible with standard microfabrication technologies, limiting large-scale and economic production and integration of moth-eye AR coatings into practical devices.

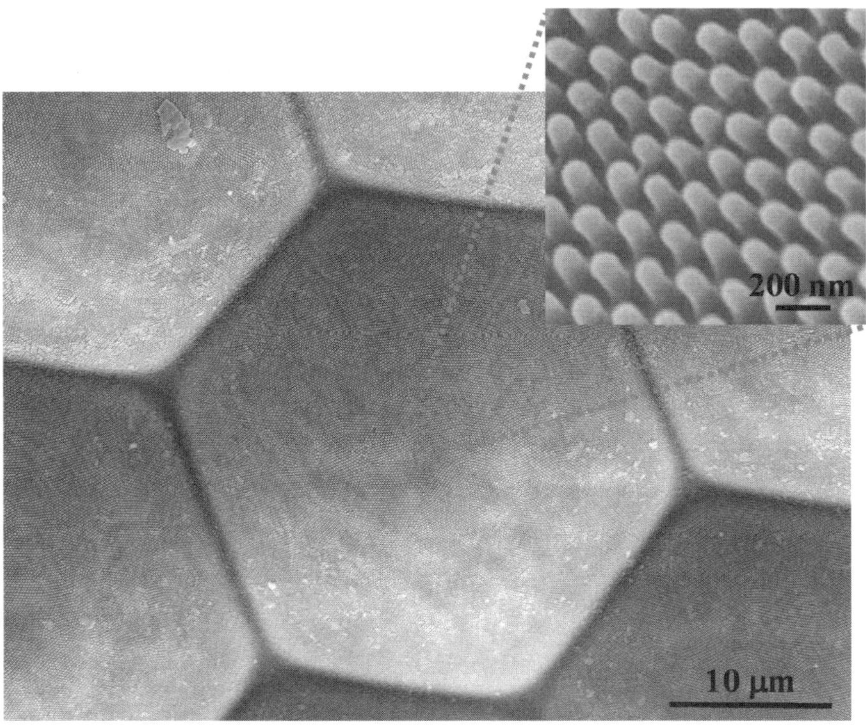

Figure 1. Typical top-view SEM image of a moth eye. The inset shows a higher magnification SEM image.

In this chapter, we will discuss two scalable nanomanufacturing platforms for large-scale production of moth-eye AR coatings on a variety of semiconductor substrates. In Section 2, a batch spin-coating platform that enables wafer-scale assembly of colloidal crystals with unusual non-close-packed crystal structure[83], and a continuous Langmuir-Blodgett platform for creating monolayer close-packed colloidal crystals on both planar and nonplanar substrates (e.g., a multicrystalline silicon wafer with very rough surfaces)[81], will be presented. Compared with existing colloidal self-assembly technologies, such as evaporation-induced convective assembly [84-87], assembly at liquid-liquid or liquid-air interface [88,89], gravitational sedimentation [90,91], electrostatic repulsion [92-95], and template-assisted assembly [96-99], these two platforms combine the simplicity and cost effectiveness of bottom-up self-assembly with the scalability and compatibility of standard microfabrication. In Section 3, the self-assembled colloidal crystals will be used as templating masks in a wet etch and a dry etch approach for fabricating broadband moth-eye AR coatings on various semiconductors including crystalline silicon, GaAs, and GaSb [100,101]. The superior antireflection performance and thermal stability of the templated AR gratings will be demonstrated by optical measurements. The experimental results are complemented by numerical simulations to pursue better understanding of the antireflection properties of moth-eye AR coatings.

10.2. Scalable Bottom-up Colloidal Self-assembly Technological Platforms

In this section, we will discuss two scalable bottom-up colloidal self-assembly nanomanufacturing platforms that enable large-scale assembly of colloidal silica microspheres into both 3-D and 2-D highly ordered colloidal crystals. The assembled silica particles will be used as structural templates for fabricating subwavelength-structured moth-eye AR coatings on various semiconductors (see Section 3). In the first spin-coating platform, which is based on a simple batch process, shear-induced crystallization of colloidal microspheres dispersed in viscous, nonvolatile monomers leads to rapid production of wafer-scale colloidal crystals with remarkably large domain sizes and unusual non-close-packed crystal structures [83,102]. The second Langmuir-Blodgett (LB) platform enables the continuous transfer of self-assembled monolayer colloidal crystals floating at an air/water interface onto both planar and nonplanar substrates (e.g., a solar-grade multicrystalline silicon wafer with very rough surfaces) [80,81].

10.2.1. Spin-coating Nano-manufacturing Platform

Spin-coating is a standard semiconductor microfabrication technology for spreading highly uniform thin films (e.g., a thin photoresist layer for subsequent photolithography) with adjustable thicknesses over large areas [103]. It has also been extensively used in assembling colloidal monolayers for unconventional nanosphere lithography [104-107]. However, traditional spin-coating approaches usually result in polycrystalline samples with poor crystalline qualities, mainly due to rapid evaporation of solvents used in dispersing colloidal particles [104]. The microspheres cannot be organized into thermodynamically stable crystalline lattices before they are immobilized on the substrates. By contrast, in the spin-coating platform, uniform silica microspheres with a wide diameter range (from ~70 nm to ~ 2 μm) are dispersed in a viscous, nonvolatile monomer, ethoxylated trimethylolpropane triacrylate (ETPTA, SR 454, Sartomer), with 1 wt% Darocur 1173 (2-hydroxy-2-methyl-1-phenyl-1-propanone) as photoinitiator [83,102]. The particle volume fraction is controlled to 0.20. Due to the strong electrostatic repulsion between silica microspheres in ETPTA [108] and the refractive index matching between silica particles and ETPTA (~ 1.46), the silica/ETPTA suspensions are transparent and stable for a long period of time (at least 3 months).

The silica-ETPTA suspension is then dispensed on a substrate and spin-coated using a typical spin coater. Formation of wafer-sized colloidal crystals occurs within 5 seconds as indicated by the appearance of a striking six-arm diffraction star (Figure 2A). This diffraction pattern is characteristic of long-range hexagonal ordering [93,109]. The ETPTA monomer can be rapidly photo-cured to make 3-D highly ordered colloidal crystal-polymer nanocomposites (Figure 2A). The polymer matrix can then be selectively removed by conducting a brief oxygen reactive ion etching (RIE) to release the embedded silica colloidal crystals.[110] The long-range hexagonal ordering of a multilayer colloidal crystal consisting of 325 nm silica microspheres is shown by the top-view SEM image in Figure 2B. Interestingly, the spin-coated colloidal crystals have unusual non-close-packed crystal structures. The top layer microspheres only fill in the triangularly arranged crevices made by the non-touching particles of the underneath layer. Systematic characterization reveals that the inter-particle distance between neighboring microspheres is ~1.41D, where D is the diameter of colloids, regardless of particles sizes and spin-coating conditions [83,102,108,111]. By simply controlling the spin speed and coating time, monolayer non-close-packed colloidal crystals can be routinely assembled over wafer-sized areas using this scalable bottom-up technology. The long-range hexagonal ordering and the non-close-packed structure of the self-assembled crystal are clearly evident by the six-arm Bragg diffraction pattern (Figure 2C) and the top-view SEM image (Figure 2D).

In addition to submicrometer-scale silica microspheres, sub-100 nm silica nanoparticles synthesized by a microemulsion method can also be scalably assembled using the spin-coating platform.[108] Figure 3A shows a top-view SEM image of a multilayer colloidal crystal consisting of ~70 nm silica nanoparticles prepared by spin-coating at 10000 rpm for 10 min. The long-range hexagonal, non-close-packed arrangement of silica nanoparticles is evident from the magnified SEM image (Figure 3B) and the fast-Fourier transform of the SEM image (inset of Figure 3A). Monolayer, non-close-packed crystals can also be assembled by extending the spin coating time to ~ 26 minutes (Figure 3C). The inter-particle distance of the neighboring silica nanoparticles is determined to be ~1.41D, by the first peak of the calculated pair correlation function as shown in Figure 3D

Figure 2. (A) Photograph of a 3-D highly ordered colloidal crystal-polymer nanocomposite prepared by the spin-coating technological platform. (B) Typical top-view SEM image of a spin-coated multilayer colloidal crystal after removing the polymer matrix by oxygen RIE. (C) Photograph of a non-close-packed monolayer colloidal crystal prepared by spin-coating. (D) Typical top-view SEM image of the sample in (C) [83] and [102]. Reprinted with permission from P. Jiang, M. J. McFarland, *J. Am. Chem. Soc.* **2004**, *126*, 13778 and P. Jiang, T. Prasad, M. J. McFarland, V. L. Colvin, *Appl. Phys. Lett.* **2006**, *89*, 011908.

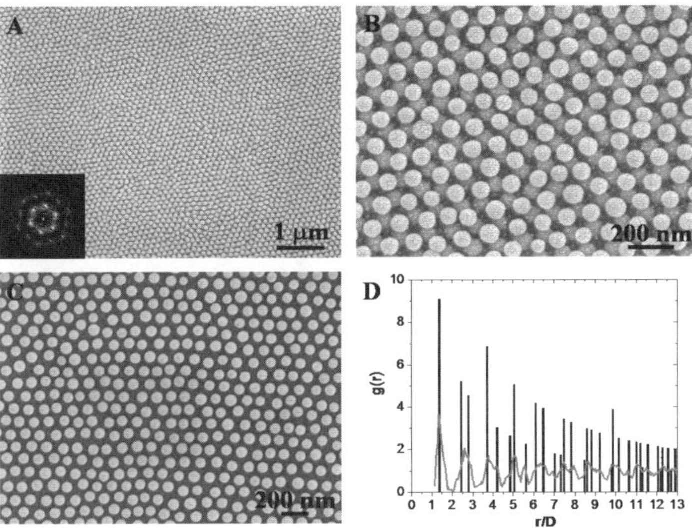

Figure 3. (A) Typical top-view SEM image of a 3-D nanoparticle colloidal crystal prepared by the spin-coating technological platform. The inset shows a Fourier transform of the image. (B) Higher magnification SEM image of the sample in (A). (C) Non-close-packed monolayer nanoparticle colloidal crystal prepared by spin-coating. (D) Pair correlation function calculated from a lower-magnification SEM image of the monolayer sample in (C). For comparison, the pair correlation function for an ideal hexagonal lattice with $\sqrt{2}D$ inter-particle distance is also shown (black lines) [108]. Reprinted with permission from W. L. Min, P. Jiang, B. Jiang, *Nanotechnology* **2008**, *19*, 475604.

The spin-coating technological platform enables rapid assembly of wafer-sized (up to 12-in.), non-close-packed colloidal crystals, which is a length scale nearly two-orders of magnitude larger than crystals assembled by other bottom-up approaches.[84-87] Additionally, the entire crystal is formed within minutes, as compared to days or even weeks needed to produce a centimeter-sized crystal using traditional colloidal self-assembly techniques. Importantly, the spin-coating platform is compatible with standard microfabrication, allowing for the nanomanufacturing of complex micropatterns for on-chip device integration. Furthermore, in addition to energetically favorable hexagonally ordered colloidal crystals, the spin-coating platform also allows for the scalable assembly of colloidal crystals with metastable square ordering [112]. Squarely ordered micropatterns are of great technological importance as they comply with the industry-standard rectilinear coordinate system for simplified addressing and circuit interconnection.[113] During spin-coating, shear-induced melting [109,114-120] of the energetically favorable hexagonal crystals lead to the formation of metastable squarely ordered crystals when the spin speed is higher than 6000 rpm. Figure 4A shows a photograph of a squarely ordered colloidal crystal consisting of 380 nm silica microspheres. The sample exhibits a distinctive four-arm diffraction pattern under white light illumination and the angles between the neighboring diffraction arms are 90°. The non-close-packed structure and the square ordering of silica microspheres are confirmed by the SEM image in Figure 4B.

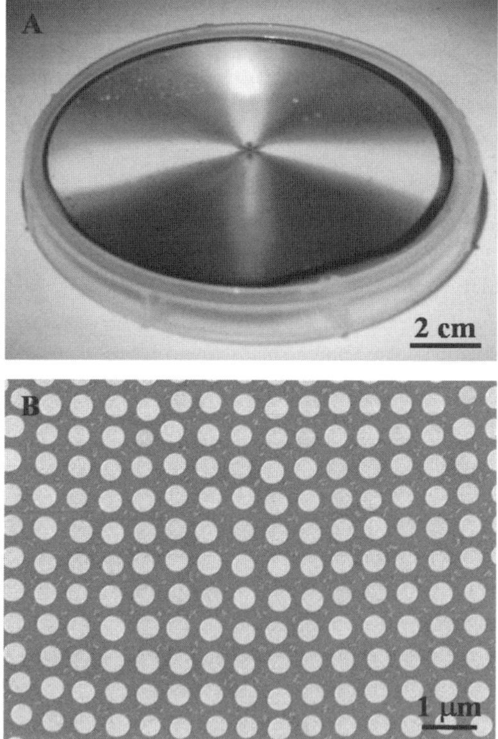

Figure 4. (A) Photograph of a spin-coated, squarely ordered monolayer silica colloidal crystal illuminated with white light. (B) Typical top-view SEM image of the sample in (A) [112]. Reprinted with permission from C. H. Sun, N. C. Linn, W. L. Min, B. Jiang, P. Jiang, *J. Vac. Sci. Technol. B* **2009**, 1043.

10.2.2 Langmuir-Blodgett Nano-Manufacturing Platform

As demonstrated above, the spin-coating platform is scalable, compatible with standard microfabrication technologies, and enables unconventional crystal structures. Unfortunately, spin-coating operation can only be performed on planar substrates with flat surfaces. This limitation is also suffered by many other bottom-up technologies, preventing the formation of high-quality colloidal crystals on nonplanar surfaces [90,91]. A simple and scalable Langmuir-Blodgett-based colloidal self-assembly technology is therefore developed to enable simultaneous deposition of monolayer colloidal crystals on both planar and nonplanar substrates [80,81]. As illustrated by the scheme in Figure 5, this technique is based on capillary force-induced colloidal self-assembly at an air/water interface. Importantly, it does not require sophisticated equipment (e.g., a LB trough). Monodispersed silica microspheres with diameter ranging from ~100 nm to over 30 μm are first cleaned and dispersed in ethylene glycol with a particle volume fraction of 0.20. A substrate, such as a glass slide or a silicon wafer, is pre-immersed in a crystallizing dish containing de-ionized water. The silica-ethylene glycol suspension is then added dropwise to the water surface. The floating silica microspheres momentarily assemble into ordered monolayer colloidal crystal induced by the high surface tension of water. The iridescent monolayer crystal floating at the air/water interface is finally transferred onto the substrate surfaces by steadily withdrawing the substrate at a rate of ~12.5 mm/min controlled by a syringe pump.

Figure 6A shows a uniform monolayer colloidal crystal consisting of 200 nm silica microspheres covering a 5-in.-diameter single-crystalline silicon wafer. The high crystalline quality of the LB-assembled crystal is demonstrated by the top-view SEM image in Figure 6B. Contrary to the crystals assembled by the spin-coating platform, the monolayer colloidal crystals fabricated by the LB technology are close-packed. Critically, this technology has a high flexibility in scalably assembling conformal particle coatings on a large variety of substrates with very complex geometries that cannot be used in traditional bottom-up approaches. One preeminent example that illustrates the capability of the LB platform is to assemble ordered colloidal monolayers on commercial solar-grade multicrystalline silicon (mc-Si) wafers with very rough surfaces (root mean square roughness of ~0.89 µm) [81]. Figure 6C shows a photograph of a 5-in. mc-Si wafer with the right half covered by a LB-assembled monolayer of 250 nm silica particles. The top-view SEM image in Figure 6D demonstrates that monolayer colloidal arrays with short-range hexagonal ordering evenly cover the rough mc-Si surface, even on the steep sidewalls of the randomly distributed, micrometer-sized pits.

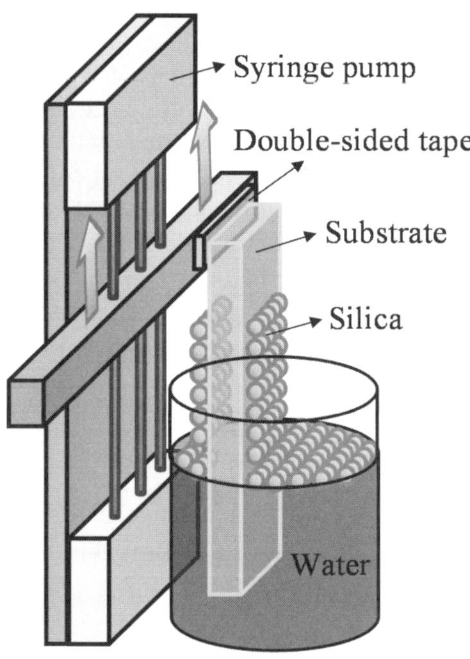

Figure 5. Schematic illustration of the Langmuir-Blodgett process for assembling and transferring monolayer colloidal crystals on a substrate [80]. Reprinted with permission from K. Askar, B. M. Phillips, X. Dou, J. Lopez, C. Smith, B. Jiang, P. Jiang, *Opt. Lett.* **2012**, *37*, 4380.

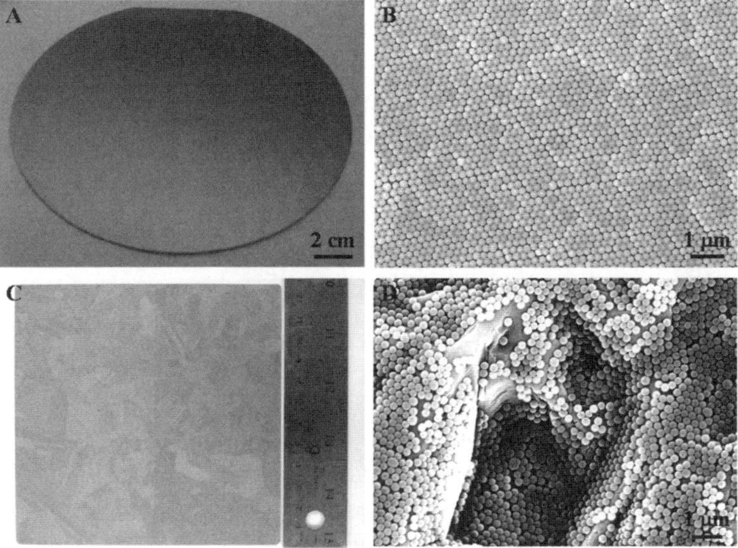

Figure 6. (A) Photograph of a monolayer silica colloidal crystal assembled by the LB technological platform on a 5-in.-sized silicon wafer. (B) Typical top-view SEM image of the sample in (A). (C) Photograph of a solar-grade multicrystalline silicon wafer with the right half covered by a monolayer silica colloidal crystal. (D) Typical top-view SEM image of the right half of the sample in (C) [81]. Reprinted with permission from B. M. Phillips, P. Jiang, B. Jiang, *Appl. Phys. Lett.* **2011**, *99*, 191103.

10.3. Templating Nano-manufacturing of Broadband Subwavelength Moth-Eye AR Coatings on Semiconductors

The scalable spin-coating and Langmuir-Blodgett bottom-up platforms enable large-scale assembly of colloidal particles into both non-close-packed and close-packed colloidal crystals. The assembled particles can then be used as structural templates in producing a myriad of nanostructured functional materials, such as plasmonic nanostructures (e.g., nanoholes and nanopyramids) [121,122], macroporous polymers [83], attoliter microvial arrays [123], 2-D magnetic nanodots [124], and subwavelength-structured broadband moth-eye antireflection coatings on a large variety of technologically important substrates (e.g., crystalline silicon, GaAs, GaSb, glass, and plastics) [12]. In this section, we will first discuss the fabrication, characterization, and modeling of moth-eye AR coatings on crystalline silicon wafers, including both single-crystalline and multicrystalline silicon. The templating nanomanufacturing technology will then be extended to GaAs and GaSb wafers.

10.3.1 Templating Nano-manufacturing of Broadband Moth-eye AR Coatings on Crystalline Silicon Wafers

Crystalline silicon is the most commonly used semiconductor in microelectronic and optoelectronic devices (e.g., computer chips, photodiodes, and solar cells) [103]. The current production of photovoltaic panels is dominated by crystalline silicon solar cells with 98% of the market share [125,126]. In addition, by leveraging the very high refractive index (~3.4) of silicon and its excellent transmission over the entire infrared to millimeter wavelengths (except for ~7–25 μm, where silicon has strong absorption), silicon is one of the best materials for manufacturing refractive optics, such as lenses, filters, and beamsplitters for compact, high resolution, and high efficiency IR spectroscopy, which is widely used in aerospace exploration [127]. Unfortunately, the large refractive index of silicon leads to high reflection loss (~30%) from a single silicon/air interface if an AR coating is not applied [16,53]. This detrimental optical reflection could greatly reduce the conversion efficiencies of crystalline silicon solar cells and impact the sensitivity of IR spectroscopy for observing faint objectives. Traditional quarter-wavelength or multilayer dielectric AR coatings exploit constructive or destructive interference for a given wavelength at the substrate/air interface. Unfortunately, it is extremely difficult to achieve low reflection over a very broad range of wavelengths using traditional AR coatings. In addition, the stringent layer thickness control, the limited material selection, and the poor mechanical/environmental stability caused by thermal-mismatch-induced delamination greatly impede the fabrication and the final durability of dielectric AR coatings on silicon substrates.

To overcome the key scientific and technical barriers faced by traditional dielectric AR coatings, two templating nanomanufacturing technologies based on simple and scalable wet-etching and dry-etching methods have been developed. As shown by the schematic illustrations in Figure 7, these techniques utilize the self-assembled monolayer colloidal crystals as structural templates for fabricating wafer-sized moth-eye AR gratings on crystalline silicon substrates.[43,44] In the wet etch approach, the non-close-packed monolayer colloidal crystals assembled by the spin-coating technology is firstly used as shadow masks during electron-beam evaporation of a 30 nm chromium layer. After lifting off the templating silica microspheres, a periodic array of chromium nanoholes is formed on the (100) silicon wafer [121]. The resulting circular nanoholes are finally used as etching masks during a standard KOH anisotropic etching process to generate periodic arrays of inverted pyramidal pits on the wafer surface.[122] By simply controlling the size of the templating silica microspheres and the templating conditions, the geometries of the final inverted pyramidal pits can be easily tuned. In the dry etch approach, the self-assembled monolayer silica colloidal crystals (either non-closed-packed or close-packed) are used as etching masks during a standard SF_6 or Cl_2 reactive ion etching (RIE) process to pattern periodic arrays of nanopillars with high aspect ratios on the wafer surface.[101] The templating silica microspheres are finally removed by a brief hydrofluoric acid wash.

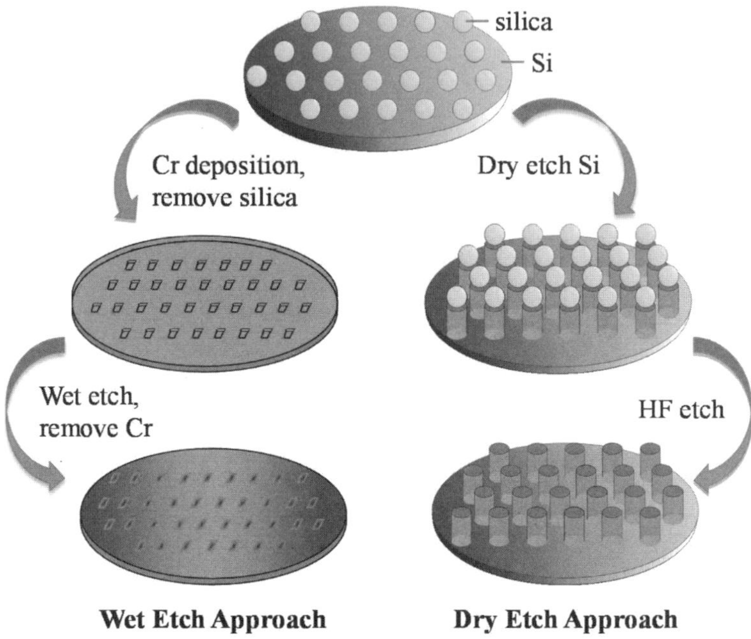

Figure 7. Schematic illustrations of the wet-etching-based and dry-etching-based templating nanomanufacturing procedures for fabricating inverted pyramidal and nanopillar antireflection gratings on the surfaces of crystalline silicon wafers.

Figure 8A shows the top-view SEM image of a periodic array of wet-etched inverted pyramidal pits templated from 320 nm silica microspheres [128]. The specular optical reflection of the templated inverted pyramidal grating is compared with that of a flat silicon wafer in Figure 8B. The polished silicon control sample exhibits shows a high level of reflection (> 30%) for both visible and near-IR wavelengths. By contrast, the templated inverted pyramidal grating exhibits a reduced reflection of ~10% for long wavelengths (> 600 nm). The reflection is further reduced to ~2% around 400 nm. The antireflection performance of the templated silicon inverted pyramidal arrays is okay, but far from being outstanding. The optical simulations based on a rigorous coupled wave analysis (RCWA) model [129,130] indicated that the limited AR performance is mainly caused by the shallow optical depth of the templated inverted pyramidal pits. Due to the characteristic 54.7° sidewalls of the anisotropically etched silicon pits [103], the maximal depth of the V-shaped inverted pyramids is limited by the inter-particle distance of the templating microspheres.

Based on extensive RCWA optical modeling, we conclude that subwavelength-structured gratings with high aspect ratios are need to produce broadband AR coatings with much improved AR performance than the above inverted pyramidal pits. We therefore have developed the dry etch approach in order to greatly increase the optical depth of the templated gratings [43,101]. In this approach, the self-assembled silica microspheres, which has a much smaller etching rate than that of silicon during a traditional Cl_2-RIE (5 mTorr pressure, 20 SCCM chlorine flow rate, and 80 W) or a SF_6-RIE (40 mTorr pressure, 26 SCCM SF_6, 5 SCCM O_2, and 25 W) process, can protect the silicon substrate immediately underneath them from being etched, leading to the formation of nanopillar arrays directly patterned on the wafer surface. By simply controlling the templating microsphere diameter and the RIE conditions (e.g., etching power and time, gas composition, and chamber pressure), the geometries of the final moth-eye nanopillars (e.g., depth, aspect ratio, and shape) can be easily tuned. Figure 9 shows the typical SEM images during the templating nanomanufacturing of silicon moth-eye nanopillars with an aspect ratio of ~6.

A

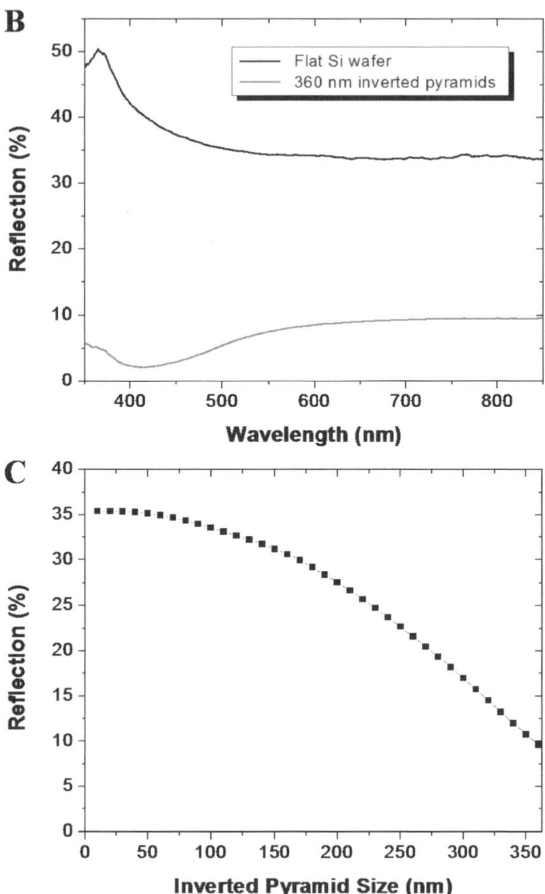

Figure 8. (A) Typical top-view SEM image of templated inverted pyramids wet-etched on the surface of a (100) silicon wafer. (B) Normal-incidence specular optical reflection spectra obtained from a flat silicon wafer and the inverted pyramidal sample in (A). (C) Simulated normal-incidence optical reflection at 600 nm wavelength for periodic arrays of inverted pyramidal pits with different sizes [44]. Reprinted with permission from C.-H. Sun, W.-L. Min, N. C. Linn, P. Jiang, B. Jiang, *Appl. Phys. Lett.* **2007**, *91*, 231105.

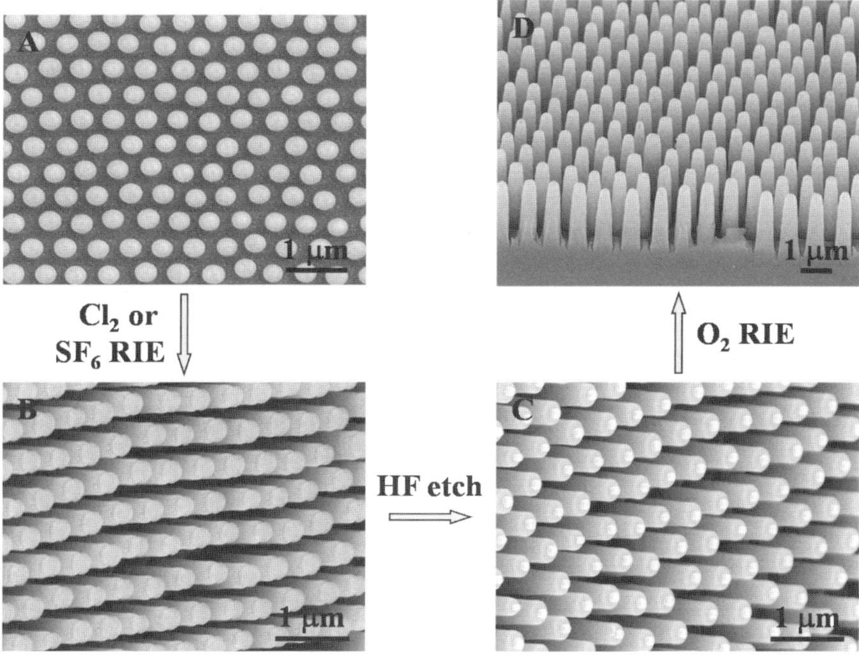

Figure 9. Typical SEM images during the dry-etching-based templating nanomanufacturing process for fabricating subwavelength-structured moth-eye antireflection gratings on a single-crystalline silicon wafer [101]. Reprinted with permission from W. L. Min, B. Jiang, P. Jiang, *Adv. Mater.* **2008**, *20*, 3914.

In addition to submicrometer-scale silica nanoparticles, the spin-coated, non-close-packed monolayer colloidal crystals consisting of sub-100 nm silica nanoparticles (see Figure 3C) can also be used as sacrificial templates in fabricating nanostructured moth-eye AR gratings.[108] Figure 10A shows a typical cross-sectional SEM image of a Cl_2-RIE-processed moth-eye grating using 70 nm silica nanoparticles as etching masks. The taper geometry, large aspect ratio, and sub-100 nm lattice spacing of the templated nanopillars enable excellent broadband antireflection properties as demonstrated by the specular reflection spectra shown in Figure 10B. Very low reflection ($< 2.5\%$) over a wide range of visible and near-IR wavelengths is demonstrated for the templated moth-eye grating. By contrast, a flat silicon control sample shows high single-side reflection (> 30%) over the entire spectrum. A silicon nitride AR coating deposited on a commercial single-crystalline silicon solar cell (with a conversion efficiency of ~13.5%) is also compared with the templated moth-eye grating. Apparently, the commercial AR coating can only effectively reduce optical reflection over a limited range of wavelengths around 700 nm. Moreover, the experimental reflection measurements match quite well with the optical simulations using the RCWA model.

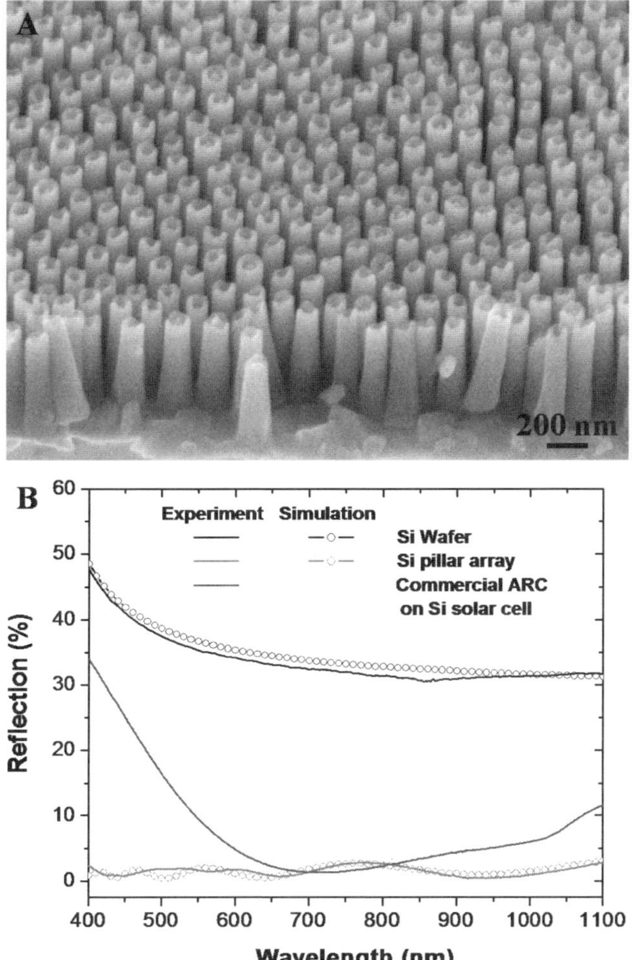

Figure 10. (A) Typical side-view SEM image of a subwavelength moth-eye grating templated from 70 nm non-close-packed monolayer silica colloidal crystal. (B) Experimental (solid) and simulated (dotted) normal-incidence specular reflection spectra from a flat silicon wafer, a templated moth-eye nanopillar array, and a commercial AR coating on a single-crystalline silicon solar substrate [108]. Reprinted with permission from W. L. Min, P. Jiang, B. Jiang, *Nanotechnology* **2008**, *19*, 475604.

As demonstrated in Section 2.1, the scalable spin-coating platform enables wafer-scale assembly of non-close-packed monolayer colloidal crystals with unusual metastable square order (see Figure 4). This provides us a unique system to conduct fundamental investigations on the effects of crystal structure of the templated moth-eye gratings on their AR properties [111]. Figure 11A shows a side-view SEM image of a moth-eye AR grating templated from a squarely ordered monolayer colloidal crystal using the same dry etch approach as described above. The specular optical reflection spectra in Figure 11B compare the AR performance of the squarely ordered moth-eye array with a hexagonal pillar array fabricated under the same conditions (e.g., same templating particle size and etched together in the same RIE chamber). It is interesting to note that the average reflection of the squarely ordered array is apparently lower than that of the nature-inspired hexagonal array with very similar structural parameters (such as shape, lattice space, and nanopillar aspect ratio).

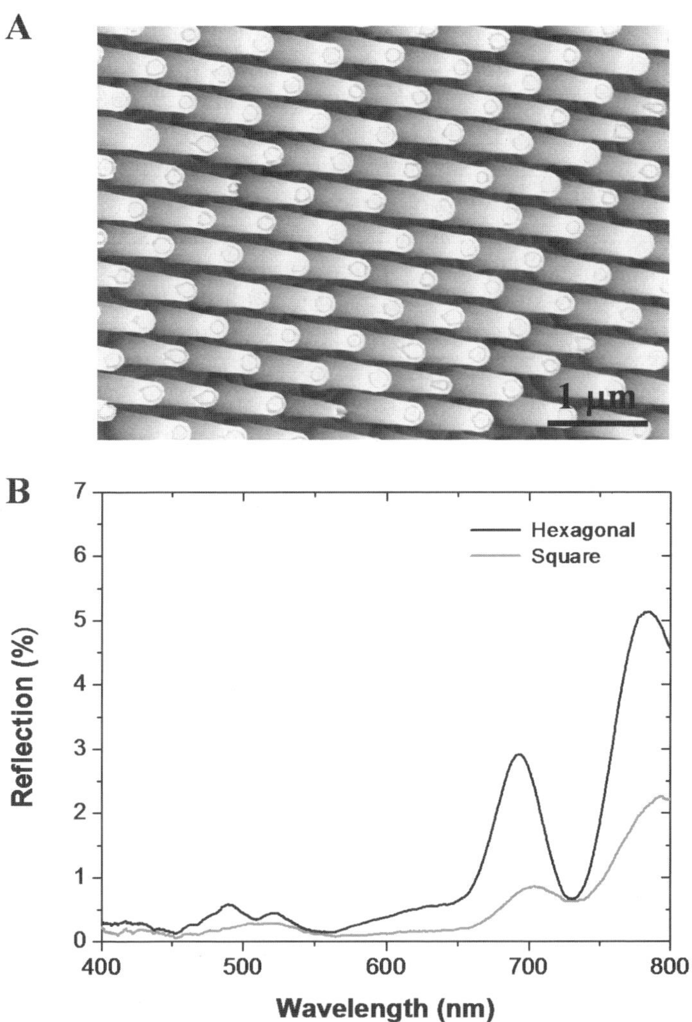

Figure 11. (A) Typical side-view SEM image of a silicon moth-eye nanopillar array templated from a non-close-packed, squarely ordered monolayer colloidal crystal prepared by spin-coating. (B) Comparison of the normal-incidence specular optical reflection spectra between a squarely ordered silicon moth-eye array and a hexagonal array RIE-processed in the same chamber [112]. Reprinted with permission from C. H. Sun, N. C. Linn, W. L. Min, B. Jiang, P. Jiang, *J. Vac. Sci. Technol. B* **2009**, 1043.

The dry etch-based templating nanomanufacturing technology is a general approach that works for both non-close-packed colloidal crystals assembled by the spin-coating platform and close-packed colloidal monolayers fabricated by the LB platform (Figure 6). Figure 12A shows a typical side-view SEM image of a templated moth-eye grating on a commercial solar-grade multicrystalline silicon wafer using LB-assembled, close-packed 250 nm silica microspheres as etching masks in a Cl_2-RIE process [81]. High aspect ratio nanopillars with short-range ordering which is similar to that of the templating colloidal monolayer (see Figure 6D) are clearly shown. The specular hemispherical reflection spectra in Figure 12B compares the AR performance of a polished single crystalline silicon (sc-Si) wafer, a solar-grade mc-Si wafer, and the templated mc-Si moth-eye grating. The flat sc-Si wafer shows 30-50% hemispherical reflectance for wavelengths from 400 to 900 nm, matching with early measurements in the literature [10,131]. The rough surface of the commercial mc-Si wafer helps to reduce its hemispherical reflectance to 20-30%. In sharp contrast, the templated mc-Si moth-eye grating exhibits excellent broadband AR property and the hemispherical reflectance is nearly zero for a wide range of wavelengths.

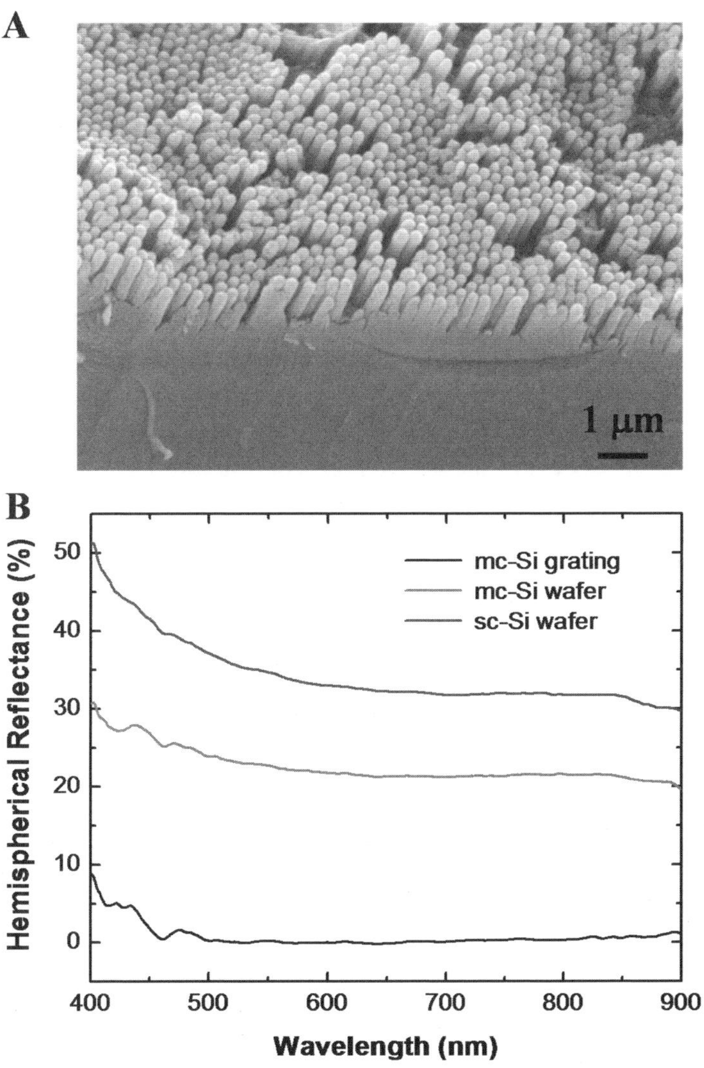

Figure 12. (A) Typical side-view SEM image of a close-packed moth-eye grating on a multicrystalline silicon substrate templated from LB-assembled colloidal monolayers. (B) Comparison of hemispherical reflectance spectra obtained from a polished single-crystalline silicon wafer, a solar-grade multicrystalline silicon wafer, and the templated multicrystalline silicon moth-eye grating in (A) [81]. Reprinted with permission from B. M. Phillips, P. Jiang, B. Jiang, *Appl. Phys. Lett.* **2011**, *99*, 191103.

10.3.2 Templating Nano-manufacturing of Broadband Moth-eye AR Coatings on GaAs and GaSb Wafers

Besides crystalline silicon substrates, the self-assembled silica microsphere-based templating nanomanufacturing technology can be easily extended to many technologically important semiconductors, provided that the RIE-etching rate of silica is significantly smaller than those of the semiconductors. Gallium arsenide has been widely used in optoelectronics, such as vertical cavity surface-emitting laser (VCSEL), near-infrared photodetector, and highly efficient concentrator solar cells [58,132,133]. Gallium antimonide is one of the most appropriate materials for thermophotovoltaic (TPV) generation due to the excellent match between the radiation spectrum and the spectral response of GaSb cells [79,134]. Monolayer silica colloidal crystals can be assembled on GaAs and GaSb wafers using the above spin-coating and LB platforms. Figure 13A shows a side-view SEM image of a templated array of conical GaAs nanopillars using spin-coated 320 nm silica microspheres as etching masks in a BCl_3-RIE process (7.5 mTorr pressure, 20 sccm BCl_3, RIE = 100 W, and ICP = 300 W) [134]. Although the aspect ratio of the templated GaAs nanopillars is only about 1.0, the tapered conical shape of the nanopillars leads to a smooth refractive index gradient from air to the bulk GaAs substrate, resulting in the excellent broadband AR performance shown by the normal-incidence specular reflection spectra in Figure 13B. Similar to crystalline silicon, the high refractive index of GaAs (> 3.6) leads to a high optical reflection loss (> 30%) for a bare GaAs control sample.

Using a spin-coated monolayer colloidal crystal as an etching mask and a Cl_2-RIE process (5.3 mTorr pressure, 8 SCCM Cl_2, 4 SCCM Ar, and 200 W), periodic arrays of subwavelength moth-eye nanopillars with large aspect ratios have been successfully fabricated on a 2-in. GaSb wafer (Figure 14A) [79]. The as-fabricated AR grating exhibits low optical reflection (< 4%) for a wide range of visible and near-IR wavelengths (black line in Figure 14B). This represents a > 90% reduction of reflective loss when compared with a flat GaSb wafer which shows > 40% specular reflection for the same wavelength range. Besides broadband antireflection, another major merit of the templated subwavelength-structured moth-eye AR gratings is their high environmental stability. This is because the templated moth-eye AR gratings are directly patterned on the wafer surface and no foreign materials need to be deposited as in traditional dielectric AR coatings. Figure 14B compares the normal-incidence specular reflection spectra of a templated GaSb moth-eye grating before and after annealing at 200°C for 6 hours. The change in the broadband AR performance is quite small. This is in sharp contrast with conventional quarter-wavelength AR coatings which exhibit significant AR performance degradation even at temperature as low as 100°C [33]. This unique thermal stability of the templated moth-eye gratings is particularly useful for fabricating high-efficiency concentrating GaAs solar cells [132-134] and GaSb TPV cells [135] as their surface temperatures are usually high. The mismatch of the coefficient of thermal expansion between common dielectric AR coatings and the semiconductor substrates greatly limits the durability and performance of the final devices.

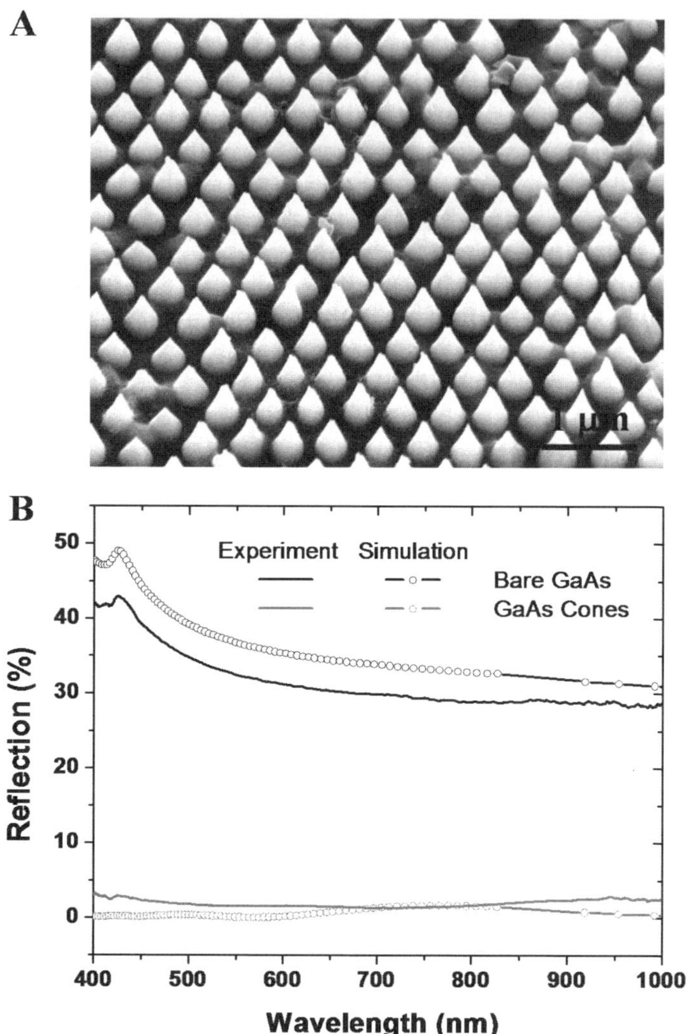

Figure 13. (A) Typical side-view SEM image of a templated GaAs conical nanonipple array. (B) Experimental (solid lines) and simulated (dotted lines) normal-incidence specular reflection spectra obtained from a bare GaAs wafer and the conical subwavelength grating in (A) [134]. Reprinted with permission from C. H. Sun, B. J. Ho, B. Jiang, P. Jiang, *Opt. Lett.* **2008**, *33*, 2224.

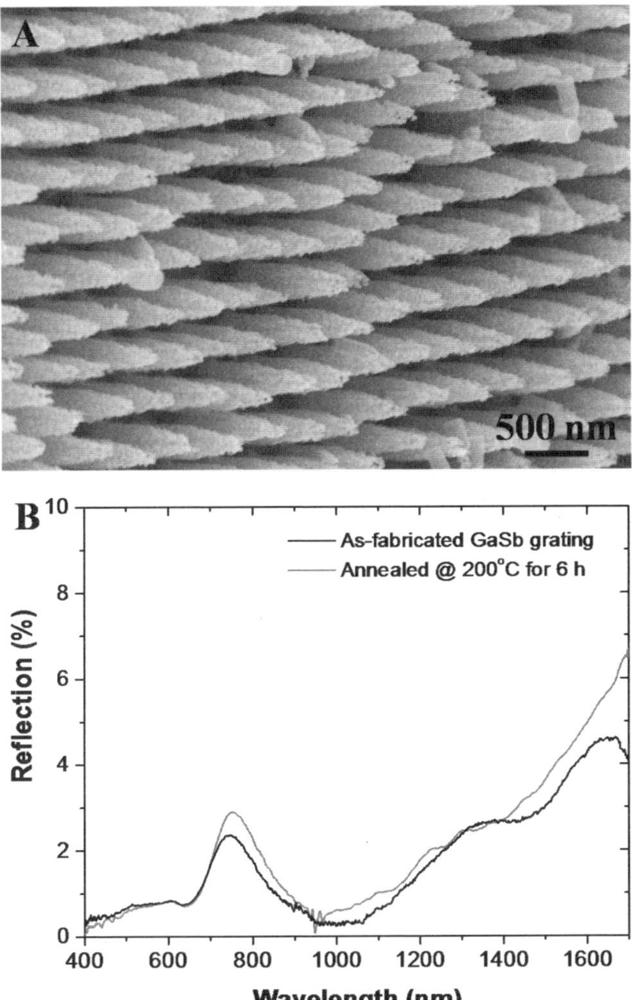

Figure 14. (A) Typical side-view SEM image of a templated GaSb moth-eye grating. (B) Comparison of normal-incidence specular reflection spectra obtained from a GaSb moth-eye grating and the same sample after annealing at 200°C for 6 hours [79]. Reprinted with permission from W. L. Min, A. P. Betancourt, P. Jiang, B. Jiang, *Appl. Phys. Lett.* **2008**, *92*, 141109.

10.4. Conclusions

We have developed two scalable bottom-up nano-manufacturing platforms for large-scale assembly of colloidal crystals, which can be used as structural templates for producing broadband subwavelength-structured moth-eye AR coatings on a large variety of semiconductor wafers. These unconventional approaches combine the simplicity and cost benefits of bottom-up colloidal self-assembly with the scalability and compatibility of standard industrial manufacturing. The spin-coating technology enable wafer-scale production of both monolayer and multilayer colloidal crystals with unusual non-close-packed structures and metastable squarely ordered crystals. The LB platform allows for continuous assembly and transfer of close-packed monolayer colloidal crystals on both planar and nonplanar substrates. The dry-etch-based templating nanofabrication technique is more versatile than the wet etch approach for patterning moth-eye gratings on different substrates, provided there is a large etch rate difference between the templating silica microsphere and the substrate material. Our experimental results and theoretical calculations show that the bioinspired moth-eye AR coatings exhibit much improved broadband AR and thermal stability than traditional quarter-wavelength AR coatings. These high-performance AR coatings could find important technological applications ranging from improving the conversion efficiency of crystalline silicon solar cells to enhancing light extraction efficiency of semiconductor light emitting diodes.

Acknowledgements. This work was partially supported by an Early Stage Innovations grant (Award No. NNX14AB07G) from NASA's Space Technology Research Grants Program and the US Defense Threat Reduction Agency, Basic Research Award # HDTRA1-15-1-0022, to University of Florida. Acknowledgments were also made to the US National Science Foundation (NSF) under Award No. CMMI-1300613.

References

1.   V. R. Almeida, C. A. Barrios, R. R. Panepucci, M. Lipson, *Nature* **2004**, *431*, 1081.

2.   J. M. Senior, *Optical Fiber Communications: Principles and Practice*, Prentice-Hall, Englewood Cliffs 1992.

3.   G. R. Fowles, *Introduction to Modern Optics*, New York 1989.
4.   M. L. Brongersma, Y. Cui, S. Fan, *Nat. Mater.* **2014**, *13*, 451.
5.   X. Liu, P. R. Coxon, M. Peters, B. Hoex, J. M. Cole, D. J. Fray, *Energy Environ. Sci.* **2014**, *7*, 3223.
6.   L. Yao, J. He, *Prog. Mater. Sci.* **2014**, *61*, 94.
7.   Y. Du, L. E. Luna, W. S. Tan, M. F. Rubner, R. E. Cohen, *ACS Nano* **2010**, *4*, 4308.
8.   D. Chen, *Sol. Energ. Mater. Sol. C.* **2001**, *68*, 313.
9.   J. A. Hiller, J. D. Mendelsohn, M. F. Rubner, *Nat. Mater.* **2002**, *1*, 59.
10.   Y.-F. Huang, S. Chattopadhyay, Y.-J. Jen, C.-Y. Peng, T.-A. Liu, Y.-K. Hsu, C.-L. Pan, H.-C. Lo, C.-H. Hsu, Y.-H. Chang, C.-S. Lee, K.-H. Chen, L.-C. Chen, *Nat. Nanotechnol.* **2007**, *2*, 770.
11.   R. Prado, G. Beobide, A. Marcaide, J. Goikoetxea, A. Aranzabe, *Sol. Energ. Mater. Sol. C.* **2010**, *94*, 1081.
12.   K. Askar, B. M. Phillips, Y. Fang, B. Choi, N. Gozubenli, P. Jiang, B. Jiang, *Colloid Surf. A* **2013**, *439*, 84.
13.   Z. Chen, P. Sana, J. Salami, A. Rohatgi, *IEEE Trans. Electron Dev.* **1993**, *40*, 1161.
14.   H. Nagel, A. G. Aberle, R. Hezel, *Prog. Photovoltaics* **1999**, *7*, 245.
15.   C. Martinet, V. Paillard, A. Gagnaire, J. Joseph, *J. Non-Cryst. Solids* **1997**, *216*, 77.
16.   P. Doshi, G. E. Jellison, A. Rohatgi, *Appl. Opt.* **1997**, *36*, 7826.
17.   R. Kishore, S. N. Singh, B. K. Das, *Sol. Energ. Mater. Sol. C.* **1992**, *26*, 27.
18.   H. Nagel, A. Metz, R. Hezel, *Sol. Energ. Mater. Sol. C.* **2001**, *65*, 71.
19.   C. C. Johnson, T. Wydeven, K. Donohoe, *Sol. Energ.* **1983**, *31*, 355.
20.   L. I. Maissel, R. Glang, *Handbook of Thin Film Technology*, McGraw-Hill Book Company, New York, NY 1970.
21.   W. H. Southwell, *Opt. Lett.* **1983**, *8*, 584.
22.   D. Hocine, M. S. Belkaid, M. Pasquinelli, L. Escoubas, J. J. Simon, G. A. Rivière, A. Moussi, *Mat. Sci. Semicon. Proc.* **2013**, *16*, 113.
23.   S. Chattopadhyay, Y. F. Huang, Y. J. Jen, A. Ganguly, K. H. Chen, L. C. Chen, *Mater. Sci. Eng. R* **2010**, *69*, 1.
24.   M. Kursawe, R. Anselmann, V. Hilarius, G. Pfaff, *J Sol-Gel Sci Technol* **2005**, *33*, 71.
25.   Y.-J. Lee, D. S. Ruby, D. W. Peters, B. B. McKenzie, J. W. P. Hsu, *Nano Lett.* **2008**, *8*, 1501.
26.   S. Walheim, E. Schäffer, J. Mlynek, U. Steiner, *Science* **1999**, *283*, 520.
27.   X.-T. Zhang, O. Sato, M. Taguchi, Y. Einaga, T. Murakami, A. Fujishima, *Chem. Mater.* **2005**, *17*, 696.
28.   D. Ristau, S. Günster, S. Bosch, A. Duparré, E. Masetti, J. Ferré-Borrull, G. Kiriakidis, F. Peiró, E. Quesnel, A. Tikhonravov, *Appl. Opt.* **2002**, *41*, 3196.
29.   H. Selhofer, E. Ritter, R. Linsbod, *Appl. Opt.* **2002**, *41*, 756.
30.   T.-S. Yang, C.-B. Shiu, M.-S. Wong, *Surf. Sci.* **2004**, *548*, 75.
31.   S.-H. Jeong, J.-K. Kim, B.-S. Kim, S.-H. Shim, B.-T. Lee, *Vacuum* **2004**, *76*, 507.

32. S. Ray, R. Banerjee, N. Basu, A. K. Batabyal, A. K. Barua, *J. Appl. Phys.* **1983**, *54*, 3497.

33. Y. Kanamori, K. Kobayashi, H. Yugami, K. Hane, *Jpn. J. Appl. Phys.* **2003**, *42*, 4020.

34. P. B. Clapham, M. C. Hutley, *Nature* **1973**, *244*, 281.

35. F. Galeotti, F. Trespidi, G. Timo, M. Pasini, *ACS Appl. Mater. Interface* **2014**, *6*, 5827.

36. F. L. Gonzalez, L. Chan, A. Berry, D. E. Morse, M. J. Gordon, *J. Vac. Sci. Technol. B* **2014**, *32*.

37. P. Vukusic, J. R. Sambles, *Nature* **2003**, *424*, 852.

38. M. Srinivasarao, *Chem. Rev.* **1999**, *99*, 1935.

39. D. G. Stavenga, S. Foletti, G. Palasantzas, K. Arikawa, *Proc. R. Soc. B* **2006**, *273*, 661.

40. Y. Ono, Y. Kimura, Y. Ohta, N. Nishida, *Appl. Opt.* **1987**, *26*, 1142.

41. J. Zhao, M. A. Green, *IEEE Trans. Electron Dev.* **1991**, *38*, 1925.

42. N. C. Linn, C.-H. Sun, P. Jiang, B. Jiang, *Appl. Phys. Lett.* **2007**, *91*, 101108.

43. C.-H. Sun, P. Jiang, B. Jiang, *Appl. Phys. Lett.* **2008**, *92*, 061112.

44. C.-H. Sun, W.-L. Min, N. C. Linn, P. Jiang, B. Jiang, *Appl. Phys. Lett.* **2007**, *91*, 231105.

45. J. W. Leem, X.-Y. Guan, M. Choi, J. S. Yu, *Sol. Energ. Mater. Sol. C.* **2015**, *134*, 45.

46. X. Li, P.-C. Li, L. Ji, C. Stender, S. R. Tatavarti, K. Sablon, E. T. Yu, *Sol. Energ. Mater. Sol. C.* **2015**, *143*, 567.

47. M. Z. Pakhuruddin, J. Huang, J. Dore, S. Varlamov, *IEEE J. Photovol.* **2016**, *6*, 159.

48. C. Zhang, P. Yi, L. Peng, X. Lai, J. Ni, *IEEE Trans. Nanotechnol.* **2015**, *14*, 1127.

49. C. Heine, R. H. Morf, *Appl. Opt.* **1995**, *34*, 2476.

50. Y. Kanamori, M. Sasaki, K. Hane, *Opt. Lett.* **1999**, *24*, 1422.

51. Q. Chen, G. Hubbard, P. A. Shields, C. Liu, D. W. E. Allsopp, W. N. Wang, S. Abbott, *Appl. Phys. Lett.* **2009**, *94*, 263118.

52. U. Schulz, *Appl. Opt.* **2006**, *45*, 1608.

53. A. Gombert, W. Glaubitt, K. Rose, J. Dreibholz, B. Bläsi, A. Heinzel, D. Sporn, W. Döll, V. Wittwer, *Thin Solid Films* **1999**, *351*, 73.

54. K. M. Baker, *Appl. Opt.* **1999**, *38*, 352.

55. V. Auzelyte, V. Flauraud, V. J. Cadarso, T. Kiefer, J. Brugger, *Microelectron. Eng.* **2012**, *97*, 269.

56. B. J. Bae, S. H. Hong, E. J. Hong, H. Lee, G. Y. Jung, *Jap. J. Appl. Phys.* **2009**, *48*, 010207.

57. R. Brunner, B. Keil, C. Morhard, D. Lehr, J. Draheim, U. Wallrabe, J. Spatz, *Appl. Opt.* **2012**, *51*, 4370.

58. X. Chen, Z. C. Fan, Y. Xu, G. F. Song, L. H. Chen, *Microelectron. Eng.* **2011**, *88*, 2889.

59. K. Choi, S. H. Park, Y. M. Song, C. Cho, H. S. Lee, *J. Mater. Chem.* **2012**, *22*, 17037.

60.   H. Deniz, T. Khudiyev, F. Buyukserin, M. Bayindir, *Appl. Phys. Lett.* **2011**, *99*, 183107.

61.   K. S. Han, J. H. Shin, H. Lee, *Sol. Energ. Mater. Sol. C.* **2010**, *94*, 583.

62.   K. S. Han, J. H. Shin, W. Y. Yoon, H. Lee, *Sol. Energ. Mater. Sol. C.* **2011**, *95*, 288.

63.   S. Y. Han, B. K. Paul, C. H. Chang, *J. Mater. Chem.* **2012**, *22*, 22906.

64.   G. Hubbard, M. E. Nasir, P. Shields, C. R. Bowen, A. Satka, K. P. Parsons, N. H. Holmes, D. W. E. Allsopp, *Nanotechnology* **2012**, *23*.

65.   J. K. Kim, S. J. Park, S. Kim, H. H. Park, K. D. Kim, J. H. Choi, J. Lee, D. G. Choi, K. Y. Suh, J. H. Jeong, *Microelectron. Eng.* **2012**, *100*, 12.

66.   D. H. Ko, J. R. Tumbleston, K. J. Henderson, L. E. Euliss, J. M. DeSimone, R. Lopez, E. T. Samulski, *Soft Matter* **2011**, *7*, 6404.

67.   Y. H. Ko, L. H. Jin, Y. H. Cho, *Thin Solid Films* **2011**, *519*, 2251.

68.   C. Morhard, C. Pacholski, D. Lehr, R. Brunner, M. Helgert, M. Sundermann, J. P. Spatz, *Nanotechnology* **2010**, *21*.

69.   K. Nakata, M. Sakai, T. Ochiai, T. Murakami, K. Takagi, A. Fujishima, *Langmuir* **2011**, *27*, 3275.

70.   Y. Ou, D. D. Corell, C. Dam-Hansen, P. M. Petersen, H. Y. Ou, *Opt. Express* **2011**, *19*, A166.

71.   H. Park, D. Shin, G. Kang, S. Baek, K. Kim, W. J. Padilla, *Adv. Mater.* **2011**, *23*, 5796.

72.   T. Senn, O. Kutz, C. Weniger, J. M. Li, M. Schoengen, H. Lochel, J. Wolf, P. Gottert, B. Lochel, *J. Vac. Sci. Technol. B* **2011**, *29*.

73.   B. K. Shin, T. I. Lee, J. Xiong, C. Hwang, G. Noh, J. H. Cho, J. M. Myoung, *Sol. Energ. Mater. Sol. C.* **2011**, *95*, 2650.

74.   M. Silvennoinen, K. Paivasaari, J. J. J. Kaakkunen, V. K. Tikhomirov, A. Lehmuskero, P. Vahimaa, V. V. Moshchalkov, *Appl. Surf. Sci.* **2011**, *257*, 6829.

75.   Y. M. Song, Y. Jeong, C. I. Yeo, Y. T. Lee, *Opt. Express* **2012**, *20*, A916.

76.   S. Tanaka, S. Fujihara, *Langmuir* **2011**, *27*, 2929.

77.   J. Tommila, V. Polojarvi, A. Aho, A. Tukiainen, J. Viheriala, J. Salmi, A. Schramm, J. M. Kontio, A. Turtiainen, T. Niemi, M. Guina, *Sol. Energ. Mater. Sol. C.* **2010**, *94*, 1845.

78.   N. Yamada, O. N. Kim, T. Tokimitsu, Y. Nakai, H. Masuda, *Prog. Photovoltaics* **2011**, *19*, 134.

79.   W. L. Min, A. P. Betancourt, P. Jiang, B. Jiang, *Appl. Phys. Lett.* **2008**, *92*, 141109.

80.   K. Askar, B. M. Phillips, X. Dou, J. Lopez, C. Smith, B. Jiang, P. Jiang, *Opt. Lett.* **2012**, *37*, 4380.

81.   B. M. Phillips, P. Jiang, B. Jiang, *Appl. Phys. Lett.* **2011**, *99*, 191103.

82.   E.-M. Park, S.-K. Lim, S.-H. Ra, S.-J. Suh, *J. Nanosci. Nanotech.* **2013**, *13*, 7586.

83.   P. Jiang, M. J. McFarland, *J. Am. Chem. Soc.* **2004**, *126*, 13778.

84.   N. D. Denkov, O. D. Velev, P. A. Kralchevsky, I. B. Ivanov, H. Yoshimura, K. Nagayama, *Nature* **1993**, *361*, 26.

85.   P. Jiang, J. F. Bertone, K. S. Hwang, V. L. Colvin, *Chem. Mater.* **1999**, *11*, 2132.

86.   Y. A. Vlasov, X.-Z. Bo, J. C. Sturm, D. J. Norris, *Nature* **2001**, *414*, 289.

87.   S. Wong, V. Kitaev, G. A. Ozin, *J. Am. Chem. Soc.* **2003**, *125*, 15589.

88.   Y. Lin, H. Skaff, T. Emrick, A. D. Dinsmore, T. P. Russell, *Science* **2003**, *299*, 226.

89.   A. Dong, J. Chen, P. M. Vora, J. M. Kikkawa, C. B. Murray, *Nature* **2010**, *466*, 474.

90.   A. Blanco, E. Chomski, S. Grabtchak, M. Ibisate, S. John, S. W. Leonard, C. Lopez, F. Meseguer, H. Miguez, J. P. Mondia, G. A. Ozin, O. Toader, H. M. van Driel, *Nature* **2000**, *405*, 437.

91.   R. Mayoral, J. Requena, J. S. Moya, C. López, A. Cintas, H. Miguez, F. Meseguer, L. Vázquez, M. Holgado, Á. Blanco, *Adv. Mater.* **1997**, *9*, 257.

92.   J. M. Jethmalani, W. T. Ford, G. Beaucage, *Langmuir* **1997**, *13*, 3338.

93.   P. Picranski, *Contemp. Phys.* **1983**, *24*, 25.

94.   S. A. Ashert, *Science* **1996**, *274*, 959.

95.   A. Yethiraj, A. van Blaaderen, *Nature* **2003**, *421*, 513.

96.   Y. Cui, M. T. Björk, J. A. Liddle, C. Sönnichsen, B. Boussert, A. P. Alivisatos, *Nano Lett.* **2004**, *4*, 1093.

97.   G. A. Ozin, S. M. Yang, *Adv. Funct. Mater.* **2001**, *11*, 95.

98.   A. van Blaaderen, R. Ruel, P. Wiltzius, *Nature* **1997**, *385*, 321.

99.   Y. Yin, Y. Lu, B. Gates, Y. Xia, *J. Am. Chem. Soc.* **2001**, *123*, 8718.

100.   C. H. Sun, P. Jiang, B. Jiang, *Appl. Phys. Lett.* **2008**, *92*, 061112.

101.   W. L. Min, B. Jiang, P. Jiang, *Adv. Mater.* **2008**, *20*, 3914.

102.   P. Jiang, T. Prasad, M. J. McFarland, V. L. Colvin, *Appl. Phys. Lett.* **2006**, *89*, 011908.

103.   M. J. Madou, *Fundamentals of Microfabrication: the Science of Miniaturization*, CRC Press, Boca Raton, FL 2002.

104.   H. W. Deckman, J. H. Dunsmuir, *Appl. Phys. Lett.* **1982**, *41*, 377.

105.   J. C. Hulteen, R. P. Vanduyne, *J. Vac. Sci. Technol. A* **1995**, *13*, 1553.

106.   C. L. Haynes, R. P. Van Duyne, *J. Phys. Chem. B* **2001**, *105*, 5599.

107.   A. Kosiorek, W. Kandulski, P. Chudzinski, K. Kempa, M. Giersig, *Nano Lett.* **2004**, *4*, 1359.

108.   W. L. Min, P. Jiang, B. Jiang, *Nanotechnology* **2008**, *19*, 475604.

109.   R. L. Hoffman, *Transactions of the Society of Rheology* **1972**, *16*, 155.

110.   P. Jiang, *Chem. Commun.* **2005**, 1699.

111.   C. H. Sun, W. L. Min, N. C. Linn, P. Jiang, B. Jiang, *J. Vac. Sci. Technol. B* **2009**, *27*, 1043.

112.   C. H. Sun, N. C. Linn, W. L. Min, B. Jiang, P. Jiang, *J. Vac. Sci. Technol. B* **2009**, 1043.

113.   C. B. Tang, E. M. Lennon, G. H. Fredrickson, E. J. Kramer, C. J. Hawker, *Science* **2008**, *322*, 429.

114.   B. J. Ackerson, N. A. Clark, *Phys. Rev. A* **1984**, *30*, 906.

115.   B. J. Ackerson, *J. Phys.-Condes. Matter* **1990**, *2*, SA389.

116.   L. B. Chen, B. J. Ackerson, C. F. Zukoski, *J. Rheol.* **1994**, *38*, 193.

117.   R. M. Amos, T. J. Shepherd, J. G. Rarity, P. Tapster, *Electron. Lett.* **2000**, *36*, 1411.

118.   C. Dux, H. Versmold, *Phys. Rev. Lett.* **1997**, *78*, 1811.

119. H. Versmold, *Phys. Rev. Lett.* **1995**, *75*, 763.
120. Y. L. Wu, D. Derks, A. van Blaaderen, A. Imhof, *Proc. Natl. Acad. Sci. USA* **2009**, *106*, 10564.
121. P. Jiang, M. J. McFarland, *J. Am. Chem. Soc.* **2005**, *127*, 3710.
122. C. H. Sun, N. C. Linn, P. Jiang, *Chem. Mater.* **2007**, *19*, 4551.
123. P. Jiang, *Angew. Chem. Int. Ed.* **2004**, *43*, 5625.
124. P. Jiang, *Langmuir* **2006**, *22*, 3955.
125. A. Luque, S. Hegedus, Eds., *Handbook of Photovoltaic Science and Engineering*, John Wiley & Sons, West Sussex 2003.
126. J. Poortmans, V. Arkhipov, Eds., *Thin Films Solar cells: Fabrication, Characterization and Applications*, John Wiley & Sons, Chichester 2006.
127. J. M. Dai, J. Q. Zhang, W. L. Zhang, D. Grischkowsky, *J. Opt. Soc. Am. B* **2004**, *21*, 1379.
128. J. Henzie, E. S. Kwak, T. W. Odom, *Nano Lett.* **2005**, *5*, 1199.
129. M. G. Moharam, T. K. Gaylord, *J. Opt. Soc. Am.* **1981**, *71*, 811.
130. M. G. Moharam, D. A. Pommet, E. B. Grann, T. K. Gaylord, *J. Opt. Soc. Am. A* **1995**, *12*, 1077.
131. S. H. Zaidi, D. S. Ruby, J. M. Gee, *IEEE Trans. Electron Devices* **2001**, *48*, 1200.
132. E. Oliva, F. Dimroth, A. W. Bett, *Prog. Photovoltics* **2008**, *16*, 289.
133. T. Yamada, A. Moto, Y. Iguchi, M. Takahashi, S. Tanaka, T. Tanabe, S. Takagishi, *J. J. Appl. Phys.* **2005**, *44*, L985.
134. C. H. Sun, B. J. Ho, B. Jiang, P. Jiang, *Opt. Lett.* **2008**, *33*, 2224.
135. S. Luca, J. L. Santailler, J. Rothman, J. P. Belle, C. Calvat, G. Basset, A. Passero, V. P. Khvostikov, N. S. Potapovich, R. V. Levin, *J. Solar Energy Eng.* **2007**, *129*, 304.

Chapter 11

Breath Biomarker Detection by Chemical Sensors

J. Huang, Y. Li, S. Sood, P.I. Gouma

Center for Nanomaterials and Sensor Development
204 Engineering Building, SUNY at Stony Brook
Stony Brook, NY 11794-2275

11.1. Breath Biomarkers

11.1.1. Biomarker

A clinical diagnosis of a disease is usually based on medical signs and symptoms reported by the patients. The latter are the indications of health and illness perceived by the patients, medical signs are objective indications of the state of health or disease of the patient observed from outside of the patient [1].

Such indications rely on 'biological markers', or biomarkers, the definition given by the National Institutes of Health Director's Initiative on Biomarkers and Surrogate Endpoints, is: 'A characteristic that is objectively measured and evaluated as an indicator of normal biological processes, patho-genic processes, or pharmacologic responses to a therapeutic intervention' [2].

Biomarkers find many applications in disease diagnosis and health status monitoring, either as diagnostic tools for a disease; as a tool for staging of disease or classification of the extent of the disease; or as an indicators of disease prognosis; and as theragnostic tool for the prediction and monitoring of the clinical response to an intervention [2].

355

Commonly used biomarkers include the concentration of glucose in the blood for diabetes monitoring, the concentration of cholesterol in blood for cardiovascular disease, the blood pressure for stoke, the tidal volume for pulmonary disease, to name a few.

11.1.2. Using Breath Constituents as Biomarkers

Breath analysis implies the identification and quantification of single or multiple compounds in the exhaled breath and their use to diagnose certain diseases. It dates back to antiquity and the time of Hippocrates of Kos [3]. Physicians from that time till now may diagnose disease with clues from the aroma in the patient's breath: "fruity odor of uncontrolled diabetes, the musty, fishy reek of advanced liver disease, the urine-like smell of failing kidneys and the putrid stench of a lung abscess" [3].

In 1784, Antoine Laurent Lavoisier, the father of the 'chemical revolution', found that animals consume oxygen and expire carbon dioxide. Together with French scientist Pierre Simon Laplace, Lavoisier measured the breath of guinea pigs. Theirs was a pioneering work on breath analysis by a chemical method [3]. In 1971, Linus Pauling first analyzed approximately 250 compounds in human breath by gas chromatography and mass spectroscopy (GC-MS) [4].

Breath analysis is superior to other biomarker sampling methods, such as blood, urine, biopsy, endoscopy and imaging as it is completely non-invasive, and with respect to frequency, access and cost [5] and it easy for any patient, from the pre-maturely born to the elderly and incapacitated to provide a breath sample. The comparison could be found in Table I.

However, there are two major challenges for clinical breath analysis: identification of biomarkers and confounding factors. Even though it has been measured nowadays that at least 400 different compounds could be found in human breath, it is difficult to find a direct relationship between biomarkers and specific diseases [5].

Table I Current widely-used medical diagnostic techniques [5]

	Blood	Urine	Biopsy	Endoscopy
Non-invasive	No	Yes	No	No
Limitlessly repeatable	No	No	No	No
Reason	Invasive: painful Non-repeatable: finite, expense, logistics	Non-repeatable: finite, timing, expense, logistics	Highly invasive Non-repeatable: major procedure, expense	Invasive procedure Non-repeatable: major procedure, expense

Table II Current widely-used medical diagnostic techniques [5]

	Examples of imaging			Breath analysis
	X-ray/CT scan	MRI	Sonography	
Non-invasive	No	Maybe	Yes	Yes
Limitlessly repeatable	No	No	No	Yes
Reason	Invasive: ionizing radiation Non-repeatable: radiation exposure, expense, logistics	Possibly invasive: injected dyes Non-repeatable: expense, logistics	Non-repeatable: expense, logistics	

The second challenge involves the sampling method and its standardization. Because the concentration of biomarkers in the breath is usually in trace amounts (parts per trillion (ppt) or parts per billion (ppb) level), any inaccuracy from the sampling process would change the output dramatically and fail the test. The error during sampling process might come from the breakdown of the molecules at high working temperature; the result might be flow rate dependent; the molecules might be adsorbed on the wall of the measuring tube or chamber, etc. The dynamics of exogenous compounds within the body, which may release biomarkers, constitutes a confounding factor. At present, standardization is not achieved because the sampling and instrumentation techniques are under development. However, it's almost impossible to compare the data if there's no standard in the field. Examples of biomarkers of some disease are listed in Table III

Table III Examples of some diseases' biomarkers [6]

Biomarker	Chemical Formula	Disease
Nitric oxide, carbon monoxide	NO, CO	Asthma
Formaldehyde, pentane	CH_2O, C_5H_{12}	Breast cancer
Acetone	$(CH_3)_2CO$	Diabetes
Carbon monoxide	CO	Angina, hyperbilirubinermia
Carbonyl sulphide, ammonia	OCS, NH_3	Liver disease
Sulphide carbonyl	OCS	Transplant rejection
Hydrogen cyanide	HCN	Cystic fibrosis

11.1.3. Nitric Oxide as Biomarker of Asthma

Nitric oxide (NO), molecule of the year 1992 [7], was considered one of the most noxious gases. It's a simple free radical gas, colorless at room temperature and can dissolve in the water in the absence of oxygen. It's released in the exhaust of cars, it could degrade ozone and it's found in the acid rain. However, it is a signaling molecule in a variety of biological functions in the human body [8]. It also acts as vasodilator, bronchodilator, neurotransmitter and mediator of the inflammatory response [9]. In the 1990s NO was found to be the biomarker for asthma in human breath [10-12]. NO in exhaled breath was studied in patients with asthma and high fractional concentration of orally exhaled nitric oxide (FE_{NO}) in these patients compared with healthy subjects [11-15] and a fall after treatment with corticosteroids [16-18] were reported.

Inflammation of the airways is a central process of asthma and other diseases. Invasive techniques, such as bronchoscopy with lavage and biopsy, or by the analysis of induced sputum, are used to analyze the inflammation in the airway [19].

It has been proposed using FE_{NO} as a biomarker to diagnose [20, 21]; to monitor the response to anti-inflammatory medications [17, 22]; to monitor adherence to steroids [23, 24]; and to predict asthma exacerbation [25-27]. Thus, FE_{NO} has become more and more recognized as very useful in the field of pulmonary disease.

In 2005, American Thoracic Society (ATS) and European Respiratory Society (ETS) jointly stated recommendations for online and offline measurement of exhaled lower respiratory nitric oxide and nasal nitric oxide [19]. In 2011, ATS published an official clinical practice guideline in order to interpret exhaled nitric oxide level for clinical application [28].

11.1.4. Two Major VOCs in Human Breath

In recent work, researchers have identified more than 3000 VOCs in breath. About 200 of them are widely detected in almost all breath samples, while the existence of others varies from individuals to individuals [29].

11.1.4.1. Acetone in Human Breath

As mentioned above, there are thousands of volatile organic compounds contained in human breath. Among all those VOCs, breath acetone is one of the most abundant ones. Acetone is generally regarded as a biomarker for diagnosis of diabetes. Diabetic patients typically have higher concentration of exhaled acetone and this may result in a sweet odor in their breath [30]. For most VOCs, the measured values in human breath vary a lot. This may due to the variability between (and within) individuals as well as from the sampling and analysis methods. In the work of Fenske et al., the concentration range of acetone was summarized and it lied in 1.2-1880 ppb [31].

11.1.4.2. Isoprene in Human Breath

Isoprene is another major VOC in human exhaled breath. Usually, the concentration of isoprene is in the sub-ppm level [31]. In 1984, Denise et al. suggested by experiments that breath isoprene was possibly linked to cholesterol synthesis [32]. Its formation occurs in human body as a by-product of the biosynthesis of cholesterol. This finding raised the possibility that measurements of breath isoprene might become a noninvasive and simple way for assessing body cholesterol status. The effectiveness of lipid-lowering therapy can also be evaluated by monitoring the concentration of breath isoprene in patients. Although breath isoprene seems to be a promising candidate for breath analysis, there are some factors (age, gender, heart rate, etc.) [33, 34] that may affect its concentration in exhale. A standard is needed for sampling and measurement.

11.2. Detection Methods for Breath Analysis

In this section, we will mainly focus on nitric oxide detection. The need for clinical breath analysis drives the development of sensor technology. An ideal sensor for FE_{NO} measurement in clinical use needs to be sensitive and selective, compact, portable, robust, reliable, real time operative, easy to operate and inexpensive. Several available and developing technologies are discussed here.

11.2.1. Chromatography and Spectroscopy

11.2.1.1. Off-line Analysis: Gas Chromatography

Gas chromatography (GC) is a common technique in the breath analysis community for biomarker detection. The basic process is injecting the gas sample into the chromatographic column where the separation process is carried out. Chemical compounds with different affinities to the stationary phase (liquid or solid) move alone the column with various rates with an inert gas. For most GC, identification of the chemicals is based solely on retention time and since many compounds have the same retention time, we need extra detector for advanced detection [35]. At the output of the column, a few methods are coupled to identify the chemicals. These methods include but not limited to:

1) ***Mass Spectroscopy (MS)***: the mass spectrometer takes injected gas from the output of GC, ionizes it in high vacuum, propels and focuses these ions and their fragmentation products in magnetic field, then collects and measures the amounts of each selected ion [35], and quantifies the chemicals, which is based on determination of mass to charge ratio of ionized atoms or molecules. The sensitivity could be down to few ppb.

2) ***Flame Ionization Detection (FID)***: in FID VOCs are burned, producing ions and electrons that conduct the current which is used for detection and quantification of the institutes. FID has high sensitivity, wade range linear response and low level of noise, that's the reason it is common in breath analysis [6]. Acetone, pentane, isoprene, hexanal and other VOCs has been detected by GC-FID [36].

3) ***Ion Mobility Spectrometry (IMS)***: the principle of IMS is a time-of-flight measurement. The gaseous sample enter the spectrometer and become ionized by the radioactive source. The charged species would be accelerated over a short distance and the time of flight would be measured and the identities of the chemicals would be determined.

   The IMS is different from the MS in that it operates under atmospheric environment so there's no need for large and expensive vacuum pumps to reduce the expense and increase the portability [37]. IMS could provide a combination of detection features such as high sensitivity (down to ppb level), fast response (a few seconds), small size (wearable), low power consumption (about 4 AA batteries) and low cost [38].

All of these technologies mentioned require analyte pre-concentration, thermal deposition or other significant delays between breath collection and measurement [5]. There is potential for introduction of confounding variables and errors with each additional step in sampling and instrumentation which increase the complexity of the process since the concentration of the analyte is so minute.

11.2.1.2. Real-Time Analysis

Online real-time analysis possesses several advantages over off-time analysis with no requirement for sample collection and storage so as to eliminate a major source of errors and confounding factors in the analysis, through mechanisms like rapid degradation of unstable compounds, leaking of gases, permeation and adhesion on the surface [39-41]. Since delays and processing requirements make ideal clinical implementation unlikely [5], an online sampling instrument is always preferable if else being equal with off-line approaches [41]. Proton transfer reaction mass spectrometry (PTR-MS) and selected ion flow tube mass spectrometry (SIFT-MS) are both well-studied real-time online analysis techniques for VOCs, details could be found in [42-45] for PTR-MS and in [46-48] for SIFT-MS.

11.2.2. Chemical Sensors

A definition of chemical sensors by Wolfbeis in 1990 [49] is as follows, 'Chemical sensors are small-sized devices comprising a recognition element, a transduction element, and a signal processor capable of continuously and reversibly reporting a chemical concentration.'
It's well accepted that chemical sensors should have the following characteristics:
- Transform chemical signals to electrical signals
- Be specific to one analyte or a group of analytes
- Respond and recover quickly

- Be small
- Be cheap
- Be stable in a definite length of time

Chemical sensors convert chemical information, like composition of compound, concentration of single molecule, mass change, chemical reaction status, etc. into measurable signals, mostly electrical signals.

Two basic components of a chemical sensor are receptor and transducer. A receptor is the surface layer which is in contact with the sample and have reaction with the analyte, either in physical or chemical ways. A transducer transforms the physical change of the receptor into electrical signals, in most sensors.

According to the transducer operating principle, sensors could be classified in such a way, according to International Union of Pure and Applied Chemistry (IUPAC) in 1991 [50]:

1. Optical sensors transform changes of optical phenomena, following absorbance, reflectance, luminescence, fluorescence, refractive index, optothermal effect and light scattering.
2. Electrochemical sensors transform the effect of the electrochemical interaction analyte-electrode into a useful signal; among them are voltammetric sensors, potentiometric sensors, chemically sensitized field effect transistor and potentiometric solid electrolyte gas sensors.
3. Electrical sensors based on measurements of the signal that arises from the change of electrical properties caused by the interaction of the analyte where no electrochemical processes take place. These included metal oxide semiconductor sensors used principally as gas phase detectors which will be discussed later, based on reversible redox processes of analyte gas components, organic semiconductor sensors, electrolyte conductivity sensors and electric permittivity sensors.
4. Mass sensitive sensors transform the mass change at a specially modified surface into a change of a property of the support material. These include piezoelectric sensors and surface acoustic wave sensors.

5. Magnetic sensors based on the change of paramagnetic properties of a gas being analyzed.
6. Thermometric sensors based on the measurement of the heat effects of a specific chemical reaction or absorption that involve the analyte.
7. Other sensors, mainly based on emission or absorption on radiation.

11.2.3. Chemiresistor

'Chemiresistor' is a term used for gas sensor elements based on resistance/conductance measurements. One of the first and most common used chemiresistor is the 'Taguchi sensor' [51]. Taguchi sensors are made by pressing and sintering powdered solid ceramic. Tin oxide and zinc oxide are the most common materials used as Taguchi sensors. Such sensors could be used for detecting reducing gases such as leakage detection in gas pipeline and alcohol breathalyzer. There are many companies in the market of chemiresistor now, such as Figaro, FIS, CityTech, Applied-Sensors, Synkera, etc. [52].

The gas sensing mechanism for metal oxide semiconductor gas sensor is not completely clear yet; however there are possible pathways reported. The first model is that when the metal oxide semiconductor sensor is in air, oxygen absorbs on the surface and dissociates to form O^-, which extracts the electron from the semiconductor to the oxygen. This act of electron extraction tends to increase the resistance (assuming the semiconductor is an n-type one whose majority charge carries are electrons) of the sensor. In the presence of combustible gases, the absorbed O^- reacts with them and electrons are re-injected to the semiconductor that tends to decrease the resistance of it. A competition between oxygen removing the electrons and combustible gases re-injecting the electrons comes to a steady state which the resistance of the metal oxide is determined [53].

Another model that exists or co-exists is that the combustible gas would extract a lattice-oxygen from the metal oxide, leaving vacancies that act as donors. The oxygen from the air tends to re-oxidize the metal oxide and remove the donor vacancies. A competition between the combustible gas creating vacancies as donors and the oxygen removing the vacancies exists [54].

For NO detection, many metal oxides have been employed as gas sensing materials. Some researchers have also investigated the effect of doping or decoration with metals [55-61]. Table 1 is a summary of sensing properties of some chemiresistor to NO.

Table 1 Summary of sensing properties of some chemiresistor to NO

Sensing material	Minimum saturation detection limit (MDL/SDL)	Response/ recovery time	Optimal working temperature	Sensitivity	Ref.
Hierarchically porous structured	N/A	response 63 s	200 °C	46.2 for 100 ppm NO	[55]
Monoclinic WO3		recovery 223 s			
β-WO3	N/A	response ~ 40 s	180 °C	13 for 10 ppm NO	[56]
Villi-like monoclinic WO3	MDL 88 ppt	N/A	200 °C	278 for 1 ppm NO	[57]
Au/δ-WO3	MDL 500 ppb	response 15 s recovery 24 s	170 °C	~55 for 10 ppm NO	[58]
Ag/δ-WO3	MDL 500 ppb	response <10 s recovery <100 s	170 °C	~90 for 2 ppm NO	[59]
WO3/Cr2O3 mixture	MDL 18 ppb	N/A	300 °C	~1.3 for 200 ppb	[60]

Pt/In2O3-WO3	MDL 25 ppb	response <20min recovery <120min	room temp.	15.2 for 25 ppb	[61]

11.2.4. Tungsten Oxide

Tungsten oxide is a transition metal oxide with wide range of applications. The crystal structure of WO_3 is a distortion of rhenium oxide cubic structure in which tungsten atoms are located in cube corners and the oxygen atoms are located on the cube edges [62]. Each tungsten atom is surrounded by six oxygen atoms, forming an octahedron. The slight rotation of these octahedral with respect to each other, as well as unequal bond lengths in the octahedral coordination, causes lattice distortion and reduces the symmetry. The distorted structure is stable in several forms giving rise to different phases depending on the temperature. Figure 1 (adapted from [63]) shows the 7 known polymorphic transformations of WO_3 in bulk.

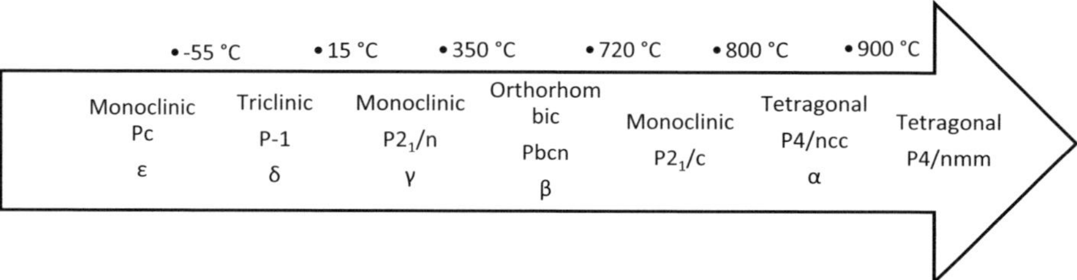

Figure 1. Phase transformations in WO_3 in bulk.

In 1991, Yamazoe et al. [64] first discovered the sensing properties of WO_3 towards NO and NO_2. The sensitivity towards 200 ppm NO and 80 ppm NO_2 was 31 and 97 respectively at 300 °C. This research marked the beginning of a new era for NO sensing, after which many studies were focused on this specific topic by using different methods to produce WO_3 sensors.

Earlier work by Gouma's group [65] introduced the polymorph-gas selection map of resistive gas sensors by studying tungsten oxides and molybdenum oxides. It provides the direction to develop chemo-resistive gas sensors by size control and operation temperature selection so as to attain the polymorph required for the gas tested.

Penza et al. [66] used radio frequency sputtering system to produce WO_3 film. The substrate they used was a 15×15 mm glass whose thickness was 5mm. The thickness of films produced varied from 500 to 5000 Å according to the deposition time and input power that changed between 100 and 300 W. The films were annealed at 400 °C in dry air for at least 6 hours to stabilize the sensing properties. They compared the XRD results from two films with different grown rates, the phase of both were tetragonal and it showed that the film deposited at lower rate had a better structural quality. They found out the optimum operating temperature was 250 °C after testing the sensing with NO_x, NH_3, H_2, H_2S, SO_2, CO_x and CH_4. The sensor showed good selectivity of NO_x against other gases. The threshold limit of the sensor at 250 °C for NO_2 and NO was 2 and 500 ppm.

Polleux et al. [67] used tungsten isopropoxide and benzyl alcohol to build a versatile reaction system for the preparation of crystalline tungsten oxide nanowires performed in a glass beaker at low temperature. The length of the nanowire bundles ranged from 300 to 1000 nm and the width was between 20 and 100 nm. The diameter of all nanowires was very uniform being in the range of (1 ± 0.1) nm and they were mostly oriented parallel to each other with an equal distance. To make a stable sensor, a suspension of the tungsten oxide nano-bundles in ethanol was drop coated onto the aluminum substrate and then the sensor was heat-treated at 500 °C to be stabilized. The sensor showed high sensitivity (≈9) at very low concentration of NO_2 (about 500 ppb), which might be attributed to the high surface area and high crystallinity with appropriate porosity of the nanowire film. The response of the sensor to NO_2 decreased with increasing temperature and almost lost their sensitivity at 300 °C higher than which it could still be sensitive to reducing gases such as hydrocarbons and volatile organic compounds.

Marquis and Vetelino [68] tried to build a sensor arrays for fossil burning gases emission. They controlled substrate material, film thickness, film doping, deposition temperature and operation temperature in radio frequency sputtering to produce tungsten trioxide gas sensor. They chose 1000 Å WO_3 un-doped film sputtered at 400 °C on sapphire operating at 300 °C as NO_x sensor and 5000 Å WO_3 film co-sputtered with 16 Å Au at 200 °C on alumina operating at 350 °C as NH_3 sensor after a large number of tests. The response time for the NH_3 film ranged from 7 min for 1 ppm NH_3 to about 4 min for 50 ppm NH_3 and the response time for the NO_x film ranged from about 6 min for 1 ppm NO_x to less than 2 min for 75 ppm NO_x. They also tested the short and long-term stability of each film exposed to 15% relative humidity in the air. The baseline variation was in 3% over a 3 min period and in 8 % over a 10-hour period for the NH_3 sensor while the short-term stability of NO_x film was in less than 0.6 % and the long-term stability was about 20 %. The sensor array showed good selectivity to 10 ppm NH_3 and 75 ppm NO_x and H_2S that is a gaseous emission in fossil burning systems.

Lee et al. [69] made NO_x sensors with TiO_2 doped tungsten trioxides by sol-coprecipation of WCl_6 and $TiCl_4$ solution with ammonium hydroxide and surfactant and screen-printing after calcination at high temperature. The grain size of TiO_2 doped WO_3 was smaller than pure WO_3 or the simple mixture of WO_3 and TiO_2 powder. The TiO_2 doped WO_3 showed excellent sensitivity to low concentration of NO_x (0.5-30 ppm), short response time (1-2 min) and short recovery time (2-3 min) at 350 °C which was the optimum temperature for NO_x testing. It was illustrated that the size effect introduced by TiO_2 enhanced the sensing performance of WO_3.

In the study of Yu-De et al. [70], The electrical and gas-sensing properties of calcined tungsten trioxide were investigated. WO_3 with SiO_2 (4 wt. %) were fabricated on an alumina tube with Au electrodes and Platinum wires. A Ni-Cd alloy crossing alumina tube as a resistor was used for the heating. The sensors were calcined at different temperatures (350-800 °C) and then were aged at 250 °C for 150 hours in air to improve the stability and repeatability. When operating at the same temperature (200 and 250 °C), sensors that were calcined at higher temperature had smaller resistance baseline. It was showed that the sensor had best thermal stability in the operating temperature range of 175 to 225 °C, where the resistance was low and had little variation. By testing the sensitivity of the sensors to different gases, such as ethanol, petrol, methane and butane, at different calcining temperature, the optimum one was 500 °C. The sensor showed highest sensitivity (about 40 at 80 °C) to ethanol (100 ppm) among those gases tested. At calcining temperature of 500 °C and operating temperature of 200 °C, the WO_3 gas sensors which were compounds of monoclinic and triclinic phase (by XRD results) had good thermal stability and good sensing properties to ethanol.

Yin et al. [71] prepared Au-modified WO_3 nanocomposites by chemically reducing $HAuCl_4$ on the surface of two-dimensional WO_3 nanoplates, which were made by intercalation-topochemical process. WO_3 with 1 wt. % Au showed best sensing performance at 170 °C to NO (0.5-10 ppm), which was higher than that of pure plate like WO_3. It showed high selectivity to NO among various inorganic gases (H_2, SO_2, and CO) and volatile organic compound (ethanol, acetone, methanol and benzene) at temperature lower than 200 °C.

Ponzoni et al. [72] prepared WO_{3-x} nanowires by thermal evaporation of tungsten powders at 1400-1450 °C in oxygen for 10 minutes. The width of the nanowires ranged from 0.7 to 200 nanometers and a 3D networks was formed by nanowires intercrossing each other. After the nanowire networks were drop coated onto alumina substrates and annealed in air at 300 °C for six days for stabilization, the sensors were measured for the response towards NO_2, H_2S, CO and NH_3. The sensor had excellent response (about 6) to NO_2 at low concentration (50 ppb) at 300 °C. The sensor was found to have better response intensity but longer response and recovery time at lower temperature (more than 40 minutes at 200 °C) at the testing range of 100-500 °C for NO_2. It appeared that the optimum temperature for sensing was 300 °C to achieve the balance between response intensity and dynamics. They also found that the sensor had weak responses towards NH_3 and CO at any working temperature, which was evidence for the good selectivity for the WO_{3-x} nanowire systems being good NO_2 sensor.

Kim et al. [73] made gas sensor by drop coating $WO_{2.72}$ nanorod solution on the Si bulk-micromachined membrane with a hotplate for temperature control. The resistance of the sensor decreased when exposed to 100 ppm ammonia (NH_3) at the temperature range of 150-250 °C but increased at temperature lower than 70 °C. Further sensing tests to gases like 2% N_2 (or air), 1000 ppm ethanol, 10 ppm NH_3 and 3 ppm NO_2 were performed at room temperature. All of them had increase in resistance except for N_2 that had little change and their response sensitivity ascended in the order of air, ethanol, ammonia and nitric dioxide when the recovery time ascended in the opposite sequence.

Prasad [74] used acid precipitation route to produce tungsten oxide thin films. Two different phases (orthorhombic and monoclinic phases) were obtained by calcination at different temperature. Phase transformation from monoclinic to orthorhombic was found to occur in the temperature range of 350-400 °C. The sensing test was performed at the temperature range of 200-500 °C to nitrogen dioxide and ammonia and it was found that both phases of WO_3 had good selectivity of nitrogen dioxide against ammonia and monoclinic phase WO_3 had best sensing performance to nitrogen dioxide in the temperature range of 200-300 °C.

Krithika [75] used sol-gel method to prepare WO_3 nanoparticles and thin films of WO_3 were prepared by spin coating and drop coating of the prepared sol on the 3 mm × 3 mm Al_2O_3 substrates. The films attained were then calcined at 400 °C for 6 hours and 515 °C for 8 hours to get monoclinic and orthorhombic phases respectively according to the result from differential scanning calorimetry (DSC) test. The average grain size of the particles which calcined at 400 °C was around 10-20 nm. WO_3 with either monoclinic or orthorhombic phase was tested for the response to NO_2, isoprene, acetone and ethanol at different ppm levels at 400 and 500 °C in respect. It showed that the monoclinic phase showed no cross-sensitivity toward reducing gases like hydrocarbons and VOCs while the orthorhombic phase had higher sensitivity for similar concentration of NO_2 but cross-sensitivity towards reducing gases. She also tested sensing using both polymorphs for ppb level sensing of NO and NO_2 at 400 and 200 °C for monoclinic and orthorhombic phase respectively. The sensitivity of the monoclinic phase for NO was a little greater than NO_2 at the same concentration while the sensitivity of the orthorhombic phase to

NO_2 was much higher than NO. In the conclusion of her thesis, that author suggested that the similarity of sensing response to oxidizing gases like NO and NO_2 of both monoclinic and orthorhombic phases was because they were isostructural and the mechanism of such sensing was absorption based that didn't affect the bond on the metal oxide surface. However the sensing behavior towards reducing gases was different. 'Magneli phase' or crystallographic shear (CS) planes existed in orthorhombic phase which played a critical role in inserting oxygen in to the hydrocarbon molecule as catalysts for selective oxidation. The sensing mechanism of such kind was reaction based. Two different sensing mechanisms exhibit from two polymorphs that are isostructural.

Wang [76] used flame spray pyrolysis method to synthesis ε-WO_3. The percentage of ε-WO_3 was related to the particle size of the particles and the smaller they are the higher percentage of ε-WO_3. However, after calcination with as-received WO_3 the ε-WO_3 vanished. So he came up with an idea of doping chromium in the material by forming a surface layer, which could prevent the ε to γ transition. The WO_3 with 10 at% Cr doped sensor was found to have good sensitivity and selectivity to low concentrations of acetone (from 0.2 to 1 ppm) compared to other interfering gases at 400 °C. The mechanism behind the high sensitivity to acetone of ε-WO_3 was believed to be related to the surface dipole of ferroelectric ε-WO_3 and the highly polar acetone gas molecules. Wang [77] built a prototype breathalyzer to detect acetone in human breath for diagnosing diabetes. The sensor was made of ε-tungsten trioxide, which is selective to acetone at the operating temperature. The concentration of gases tested in the study was down to 200 ppb and it showed adequate selectivity at the threshold concentration (1.8 ppm) of diabetes diagnosis. It was suggested that the acentric structure of ε-tungsten trioxide played an important role in the selective detection of acetone that has a high dipole moment than any other gases tested.

11.3. Processing Methods for Breath Gas Sensing Materials

In Gouma's research group, many processing methods have been applied for the production of gas sensing materials. In this section, three of the most used one are being introduced.

11.3.1. Sol-Gel Method

Sol-gel process [78] involves solid nanoparticles dispersed in a continuous liquid (sol) which agglomerate to form interconnected three-dimensional solid state network with isolated liquid phase (gel). This is a relatively inexpensive method and could be easily applied in the lab. The starting materials for sol are usually inorganic metal salts or organic metal compounds. The versatility of the sol-gel process is manifested in the very many ways which could be applied to attain the final products depending on the application requirements, such as spin coating, drop coating, electrospinning.

Krithika [79] prepared nanostructured WO_3 using sol-gel method and spin coated on aluminum substrates. After heat treatment at 400 °C for six hours, the amorphous phase turned into monoclinic phase by XRD result and Raman spectrum. A TEM picture of the heat treated sample is shown in

Figure 2. The sensor was tested to nitric oxide, acetone, carbon monoxide, ethanol, isoprene and methanol, showing a good selectivity for nitric oxide against other gases. Also, the concentration of nitric oxide was down to 300 ppb, which is relevant to those found in human breath level. The sensing result is shown in Figure 3. With that being tested, the materials could be a good prospect for human breath nitric oxide detection.

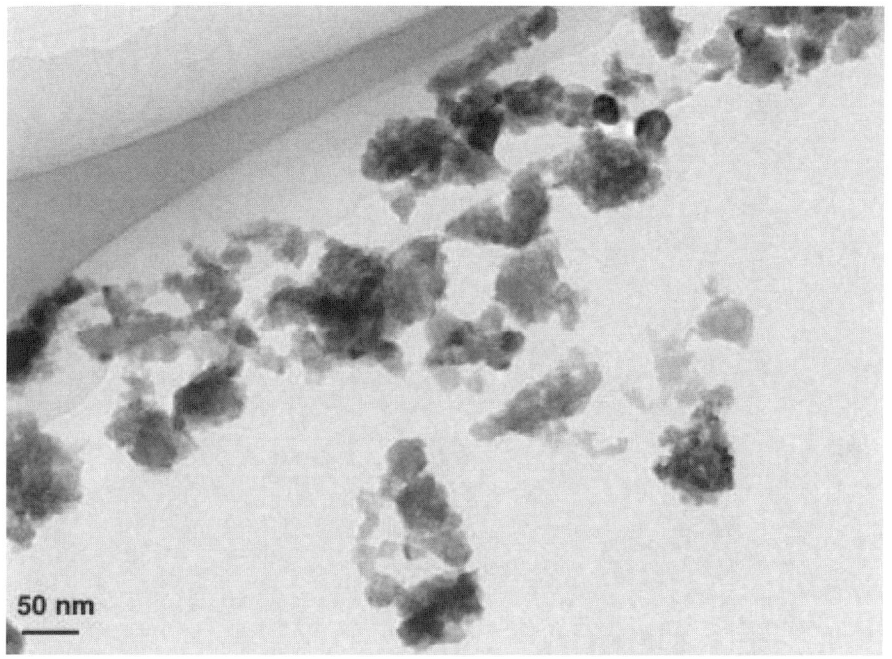

Figure 2. General TEM view of the 400°C annealed sol-gel sample [75].

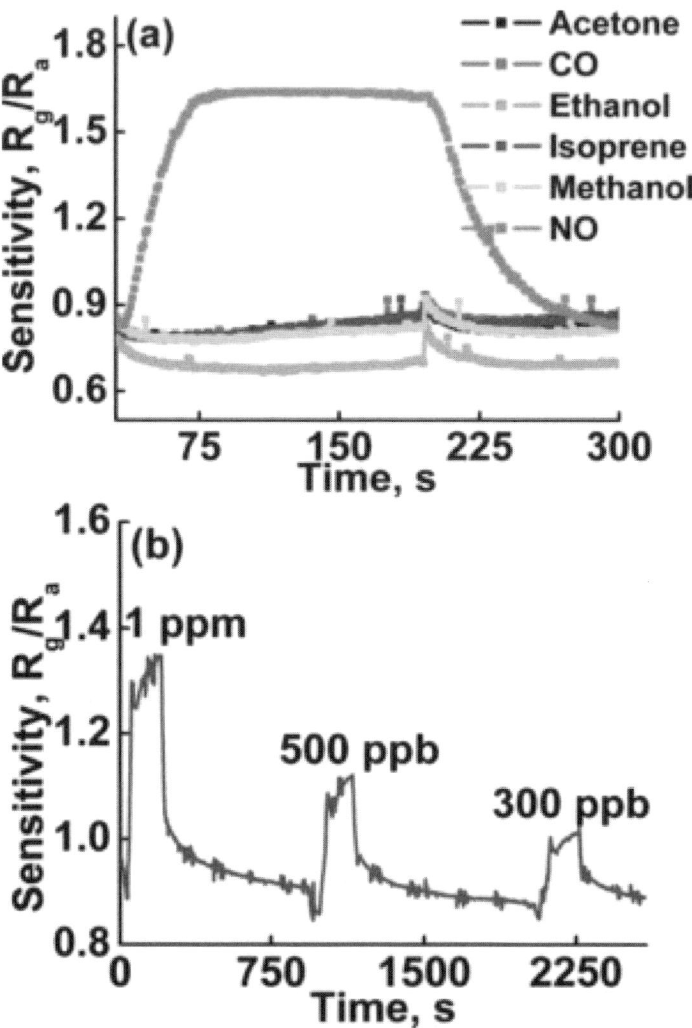

Figure 3. (a) Gas sensing response of monoclinic-WO3 to the following gases: 10 ppm NO, 10 ppm acetone, 10 ppm isoprene, 50 ppm ethanol, 50 ppm methanol, and 50 ppm CO in a background of 80/20-N_2 /O_2. (b) Low concentration gas sensing response of monoclinic-WO3 to 1 ppm, 500 ppb, and 300 ppb NO in a background of 80/20-N_2 /O_2 [79].

11.3.2. Electrospinning

Electrospinning is an easy to go, way to produce nanofiber. It uses electrical charge to extract fine fibers from liquid to deposit on the charged substrate. The basic setup for electrospinning usually consists of syringe pump, needle that is connected to power supply and grounded substrate. There are many parameters controlling the outcome of the fibers: the properties of the solution (viscosity, surface tension and conductivity), potential of the charge, flow rate, distance between needle and substrate, ambient atmosphere (temperature, humidity and air flow).

Sawicka and Prasad [80] combined metal oxide sol-gel and polymer to prepare the precursor for electrospinning and they focused on the MoO_3/WO_3 and polyvinylpyrrolidone (PVP) system. Compared with the sol-gel prepared films, the nanowires of WO_3 showed better sensitivity, fast response and lower sensing limit to nitrogen dioxide. However, the range of concentration was from 50 to 500 ppm, which is too high for human breath sample.

11.3.3. Flame Spray Pyrolysis

To produce nano-sized particles with high purity and good thermal stability in a scalable way, scientists came up with the idea of using flame in gas phase process. Compared to those wet phase process mostly used, which involves redundant moves like filtration, drying, calcination, etc., gas phase process is easier to control and ecofriendly because of water saving, just named a few advantages here.

Flame synthesis of materials doesn't need extra sources of energy from the system to converse the precursor and the energy for the particles formation is coming from the in-situ chemical reaction of the precursor in the synthesis process. Various reactors and methods have been adapted to produce different kinds of metals and metal oxides. According to the precursor state fed into the system, the flame synthesis process could be classified to two main categories: vapor-fed aerosol flame synthesis (VAFS) and liquid-fed aerosol flame synthesis (LAFS).

Vapor-fed aerosol flame synthesis is the pyrolysis of volatile precursors in flames. The particles form by nucleation in the gas phase after conversion of volatile precursors and grow by surface reaction and/or coagulation and coalescence into larger particles. The limitation of VAFS is the lack of available volatile precursors at reasonable price in the market. Fumed silica, alumina and pigmentary titania have been already produced by VAFS [81].

Liquid-fed aerosol flame synthesis is more versatile because the precursor doesn't have to be volatile. In this case, the liquid precursor is sprayed by air-assisted atomization and vaporized to form particles. Complete vaporization leads to formation of nanoparticles like VAFS but incomplete vaporization leads to micron-sized shell-like or hollow particles. Depending on the enthalpy content of the precursor and whether an extra flame is used as heat source to evaporate the precursor, liquid-fed aerosol flame synthesis could be classified to flame spray pyrolysis (FSP) and flame assisted spray pyrolysis (FASP) [81].

In FSP, metal organic precursor is dispersed by gas convection through a nozzle forming fine spray then ignited, vaporized and particles are formed by nucleation and grow from the gas phase. Because of the high maximum temperature and short residence time, the particles could be nano-sized and homogeneously formed; the schematic is shown as Figure 4 [78].

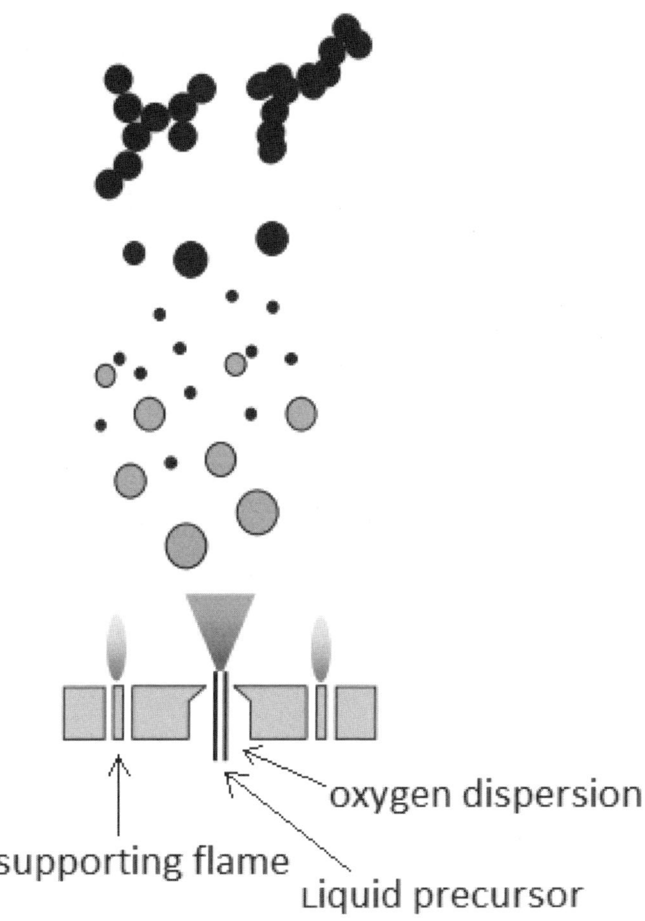

Figure 4. Schematic of FSP process.

The as-prepared particles in flame synthesis process could be collected by filters, electrostatic precipitators or cyclones. Alternatively, a major advance of phase synthesis compared to wet-phase method is direct deposition on the substrate. The thickness and morphology of the materials on substrate could be reproduced and well controlled by this method [82].

Wang [76] used flame spray pyrolysis method to synthesis ε-WO_3. The WO_3 with 10 at% Cr doped (TEM and HRTEM pictures as shown in Figure 5) sensor was found to have good sensitivity and selectivity to low concentrations of acetone (from 0.2 to 1 ppm) compared to other interfering gases at 400 °C (Figure 6).

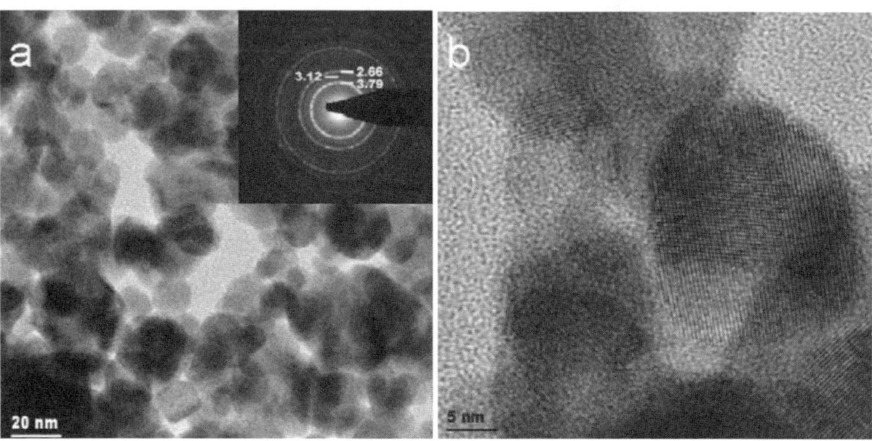

Figure 5 (a) TEM (inset: SAED pattern) and (b) HRTEM images of heat treated 10 at % Cr-doped WO_3 [76].

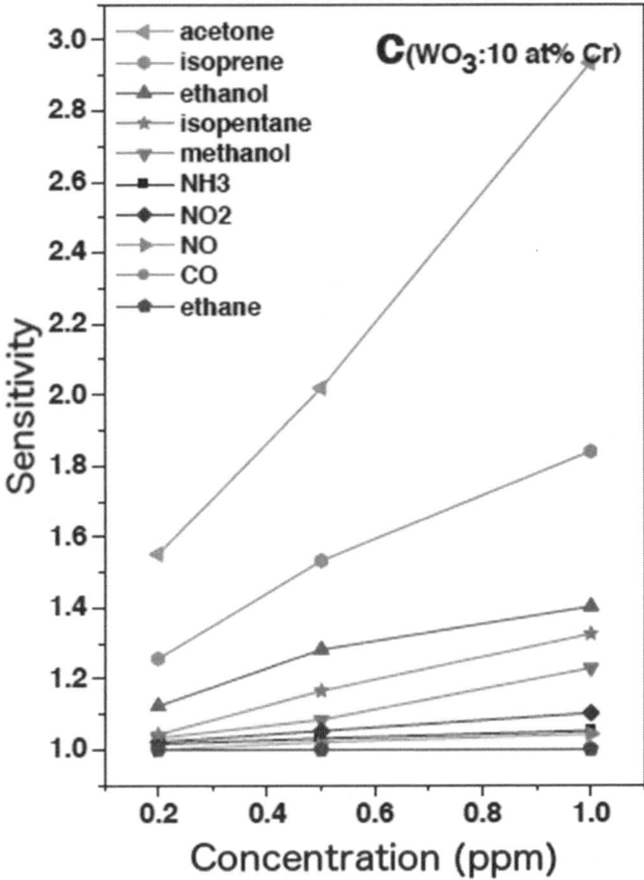

Figure 6. Sensing result of 10 at% Cr-doped WO3 to acetone, isoprene, ethanol, isopentane, methanol, NH3, NO2, NO, Co and ethane [76].

11.4. Breath Analysis Devices by Gouma's Research Group

11.4.1. Nanosensor Device for Breath Acetone Detection

Wang et al. has pioneered in the production of the nanosensor device for breath acetone detection prototype [77]. The material used in this prototype was 10 at% Cr-doped WO_3 nanoparticles, with 80% ε-WO_3

phase content. Details of sensing tests and procedure ware described in [76]. The nanostructured sensor is linear in the detection range of interest to diabetes monitoring (from 0.2 ppm to 2.0 ppm). The concentration of acetone in exhaled human breath normally falls within this range [83]. The diabetes diagnosis threshold was set at 1.8 ppm.

There is an approximately linear relationship between the concentration of acetone and sensor sensitivity, as shown in Figure 7. The sensitivity defined here is R_o/R_g.

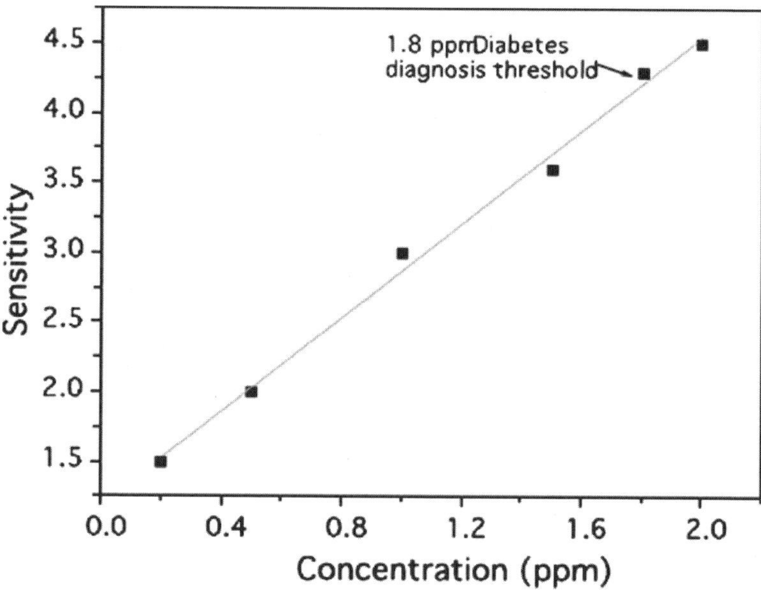

Figure 7. Relationship between concentration of acetone and sensitivity.

The basic idea of the sensing device is to compare the resistance of the sensor material in the exposure to human breath to a comparative resistor. The resistance of the comparative resistor is determined by a pre-assumed biomarker concentration threshold (1.8 ppm in this prototype for diabetes detection) in the human breath as well as the character of the sensing material. The sensor's resistance is determined by the actual biomarker concentration. Assuming the biomarker is a reducing gas, if the resistance of the sensor is lower than that of the preset resistance, the concentration of the biomarker is then higher than the threshold, which implies that the patient has a subject has a high probability to be afflicted with a given disease or to suffer metabolic malfunction. The comparison result is manifested by a LED indication. The photograph of the manufactured prototype is shown in Figure 8. The dimension of the prototype is 15 cm (L) × 7.5 cm (W) which meets the requirement for "portability". The bottom left pat in the figure is the sensor. It is isolated from the environment by a chamber made of Teflon (not shown). A mouthpiece is connected with the chamber to introduce human single breath to the sensing material.

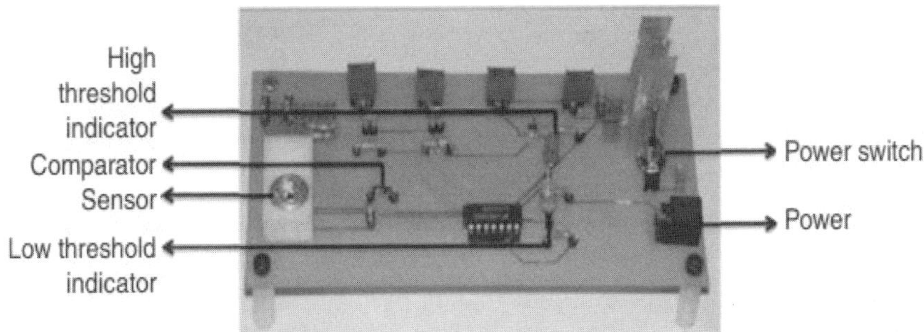

Figure 8. Designed portable device for disease diagnosis.

At 1.8 ppm acetone exposure, the resistance was 3.5 MΩ which was set as the lower threshold value of the device. The upper threshold was set to 20 MΩ, a little higher than the sensor's baseline value. Three different concentration of acetone was introduced (500 ppb, 1 ppm and 1.5 ppm) and no change of the LED light was observed until the concentration was increased to 1.8 ppm or higher.

In order to validate the selectivity of the device to potential interfering gases in human breath, several types of other gases were introduced, including NO, NH_3, CO, ethanol methanol and ethane. The device did not turn on the LED light to NO, NH_3, CO and ethane up to

10 ppm, which is much higher than the respective gas concentration found in exhaled human breath. It also did not turn on the LED light to ethanol and methanol up to 3 ppm. As elevated levels of ethanol and methanol are usually associated with alcohol ingestion and fruit consumption [84], the subject should avoid consuming alcohol or fruits prior to the test.

11.4.2. Three-Nanosensor Array Microsystem to Monitor Infections

A three-sensor array consisting of a NO, NH_3 and isoprene sensor has been developed. The sensor array interfaces to an integrated circuit for electrical readout and temperature control, thereby providing a complete microsystem capable of capturing a single exhaled breath and analyzing it with respect to the relative content of NO, NH_3, and isoprene.

With respect to the breathalyzer prototype, in order to quantify and transmit the response of the sensor array, a readout system has been designed. The sensors in the array behave electrically as resistors, the resistance of which depends on the concentrations of the specific gases in the environment. Thus, the readout circuitry converts the resistance from each sensor into a voltage signal that is then digitized. The three-sensor system is based on a single sensor design with a Bluetooth interface on a printed circuit board (PCB) with dimensions 3 x 3 inches (Figure 9).

Figure 9. The breathalyzer prototype (the NO sensor mounted).

11.5. Outlook-Single Crystal Nanowire Sensors for Breath Diagnostics

11.5.1. Nanowires of Semiconducting Oxides

Nanowires of semiconducting materials are nanostructures with high-aspect ratio, enormous surface area, and controlled crystallographic orientation that make them ideal for applications in sensing in a breath diagnostic tool [85], catalysis, and nano-electronics device [86, 87]. Such nanostructures have been widely considered in the literature since the early report of single crystal silicon by a vapor-liquid solid technique [88] and single crystal semiconducting nanowires have been synthesized by pulsed-laser ablation [89]. Monocrystalline nanowires of metal oxides (WO_3, MoO_3, CuO etc) have been produced through various routes, from thermal oxidation of a metal mesh, to template-assisted blend electrospining, to colloidal synthesis of oxides [89-93], as tabulated in *Table 2*.

Table 2 Results from earlier research in our group

	Synthesis method	Possible analyte sensed	Ref.
WO$_3$ nanowires	Electrospinning	Nitrogen dioxide	[86]
CuO nanowires	CuO nanowires Template assisted growth	--	[91]
MoO$_3$ nanowires	Electrospinning	Ammonia	[93, 94]
3D CuO nanogrids	Electrospinning and thermal oxidation	Ethanol	[95]
Hexagonal WO$_3$	Chemical synthesis	Ammonia	[75]
TiO$_2$ nanoplates	Thermal oxidation	Carbon monoxide	[96]

Single crystal nanowires are of higher value than polycrystalline/multigranular nanowires. Although this is only true when single crystal nanowires have a specific plane exposed, and the properties like gas sensitivity are enhanced [93]. In that case the polycrystalline nanowires have a disadvantage, because the planes are randomly oriented. But reverse maybe true when gas sensitivity is not dependent on specifically oriented exposed plane [97]. For example, in the case of polycrystalline nanowires of In_2O_3 [97], it was seen that porosity due to very small grain size increased the sensitivity to ethanol, as compared to single crystal nanowires, this can be the case in non-specific materials. For highly specific materials, like single crystals of monoclinic WO_3 [79] the sensitivity is highly dependent on the specific planes being exposed and the absence of defects.

WO_3 nanowires, in particular, have also been synthesized from growth directly from tungsten substrate [98] or by thermal heating of a tungsten plate [99]. Non-aqueous preparation of crystalline tungsten oxide nanowires has been carried out using a tungsten isopropoxide precursor in benzyl alcohol [99]. Nevertheless, there has been no direct observation of WO_3 nanowire growth in-situ and solely from solid oxide precursors until now, and thus the nature of the spontaneous transition of oxide nanocrystals to nanowires under the influence of an electron beam is being addressed here.

MoO_3 nanowires, in particular have also been synthesized by other researchers. Pure orthorhombic structured nanowire arrays of molybdenum trioxide was synthesized by Zhou et al. [100], these were grown on a silicon substrate. Due to the instability of trioxide form of molybdenum oxides, many works have managed to synthesize molybdenum dioxide in conjunction with molybdenum trioxide. Such structures have been reviewed extensively by Mai et al. [101] and the readers are directed to the source for more information.

11.5.2. Polymorphic Nanowires as Sensing Materials

Nanowires nanostructures are solid, one dimensional materials with diameters in the 10-80 nm range, and are metallic or semiconducting in nature, made from metals or metal oxides most often. Wires made from molecular entities and diameter less than 5 nm are often called molecular nanowires, and these are different from nanowires. Carbon nanotube research and nanowires research are often parallel. Though carbon nanotube materials are different, the sensing mechanisms and underlying behavior are usually very similar. Nanowire have large surface area to volume ratio promising high sensitivity and the size of the nanostructures is similar to the size of species being sensed, thus nanowires make good candidate transducers for producing the signals that are then read and recorded by conventional instruments. Since they were discovered, silicone nanowires have primarily been investigated for sensing applications in biological systems [102].

Biomedical research would greatly benefit from the advent of nanowires as sensors and they have been viewed as one of most promising areas [103]. Techniques for generating various types of important nanowires, nanorods, nanobelts, and nanotubes, synthetic strategies, design of sensing devices, and sensing mechanism has been reviewed [104].

Nanowire nanostructure sensors of metal oxides have enhanced properties as opposed to nanoparticle nanostructures, this is particularly true for sensing sensitivity. For example, sensitivity of nanowire structures of orthorhombic phase of MoO_3 to 100 ppm ammonia gas was 20.5 in comparison to nanoparticles where the sensitivity was 6.8 for same concentration [75, 92]. Similarly, in the case of WO_3 the sensor sensitivity analysis towards NO_2 showed that the sensitivity improved significantly when nanowires were used as opposed to nanoparticles [80]. Single crystals are of particular importance for sensing of analytes, as compared to the multigranular crystals. Instabilities may form at the grain boundaries, which cause lowering of sensitivity towards gaseous analytes. Ra et al. [105] studied such an affect for ZnO nanowires. They observed much slower recovery and response times and additionally the sensing response was much more unstable for multigranular nanowires as compared to single crystal ZnO nanowires.

Therefore, the single crystal character of well-defined nanowires is highly desirable for their use in sensing applications. Stoichiometry also plays an important role in defining the sensing properties of nanowires. The difference between a non-stoichiometric and stoichiometric nanowire may cause it to change sensing behavior significantly. Gali et al. [106] studied such behavior for In_2O_3 nanowires, and found that lower the stoichiometry of the nanowires higher was response to the oxidizing gas, whereas when the stoichiometric behavior increased it led to a more significant response to reducing gas. Therefore, a single crystal nanowire of desired and known stoichiometry would be an ideal candidate for sensing low concentrations of specific analytes. Exposed surfaces of the single crystals of nanowires is another factor that influences the sensitivity behavior of the crystals. In many instances the presence of higher surface oxygen on certain planes promote adsorption of the analytes at the surface. For example, Kaneti et al. [107] studied the gas-sensing behavior of ZnO as it varies due to differently exposed crystalline planes. It was observed that the (0001) plane had double sensitivity towards Ethanol as compared to the (1010) plane.

Acknowledgements

Research carried out in part at the Center for Functional Nanomaterials, Brookhaven National Laboratory, which is supported by the U.S. Department of Energy, Office of Basic Energy Sciences, under Contract No. DE-SC0012704. The authors also wish to acknowledge the support of the National Science Foundation through grants: NSF DMR #1106168 and NSF IIS #1231761.

References

1. Strimbu, K. and J.A. Tavel, *What are biomarkers?* Current Opinion in Hiv and Aids, 2010. **5**(6): p. 463-466.

2. Atkinson, A.J., et al., *Biomarkers and surrogate endpoints: Preferred definitions and conceptual framework.* Clinical Pharmacology & Therapeutics, 2001. **69**(3): p. 89-95.

3. Phillips, M., *Breath Tests in Medicine.* Scientific American, 1992. **267**(1): p. 74-79.

4. Pauling, L., et al., *Quantitative Analysis of Urine Vapor and Breath by Gas-Liquid Partition Chromatography.* Proceedings of the National Academy of Sciences of the United States of America, 1971. **68**(10): p. 2374-&.

5. Braun, P.X., C.F. Gmachl, and R.A. Dweik, *Bridging the Collaborative Gap: Realizing the Clinical Potential of Breath Analysis for Disease Diagnosis and Monitoring–Tutorial.* Sensors Journal, IEEE, 2012. **12**(11): p. 3258-3270.

6. Wojtas, J., et al., *Ultrasensitive laser spectroscopy for breath analysis.* Opto-Electronics Review, 2012. **20**(1): p. 26-39.

7. Culotta, E. and D.E. Koshland, Jr., *NO news is good news.* Science, 1992. **258**(5090): p. 1862-5.

8. Moncada, S. and A. Higgs, *The L-arginine-nitric oxide pathway.* N Engl J Med, 1993. **329**(27): p. 2002-12.

9.   Yates, D.H., *Role of exhaled nitric oxide in asthma.* Immunol Cell Biol, 2001. **79**(2): p. 178-90.

10.   Hamid, Q., et al., *Induction of nitric oxide synthase in asthma.* Lancet, 1993. **342**(8886-8887): p. 1510-3.

11.   Kharitonov, S.A., et al., *Increased nitric oxide in exhaled air of asthmatic patients.* Lancet, 1994. **343**(8890): p. 133-5.

12.   Massaro, A.F., et al., *Expired nitric oxide levels during treatment of acute asthma.* Am J Respir Crit Care Med, 1995. **152**(2): p. 800-3.

13.   Alving, K., E. Weitzberg, and J.M. Lundberg, *Increased amount of nitric oxide in exhaled air of asthmatics.* Eur Respir J, 1993. **6**(9): p. 1368-70.

14.   Persson, M.G., et al., *Single-breath nitric oxide measurements in asthmatic patients and smokers.* Lancet, 1994. **343**(8890): p. 146-7.

15.   Kharitonov, S.A., et al., *EXhaled nitric oxide is increased in asthma.* Chest, 1995. **107**(3_Supplement): p. 156S-157S.

16.   Silkoff, P.E., et al., *Exhaled nitric oxide and bronchial reactivity during and after inhaled beclomethasone in mild asthma.* J Asthma, 1998. **35**(6): p. 473-9.

17.   Yates, D.H., et al., *Effect of a nitric oxide synthase inhibitor and a glucocorticosteroid on exhaled nitric oxide.* Am J Respir Crit Care Med, 1995. **152**(3): p. 892-6.

18.   Massaro, A.F. and J.M. Drazen, *"Exhaled nitric oxide during exercise: site of release and modulation by ventilation and blood flow".* J Appl Physiol (1985), 1996. **80**(6): p. 1863-4.

19.   American Thoracic, S. and S. European Respiratory, *ATS/ERS recommendations for standardized procedures for the online and offline measurement of exhaled lower respiratory nitric oxide and nasal nitric oxide, 2005.* Am J Respir Crit Care Med, 2005. **171**(8): p. 912-30.

20.   Ludviksdottir, D., et al., *Clinical aspects of using exhaled NO in asthma diagnosis and management.* Clin Respir J, 2012. **6**(4): p. 193-207.

21.   Smith, A.D., et al., *Diagnosing asthma: comparisons between exhaled nitric oxide measurements and conventional tests.* Am J Respir Crit Care Med, 2004. **169**(4): p. 473-8.

22.   Yates, D.H., et al., *Endogenous nitric oxide is decreased in asthmatic patients by an inhibitor of inducible nitric oxide synthase.* Am J Respir Crit Care Med, 1996. **154**(1): p. 247-50.

23.   Price, D., et al., *Using fractional exhaled nitric oxide (FeNO) to diagnose steroid-responsive disease and guide asthma management in routine care.* Clin Transl Allergy, 2013. **3**(1): p. 37.

24.   Beck-Ripp, J., et al., *Changes of exhaled nitric oxide during steroid treatment of childhood asthma.* Eur Respir J, 2002. **19**(6): p. 1015-9.

25.   Pijnenburg, M.W., et al., *Exhaled nitric oxide predicts asthma relapse in children with clinical asthma remission.* Thorax, 2005. **60**(3): p. 215-8.

26.   Harkins, M.S., K.L. Fiato, and G.K. Iwamoto, *Exhaled nitric oxide predicts asthma exacerbation.* J Asthma, 2004. **41**(4): p. 471-6.

27.   Jones, S.L., et al., *The predictive value of exhaled nitric oxide measurements in assessing changes in asthma control.* Am J Respir Crit Care Med, 2001. **164**(5): p. 738-43.

28. Dweik, R.A., et al., *An official ATS clinical practice guideline: interpretation of exhaled nitric oxide levels (FENO) for clinical applications.* Am J Respir Crit Care Med, 2011. **184**(5): p. 602-15.

29. Phillips, M., et al., *Variation in volatile organic compounds in the breath of normal humans.* Journal of Chromatography B, 1999. **729**(1-2): p. 75-88.

30. Deng, C.H., et al., *Determination of acetone in human breath by gas chromatography-mass spectrometry and solid-phase microextraction with on-fiber derivatization.* Journal of Chromatography B-Analytical Technologies in the Biomedical and Life Sciences, 2004. **810**(2): p. 269-275.

31. Fenske, J.D. and S.E. Paulson, *Human breath emissions of VOCs.* Journal of the Air & Waste Management Association, 1999. **49**(5): p. 594-598.

32. Deneris, E.S., R.A. Stein, and J.F. Mead, *In vitro biosynthesis of isoprene from mevalonate utilizing a rat liver cytosolic fraction.* Biochem Biophys Res Commun, 1984. **123**(2): p. 691-6.

33. Cikach, F.S., Jr. and R.A. Dweik, *Cardiovascular biomarkers in exhaled breath.* Prog Cardiovasc Dis, 2012. **55**(1): p. 34-43.

34. Karl, T., et al., *Human breath isoprene and its relation to blood cholesterol levels: new measurements and modeling.* J Appl Physiol (1985), 2001. **91**(2):p. 762-70.

35. McMaster, M., *GC/MS: A Practical User's Guide.* 2008: Wiley.

36. Mueller, W., et al., *Method for analysis of exhaled air by microwave energy desorption coupled with gas chromatography-flame ionization detection-mass spectrometry.* J Chromatogr B Biomed Sci Appl, 1998. **716**(1-2): p. 27-38.

37. Ho, C.K., et al., *Review of Chemical Sensors for In-Situ Monitoring of Volatile Contaminants.* 2001.

38. Kanu, A.B., et al., *Ion mobility-mass spectrometry.* Journal of Mass Spectrometry, 2008. **43**(1): p. 1-22.

39. Amorim, L.C. and L.C.Z. de, *Breath air analysis and its use as a biomarker in biological monitoring of occupational and environmental exposure to chemical agents.* J Chromatogr B Analyt Technol Biomed Life Sci, 2007. **853**(1-2): p. 1-9.

40. Pleil, J.D., et al., *Breath analysis science at PittCon 2012, Orlando, Florida.* J Breath Res, 2012. **6**(3): p. 039001.

41. Pleil, J.D. and A. Hansel, *Submarines, spacecraft and exhaled breath.* J Breath Res, 2012. **6**(1): p. 019001.

42. Beauchamp, J., et al., *On the performance of proton-transfer-reaction mass spectrometry for breath-relevant gas matrices.* Measurement Science & Technology, 2013. **24**(12).

43. Hansel, A., et al., *Proton-Transfer Reaction Mass-Spectrometry - Online Trace Gas-Analysis at the Ppb Level.* International Journal of Mass Spectrometry, 1995. **149**: p. 609-619.

44. Lindinger, W., A. Hansel, and A. Jordan, *Proton-transfer-reaction mass spectrometry (PTR-MS): on-line monitoring of volatile organic compounds at pptv levels.* Chemical Society Reviews, 1998. **27**(5): p. 347-354.

45. Blake, R.S., P.S. Monks, and A.M. Ellis, *Proton-Transfer Reaction Mass Spectrometry.* Chemical Reviews, 2009. **109**(3): p. 861-896.

46. Smith, D. and P. Spanel, *Selected ion flow tube mass spectrometry (SIFT-MS) for on-line trace gas analysis.* Mass Spectrometry Reviews, 2005. **24**(5): p. 661-700.

47. Spanel, P. and D. Smith, *Progress in Sift-Ms: Breath Analysis and Other Applications.* Mass Spectrometry Reviews, 2011. **30**(2): p. 236-267.

48. Spanel, P. and D. Smith, *Advances in On-line Absolute Trace Gas Analysis by SIFT-MS.* Current Analytical Chemistry, 2013. **9**(4): p. 525-539.

49. Wolfbeis, O.S., *Chemical Sensors - Survey and Trends.* Fresenius Journal of Analytical Chemistry, 1990. **337**(5): p. 522-527.

50. Hulanicki, A., S. Glab, and F. Ingman, *Chemical Sensors Definitions and Classification.* Pure and Applied Chemistry, 1991. **63**(9): p. 1247-1250.

51. Seiyama, T., et al., *A New Detector for Gaseous Components Using Semiconductive Thin Films.* Analytical Chemistry, 1962. **34**(11): p. 1502-1503.

52. Barsan, N., D. Koziej, and U. Weimar, *Metal oxide-based gas sensor research: How to?* Sensors and Actuators B-Chemical, 2007. **121**(1): p. 18-35.

53. Moseley, P., *New trends and future prospects of thick-and thin-film gas sensors.* Sensors and Actuators B: Chemical, 1991. **3**(3): p. 167-174.

54. Sze, S.M., *Semiconductor sensors.* 1994: Wiley-Interscience.

55. Xia, Y., et al., *Biotemplated fabrication of hierarchically porous NiO/C composite from lotus pollen grains for lithium-ion batteries.* Journal of Materials Chemistry, 2012. **22**(18): p. 9209-9215.

56. Manno, D., et al., *Physical and structural characterization of tungsten oxide thin films for NO gas detection.* Thin Solid Films, 1998. **324**(1-2): p. 44-51.

57. Moon, H.G., et al., *Extremely Sensitive and Selective NO Probe Based on Villi-like WO3 Nanostructures for Application to Exhaled Breath Analyzers.* Acs Applied Materials & Interfaces, 2013. **5**(21): p. 10591-10596.

58. Yin, L., et al., *Enhanced selective response to nitric oxide (NO) of Au-modified tungsten trioxide nanoplates.* Materials Chemistry and Physics, 2013. **143**(1): p. 461-469.

59. Chen, D.L., et al., *Low-temperature and highly selective NO-sensing performance of WO3 nanoplates decorated with silver nanoparticles.* Sensors and Actuators B-Chemical, 2013. **185**: p. 445-455.

60. Sun, C.H., G. Maduraiveeran, and P. Dutta, *Nitric oxide sensors using combination of p- and n-type semiconducting oxides and its application for detecting NO in human breath.* Sensors and Actuators B-Chemical, 2013. **186**: p. 117-125.

61. Chang, B.Y., et al., *Evaluation of Pt/In2O3-WO3 nano powder ultra-trace level NO gas sensor.* Journal of the Taiwan Institute of Chemical Engineers, 2014. **45**(3): p. 1056-1064.

62. Buseck, P. and M.S.o. America, *Minerals and Reactions at the Atomic Scale: Transmission Electron Microscopy.* 1992: Mineralogical Society of America.

63. Gurlo, A., *Nanosensors: towards morphological control of gas sensing activity. SnO2, In2O3, ZnO and WO3 case studies.* Nanoscale, 2011. **3**(1): p. 154-165.

64. Akiyama, M., et al., *Tungsten Oxide-Based Semiconductor Sensor Highly Sensitive to NO and NO2.* Chemistry Letters, 1991. **20**(9): p. 1611-1614.

65. Gouma, P., A. Prasad, and K. Iyer, *Selective nanoprobes for'signalling gases'.* Nanotechnology, 2006. **17**(4): p. S48.

66. Penza, M., et al., *Tungsten trioxide (WO₃) sputtered thin films for a NOₓ gas sensor.* Sensors and Actuators B: Chemical, 1998. **50**(1): p. 9-18.

67. Polleux, J., et al., *Template-Free Synthesis and Assembly of Single-Crystalline Tungsten Oxide Nanowires and their Gas-Sensing Properties.* Angewandte Chemie, 2006. **118**(2): p. 267-271.

68. Marquis, B.T. and J.F. Vetelino, *A semiconducting metal oxide sensor array for the detection of NOₓ and NH₃.* Sensors and Actuators B: Chemical, 2001. **77**(1): p. 100-110.

69. Lee, D.-S., et al., *Nitrogen oxides-sensing characteristics of WO₃-based nanocrystalline thick film gas sensor.* Sensors and Actuators B: Chemical, 1999. **60**(1): p. 57-63.

70. Yu-De, W., et al., *Electrical and gas-sensing properties of WO₃ semiconductor material.* Solid-State Electronics, 2001. **45**(5): p. 639-644.

71. Chen, D., et al., *Low-temperature and highly selective NO-sensing performance of WO3 nanoplates decorated with silver nanoparticles.* Sensors and Actuators B: Chemical, 2013. **185**(0): p. 445-455.

72. Ponzoni, A., et al., *Ultrasensitive and highly selective gas sensors using three-dimensional tungsten oxide nanowire networks.* Applied Physics Letters, 2006. **88**(20): p. 3.

73. Kim, Y.S., et al., *Room-temperature semiconductor gas sensor based on nonstoichiometric tungsten oxide nanorod film.* Applied Physics Letters, 2005. **86**(21): p. -.

74. Prasad, A.K., *Study of gas specificity in molybdenum trioxide/tungsten oxide thin film sensors and their arrays.* 2005.

75. Kalyanasundaram, K., *Biomarker sensing using nanostructured metal oxide sensors.* 2007: ProQuest.

76. Wang, L., *Tailored synthesis and characterization of selective metabolite-detecting nanoprobes for handheld breath analysis.* 2008.

77. Wang, L., et al., *Nanosensor device for breath acetone detection.* Sensor Letters, 2010. **8**(5): p. 709-712.

78. Strobel, R., A. Alfons, and S.E. Pratsinis, *Aerosol flame synthesis of catalysts.* Advanced Powder Technology, 2006. **17**(5): p. 457-480.

79. Gouma, P.I. and K. Kalyanasundaram, *A selective nanosensing probe for nitric oxide.* Applied Physics Letters, 2008. **93**(24).

80. Sawicka, K.M., A.K. Prasad, and P.I. Gouma, *Metal oxide nanowires for use in chemical sensing applications.* Sensor Letters, 2005. **3**(1-4): p. 31-35 %@ 1546-198X.

81. Strobel, R. and S.E. Pratsinis, *Flame aerosol synthesis of smart nanostructured materials.* Journal of Materials Chemistry, 2007. **17**(45): p. 4743-4756.

82. Madler, L., et al., *Direct formation of highly porous gas-sensing films by in situ thermophoretic deposition of flame-made Pt/SnO2 nanoparticles.* Sensors and Actuators B-Chemical, 2006. **114**(1): p. 283-295.

83. Diskin, A.M., P. Spanel, and D. Smith, *Time variation of ammonia, acetone, isoprene and ethanol in breath: a quantitative SIFT-MS study over 30 days.* Physiol Meas, 2003. **24**(1): p. 107-19.

84. Taucher, J., et al., *Methanol in human breath.* Alcohol Clin Exp Res, 1995. **19**(5): p. 1147-50.

85.   Gouma, P. and S. Sood, *3-Sensor array for hand held breath diagnostic tool.* MRS Online Proceedings Library, 2013. **1533**(Business of Nanotechnology IV): p. 2013.146/1-2013.146/5.

86.   Gouma, P. and P.I. Gouma, *Nanomaterials for chemical sensors and biotechnology.* 2009: Pan Stanford Publishing.

87.   Wang, L. and P. Gouma, *Selective Crystal Structure Synthesis and Sensing Dependencies*, in *Metal Oxide Nanomaterials for Chemical Sensors*, M.A. Carpenter, S. Mathur, and A. Kolmakov, Editors. 2013, Springer New York. p. 167-188.

88.   Wagner, R.S. and W.C. Ellis, *Vapor-liquid-solid mechanism of single crystal growth.* Applied Physics Letters, 1964. **4**(5): p. 89-90.

89.   Zhang, Y.F., et al., *Silicon nanowires prepared by laser ablation at high temperature.* Applied Physics Letters, 1998. **72**(15): p. 1835-1837.

90.   Frey, G.L., et al., *Investigations of Nonstoichiometric Tungsten Oxide Nanoparticles.* Journal of Solid State Chemistry, 2001. **162**(2): p. 300-314.

91.   Sawicka, K., et al., *Oxidation Synthesized CuO Nanowires for Gas Sensing Applications.* Microscopy and Microanalysis, 2004. **10**(SupplementS02): p. 360-361.

92.   Gouma, P., K. Kalyanasundaram, and A. Bishop, *Electrospun single-crystal MoO3 nanowires for biochemistry sensing probes.* Journal of Materials Research, 2006. **21**(11): p. 2904-2910.

93.   Gouma, P.I., A.S. Haynes, and K. Kalyanasundaram, *Electrospun single crystal MoO3 nanowires for bio-chem sensing probes.* 2011, Google Patents.

94.   Sawicka, K., P. Gouma, and S. Simon, *Electrospun biocomposite nanofibers for urea biosensing.* Sensors and Actuators B: Chemical, 2005. **108**(1–2): p. 585-588.

95.   Lee, J. and P.I. Gouma, *Tailored 3D CuO Nanogrid Formation.* Journal of Nanomaterials, 2011.

96.   Gouma, P.I. and M.J. Mills, *Anatase-to-rutile transformation in titania powders.* Journal of the American Ceramic Society, 2001. **84**(3): p. 619-622.

97.   Liu, J., et al., *Highly porous metal oxide polycrystalline nanowire films with superior performance in gas sensors.* Journal of Materials Chemistry, 2011. **21**(30): p. 11412-11417.

98.   Xu, F., et al., *Flame synthesis of aligned tungsten oxide nanowires.* Applied Physics Letters, 2006. **88**(24).

99.   Chen, G.Y., et al., *Growth of tungsten oxide nanowires using simple thermal heating.* 2006 IEEE Conference on Emerging Technologies - Nanoelectronics, 2006: p. 376-378.

100.   Zhou, J., et al., *Large-area nanowire arrays of molybdenum and molybdenum oxides: synthesis and field emission properties.* Advanced Materials, 2003. **15**(21): p. 1835-1840.

101.   Mai, L.Q., et al., *Molybdenum oxide nanowires: synthesis & properties.* Materials Today, 2011. **14**(7-8): p. 346-353.

102.   Cui, Y., et al., *Nanowire nanosensors for highly sensitive and selective detection of biological and chemical species.* Science, 2001. **293**(5533): p. 1289-1292.

103.   Hood, L., et al., *Systems biology and new technologies enable predictive and preventative medicine.* Science, 2004. **306**(5696): p. 640-643.

104. Huang, X.J. and Y.K. Choi, *Chemical sensors based on nanostructured materials.* Sensors and Actuators B-Chemical, 2007. **122**(2): p. 659-671.

105. Ra, H.W., et al., *The effect of grain boundaries inside the individual ZnO nanowires in gas sensing.* Nanotechnology, 2010. **21**(8): p. 85502.

106. Gali, P., et al., *Stoichiometry dependent electron transport and gas sensing properties of indium oxide nanowires.* Nanotechnology, 2013. **24**(22).

107. Kaneti, Y.V., et al., *Crystal plane-dependent gas-sensing properties of zinc oxide nanostructures: experimental and theoretical studies.* Physical Chemistry Chemical Physics, 2014. **16**(23): p. 11471-11480.

Chapter 12

Gallium Nitride Microelectronics for High-Temperature Environments

Debbie G. Senesky[1,2], Hongyun So[1], Ateeq J. Suria[3], Ananth Saran Yalamarthy[3], Sambhav R. Jain[2], Caitlin A. Chapin[3], Heather C. Chiamori[1], and Minmin Hou[2]

[1]Department of Aeronautics and Astronautics, Stanford University, Stanford, California, USA
[2]Department of Electrical Engineering, Stanford University, Stanford, California, USA
[3]Department of Mechanical Engineering, Stanford University, Stanford, California, USA

12.1. Introduction

Sensors and electronics with robust operation under extreme harsh environments are required for a host of chemical, optical, physical and radiative applications. Gallium nitride (GaN) based sensors and electronics, particularly the high electron mobility transistor (HEMT) configuration, are emerging as candidates for gathering and processing local situational data from such high temperature, high radiation and corrosive environments. When combined with ternary alloys, such as aluminum, GaN-based heterostructures can provide high efficiency operation due to high operating voltages, high currents and low on-resistance. [1] High power per unit width of AlGaN/GaN devices [1] results in smaller device footprints, higher impedances for matching and capability of large sensor-node dissemination for real-time monitoring systems. With sub-millimeter length scales, a multitude of GaN-based devices (electronics, sensors, energy harvesters and electronics) can be incorporated onto the same chip lending multifunctionality to each die (Figure 1) [2]. This supports Internet of Things (IoT) applications where high-temperature data streams (e.g., industrial gas turbines, automotive and downhole) enable new architectures for predictive monitoring.

GaN HEMTs have been developed for high power electronics and high frequency applications, such as high power radar systems, although other viable commercial needs are driving sensor growth, including the energy and space sectors. High power silicon devices, experiencing higher voltages and higher electric fields, can be subject to increased leakage currents if internal junction temperatures are not maintained at less than 200°C and becomes problematic as the ambient temperature increases. [3]. However, advancements in metal contact reliability, packaging and interconnects, a deeper understanding of the sensor and electronic transduction mechanisms are required to enable operation within high temperature environments. Here, a discussion of the material properties of the GaN platform and state-of-the-art sensor and electronics is provided, as well as future outlook for monolithic integration of components for high-temperature working environments.

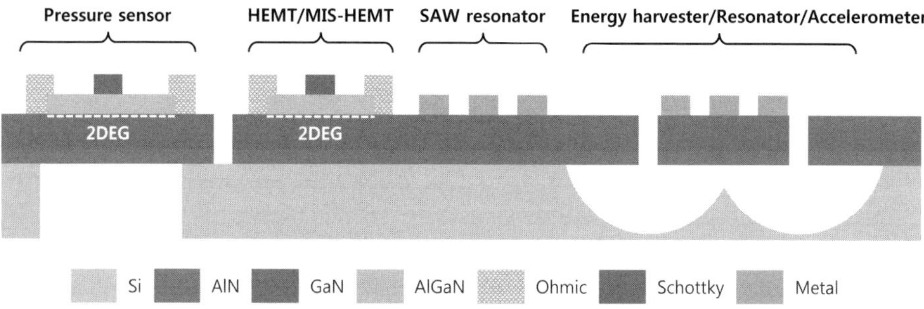

Figure 1. Schematic illustration of multifunctional electromechanical devices monolithically integrated on a single substrate by leveraging the AlGaN/GaN platform.

12.2. Material Properties of GaN

GaN is semiconductor material well-suited for harsh environments applications due to its wide bandgap (3.4 eV), with potential for high temperature operation (600°C and beyond) [1, 2]. GaN based devices can operate under high applied voltages (up to 42 V) and high frequencies with high linearity, high efficiency and low noise [1]. In addition, GaN is already expected to surpass state of the art GaAs and InP for millimeter wave applications [4]. As shown in Table 1, wide bandgap semiconductor materials intrinsic properties that support for high temperature and high power applications, as well as chemically corrosive situations.

Table 1. Material properties of wide bandgap semiconductors (GaN, 6H-SiC, AlN and diamond) compared with Si.

Property	GaN	AlN	6H-SiC	Diamond	Silicon
Melting Point (°C)	1727-2227[a] [5] ~2500[6]	>2000[7] ~3200[6]	>1050 [b], ~1250 [c] [8]	4440 (12.4GPa)[d][9] 1200[e][10]	1415 [11]
Energy Gap (eV)	3.4 [7,12]	6.2 [7,12]	3.0 [8,13]	5.48[10] 5.4[14]	1.12 [11]
Critical Field ($\times 10^6$ V/cm)	3-5 [12]	1.2-1.8 [12]	2.5 [f] [13]	3.75 [10]	0.3 [11]
Thermal Conductivity (W/cm-K)	1.3[7] 2.3 [12]	2[7] 2.8-3.2[12]	2.31-4.9 [g] [15]	5-20[14]	1.24[9]
Young's Modulus (GPa)	150 [12]	308, 374 [12]	400-500 [8]	910-1250 [10]	130-188 [16]
Acoustic Velocity ($\times 10^3$ m/s)	7.96 [12]	11.27 [h] [12]	~13.3[15]	10.5[14] 20[10]	8.4[17]
Yield Strength (GPa)	0.1 [i][12]	0.3 [j][12]	21[18]	53[18]	7[18]
Coeff. of Thermal Exp. (°C $\times 10^{-6}$)	3.17[12]	5.3[7]	3.3 [18]	1 [18]	2.33 [18]
Chemical Stability	Good [7,12]	Excellent [12]	Excellent [18]	Fair [k] [10]	Fair [18]
Mobility (cm²/V-s)	300-600 [7]	14 [l][7]	400[13]	2200[10]	1450[19]

Table notes:

(a) depends on crystal direction

(b) sublimation of Si atoms from Si-rich regions, diamond-like (sp^3) bonding of carbon atoms

(c) graphene sheet sp^2 carbon formation

(d) phase change

(e) transition to graphite begins

(f) 10x the value of silicon

(g) at different nitrogen and aluminum doping concentrations

(h) wave character in longitudinal direction

(i) at 1000K

(j) at 1000°C

(k) susceptible to oxidation starting at 250°C

(l) hole mobility of impure, low resistivity AlN

12.3. Emerging High-Temperature Applications

State-of-the-art silicon devices are limited to ~ 300°C operational temperatures due to generation of thermal carriers. The wide bandgap of GaN allows potential device operation to 600°C and possibly above. The potential to have uncooled devices operating at these higher temperatures can benefit a multitude of commercial, military and space applications. Table 2 illustrates the potential applications for high temperature microelectronics.

To date, most investigation into the high temperature performance of GaN-based devices has focused on heterostructure field effect transistors (HFETs) or HEMTs. Another HFET design with high 2DEG mobilities of 100-120 cm^2/V-s was demonstrated at 400°C, indicating that 2DEG mobility could be larger than bulk mobility at these temperatures [20]. GaN based devices have been tested on sapphire and SiC substrates as well as with experimental dielectric gates up to 500°C, suggesting that off-state degradation does not occur until ~450°C [21,22] A GaN HFET with an improved Ohmic contact showed fast switching up to 225°C with a high breakdown voltage (750 V) and little degradation [23]. These results indicate that the capability of GaN-based devices for use in high-temperature environments can occur with advancements in areas such as design, materials synthesis, fabrication processing and packaging, as well as a fundamental understanding of the high-temperature transduction mechanisms (piezoelectric, pyroelectric and piezoresistive) and thermo-chemical effects.

Table 2. Various industry sectors with high-temperature working environments. Potential types of microelectronics sensors and devices include: *O=optical; P=pressure; T=temperature; H=magnetic field; R=radiation; W=wireless telemetry;* and *E=energy harvesting.*

Application/Industry	Estimated Operating Temperature (°C)	Sensors and Microelectronics
Oil & Gas Exploration	150 to 300 [24]	*O,P,T,H,W,E*
Geothermal	> 300 [24]	*O,P,T,H,W,E*
Space Exploration (e.g. Europa & Venus)	-173 to 500 [25]	*O,P,T,H,R,W,E*
Automotive Engines (e.g. exhaust)	350 to 900 [26]	*O,P,T,H,W,E*
Gas Turbines	> 600 [27]	*O,P,T,H,W,E*
Aerospace Systems	> 600 [27]	*O,P,T,H,R,W,E*
Hypersonic Flight	> 1000 [28]	*O,P,T,H,R,W,E*

12.4. Electro-mechanical and Thermo-chemical Effect in Gan Based Devices

The two-dimensional electron gas (2DEG) in the AlGaN/GaN heterostructure, which arises due to differences in strain induced piezoelectric polarization charges between the AlGaN and GaN layers [29,30] is subject to a wide variety of inter-dependent electro-thermo-mechanical [31–34] and chemical phenomena [14, 15]. These phenomena arise due to intrinsic and applied strain and the inverse-piezoelectric effect, electron scattering and hot electron effects, as well as diffusion and degradation effects in device contacts [37]. The interaction of these effects lead to changes in bulk properties such as electron mobility, threshold voltage, trans-conductance, saturation and leakage currents, current collapse, as well as eventual failure [10, 14, 15] in AlGaN/GaN devices. The typical composition of AlGaN is 15% to 30% [8, 9], resulting a strain of ~ 0.4-0.7% in the AlGaN alloy layer, leading to a charge density of ~ 10^{13} cm^{-2} at the AlGaN/GaN interface.

Additional strain in the AlGaN/GaN layers can be induced by external application (such as in pressure sensors [38–40] and electronics packaging), residual strain arising from thermal expansion [41], relaxation in the AlGaN layer at high temperatures [42], hot-spot lattice heating [43] and from application of electric fields due to the inverse piezoelectric effect at high reverse bias gate voltages [44]. The piezoelectric polarization fields and the conduction-valence band offset are modified due to applied strain [45,46], leading to a change in the charge density and mobility in the 2D electron gap. Using semi-empirical models, a gauge factor of -7.9-/+5.2 [47] was estimated at 25°C, and an effective mass change of ~ -2.7% per 0.1% applied biaxial strain is determined [48]. At high strain levels above 0.2%, the dependence of the 2DEG charge on the piezoelectric polarization fields and the band-gap is non-linear, thus more accurate calculations are required [24, 25, 28]. In addition to the change in the amount of charge, temperature rise decreases the electron-optical phonon scattering time, leading to a $T^{-1.5}$ dependence in the electron mobility in the 2DEG [29, 30]. In order to illustrate the coupling of electrical, thermal and applied mechanical effects (excluding inverse piezo-electric effect), the modeled current-voltage characteristics of an AlGaN/GaN HEMT with gate length L = 10 μm and width W = 100 μm for a biaxial applied strain level from -0.5% to 0.5% on the AlGaN and GaN layers at a uniform temperature of 25°C and 225°C is shown in Fig. 2 at drain-source voltage V_D = 1 V for an interface state density of D_{it} of 5×10^{13} cm^{-2}eV^{-1} at the Schottky contact. At a fixed temperature, applied tensile strain leads to a change in 2DEG electron sheet density, causing a shift in the threshold voltage (V_{th}) and an increase in the 2DEG mobility, leading to a change in the drain-source current (I_D) as can be observed in Fig. 2(c). As the device temperature rises, I_D decreases due to a combination of increase in electron-optical phonon scattering rate [29, 30], as well as positive shift in V_{th} arising from strain relaxation due to reduction in 2DEG charge density [42] at the AlGaN/GaN interface, although distinct current-voltage curves are still observed for each applied strain level, as can be seen in Fig. 2(c). At very high temperatures above 600°C, irreversible failure of the 2DEG channel is observed [42].

At high gate reverse bias, typically between 20 V and 25 V from gate to drain [16, 31], large electric fields above ~ 1.8 MV/cm lead to critical levels of strain in the AlGaN/GaN layers due to the inverse piezoelectric effect [53]. This leads to the irreversible generation of defect sites at which electrons get trapped near the AlGaN/GaN interface, which reduces the 2DEG interface charge leading to a positive shift in V_{th} [37], as well as decrease in the drain current I_D and several orders of magnitude increase in the gate current I_G, as illustrated in Fig. 3 [54–56]. Several groups have postulated that the increase in I_G and the decrease in I_D is due to the generation of defect sites by hot electrons at high gate to drain voltage levels [57–59], however, the inverse-piezo-electric hypothesis seems to be more accurate in the light of recent experiments. The presence of these defect sites is also correlated with the appearance of pit-like defects near the drain side of the gate in the AlGaN layer which indicate strain due to inverse piezo-electric effect [60]. Lastly, temperature rise leads to chemical alloying, diffusion, and degradation effects in AlGaN-GaN device contacts. An extensive discussion of these thermo-chemical effects is presented in a later section of this chapter.

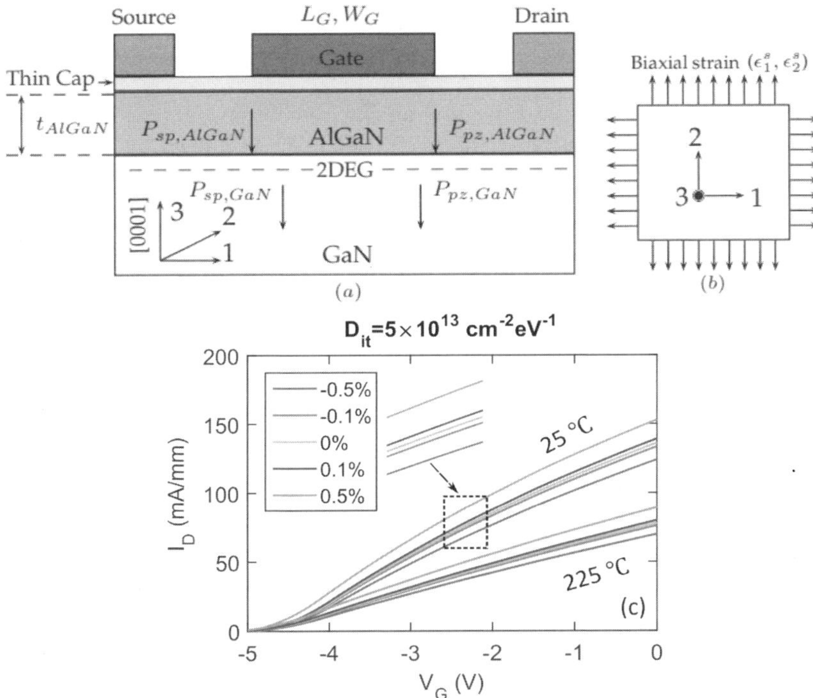

Figure 2. Effect of applied strain and temperature on AlGaN-GaN HEMTs. (a) Cross-sectional image of a typical Ga-face AlGaN/GaN heterostructure used in HEMT device microfabrication. The shorthand 1, 2, and 3 corresponds to <11$\bar{2}$0>, <10$\bar{1}$0>, and <000$\bar{1}$> directions in the Wurtzite AlGaN and GaN crystals, respectively. (b) Top-view of the AlGaN/GaN HEMT under biaxial applied strain in the 1- and 2-directions (basal plane). (c) Current - voltage response of AlGaN/GaN HEMTs due to a biaxial strain level of -0.5% – 0.5% (shown in the legend) at 25°C and 225°C for an interface state density D_{it} of 5×10^{13} cm^{-2}eV^{-1} at the Schottky contact. The drain voltage is fixed at 1 V.

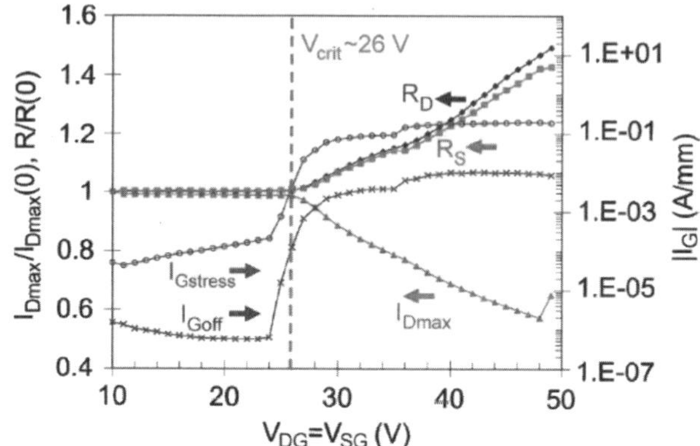

Figure 3. Increase in I_G and reduction in I_D arising from the inverse piezoelectric effect due to high reverse gate voltages. V_{GS} from -10 to -50 V is applied for durations of 1 minute and steps of 1 V with at V_{DS} = 0 to ensure negligible channel current to exclude degradation due to hot-electron effects. After each 1 minute step, I_{Dmax} is measured at V_{DS} = 5 V and V_{GS} = 2 V, while I_G is measured at V_{DS} = 0.1 V and V_{GS} = -5 V. Reprinted from IEEE Electron Device Letters, Joh et al., Critical Voltage for Electrical Degradation of GaN High-Electron Mobility Transistors, 289, Copyright 2008, with permission from IEEE [56].

12.5. Physical Sensors

12.5.1. Strain Sensing

As mentioned above, the geometry of the AlGaN/GaN heterostructure, spontaneous polarization, piezoelectric polarization, and band bending contribute to the formation of the 2DEG quantum well. Unlike traditional piezoelectric materials, the charge that accumulates upon mechanical stress is stable since the electrons are trapped at the interface of the heterostructure. The fixed polarization at the interface as a function of the aluminum concentration, $\sigma(x)$, is described by Equation 1, where P_{SP} is the the spontaneous polarization and P_{PE} is the piezoelectric polarization [30].

$$\sigma(x) = P_{SP,GaN} + P_{PE,GaN} - P_{SP,AlGaN} - P_{PE,AlGaN} \tag{1}$$

The piezoelectric polarization depends on the strain the material experiences. The stability of GaN at high temperatures and in hostile environments, in addition to the strain sensitivity, has provided motivation for the development of physical sensors made from GaN.

One of the first papers to study how mechanical stimuli affected the 2DEG did so in a thickness optimization study of 2DEG mobility and carrier concentration. By varying the thickness of the AlGaN grown on the bulk GaN one can vary the plain stress across the wafer. Thickness on the film had a direct correlation to mobility and carrier concentration because the strain experience by the film increase up to a critical point [61]. The authors did not model or calculate the strain induced, however, to our knowledge this is the first indication in the literature that shows the 2DEG is sensitive to mechanical stimuli.

Several researchers have further explored the strain sensitivity of GaN and AlGaN heterostructures via traditional beam bending experiments [39,47,62–67]. The gauge factor (GF) is the primary figure of merit for strain sensing and can be used to compare various works. The GF is a measure of the device's sensitivity to strain (ε); generally, it is described as the percent change in resistance ($\Delta R/R_0$) per strain as shown in Equation 2,

$$ \mathrm{GF} = \frac{\Delta R/R_0}{\varepsilon}. \tag{2} $$

It should be noted that in the literature, the change in different electrical parameters (i.e., current or capacitance) to strain are often used and therefore cannot be directly compared to the piezoresistive GF without knowledge of the interface circuit. Eickhoff, *et al.* compared the piezoresistive GF of AlGaN to the piezoelectric GF of the AlGaN/GaN heterostructures via a 4-point bending experiment. It was shown that sensitivity of the heterostructure device increased the gauge factor to -85 from -26 for bulk AlGaN [62]. These results are depicted in Figure 4. Zimmerman reported a GF of ~90 [63]. Tilak *et al.* reports a capacitive gauge factor of 575 at room temperature and 361 at 400°C [64]. Yilmazoglu *et al.* reported varying a gauge factor base on bias conditions from 19 to 350 [65]. Yilmazoglu *et al.* attributes this change to variations in channel temperature which would support what Chapin *et al.* observed when (absolute value) of gauge factor increases from 18 to 83 when temperature increase from room temperature to 90°C [39]. In fact in one work comparing DC and high frequency strain responses, Talukdar *et al.* reported gauge factors varying from -3500 to 1350

[66]. According to Koeler *et al.*, the repeatability of GF could be at issue because of parasitic trapped charges; under Koehler's model the GF simulated was -7.9+/-5.2. Koehler *et al.* attempted to eliminate the trap charges by optically exciting them under UV and experimentally found a gauge factor of —2.8 ± 0.4 [47]. This could even explain temperature variations in GF since increased temperature could reduce trapped charges. As further proof that presences of interface/surface trap states might be inflating the presented gauge factors Yao *et al.* demonstrated that the Schottky barrier height is effected by strain and the defect density [67].

Piezoeresitive transductive properties for silicon and silicon carbide are well understood in terms of piezoresistive coefficient dependent on crystal orientation, temperature, and doping [68–71]. N-type silicon and silicon carbide has a GF varying from -100 to 50 and from -32 to 20 based on orientation, respectively [69,70]. Alternatively, the piezoelectric GF of AlGaN/GaN heterostructures is not well understood. An extremely wide range in sensitivity to strain, i.e. from 1350 to -3500, as reported in a single paper [66]. These inconsistencies may be caused by various material and electrical defects, as well as environmental conditions. For example, surface charge (causing floating gate/bias) due environmental or passivation conditions, trapped charge at the AlGaN/GaN interface, interfacial and hot electron based defect sites, and temperature dependencies caused by self-heating during operation. These devices show promise for pressure sensing in high temperature and extreme environments. However, commercialization requires further study of the material-, electrical-, and environmental-dependent sensor properties, as well as viable packaging schemes. Based on the variation in the literature, there is still a need to study GaN heterostructures sensitivity and response to strain. Authors have proven the potential of GaN to be a highly sensitive to strain, however, for GaN heterostructures to become viable for high temperature sensing more needs to be done to understand the variation in GF values reported in the literature.

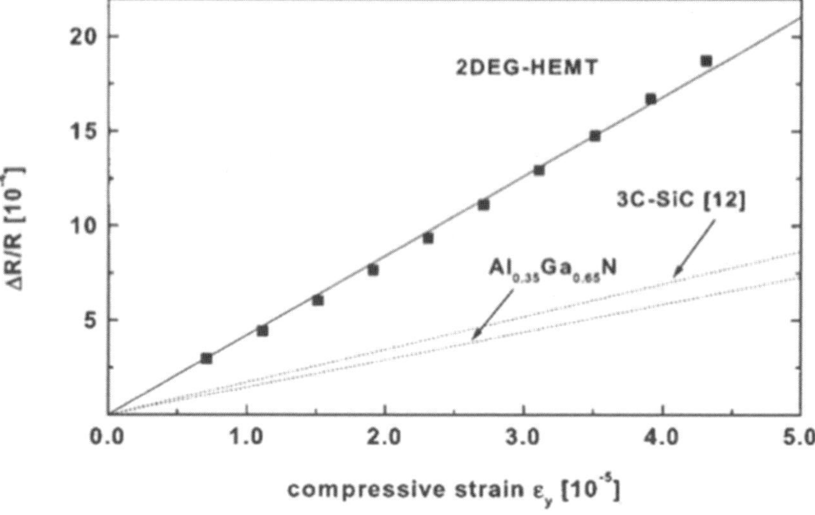

Figure 4. Relative change in conductivity of GaN 2DEG HEMT compared to bulk $Al_{0.35}Ga_{0.65}N$ and 3C-SiC. Reprinted from Journal of Applied Physics, 90, Eickhoff et al., Piezoresistivity of $Al_xGa_{1-x}N$ layers and $Al_xGa_{1-x}N$/GaN heterostructures, 2283, Copyright 2001, with permission from AIP Publishing. Ref. [62].

12.5.2. Pressure Sensing

In addition to strain sensing, pressure sensors have been demonstrated using GaN heterostructures. Figure 1 illustrates a concept for monolithically integrating AlGaN/GaN HEMT pressure sensors with HEMT electronics (e.g., power electronics and interface circuits). To create the pressure sensor, strain-sensitive elements such as 2DEG-based piezoresistors or strain-sensitive HEMTS are positioned on a released, pressure-sensitive diaphragm. Similar to Si-based pressure sensors, the underlying Si can be removed using deep reactive ion etching (DRIE) or KOH etching. However, AlGaN/GaN pressure sensors leverage the piezoelectric strain-sensitivity of the 2DEG as opposed to doped-Si piezoresistive elements [63,69]. To transfer the in-plane strains, sense structures are periodically placed across the pressure-sensitive diaphragm [38,72], are configured as interdigitated fingers across the diaphragm [73–75], or leverage circular-shaped ring designs placed at the clamped-edge of the diaphragm [76–78]. These pressure sensors leverage several types of sensing mechanisms including capacitive [76], conductance without external gate [73,74,79,80], and conductance with external gate [38,72,77,78].

While many pressure sensor designs have been presented previously very few present experimental results looking at the GaN transduction [38,72–74,79,80]. Of these works, there are only three different device designs; only one use micromachining techniques to release the membranes [74,80] and only one looks at the HEMT response under pressure [38,72]. Son *et al.* and Le Boulbar *et al.* shows that sensitivity of the device is dependent on bias conditions [38,80]. Son's team demonstrated changes in sensitivity of n-I-n AlGaN/GaN circular pressure sensors. The sensitivity (2 µA/MPa) was highest near the turn on region, 0 V, but also effected by noise due to the small magnitude currents measured [80]. By using rectangle-shaped HEMTs on sapphire, located across a 8 mm by 8 mm chip clamped to a steel diaphragm, Le Boulbar *et al.* was able to show that the region the transistor was operated in affected the sensitivity of the device. Le Boulbar demonstrated the sensitivity of the devices changed from about 2%/MPa to 80%/MPa by operating the device in the weak inversion region ($V_{GS} \approx V_T$) rather than when the device was in strong inversion ($V_{GS} \approx 0$ V).

12.6. Optical and Radiating Sensors

12.6.1. Optical Sensing

In recent years, photodetectors using GaN have attracted great interest because GaN offers stable electrical performance at elevated temperatures [81–83] and tunable optical properties by regulating Al fraction in $Al_xGa_{1-x}N$ [84]. In addition, GaN provides higher resistance to radiation damage, allowing for flame detection and space exploration applications [85–88]. Particularly, direct band gap of GaN (3.42 eV) covers an ultraviolet (UV) photon energy region and can get high internal gain upon expose to UV lights [89–92]. Based on the device structure, photodetectors can be categorized as: photoconductors [93–95], Schottky diodes [90,96–100], metal-semiconductor-metal structures [91,101–104], p-n junctions [92,105], p-i-n junctions [88,89,106–109], and phototransistors [110–114] type, as shown in Figure 5 [115]. Among these types of photosensitive devices, photodetectors using AlGaN/GaN heterostructure provides higher response through highly conductive channel (called the two-dimensional electron gas, 2DEG) created at the interface between AlGaN/GaN [89,104,107,110,111,114–116]. In GaN p-n junction type devices, electron-hole pairs are generated by absorbed UV light and minority carriers created in junction move into space-charge region by electric field, inducing current change upon exposure to UV light [92,114]. On the other hand, in photodetectors operated by the 2DEG, electrons in the AlGaN/GaN films excited by UV light move toward the 2DEG layer, whereas holes drift toward AlGaN surface due to a piezoelectric polarization field [29,30,117,118]. Although photodetectors are generally fabricated on sapphire and silicon carbide substrates, these common substrates for III-nitride semiconductor materials have poor thermal conductivity and high cost, respectively [119–122]. Consequently, the development of photodetectors on silicon (111) substrates with molecular beam epitaxy has emerged [123–125].

Photodetectors made of III-V semiconductors have a persistent photoconductivity (PPC) effect [115,126–136]. The PPC effect shows extremely slow decay time (takes several minutes or even hours) when UV excitation is removed from the detectors, as shown in Figure 6 [126], which might be a chronic drawback to many applications due to slow recovery time [115]. This non-exponential decay behavior mainly resulted from dislocations, impurities, intrinsic defects, or deep level defects (DX centers), forming deep trap sites and potential energy barriers [126–136]. The PPC effect can be mitigated with elevated temperature because of accelerated recombination rate of electron-hole pair at high temperature region [126,130,131,133,137–139]. High doping of AlGaN and GaN can further attenuate the PPC effect by enhanced tunneling effects across barriers [129,133].

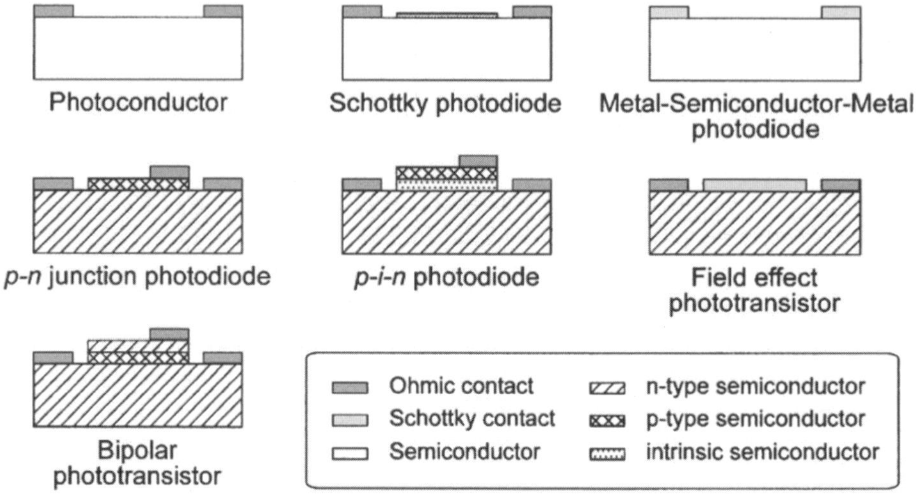

Figure 5. Schematic device structure of GaN-based photodetectors. Reprinted from Journal of Crystal Growth, 230, Monroy et al., AlGaN-bsed UV photodetctors, 537, Copyright 2001, with permission from Elsevier. Ref. [115].

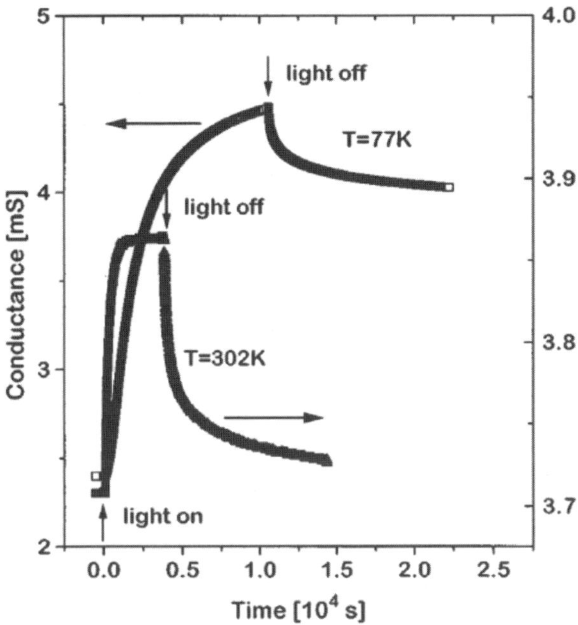

Figure 6. Persistent photoconductivity (PPC) effect in n-type GaN photodetectors. Reprinted from Applied Physics Letters, 71, Hirsch et al., Persistent photoconductivity in n-type GaN, 1098, Copyright 1997, with permission from AIP Publishing LLC. Ref. [126].

12.6.2. Radiating Sensing

There has been considerable interest in developing of devices and electronics with resistance to high-energy ionizing radiation. GaN has emerged as a viable material in this space, as well as within nuclear and particle-physics reactors, due to its radiation hardness and ability to withstand very high doses of ionizing energy, wide bandgap, thermal stability and large displacement energy [140–155]. There have been several studies and review articles published on radiation damage and use of GaN as an ionizing radiation detector. The impact of alpha particles, protons, X-rays, ultraviolet radiation, neutrons and electrons on n-GaN, p-GaN and heterostructures of GaN such as AlGaN/GaN, AlGaN/AlN/GaN, AlInN/AlN/GaN, etc. were part of these investigations [141–145,156–163]. A large majority of the studies are focused on studying the radiation damage characteristics of these devices, but the response can be evaluated for use of these devices for radiation sensing.

The need for development of GaN based ionizing radiation detectors was explored by Grant *et al.* for use in the CERN Large Hadron Collider (CLHC). The intention was to explore wide bandgap semiconductors to develop radiation hard detectors that can survive fluences up to 10^{16} fast hadrons/cm^2. A non-linear increase in leakage current was observed with neutron irradiation, and charge collection efficiency (CCE) of ~5% was determined for protons and neutrons [161]. P. Sellin and J. Vaitkus demonstrated CCE values close to 100% by for alpha particles and protons using Schottky pad detectors fabricated on epitaxial GaN on sapphire wafers. The CCE response is shown in Figure 7 for these detectors to alpha particles [145]. Another study by Grant *et al.* developed pad structures built on GaN epitaxial wafers and irradiation with 24 GeV/c protons and 1 MeV neutrons, with a fluence of 10^{16} cm^{-2} for both particle types. The maximum CCE for unirradiated devices was 55%. Post irradiation with protons the CCE was 26%, and 20% after neutron irradiation, along with a decrease in leakage current with increasing radiation level [144]. Such measurements can be utilized for ionizing radiation detection. Howgate *et al.* developed ultrahigh gain AlGaN/GaN radiation detectors utilizing the two-dimensional electron gas (2DEG). Their work demonstrated large sensitivities in a very low dose regime for X-rays in 1-20 keV range and up to 20 MeV protons [160]. Gu *et al.* irradiated AlGaN/GaN HEMTs with 1 MeV neutron irradiation with fluence of 10^{15} cm^{-2}. Their work demonstrated minimal change degradation in drain saturation current ($I_{ds,sat}$), maximum extrinsic transconductance (g_m), and threshold voltage (V_{th}) thus making it an excellent candidate for neutron irradiation hardness, but not detection. Figure 8 shows the I_{ds}-V_{ds} characteristics for these HEMTs [159]. Ahn *et al.* investigated the response of AlGaN/GaN MIS-HEMTs with an ALD Al2O3 gate dielectric to proton irradiation. Proton energies of 5, 10 and 15 MeV resulted in a 95.3%, 68.3%, and 59.8% decrease in $I_{ds,sat}$, respectively. The forward and reserve gate leakage current were reduced by more than 10 times, along with shifts in V_{th} [157]. These and other changes are represented in Figures 9a and 9b.

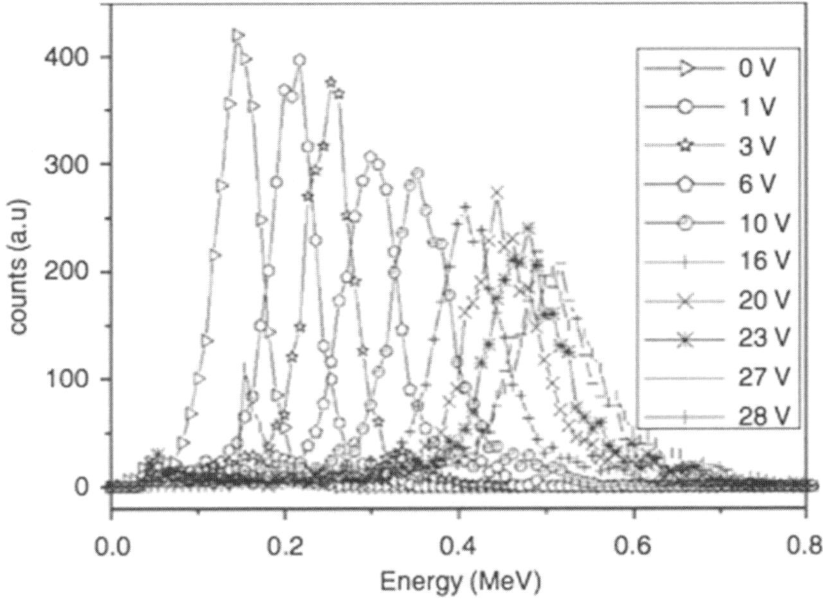

Figure 7. Alpha particle pulse height spectra obtained from a 2 μm thick semi-insulating GaN Schottky diode. The data from this figure is used to calculate the charge capture efficiency (CCE). Reprinted from Nuclear Instruments and Methods Physics Research A, 557, Sellin *et al.*, New materials for radiation hard semiconductor detectors, 481, Copyright 2006, with permission from Elsevier Limited. Ref.[145].

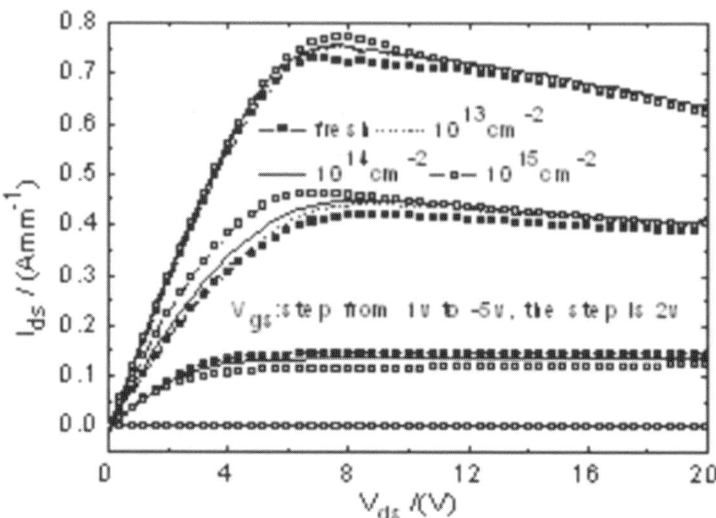

Figure 8. I_{ds}-V_{ds} characteristics of HEMTs before and after 1 MeV neutron irradiation for three fluence levels is shown. Reprinted from Physics Status Solidi, 7, Gu *et al.*, The effect of neutron irradiation on the AlGaN/GaN high electron mobility transistors, 1992, Copyright 2010, with permission from John Wiley and Sons. Ref. [159].

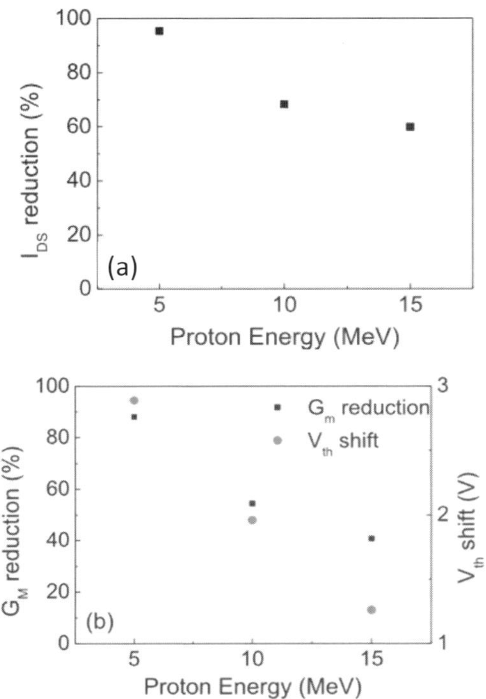

Figure 9. (a) Ids-Vds reduction of AlGaN/GaN HEMTs before and after 10 MeV proton irradiation with a fluence of 5 x 1015 cm-2. (b) Reduction of gm and Vth shift as function of proton irradiation energies at fixed fluence of 5 x 1015 cm-2. Reprinted from Journal of Vacuum Science and Technology B, 33, Ahn et al., Effect of proton irradiation energy on AlGaN/GaN metal-oxide semiconductor high electron mobility transistors, 0501208-2,3, Copyright 2015, with permission from AIP Publishing LLC. Ref. [157].

There are several review papers published on the use of GaN based devices for ionizing radiation detection. Polyakov *et al.* discussed that irradiation of AlGaN/GaN HEMTs by 1.8 MeV protons results in decrease of $I_{ds,sat}$, g_m and V_{th} after a fluence of 10^{14} cm^{-2}. The change in I_{ds}-V_{ds} characteristics and $I_{ds,sat}$ are shown in Figure 10. The impact of high proton energy (~40 MeV) was minimal, thus making AlGaN/GaN HEMTs candidates for low proton energy detection. Gamma radiation up to 600 MRad also demonstrated no substantial change in characteristics [158]. Wang *et al.*

413

reviewed the use of the GaN ionizing radiation detectors for alpha particles, X-ray, thermal neutron and electrons [143]. In 2015, Pearton *et al.* reviewed a wide range of studies on GaN devices and concluded there was decreased in $I_{ds,sat}$, g_m and V_{th} after 40 MeV proton doses equivalent to decades in lower-Earth orbit. For $I_{ds,sat}$ 15-20% degradation was determined for fluence of 5×10^9 cm^{-2}, 30-50% for 5×10^{10} cm^{-2}. Additionally, 50% degradation was observed with 10 MeV electron irradiation with fluence of 6.5×10^{15} cm^{-2}.

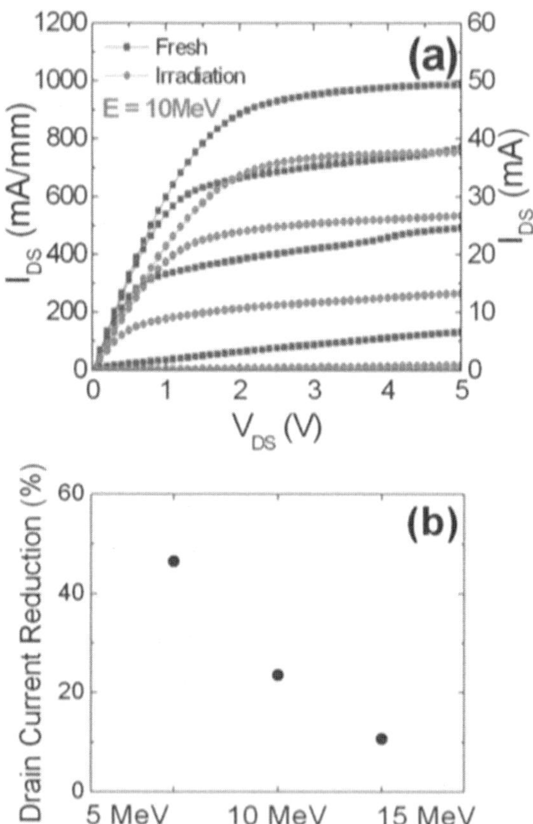

Figure 10. (a) Ids-Vds characteristics of AlGaN/GaN HEMTs before and after 10 MeV proton irradiation. (b) Percentage change of saturation current at Vds = 5V is shown as a function of irradiation energies of 5, 10 and 15 MeV. Reprinted from Journal of Material Chemistry C, 1, Polyakov et al., Radiation effects in GaN materials and devices, 882, Copyright 2013, with permission from Royal Society of Chemistry. Ref. [158].

As discussed above, there are numerous reports that describe the response of AlGaN/GaN and GaN devices around their radiation response to proton, neutron, gamma, electron, X-ray, etc. To enable their use onboard spacecraft and satellites, further studies are needed to study the combined effect multiple ionizing radiation sources. Additionally, the effect of irradiation with multiple sources of ionizing radiation at elevated and subzero temperatures needs to be investigated. The impact of annealing of AlGaN/GaN and GaN devices through room temperature and high temperature annealing [141,143,145,158,159] has been investigated, and varies based on the type of radiation source. To enable AlGaN/GaN and GaN devices devices to be used in space environments, the impact of annealing due to temperature variations need to be explored, in addition to the impact of multiple radiation sources. All together with multiple ionizing radiation sources and large swings in temperatures in space environments due to exposure to the sun presents a complex coupled problem that will need to be further explored.

12.7. High-temperature Electronics

12.7.1. High-temperature Transistors

Gallium nitride (GaN) based high electron mobility transistors (HEMTs) are attractive candidates for operation at high-temperature conditions [29, 56, 129–132]. The HEMTs are formed by growing a heterostructure of GaN with its alloys. Examples of heterostructures include AlGaN/GaN, AlInN/GaN and InAlN/GaN. Previous work has shown transistor like operation of HEMT devices to temperatures as high as 900°C [166]. Additionally, they are used in high voltage and high power applications [23,167,168]. However, these HEMTs face issues of high leakage current, gate and drain lag during operation, and current collapse [82,169–171]. This has been addressed by surface passivation and addition of a gate dielectric to form metal-insulator-semiconductor HEMTs (MIS-HEMTs) [172,173].

Several studies have been carried out to investigate the electrical response of HEMTs and MIS-HEMTs over a wide range of temperatures. Stable operation up to 340°C was demonstrated by Gaska *et al.* for

AlGaN/GaN HEMTs [81]. Additionally, Daumiller *et al.* reported an order of magnitude decrease after operation of 800°C and upon further investigation concluded that AlGaN/GaN heterostructure undergoes an irreversible degradation above 600°C [165]. Son *et al.* demonstrated that for MIS-HEMTs no off-state degradation up to 400°C and excellent pinch off behavior up to 450°C. Above 500°C the off-state current was observed to increase [21]. Maier *et al.* have shown operation of InAlN/GaN HEMTs up to 900°C for 50 hours under vacuum. Contact metallization stability is thought to be reason of the failure [174]. Wang *et al.* developed and experimentally validated an analytical model for AlGaN/GaN HEMTs to predict temperature dependence of 2DEG sheet density, threshold voltage, I_D-V_G, and I_D-V_D characteristics [51]. Additionally, Vitanov *et al.* investigated high temperature modeling of AlGaN/GaN HEMTs [32]. The temperature dependence of maximum current and cut-off frequency was investigated, as shown in Figures 11a and 11b, respectively.

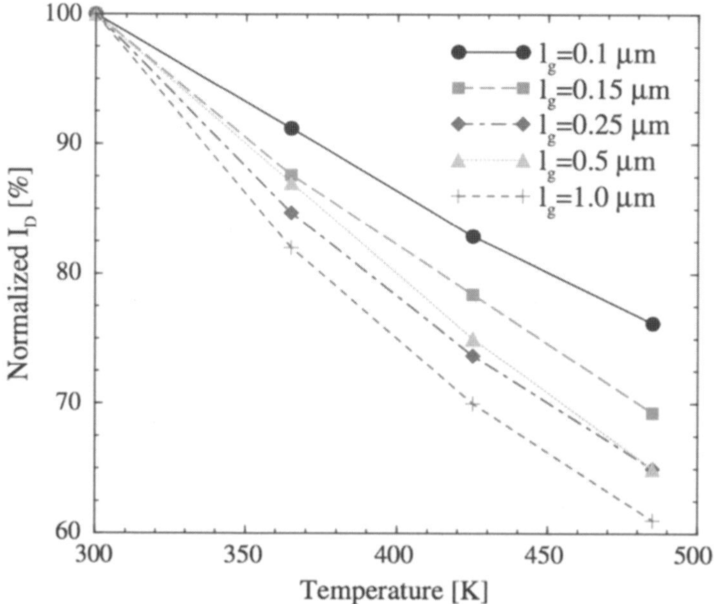

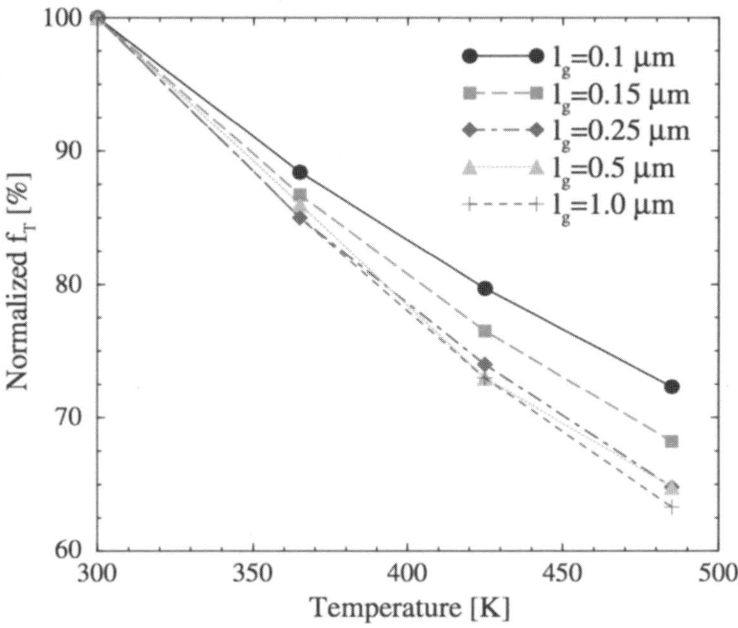

Figure 11. Simulated maximum (a) I_{ds} (V_{ds} = 7 V) and (b) f_T (V_{ds} = 7 V) as a function of ambient temperature normalized to 300 K values. Reprinted from Solid State Electronics, 54, Vitanov et al., High-temperature modeling of AlGaN/GaN HEMTs, 1112, Copyright 2010, with permission from Elsevier Limited. Ref. [32].

12.7.2. High-temperature Contacts

Low-resistance Ohmic contacts to GaN-based HEMTs have received much attention in the past decade. The metallization schemes to contact the 2DEG at the AlGaN/GaN interface have been adopted from the optimized Ohmic contacts to n-type GaN [175–179]. Low work-function metals such as Ti, Al, and Ti/Al react with GaN at high temperatures to form an Ohmic contact, the mechanisms of which are detailed in [180,181]. Several different capping layers (e.g., Pt/Au, Ni/Au, Mo/Au) are added to the Ti/Al bilayer system to prevent the oxidation of the underlying metals and intermixing at high temperatures [182,183].

Thus variants of the standard Ti/Al/X/Au (where X is Pt, Ni, Mo) metallization scheme have been used as Ohmic contacts to n-GaN and AlGaN/GaN heterostructures [183–201]. On the other hand, metals with large work-functions (Au, Pd, Ni, Ir, or Pt) are better suited as Schottky contacts to n-GaN, because the effect of Fermi level pinning at the metal-semiconductor interface is much reduced by the ionic nature of Ga-N and Al-N bonds and the barrier height has a direct dependence on the metal work-function [202–205]. I-V and C-V measurements were used to compare the Schottky barrier heights, ideality factors and reverse breakdown voltages of Ni, Pt, Pd, Au and Cr Schottky contacts to n-GaN with Ni and Pt showing superior properties [206,207]. Hence, the most widely used Schottky contacts to GaN consist of Ni/Au [101,208] or Pt/Au [209,210] bilayers. Further reduction of gate leakage current by four orders of magnitude was observed through the insertion of high work-function metal (Pt) in the Ni/Au bilayer, followed by thermal annealing (500°C, 5 minutes) in N_2 ambient [1].

Thermally activated degradation of contacts is one of the major reliability concerns in GaN-based HEMTs [211]. Schottky (gate) and Ohmic contacts are stable below 300°C [212]. Few studies have been reported on the long-term thermal stability and electrical characteristics of multilayer metallization schemes at elevated temperatures and even less so in oxidizing environments. For instance, Lee and Kao studied the long-term drift in specific contact resistivity of Ti/Al/Pt/Au contacts to n-GaN under different annealing conditions in N_2 ambient [198]. The Ohmic properties were found to be stable for longer durations at lower anneal temperatures (600 minutes at 750°C, 540 minutes at 850°C, and 60 minutes at 950°C). The significance of the Pt barrier in preventing Au in-diffusion and intermixing was highlighted. Kumar *et al.* observed no degradation of Ti/Al/Mo/Au contacts to n-GaN for up to 360 hours at 500°C in N_2 ambient [199]. The use of Mo as the diffusion barrier is justified over Ni, Ti, Pd or Pt, owing to its higher melting point and lower Au solubility in Mo (< 1%) at the annealing temperatures of 850°C. Selvanathan *et al.* further compared the thermal stability of Ti/Al/Mo/Au, Mo/Al/Mo/Au and V/Al/Mo/Au contacts to AlGaN/GaN HEMTs up to 700°C in N_2 ambient [201]. Although Mo/Al/Mo/Au showed the lowest contact resistance of the three schemes, its thermal stability was poorer than Ti/Al/Mo/Au. Piazza *et al.* reported a large increase in contact resistance of Ti/Al/Ni/Au contacts to GaN-HEMTs after 100 hour thermal storage at 340°C [213]. Through AES and EDS observations, Ga out-diffusion and Au inter-diffusion were identified as the degradation mechanisms. Recently, Hou and Senesky reported the time evolution of Ti/Al/Pt/Au contacts to n-GaN at 600°C in air over a 10 hour thermal storage test [200]. Even within an oxidizing air ambient (instead of inert ambient or vacuum), the contact resistance remained relatively stable during the thermal storage, as shown in Figure 12 [200]. From AFM and AES depth analysis, minimal physical change in the Ti/Al/Pt/Au contacts was revealed after the thermal storage. Thermally stable Ti/Al/Ni (no Au overlay) Ohmic contacts to ultrathin body InAlN/GaN HEMTs were reported for a testing period of 30 minutes at 600°C in vacuum (p ~ 10^{-6} mbar) [214]. Maier *et al.* further evaluated the Ti/Al/Ni/Au stack on InAlN/GaN HEMTs at temperatures of about 800°C in nitrogen [174]. Intermixing and Au accumulation at the interface was reported, as seen from the cross section TEM in Figure 13.

On the other hand, metals used for Schottky contacts react with GaN on prolonged thermal treatments at 500 to 600°C, thus forming gallides and nitrides at the interface which can affect the Schottky properties [148,215]. The reaction between Ni and GaN was studied by Venugopalan *et al.* [216], and reasonably stable contacts up to 500°C were observed [217,218]. Liu *et al.* reported failure of Ni Schottky contact to n-GaN at 600°C due to Ga migration into the Ni layer, however NiSi Schottky contact remained stable after an hour at 600°C, possibly due to a thin insulation layer formed at the NiSi/GaN interface [217]. Similarly, Schottky contacts on n-GaN using W were reported to degrade at temperatures $\geq$ 500°C, but those using WSi_x showed improved thermal stability up to 600°C [75,219]. Pd as a gate contact to GaN was found to degrade at 300°C within 2 minutes [203,220]. Re Schottky contact to GaN was reported to be stable up to 600°C, due to the metallurgical stability of Re on GaN [220]. Wu *et al.* have studied the electrical and structural degradation in AlGaN/GaN HEMTs under high-power and high temperature DC stress [221]. Device self-heating due to large electric fields [222] is suspected to be (in part) the reason for the structural degradation of the gate closer to the drain contact. Large gate currents under a strong positive gate bias have also shown to thermally degrade the

Schottky contact through an observed reduction of $V_{G,on}$ and ideality factor degradation [223].

Thus robust high-temperature tolerant contacts to GaN-based HEMTs are crucial for their long-term stability in hot oxidizing environments. Novel material stacks using a combination of refractory metals (resistant to heat and water) and noble metals (resistant to oxidation and corrosion) should be identified and developed to achieve low resistance Ohmic contacts. Further work is needed to minimize drift in contact resistance with temperature and duration of exposure (aging). New barrier mechanisms are needed to prevent high temperature diffusion and oxidation of Schottky contact materials. Essentially, advancement both at the materials front and new contact topologies is essential in exploiting GaN technology for high temperature electronics and sensors.

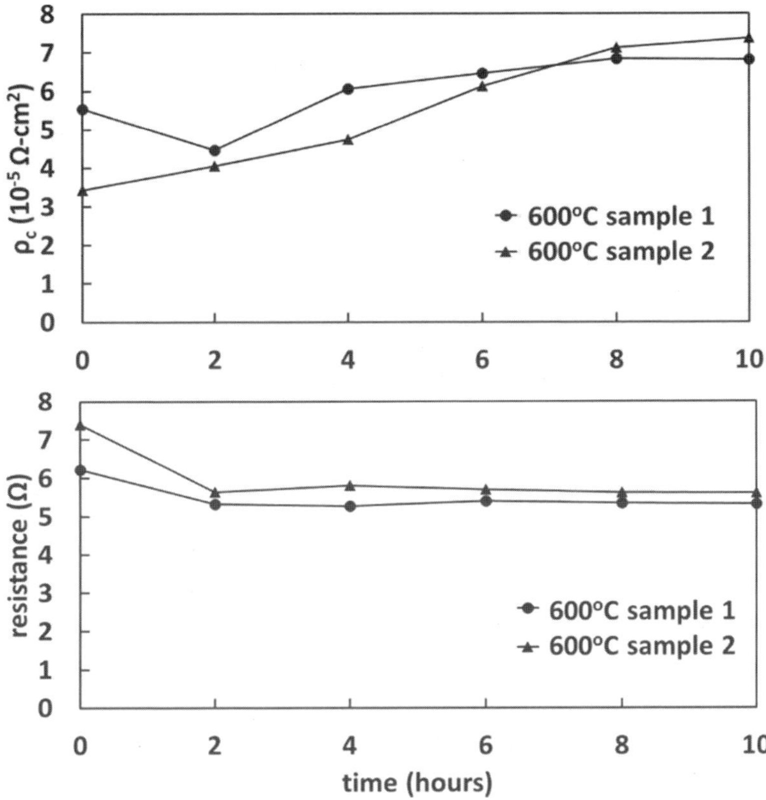

Figure 12. Experimentally measured specific contact resistivity and the total resistance of Ti/Al/Pt/Au multilayer contacts to GaN over 10 hours thermal storage at 600°C. Reprinted from Applied Physics Letters, 105, Hou et al., Operation of Ohmic Ti/Al/Pt/Au multilayer contacts to GaN at 600°C in air, 081905, Copyright 2014, with permission from AIP Publishing LLC. Ref. [200].

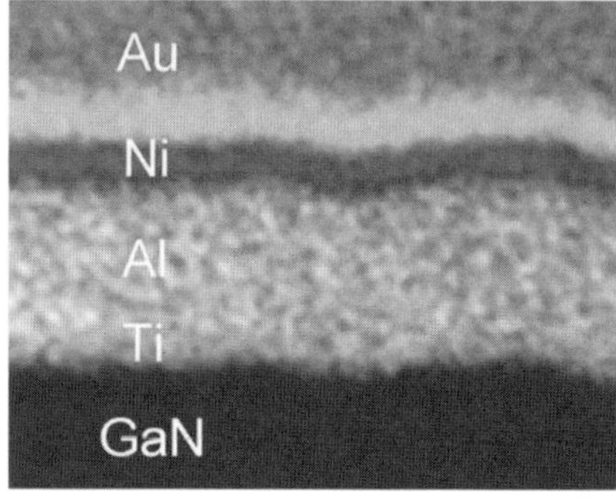

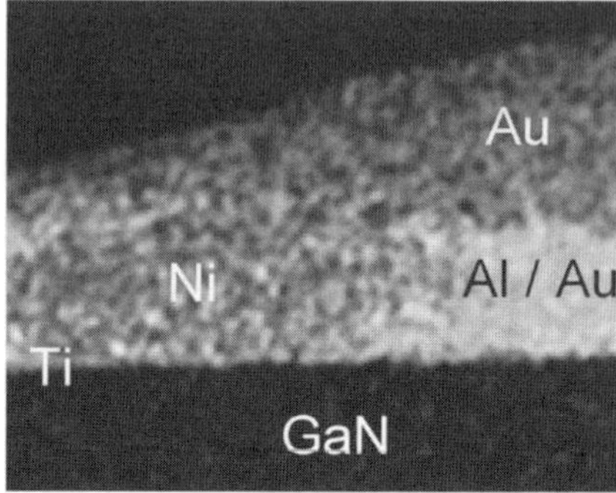

Figure 13. Elemental energy dispersive X-ray spectroscopy maps of cross sections of Ohmic contact stack (Ti/Al/Ni/Au, 10 nm/200 nm/40 nm/100 nm) on 8-nm InAlN/GaN without AlN spacer, after (left) RTA and (right) 30-min testing at 800°C. Reprinted from *IEEE Trans. Device Mater. Reliab.*, **10,** Maier et al., Testing the Temperature Limits of GaN-Based HEMT Devices, 427–436, Copyright 2010, with permission from IEEE [174].

12.8. Future Outlook

Although there have been many compelling demonstrations and advancements using the AlGaN/GaN material system, the realization of a monolithic AlGaN/GaN platform, in which micro-scale sensors, electronics and energy harvesters are integrated on a single chip, requires further development. For example, approaches for sensing additional measurands, such as external accelerations and acoustic signatures complements the existing set of sensors and electronics (resonators and transistor circuits) that leverage the AlGaN/GaN platform for its temperature-tolerance, strain-sensitivity and radiation hardness. In addition, high-temperature and commercially viable packaging methods must be further developed for integration, environmental/chemical shielding, and reliable operation of multifunctional electronic components. More specifically, AlGaN/GaN die attach to ceramics, wire bonding, and multilevel interconnect technology, as well as additional diaphragm etch technology using SiC substrates should be developed for high-temperature operation which creates additional technical challenges and the need for new reliability protocols and standards. High-temperature capable wireless read-out circuitry, as well as energy harvesting, low-power operation and power management methods should be developed for integration within complex environments using the AlGaN/GaN platform. This will enable autonomous sensor units for the advancement to support IoT applications that demand data from high-temperature environments (e.g., smart engines and oilfields).

12.9. Conclusion

Many industries can benefit from multifunctional sensor nodes that can operate at temperatures beyond 200°C where traditional Si-based sensing platforms are limited. The GaN platform is currently being developed as an alternative to Si and SiC-based sensors and other microelectronic components due to its thermal-tolerance and that ability to monolithically integrate with AlGaN/GaN electronics. The current status of numerous high-temperature GaN-based microelectronics, along with and explanation of compelling demonstrations of high-temperature AlGaN/GaN transistors and metal contacts to GaN, have been discussed in this chapter. These data from various academic and industrial research groups support the use of AlGaN/GaN as a high-temperature platform material to advance current sensing approaches. However, the interdependent material-, electrical- and environmental-properties of the AlGaN/GaN platform including interface metals and dielectric materials under physical stimulus should be fully examined.

References

1.	Miura N., Nanjo T., Suita M., Oishi T., Abe Y., Ozeki T., Ishikawa H., Egawa T. and Jimbo T., 2004, *Solid. State. Electron.*, **48**, 689.
2.	Rais-Zadeh M., Gokhale V. J., Ansari A., Théron D., Faucher M., Cordier Y. and Buchaillot L., 2014, *J. Micromechanical Syst. IEEE*, **23**, 1.
3.	Neudeck P., Okojie R. and Chen L., 2002, *Proc. IEEE*, **90**, 1065.
4.	_Itrs, 2010, *Rep. - Itrs*, 70.
5.	Harafuji K., Tsuchiya T. and Kawamura K., 2004, *J. Appl. Phys.*, **96**, 2513.
6.	Porowski S., 1997, *Mater. Sci. Eng. B*, **44**, 407.
7.	Strite S., 1992, *J. Vac. Sci. Technol. B Microelectron. Nanom. Struct.*, **10**, 1237.
8.	Maboudian R., Carraro C., Senesky D. G. and Roper C. S., 2013, *J. Vac. Sci. Technol. A Vacuum, Surfaces, Film.*, **31**, 050805.
9.	Berger L. I., 2015, In: Haynes, MW (Ed.), *CRC Handb. Chem. Physics, 96th Ed. 2015-2016, CRC Press*, Boca Raton, FL, pp. 12–80 – 12–93.
10.	Pierson H. O., 1993, *Handb. carbon, Graph. diamond, fullerenes Prop. Process. Appl., William Andrew Publishing*, pp. 244–277.
11.	Ng K. K., 2009, *John Wiley & Sons, Inc.*, New York.
12.	Morkoç H., 2009, vol. 1, *Wiley-VCH Verlag GmbH & Co. KGaA*.
13.	Casady J. B. and Johnson R. W., 1996, *Solid. State. Electron.*, **39**, 1409.
14.	Railkar T. a., Kang W. P., Windischmann H., Malshe a. P., Naseem H. a., Davidson J. L. and Brown W. D., 2000, *Crit. Rev. Solid State Mater. Sci.*, **25**, 163.
15.	Harris G. L., 1995, *Prop. Silicon Carbide, Institution of Engineering and Technology*, pp. 2–12.
16.	Hopcroft M. a, Nix W. D. and Kenny T. W., 2010, **19**, 229.
17.	Hall J. J., 1967, *Phys. Rev.*, **161**, 756.
18.	Petersen K. E., 1982, *Proc. IEEE*, vol. 70.
19.	Jacoboni C., Canali C., Otiaviani G. and Quaranta a A., 1977, *Solid State Phys.*, **20**, 77.
20.	Maeda N., Tsubaki K., Saitoh T. and Kobayashi N., 2001, *Appl. Phys. Lett.*, **79**, 1634.
21.	Son K., Liao A., Lung G., Gallegos M., Hatake T., Harris R. D., Scheick L. Z. and Smythe W. D., 2010, In: George, T, Islam, MS, Dutta, AK (Eds.), *Proc. SPIE 7679, Micro- Nanotechnol. Sensors, Syst. Appl. II.*
22.	Arulkumaran S., Egawa T., Ishikawa H. and Jimbo T., 2002, *Appl. Phys. Lett.*, **80**, 2186.
23.	Nomura T., Kambayashi H., Masuda M., Ishii S., Ikeda N., Lee J., Yoshida S., Improved A., Hfet G., Improvement a and Electrode O., 2006, *Transacations on Electron Devices*, **53**, 2908.
24.	DeBruijn G., Skeates C., Greenaway R., Harrison D., Parris M., Simon J., Mueller F., Shantonu R., Riding M., Temple L. and Wutherich K., 2008, *Oilf. Rev.*, **20**, 46.
25.	George T., Son K. -a., Powers R. a., Castillo L. Y. D. and Okojie R., 2005, *IEEE Sensors, 2005., IEEE*, Irvine, CA, pp. 1253–1258.

26. Spetz A. L., Baranzahi a, Tobias P. and Lundström I., 1997, *Phys. Status Solidi,* **162**, 493.

27. Noor A. K., Venneri S. L., Paul D. B. and Hopkins M. a., 2000, *Comput. Struct.,* **74**, 507.

28. Fidan B., Mirmirani M. and Ioannou P. a, 2003, *Proc. AIAA Int. Sp. Planes Hypersonics Syst. Technol.,* 1.

29. Ambacher O., Smart J., Shealy J. R., Weimann N. G., Chu K., Murphy M., Schaff W. J., Eastman L. F., Dimitrov R., Wittmer L., Stutzmann M., Rieger W. and Hilsenbeck J., 1999, *J. Appl. Phys.,* **85**, 3222.

30. Ambacher O., Foutz B., Smart J., Shealy J. R., Weimann N. G., Chu K., Murphy M., Sierakowski A. J., Schaff W. J., Eastman L. F., Dimitrov R. M. A. and S. M., 2000, *J. Appl. Phys.,* **87**, 334.

31. Auf der Maur M., Romano G. and Di Carlo A., 2012, *Comput. Electron. (IWCE), 2012 15th Int. Work.,* pp. 1–4.

32. Vitanov S., Palankovski V., Maroldt S. and Quay R., 2010, *Solid. State. Electron.,* **54**, 1105.

33. Venkatachalam A., James W. T. and Graham S., 2011, *Semicond. Sci. Technol.,* **26**, 8085027.

34. Duan, Y., Tang, G., Qin, L. and Shi L., 2008, *Eur. Phys. J. B,* **66**, 215.

35. Ambacher O., Eickhoff M., Link A., Hermann M., Stutzmann M., Bernardini F., Fiorentini V., Smorchkova Y., Speck J., Mishra U., Schaff W., Tilak V. and Eastman L. F., 2003, *Phys. Status Solidi,* **6**, 1907.

36. Eickhoff M., Schalwig J., Steinhoff G., Weidemann O., Görgens L., Neuberger R., Hermann M., Baur B., Müller G., Ambacher O. and Stutzmann M., 2003, *Phys. status solidi,* **6**, 1642.

37. Cheney D., Douglas E., Liu L., Lo C.-F., Gila B., Ren F. and Pearton S., 2012, *Materials (Basel).,* **5**, 2498.

38. Le Boulbar E. D., Edwards M. J., Vittoz S., Vanko G., Brinkfeldt K., Rufer L., Johander P., Lalinský T., Bowen C. R. and Allsopp D. W. E., 2013, *Sensors Actuators A Phys.,* **194**, 247.

39. Chapin C. A., Chiamori H. C., Hou M. and Senesky D. G., 2013, *Int. Work. Struct. Heal. Monit.,* vol. 9, pp. 1621–1628.

40. Dzuba J., Vanko G., Ryger I., Vallo M., Kutis V. and Lalinsky T., 2014, *10th Int. Conf.,* p. 1.

41. Steude G., Meyer B., G̈oldner, A. Hoffmann A., Kaschner A., Bechstedt F., Amano H. and Akasaki I., 1999, *Jpn. J. Appl. Phys.,* **38**, L498.

42. Chen D. J., Zhang K. X., Tao Y. Q., Wu X. S., Xu J., Zhang R., Zheng Y. D. and Shen B., 2006, *Appl. Phys. Lett.,* **88**, 102106.

43. Gao F., Lo H.-Y., Ram R. and Palacios T., 2010, *Device Res. Conf.,* pp. 127–128.

44. Sarua A., Ji H., Kuball M., Uren M. J., Martin T., Nash K. J., Hilton K. P. and Balmer R. S., 2006, *Appl. Phys. Lett.,* **88**, 103502.

45. Yan Q., Rinke P., Scheffler M. and Van De Walle C. G., 2009, *Appl. Phys. Lett.,* **95**, 3.

46. Yan Q., Rinke P., Janotti A., Scheffler M. and Van de Walle C. G., 2014, *Phys. Rev. B,* **90**, 125118.

47. Koehler A. D., Gupta A., Chu M., Parthasarathy S., Linthicum K. J., Johnson J. W., Nishida T., Thompson S. E. and Repeatable A., 2010, *IEEE Electron Device Lett.*, **31**, 665.

48. Chu M., Koehler A. D., Gupta A., Nishida T. and Thompson S. E., 2010, *J. Appl. Phys.*, **108**, 104502.

49. Ambacher O., Majewski J., Miskys C. R., Link A., Hermann M., Eickhoff M., Stutzmann M., Bernardini F., Fiorentini V., Tilak V., Schaff B. and Eastman L. F., 2002, *J. Phys. Condens. Matter*, **14**, 3399.

50. Husna F., Lachab M., Sultana M., Adivarahan V., Fareed Q. and Khan A., 2012, *IEEE Trans. Electron Devices*, **59**, 2424.

51. Wang Y.-H., Liang Y. C., Samudra G. S., Chang T.-F., Huang C.-F., Yuan L. and Lo G.-Q., 2013, *Semicond. Sci. Technol.*, **28**, 125010.

52. Kim J., Freitas J. a., Mittereder J., Fitch R., Kang B. S., Pearton S. J. and Ren F., 2006, *Solid. State. Electron.*, **50**, 408.

53. Gaska R., Osinsky a, Yang J. W. and Shur M. S., 1998, *Electron Device Lett. IEEE*, **19**, 89.

54. Roesch W. J., 2006, *Microelectron. Reliab.*, **46**, 1218.

55. Singhal S., Roberts J. C., Rajagopal P., Li T., Hanson A. W., Therrien R., Johnson J. W., Kizilyalli I. C. and Linthicum K. J., 2006, *44th Annu. IEEE Int. Reliab. Phys. Symp.*, pp. 95–98.

56. Joh J. and del Alamo J. a., 2008, *IEEE Electron. Lett.*, **29**, 5.

57. Kim H., Tilak V., Green B. M., Smart J. a., Schaff W. J., Shealy J. R. and Eastman L. F., 2001, *Phys. Status Solidi*, **188**, 203.

58. Sozza a., Dua C., Morvan E., diForte-Poisson M. a., Delage S., Rampazzo F., Tazzoli a., Danesin F., Meneghesso G., Zanoni E., Curutchet a., Malbert N., Labat N., Grimbert B. and Jaeger J.-C. De, 2005, *IEEE Int. Devices Meet. 2005. IEDM Tech. Dig.*, **00**, 4.

59. Kim H., Thompson R. M., Tilak V., Prunty T. R., Shealy J. R. and Eastman L. F., 2003, *IEEE Electron Device Lett.*, **24**, 421.

60. Meneghesso G., Verzellesi G., Danesin F., Rampazzo F., Zanon F., Tazzoli A., Meneghini M. and Zanoni E., 2008, *IEEE Trans. Device Mater. Reliab.*, **8**, 332.

61. Shen B., Someya T. and Arakawa Y., 2000, *Appl. Phys. Lett.*, **76**, 2746.

62. Eickhoff M., Ambacher O., Krötz G. and Stutzmann M., 2001, *J. Appl. Phys.*, **90**, 3383.

63. Zimmermann T., Neuburger M., Benkart P., Hernandez-Guillen F. J., Pietzka C., Kunze M., Daumiller I., Dadgar a., Krost a. and Kohn E., 2006, *IEEE Electron Device Lett.*, **27**, 309.

64. Tilak V., Jiang J., Batoni P. and Knobloch a., 2007, *J. Mater. Sci. Mater. Electron.*, **19**, 195.

65. Yilmazoglu O., Mutamba K., Pavlidis D. and Mbarga M. R., 2006, *IEICE Trans. Electron.*, **E89-C**, 1037.

66. Talukdar A., Qazi M. and Koley G., 2012, *Appl. Phys. Lett.*, **101**, 252102.

67. Yao K., Khandelwal S., Sammoura F., Kazama A., Hu C. and Lin L., 2015, *IEEE Electron Device Lett.*, **36**, 902.

68. Kanda Y., 1991, *Sensors Actuators A Phys.*, **28**, 83.

69. Kanda Y., 1982, *IEEE Trans. Electron Devices*, 64.

70.   Shor J. S., Goldstein D. and Kurtz a. D., 1993, *IEEE Trans. Electron Devices*, **40**, 1093.

71.   Smith C., 1954, *Phys. Rev.*, **94**, 42.

72.   Edwards M. J., Le Boulbar E. D., Vittoz S., Vanko G., Brinkfeldt K., Rufer L., Johander P., Lalinský T., Bowen C. R. and Allsopp D. W. E., 2012, *Phys. Status Solidi*, **9**, 960.

73.   Kang B. S., Kim S., Ren F., Johnson J. W., Therrien R. J., Rajagopal P., Roberts J. C., Piner E. L., Linthicum K. J., Chu S. N. G., Baik K., Gila B. P., Abernathy C. R. and Pearton S. J., 2004, *Appl. Phys. Lett.*, **85**, 2962.

74.   Chu S. N. G., Ren F., Pearton S. J., Kang B. S., Kim S., Gila B. P., Abernathy C. R., Chyi J.-I., Johnson W. J. and Lin J., 2005, *Mater. Sci. Eng. A*, **409**, 340.

75.   Pearton S. J., Kang B. S., Kim S., Ren F., Gila B. P., Abernathy C. R., Lin J. and Chu S. N. G., 2004, *J. Phys. Condens. Matter*, **16**, R961.

76.   Kang B. S., Kim J., Jang S., Ren F., Johnson J. W., Therrien R. J., Rajagopal P., Roberts J. C., Piner E. L., Linthicum K. J., Chu S. N. G., Baik K., Gila B. P., Abernathy C. R. and Pearton S. J., 2005, *Appl. Phys. Lett.*, **86**, 253502.

77.   Lalinský T., Držík M., Vanko G., Vallo M., Kutiš V., Bruncko J., Haščík Š., Jakovenko J. and Husák M., 2011, *Microelectron. Eng.*, **88**, 2424.

78.   Lalinský T., Hudek P., Vanko G., Dzuba J., Kutiš V., Srnánek R., Choleva P., Vallo M., Držík M., Matay L. and Kostič I., 2012, *Microelectron. Eng.*, **98**, 578.

79.   Son K. A., Liu Y., Ruden P. P., Xic J., Biyikli N., Moon Y. T., Onojima N. and Morkoc H., 2005, *2005 IEEE Sensors (IEEE Cat. No.05CH37665C)*, 4 pp.|CD.

80.   Son K., Son K., Yang B. and Yang B., 2007, *Current*, 8.

81.   Gaska R., Chen Q., Yang J., Osinsky A., Asif Khan M. and Shur M. S., 1997, *IEEE Electron Device Lett.*, **18**, 492.

82.   Mishra U. K., Parikh P. and Wu Y. F., 2002, *Proc. IEEE*, **90**, 1022.

83.   Wu Y. F., Kapolnek D., Ibbetson J. P., Parikh P., Keller B. P. and Mishra U. K., 2001, *IEEE Trans. Electron Devices*, **48**, 586.

84.   Li J., Oder T. N., Nakarmi M. L., Lin J. Y., Jiang H. X. and Al M., 2002, *Appl. Phys. Lett.*, **80**, 1210.

85.   Monroy E., Muñoz E., Sánchez F. J., Calle F., Calleja E., Beaumont B., Gibart P., Muñoz J. a and Cussó F., 1999, *Semicond. Sci. Technol.*, **13**, 1042.

86.   Davis R. F., 1991, *Proc. IEEE*, **79**, 702.

87.   Henini M., 1999, *III-Vs Rev.*, **12**, 30.

88.   Hirano A., Pernot C., Iwaya M., Detchprohm T., Amano H. and Akasaki I., 2001, *Phys. Status Solidi a-Applied Res.*, **188**, 293.

89.   Xu G. Y., Salvador A., Kim W., Fan Z., Lu C., Tang H., Morkoç H., Smith G., Estes M., Goldenberg B., Yang W. and Krishnankutty S., 1997, *Appl. Phys. Lett.*, **71**, 2154.

90.   Khan M. A., Kuznia J. N., Olson D. T., Blasingame M. and Bhattarai A. R., 1993, *Appl. Phys. Lett.*, **63**, 2455.

91.   Walker D., Monroy E., Kung P., Wu J., Hamilton M., Sanchez F. J. and Diaz J., 2001, *Appl. Phys. Lett.*, **74**, 1999.

92.   Chen Q., Asif Khan M., Sun C. J. and Yang J. W., 1995, *Electron. Lett.*, **31**, 3761.

93.   B. W. Lim, Q. C. Chen, J. Y. Yang and M. A. K., 1996, *Appl. Phys. Lett.*, **68**, 3761.

94. Khan M. A., Kuznia J. N., Olson D. T., Van Hove J. M., Blasingame M. and Reitz L. F., 1992, *Appl. Phys. Lett.*, **60**, 2917.

95. Walker D., Zhang X., Saxler A., Kung P., Xu J. and Razeghi M., 1997, *Appl. Phys. Lett.*, **70**, 949.

96. Lim B. W., Gangopadhyay S., Yang J. W., Osinsky A., Chen Q., Anwar M. Z. and Khan M. A., 1997, *IEEE Electron. Lett. Online*, **33**, 7.

97. Chen Q., Yang J. W., Osinsky A., Gangopadhyay S., Lim B., Anwar M. Z., Asif Khan M., Kuksenkov D. and Temkin H., 1997, *Appl. Phys. Lett.*, **70**, 2277.

98. Monroy E., Calle F., Muñoz E., Omnès F., Gibart P. and Muñoz J. A., 1998, *Appl. Phys. Lett.*, **73**, 2146.

99. Monroy E., Calle F., Muñoz E., Beaumont B., Omnès F. and Gibart P., 1999, *Electron. Lett.*, **35**, 1488.

100. Monroy E., Calle F., Pau J. L., Sanchez F. J., Munoz E., Omnes F., Beaumont B. and Gibart P., 2000, *J. Appl. Phys.*, **88**, 2081.

101. Monroy E., Calle F., Muñoz E. and Omnès F., 1999, *Appl. Phys. Lett.*, **74**, 3401.

102. Monroy E., Calle F., Muñoz E. and Omnès F., 1999, *Phys. Status Solidi Appl. Res.*, **176**, 157.

103. Carrano J. C., Li T., Grudowski P. A., Eiting C. J., Dupuis R. D. and Campbell J. C., 1998, *J. Appl. Phys.*, **83**, 6148.

104. Huang Y., Chen D. J., Lu H., Shi H. B., Han P., Zhang R. and Zheng Y. D., 2010, *Appl. Phys. Lett.*, **96**, 1.

105. Van Hove J. M., Hickman R., Klaassen J. J., Chow P. P. and Ruden P. P., 1997, *Appl. Phys. Lett.*, **70**, 2282.

106. Monroy E., Hamilton M., Walker D., Kung P., Sánchez F. J. and Razeghi M., 1999, *Appl. Phys. Lett.*, **74**, 1171.

107. Yang W., Nohova T., Krishnankutty S., Torreano R., McPherson S. and Marsh H., 1998, *Appl. Phys. Lett.*, **73**, 1086.

108. Walker D., Saxler A., Kung P., Zhang X., Hamilton M., Diaz J. and Razeghi M., 1998, *Appl. Phys. Lett.*, **72**, 3303.

109. Parish G., Keller S., Kozodoy P., Ibbetson J. P., Marchand H., Fini P. T., Fleischer S. B., DenBaars S. P., Mishra U. K. and Tarsa E. J., 1999, *Appl. Phys. Lett.*, **75**, 247.

110. Yang W., Nohava T., Krishnankutty S., Torreano R., McPherson S. and Marsh H., 1998, *Appl. Phys. Lett.*, **73**, 978.

111. Khan M. A., Shur M. S., Chen Q., Kuznia J. N. and Sun C. J., 1995, *Electron. Lett.*, **31**, 398.

112. McIntosh K. a., Molnar R. J., Mahoney L. J., Lightfoot a., Geis M. W., Molvar K. M., Melngailis I., Aggarwal R. L., Goodhue W. D., Choi S. S., Spears D. L. and Verghese S., 1999, *Appl. Phys. Lett.*, **75**, 3485.

113. Carrano J. C., Lambert D. J. H., Eiting C. J., Collins C. J., Li T., Wang S., Yang B., Beck a. L., Dupuis R. D. and Campbell J. C., 2000, *Appl. Phys. Lett.*, **76**, 924.

114. Morkoç H., Carlo A. Di and Cingolani R., 2002, *Solid. State. Electron.*, **46**, 157.

115. Monroy E., Calle F., Pau J. L., Muñoz E., Omnès F., Beaumont B. and Gibart P., 2001, *J. Cryst. Growth*, **230**, 537.

116. Shur M. S. and Khan M. A., 1997, *MRS Bull.*, **22**, 44.

117. Smart J. a., Schremer a. T., Weimann N. G., Ambacher O., Eastman L. F. and Shealy J. R., 1999, *Appl. Phys. Lett.*, **75**, 388.

118. Zaidi Z. H. and Houston P. A., 2013, **60**, 2776.

119. Behtash R., Tobler H., Neuburger M., Schurr A., Leier H., Cordier Y., Semond F., Natali F. and Massies J., 2003, *Electron. Lett.*, **39**, 626.

120. Dunmka D. C., Lee C., Tserng H. Q., Saunier P. and Kumar M., 2004, *Electron Lett*, **40**, 1023.

121. Schremer a. T., Smart J. a., Wang Y., Ambacher O., MacDonald N. C. and Shealy J. R., 2000, *Appl. Phys. Lett.*, **76**, 736.

122. Semond F., Lorenzini P., Grandjean N. and Massies J., 2001, *Appl. Phys. Lett.*, **78**, 335.

123. Osinsky a., Gangopadhyay S., Yang J. W., Gaska R., Kuksenkov D., Temkin H., Shmagin I. K., Chang Y. C., Muth J. F. and Kolbas R. M., 1998, *Appl. Phys. Lett.*, **72**, 551.

124. Pau J. L., Monroy E., Naranjo F. B., Munoz E., Calle F., Sanchez Garcia M. a and Calleja E., 2000, *Appl. Phys. Lett.*, **76**, 2785.

125. Stevens K. S., Kinniburgh M. and Beresford R., 1995, *Appl. Phys. Lett.*, **66**, 3518.

126. Hirsch M. T., Wolk J. A., Walukiewicz W. and Haller E. E., 1997, *Appl. Phys. Lett.*, **71**, 1098.

127. Chen H. M., Chen Y. F., Lee M. C. and Feng M. S., 1997, *J. Appl. Phys.*, **82**, 899.

128. Pernot C., Hirano A., Iwaya M., Detchprohm T., Amano H. and Akasaki I., 1999, *Japanese J. Appl. Phys. Part 2-Letters*, **38**, L487.

129. Monroy E., Calle F., Garrido J. A., Youinou P., Muñoz E., Omnès F., Beaumont B. and Gibart P., 1999, *Semicond. Sci. Technol.*, **14**, 685.

130. Johnson C., Lin J. Y., Jiang H. X., Khan M. A. and Sun C. J., 1996, *Appl. Phys. Lett.*, **1808**, 66.

131. Li J. Z., Lin J. Y., Jiang H. X., Salvador A., Botchkarev A. and Morkoc H., 1996, *Appl. Phys. Lett.*, **69**, 1474.

132. Lang D. V. and Logan R. a., 1977, *Phys. Rev. Lett.*, **39**, 635.

133. Qiu C. H. and Pankove J. I., 1997, *Appl. Phys. Lett.*, **70**, 1983.

134. Muñoz E., Monroy E., Garrido J. A., Izpura I., Sánchez F. J., Sánchez-García M. A., Calleja E., Beaumont B. and Gibart P., 1997, *Appl. Phys. Lett.*, **71**, 870.

135. Garrido J. a, Monroy E., Izpura I. and Muñoz E., 1999, *Semicond. Sci. Technol.*, **13**, 563.

136. Mooney P. M., 1990, *J. Appl. Phys.*, **67**, R1.

137. Shishehchi S., Asgari A. and Kheradmand R., 2010, *Opt. Quantum Electron.*, **41**, 525.

138. Kioupakis E., Yan Q., Steiauf D. and Van de Walle C. G., 2013, *New J. Phys.*, **15**, 125006.

139. Milani N. M., Mohadesi V. and Asgar i A., 2015, *Phys. E*, **71**, 64.

140. Nord J., Nordlund K., Keinonen J. and Albe K., 2003, *Nucl. Instruments Methods Phys. Res. Sect. B*, **202**, 93.

141. Pearton S. J., Hwang Y.-H. and Ren F., 2015, *ECS Trans.*, **66**, 3.

142. Vaitkus J., Cunningham W., Gaubas E., Rahman M., Sakai S., Smith K. M. and Wang T., 2003, *Nucl. Instruments Methods Phys. Res. Sect. A Accel. Spectrometers, Detect. Assoc. Equip.*, **509**, 60.

143. Wang J., Mulligan P., Brillson L. and Cao L. R., 2015, *Appl. Phys. Rev.*, **2**, 031102.

144. Grant J., Bates R., Cunningham W., Blue a., Melone J., McEwan F., Vaitkus J., Gaubas E. and O'Shea V., 2007, *Nucl. Instruments Methods Phys. Res. Sect. A Accel. Spectrometers, Detect. Assoc. Equip.*, **576**, 60.

145. Sellin P. J. and Vaitkus J., 2006, *Nucl. Instruments Methods Phys. Res. Sect. A Accel. Spectrometers, Detect. Assoc. Equip.*, **557**, 479.

146. Lu M., Yu G. H. and Zhang G. G., 2010, *Yuanzineng Kexue Jishu/Atomic Energy Sci. Technol.*, **44**, 750.

147. Wang G., Fu K., Yao C., Wang J. and Lu M., 2011, *Guti Dianzixue Yanjiu Yu Jinzhan/Research Prog. Solid State Electron.*, **31**.

148. Monroy E., Calle F., Ranchal R., Palacios T., Verdú M., Sánchez F. J., Montojo M. T., Eickhoff M., Omnès F., Bougrioua Z. and Moerman I., 2002, *Semicond. Sci. Technol.*, **17**, L47.

149. Narita S., Hitora T., Yamaguchi E., Sakemi Y., Itoh M., Yoshida H., Kasagi J. and Neichi K., 2013, *Nucl. Instruments Methods Phys. Res. Sect. A Accel. Spectrometers, Detect. Assoc. Equip.*, **717**, 1.

150. Mulligan P., Wang J. and Cao L., 2013, *Nucl. Instruments Methods Phys. Res. Sect. A Accel. Spectrometers, Detect. Assoc. Equip.*, **719**, 13.

151. Pearton S. J., Deist R., Ren F., Liu L., Polyakov A. Y. and Kim J., 2013, *J. Vac. Sci. Technol. A Vacuum, Surfaces, Film.*, **31**, 050801.

152. Mulligan P., Qiu J., Wang J. and Cao L. R., 2014, *IEEE Trans. Nucl. Sci.*, **61**, 2040.

153. Pearton S. J., Ren F., Patrick E., Law M. E. and Polyakov A. Y., 2016, *ECS J. Solid State Sci. Technol.*, **5**, Q35.

154. Wang J., Mulligan P. L. and Cao L. R., 2014, *Nucl. Instruments Methods Phys. Res. Sect. A Accel. Spectrometers, Detect. Assoc. Equip.*, **761**, 7.

155. Lu M., Wang G. and Yao C. S., 2011, *Adv. Mater. Res.*, **343-344**, 56.

156. Aktas O., Kuliev a., Kumar V., Schwindt R., Toshkov S., Costescu D., Stubbins J. and Adesida I., 2004, *Solid. State. Electron.*, **48**, 471.

157. Ahn S., Dong C., Zhu W., Kim B., Hwang Y.-H., Ren F., Pearton S. J., Yang G., Kim J., Patrick E., Tracy B., Smith D. J. and Kravchenko I. I., 2015, *J. Vac. Sci. Technol. B, Nanotechnol. Microelectron. Mater. Process. Meas. Phenom.*, **33**, 051208.

158. Polyakov A. Y., Pearton S. J., Frenzer P., Ren F., Liu L. and Kim J., 2013, *J. Mater. Chem. C*, 877.

159. Gu W., Hao Y., Yang L., Duan C., Duan H., Zhang J. and Ma X., 2010, *Phys. Status Solidi*, **7**, 1991.

160. Howgate J. D., Hofstetter M., Schoell S. J., Schmid M., Schäfer S., Zizak I., Hable V., Greubel C., Dollinger G., Thalhammer S., Stutzmann M. and Sharp I. D., 2012, *Phys. Status Solidi*, **209**, n/a.

161. Grant J., Cunningham W., Blue a., O'Shea V., Vaitkus J., Gaubas E. and Rahman M., 2005, *Nucl. Instruments Methods Phys. Res. Sect. A Accel. Spectrometers, Detect. Assoc. Equip.*, **546**, 213.

162. Sellin P. J., Hoxley D., Lohstroh A., Simon A., Cunningham W., Rahman M., Vaitkus J. and Gaubas E., 2004, *Nucl. Instruments Methods Phys. Res. Sect. a-Accelerators Spectrometers Detect. Assoc. Equip.*, **531**, 82.

163. Vaitkus J., Gaubas E., Shirahama T., Sakai S., Wang T., Smith K. M. and Cunningham W., 2003, *Nucl. Instruments Methods Phys. Res. Sect. a-Accelerators Spectrometers Detect. Assoc. Equip.*, **514**, 141.

164. Binari S. C., Redwing J. M., Kelner G. and Kruppa W., 1997, *Electron. Lett.*, **33**, 242.

165. Daumiller I., Kirchner C., Kamp M., Ebeling K. J. and Kohn E., 1999, *IEEE Electron Device Lett.*, **20**, 448.

166. Guhel Y., Boudart B., Hoel V., Werquin M., Gaquiere C., De Jaeger J. C., Poisson M. A., Daumiller I. and Kohn E., 2002, *Microw. Opt. Technol. Lett.*, **34**, 4.

167. Karmalkar S. and Mishra U. K., 2001, *Solid. State. Electron.*, **45**, 1645.

168. Chow T. P. and Tyagi R., 1994, *IEEE Trans. Electron Devices*, **41**, 1481.

169. Mitrofanov O. and Manfra M., 2003, *Superlattices Microstruct.*, **34**, 33.

170. Vetury R., Zhang N. Q., Keller S. and Misha U. K., 2001, *IEEE Trans. Electron Devices*, **48**, 560.

171. Ambacher O., 1998, *J. Phys. D. Appl. Phys.*, **31**, 2653.

172. Medjdoub F., Sarazin N., Tordjman M., Magis M., di Forte-Poisson M. A., Knez M., Delos E., Gaquière C., Delage S. L. and Kohn E., 2007, *Electron. Lett.*, **43**, 691.

173. Gregušová D., Stoklas R., Čičo K., Heidelberger G., Marso M., Novák J. and Kordoš P., 2007, *Phys. status solidi*, **4**, 2720.

174. Maier D., Alomari M., Grandjean N., Carlin J. F., Diforte-Poisson M. A., Dua C., Chuvilin A., Troadec D., Gaquière C., Kaiser U., Delage S. L. and Kohn E., 2010, *IEEE Trans. Device Mater. Reliab.*, **10**, 427.

175. Luther B. P., Mohney S. E., Jackson T. N., Asif Khan M., Chen Q. and Yang J. W., 1997, *Appl. Phys. Lett.*, **70**, 57.

176. Smith L. L., Davis R. F., Kim M. J., Carpenter R. W. and Huang Y., 1996, *J. Mater. Res.*, **11**, 2257.

177. Van Daele B., Van Tendeloo G., Ruythooren W., Derluyn J., Leys M. R. and Germain M., 2005, *Appl. Phys. Lett.*, **87**, 061905.

178. Fan Z., Mohammad S. N., Kim W., Aktas O., Botchkarev A. E. and Morkoç H., 1996, *Appl. Phys. Lett.*, **68**, 1672.

179. Liu Q. Z. and Lau S. S., 1998, *Solid. State. Electron.*, **42**, 677.

180. Liu Q. Z., Yu L. S., Deng F., Lau S. S., Chen Q., Yang J. W. and Khan M. A., 1997, *Appl. Phys. Lett.*, **71**, 1658.

181. Ruvimov S., Liliental-Weber Z., Washburn J., Qiao D., Lau S. S. and Chu P. K., 1998, *Appl. Phys. Lett.*, **73**, 2582.

182. Taking S., 2012, AlN/GaN MOS-HEMTs Technology. University of Glasgow, 2012.

183. Jacobs B., 2004, Towards Integrated AlGaN/GaN Based X-Band High-Power Amplifiers. Technische Universiteit Eindhoven, 2004.

184. Cai S. J., Li R., Chen Y. L., Wong L., Wu W. G., Thomas S. G. and Wang K. L., 1998, *Electron. Lett.*, **34**, 2354.

185. Soltani A., BenMoussa A., Touati S., Hoël V., De Jaeger J.-C., Laureyns J., Cordier Y., Marhic C., Djouadi M. A. and Dua C., 2007, *Diam. Relat. Mater.*, **16**, 262.

186. Mohammed F. M., Wang L., Selvanathan D., Hu H. and Adesida I., 2005, *J. Vac. Sci. Technol. B Microelectron. Nanom. Struct.*, **23**, 2330.

187. Qin Z. X., Chen Z. Z., Tong Y. Z., Ding X. M., Hu X. D., Yu T. J. and Zhang G. Y., 2004, *Appl. Phys. A Mater. Sci. Process.*, **78**, 729.

188. Krämer M. C. J. C. M., 2006, Gallium Nitride-Based Microwave High-Power Heterostructure Field-Effect Transistors. Technische Universiteit Eindhoven, 2006.

189. Heo J.-G., Sung H.-K., Lim J.-W., Kim S.-K., Park W.-K. and Ko C.-G., 2010, *IEEE MTT-S Int. Microw. Symp.*, *IEEE*.

190. Chaturvedi N., Zeimer U., Würfl J. and Tränkle G., 2006, *Semicond. Sci. Technol.*, **21**, 175.

191. Buttari D., Chini A., Meneghesso G., Zanoni E., Moran B., Heikman S., Zhang N. Q., Shen L., Coffie R., DenBaars S. P. and Mishra U. K., 2002, *IEEE Electron Device Lett.*, **23**, 76.

192. Mohammed F. M., Wang L. and Adesida I., 2007, *J. Vac. Sci. Technol. B Microelectron. Nanom. Struct.*, **25**, 324.

193. Wang L., Mohammed F. M., Ofuonye B. and Adesida I., 2007, *Appl. Phys. Lett.*, **91**, 012113.

194. Sun Y. and Eastman L. F., 2006, *J. Vac. Sci. Technol. B Microelectron. Nanom. Struct.*, **24**, 2723.

195. Kang B. S., Kim S., La Roche J. R., Ren F., Fitch R. C., Gillespie J. K., Moser N., Jenkins T., Sewell J., Via D., Crespo A., Dabiran A. M., Chow P. P., Osinsky A. and Pearton S. J., 2004, *J. Vac. Sci. Technol. B Microelectron. Nanom. Struct.*, **22**, 2635.

196. Selvanathan D., Zhou L., Kumar V. and Adesida I., 2002, *Phys. status solidi*, **194**, 583.

197. Chu K. K., Murphy M. J., Burm J., Schaff W. J., Eastman L. F., Botchkarev A., Haipeng Tang and Morkoc H., 1997, *Compd. Semicond. 1997. Proc. IEEE Twenty-Fourth Int. Symp. Compd. Semicond.*, *IEEE*, pp. 427–430.

198. Lee C.-T. and Kao H.-W., 2000, *Appl. Phys. Lett.*, **76**, 2364.

199. Kumar V., Zhou L., Selvanathan D. and Adesida I., 2002, *J. Appl. Phys.*, **92**, 1712.

200. Hou M. and Senesky D. G., 2014, *Appl. Phys. Lett.*, **105**, 081905.

201. Selvanathan D., Mohammed F. M., Tesfayesus A. and Adesida I., 2004, *J. Vac. Sci. Technol. B Microelectron. Nanom. Struct.*, **22**, 2409.

202. Kampen T. U. and Mönch W., 1997, *Appl. Surf. Sci.*, **117-118**, 388.

203. Schmitz A. C., Ping A. T., Khan M. A., Chen Q., Yang J. W. and Adesida I., 1996, *Electron. Lett.*, **32**, 1832.

204. Arulkumaran S., Egawa T., Zhao G.-Y., Ishikawa H., Jimbo T. and Umeno M., 2000, *Jpn. J. Appl. Phys.*, **39**, L351.

205. Würfl J., Abrosimova V., Hilsenbeck J., Nebauer E., Rieger W. and Tränkle G., 1999, *Microelectron. Reliab.*, **39**, 1737.

206. Schmitz A. C., Ping A. T., Khan M. A. and Adesida I., 1995, *MRS Proc.*, **395**, 831.

207. Kalinina E. V., Kuznetsov N. I., Dmitriev V. A., Irvine K. G. and Carter C. H., 1996, *J. Electron. Mater.*, **25**, 831.
208. Eastman L. F., Tilak V., Smart J., Green B. M., Chumbes E. M., Dimitrov R., Kim H., Ambacher O. S., Weimann N., Prunty T., Murphy M., Schaff W. J. and Shealy J. R., 2001, *IEEE Trans. Electron Devices*, **48**, 479.
209. Walker D., Monroy E., Kung P., Wu J., Hamilton M., Sanchez F. J., Diaz J. and Razeghi M., 1999, *Appl. Phys. Lett.*, **74**, 762.
210. Binari S. C., Ikossi K., Roussos J. A., Kruppa W., Park D., Dietrich H. B., Koleske D. D., Wickenden A. E. and Henry R. L., 2001, *IEEE Trans. Electron Devices*, **48**, 465.
211. Meneghesso G., Meneghini M., Stocco a., Bisi D., de Santi C., Rossetto I., Zanandrea a., Rampazzo F. and Zanoni E., 2013, *Microelectron. Eng.*, **109**, 257.
212. Cheney D. J., Douglas E. A., Liu L., Lo C. F., Xi Y. Y., Gila B. P., Ren F., Horton D., Law M. E., Smith D. J. and Pearton S. J., 2013, *Semicond. Sci. Technol.*, **28**, 074019.
213. Piazza M., Dua C., Oualli M., Morvan E., Carisetti D. and Wyczisk F., 2009, *Microelectron. Reliab.*, **49**, 1222.
214. Herfurth P., Maier D., Lugani L., Carlin J.-F., Rosch R., Men Y., Grandjean N. and Kohn E., 2013, *IEEE Electron Device Lett.*, **34**, 496.
215. Guo J. D., Pan F. M., Feng M. S., Guo R. J., Chou P. F. and Chang C. Y., 1996, *J. Appl. Phys.*, **80**, 1623.
216. Venugopalan H. S., Mohney S. E., Luther B. P., Wolter S. D. and Redwing J. M., 1997, *J. Appl. Phys.*, **82**, 650.
217. Liu Q. Z., Yu L. S., Deng F., Lau S. S. and Redwing J. M., 1998, *J. Appl. Phys.*, **84**, 881.
218. Arulkumaran S., Egawa T., Ishikawa H., Umeno M. and Jimbo T., 2001, *IEEE Trans. Electron Devices*, **48**, 573.
219. Kim J., Ren F., Baca A. G. and Pearton S. J., 2003, *Appl. Phys. Lett.*, **82**, 3263.
220. Venugopalan H. S., Mohney S. E., DeLucca J. M. and Molnar R. J., 1999, *Semicond. Sci. Technol.*, **14**, 757.
221. Wu Y., Chen C.-Y. and del Alamo J. A., 2015, *J. Appl. Phys.*, **117**, 025707.
222. Hosch M., Pomeroy J. W., Sarua A., Kuball M., Jung H. and Schumacher H., 2009, *CS Mantech*.
223. Joh J., Xia L. and del Alamo J. A., 2007, *IEEE Int. Electron Devices Meet.*, *IEEE*, pp. 385–388.

Chapter 13

Emerging Nanotechnology for Strain Gauge Sensor

Sheng-Po Fang, Todd Schumann, Lisdelys Garcia, and
Yong-Kyu Yoon

Electrical and Computer Engineering
University of Florida, Gainesville, FL 32611, USA

13.1 Introduction

13.1.1. Nanotechnology

Nanotechnology has become a popular subject recently with the ability to precisely manipulate materials below a few hundred nanometers in a single dimension. It has opened up numerous possibilities for the future to manufacture lighter, stronger, and programmable materials that require less energy to operate or manufacture than conventional bulky materials, possibly producing less waste than with conventional manufacturing. It also promises greater fuel efficiency in land transportation, ships, aircraft, and space vehicles. Figure 1 shows various potential applications of nanotechnology [1]. Nanocoatings for both opaque and transparent surfaces may render them resistant to corrosion, scratches, and radiation. Nanoscale electronic, magnetic, and mechanical devices and systems with unprecedented levels of information processing may be fabricated, and so may chemical, photochemical, and biological sensors for protection, health care, manufacturing, and the environment; new photoelectric materials that will enable the manufacture to fabricate efficient solar-energy panels; and molecular-semiconductor hybrid devices that may become engines for the next step in the information age. The potential for improvements in health, safety, quality of life, and conservation of the environment is vast. Meantime, fabrication difficulties in some nanomaterials and high manufacturing cost still remain as major barriers for the commercial usage of nanotechnology. In fact, the USA has invested

435

3.7 billion dollars, the European Union 1.2 billion, and Japan 750 million dollars through its National Nanotechnology Initiative until 2012 [2] to improve nanotechnology and its production.

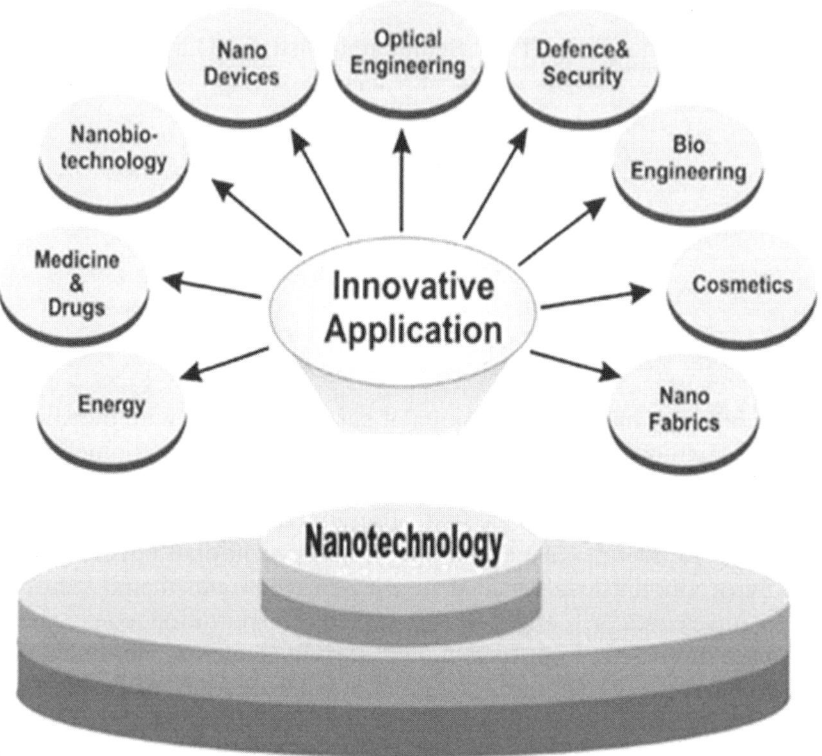

Figure 1. Various applications of nanotechnology [1].

One interesting phenomenon of nanotechnology is that materials at the nanoscale can have different qualitative properties. For example, quantum mechanical effects come to play at very small dimensions and lead to new physics and chemistry. Additionally, typical nanomaterials have a very large surface-to-volume ratio, which means that no atom is very far from a surface or interface, and the behavior of atoms at these higher-energy sites is highly influenced by the surface physics.

Quantum mechanical properties for confinement of electrons in one dimension have long been exploited in solid-state electronics. Semiconductor devices are grown with thin layers of differing composition so that electrons can be confined in specific regions of the quantum wells. Thin layers with larger energy bandgaps can serve as barriers that restrict the flow of charges to certain conditions under which they can "tunnel" through these barriers—the basis of resonant tunneling diodes. Superlattices are periodic structures of repeating wells governed by a new set of selection rules which affect the conditions for charges to flow through the structures. Superlattices have been exploited in cascade lasers to achieve far infrared wavelengths. Modern telecommunications would be based on semiconductor lasers that exploit the unique properties of quantum wells to achieve specific wavelengths and high efficiency.

The high surface-to-volume ratio is desired whenever surface chemistry needs to play a role in the performance of a device. This is most prevalently seen in chemical sensors where an ambient gas or liquid must interact with the sensor's active area to produce an electrical readout. The higher the surface-to-volume ratio, the more fluid can influence the sensor per unit measurement volume. The result is that the sensor can now be made smaller, and more importantly the sensor can have a much higher sensitivity. This allows the sensor to measure much lower concentrations of fluids and maintain an accurate readout. For example, one may detect dangerous hydrocarbons much sooner than they reach critical levels or one may use saliva for "bloodwork" instead of patient's blood.

To take advantage of nanotechnology, efficient nanofabrication methods that can manipulate nanomaterials and integrate them into micro- and macroscale systems are required. Typically fabrication approaches are categorized as the 'top-down', 'bottom-up', and intermediate approaches. Figure 2 shows the three fabrication schemes [3]. The top-down approach is analogous to removing materials from a larger mass. Examples of this kind of approach include the various types of lithographic techniques such as photo-, ion beam-, electron- or X-ray- lithography for cutting, etching and grinding. Lithography is a selective process that allows the patterning of a desired design onto the starting material. A mask is used to protect the areas of material that need to be maintained, while the rest of the material is exposed to a beam of UV, ions, electrons or X-rays. Etching and/or deposition on the surface is then performed, leaving the desired shape. On the other hand, the bottom-up approach starts with atoms or molecules to form nanomaterials. Typical fabrication techniques include chemical vapor deposition (CVD), atomic layer deposition (ALD), physical deposition, and electrodeposition. The intermediate approach is a kind of hybrid approach. These techniques have been used to produce compact and functional electronic components, such as memory chips, micro-electromechanical systems (MEMS) devices in consumer products, computer hard drives, sensors and actuators, and CD and DVD players.

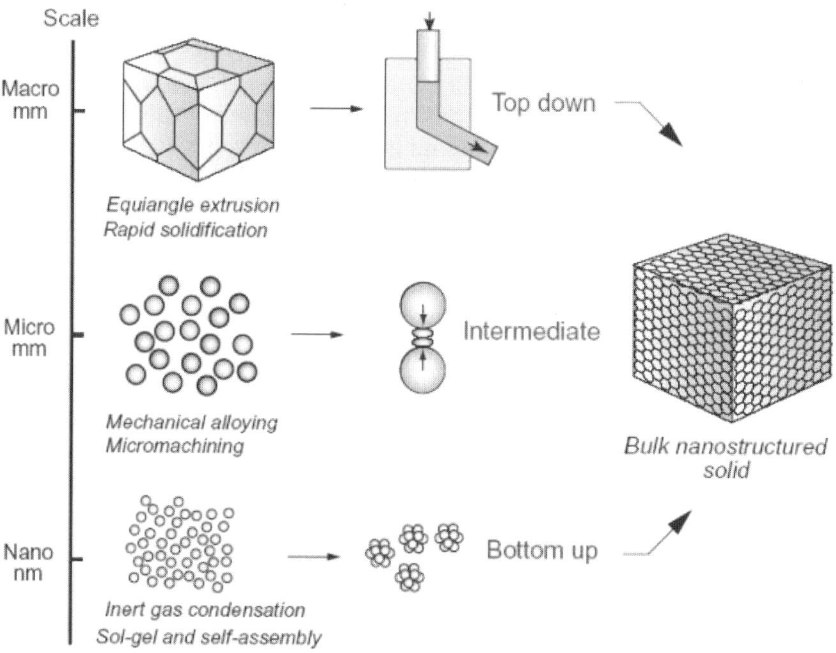

Figure 2. The three fabrication schemes, top-down, intermediate, and bottom-up approaches [3].

The impact of nanotechnology processes, machines, and products is expected to be far-reaching, affecting nearly every conceivable information technology, energy source, agricultural product, medical and pharmaceutical device, and material used in manufacturing. For example, the dimensions of electronic circuits on semiconductors continue to shrink, with minimum feature sizes now reaching the nano realm, under 100 nanometers. Likewise, magnetic memory materials, which form the basis of hard disk drives, have achieved a dramatically greater memory density as a result of nanoscale structuring to exploit new magnetic effects at nano dimensions. It is needless to say that there are immense markets taking advantage of nanomaterials and nanomanufacturing.

13.1.2. Strain Gauge Sensor

A strain gauge sensor is defined as a sensor whose electrical resistance varies with an applied force. Such a sensor converts the change of mechanical pressure or strain into electrical resistance for external electronics to recognize. When external forces are applied to a stationary object, stress and strain are resulted. Stress is defined as the object's internal resisting force responding to an applied external force, and strain is defined as the displacement and deformation that occur. In other words, if an object is experiencing an external force, it internally generates a repelling force to maintain the original shape. The repelling force is called the internal force and stress is defined as the internal force divided by the cross-sectional area of the object. Stress is expressed as a unit of Pa (Pascal) or N/m^2. On the other hand, strain is defined as the ratio of object's elongation to the original length.

In 1937, Edward Simmons of Caltech discovered that a wire with a small diameter could be used to measure surface strain by measuring its change in electrical resistance. A year later, Arthur Ruth from MIT also discovered the same phenomenon independently. Upon learning of Simmons work, Ruge worked with Simmons together for filing the patent. Later, Ruge went on to start a business with another MIT professor, Alfred deForest, to manufacture the strain gauge sensor, which was the first commercially available sensor that provides an accurate weighting scale.

The strain gauge sensor is still considered one of the most important sensors for precise mechanical quantity measurements. Strain gauge sensors utilize the piezoresistive sensing mechanism, which translates a mechanical displacement into an electrical signal, and are useful for monitoring minute structural deformations in flexible supporting layers over time. In this case, the change in electrical resistance can be measured from a specific active matrix with or without an on/off switchable transistor.

The piezoresistivity comes from both the macro- and the nano-scale. In the macro-scale, the stress in the device causes the structural deformation of the device, the strain. However, the strain itself is a reduction of the cross-sectional area and an increase in the length of the device. As a force is applied to the top of a soft rubber cube, the cube shrinks in height and bulges outwards on the lateral sides. The left-right bulge is equivalent to an increase in the length of the device. The height shrinkage as well as the front-back bulging counteract each other, but the overall cross sectional area will decrease. Assuming the resistivity of the rubber stays the same, the equation for the electrical resistance is:

$$R = \frac{\rho L}{A} \tag{1}$$

where ρ is the electrical resistivity, and A and L are the cross sectional area and length, respectively. As the length increases and the cross sectional area decreases, the overall resistance will increase. The same effect will happen if a tensile force is applied to the cube in the left-right direction.

In the nano-scale, the stress in the device is caused by a shifting of the atoms in the lattice structure. Let's consider a cube made up of a lattice of balls, 5 x 5 x 5, all connected with springs. As a force is applied to the cube, the balls, which represent atoms, shift around to compensate the stress to maintain the general shape of the device. As the atoms shift, the electron wave-functions change, which changes the band structure and energy levels in the device. The overall result is a change in the resistivity of the device, independent of any macro-scale physical change. The combination of the macro- and nano-scale effects caused by an external force on the device results in an overall absolute resistance change, which can be electrically read out.

Figure 3 shows the schematic of the bonded strain gauge sensor which was originally proposed by Dr. Arthur Ruth of MIT. The bond strain gauge consists of a thin strip of metallic film packaged in a non-conducting flexible material. Since the metallic film has piezoresistive property, the electrical resistance changes according to both tensile and compressive forces. Please note that the sensor is more sensitive to a single directional mechanical force since the length of metallic trace along the other direction is relatively shorter than in the designed direction.

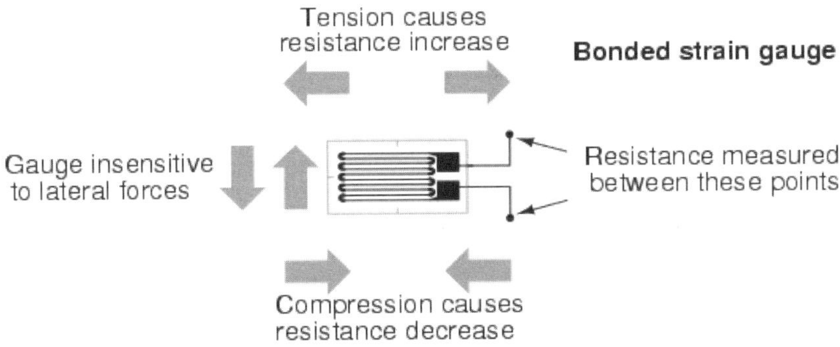

Figure 3. Schematic of bonded strain gauge.

A typical strain gauge resistance ranges from 30 Ω to 3 kΩ when unstressed and changes only a fraction of a percent when fully extended depending on the elastic limits of the gauge material and specimen. The sensitivity of the strain gauge sensor is called the gauge factor (GF), which is given in Eq. 2.

$$GF = \frac{\frac{\Delta R}{R}}{\frac{\Delta L}{L_0}} \qquad (2)$$

where ΔR is the change in strained resistance, R is the unstrained resistance, ΔL is the absolute change in length, and L_0 is the original length. Typical metallic strain gauge sensors have a gauge factor of about 2-5.

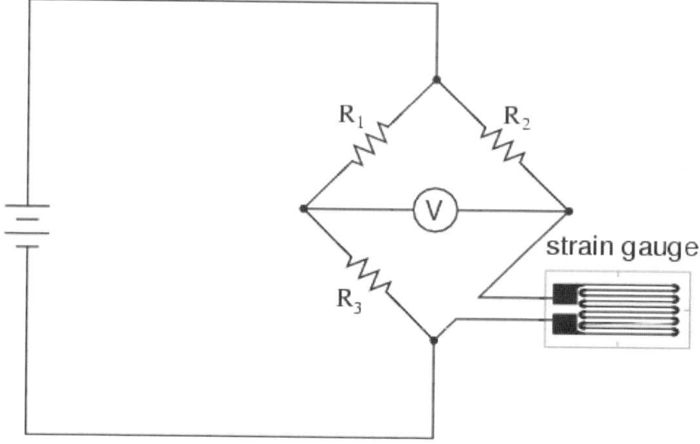

Figure 4. Wheatstone bridge configuration.

A Wheatstone bridge is typically used as the external detection circuitry to get high accuracy measurements. The Wheatstone bridge is configured as shown in Figure 4 where the arm resistances are balanced, producing zero volt reading when no strain is applied. As an external force is applied to the sensor, the voltage reading increases or decreases. Since the temperature of a device will change the resistance, the Wheatstone bridge can "drift" as the temperature changes. However, temperature compensation can be achieved by replacing R2 with an unstressed gauge meter identical to the one used in the lower right of the circuit.

Recently, technology trends have been focused on smart wearables, which requires flexible, stretchable and wearable sensors and electronics that can be easily integrated on clothing and apparel, or directly attached on the skin. Motion detection has been identified as one of the most popular applications, which requires highly stretchable and sensitive sensors. Figure 5 shows the schematic of a strain gauge sensor attached on the elbow as an example of smart wearables. Many research efforts have been exerted to explore nanotechnological approaches for the next generation of smart wearables. Since the strain gauge sensors must already be flexible due to their nature, they are the ideal choice for mechanical sensors for wearable applications.

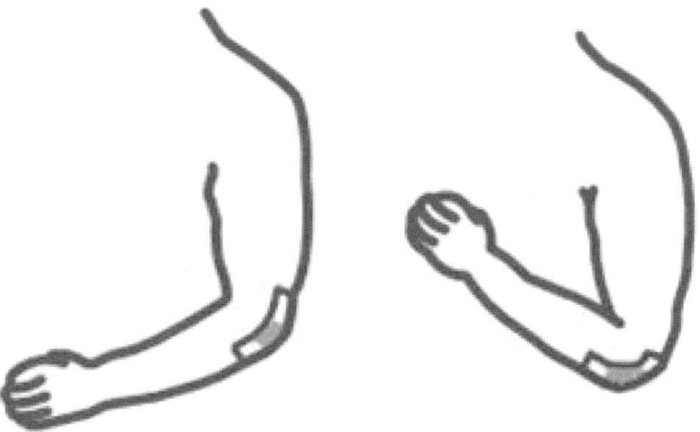

Figure 5. Schematic of the strain gauge sensor for smart wearables.

13.2. Emerging Nano Technologies

13.2.1. Nanomaterial Graphene Nanoribbon

High performance, lightweight, foldable and stretchable technologies would enable many new applications, particularly in the biomedical disciplines where personal health devices might require extensive deformation for installation and/or daily use. One of the popular nanomaterials is Graphene, which is a two dimensional monolayer of carbon atoms arranged into a hexagonal lattice with excellent electron transport performance and good mechanical elasticity. In 2011, M. Huang et al., demonstrated a suspended graphene ribbon with a linear resistive change [4]. Figure 6(top) shows the schematic of a graphene ribbon suspended by two contacts. The relative change in resistance is extracted according to the applied stain. Figure 6(bottom) shows the measurement result of the suspended graphene device, which shows proportional resistive changes to strain. A gauge factor of 1.9 is extracted from the curve. Figure 7 shows the scanning electron microscope images of the fabricated device.

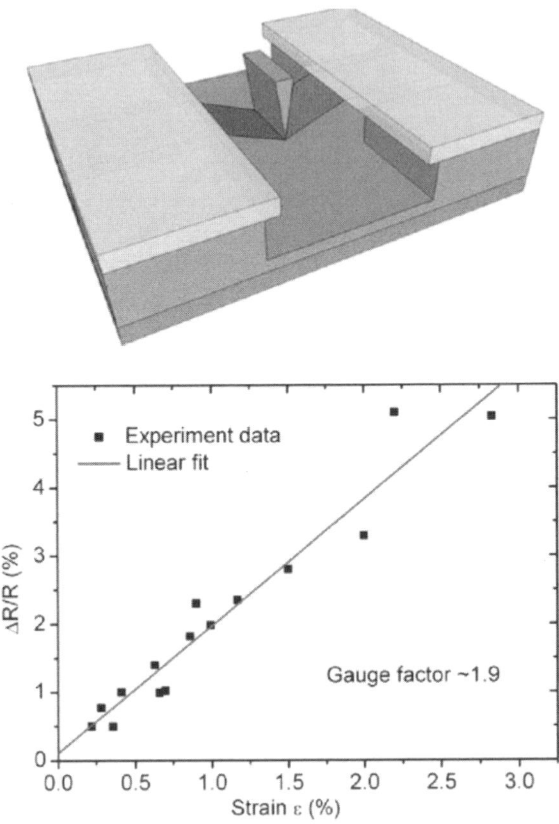

Figure 6. Schematic of a suspended graphene ribbon for a strain gauge sensor (left) and resistance change as a function of strain (right) [4].

As for graphene, the macro- and nano-scale causes of resistance changes are the same, being that it's a truly-2D material. As a point force is applied to the center of the ribbon, the length increases and the width shrinks, though this is almost entirely realized through atomic interaction. Graphene is a semiconductor with 0 eV bandgap. The hope was that the induced strain would cause the bandgap to open, which would create an exponential change in resistivity resulting in a high resistivity. However, it turned out that the bandgap did not open so that the resistance change is entirely piezoresistive, which was not significant compared to other strain gauge sensors.

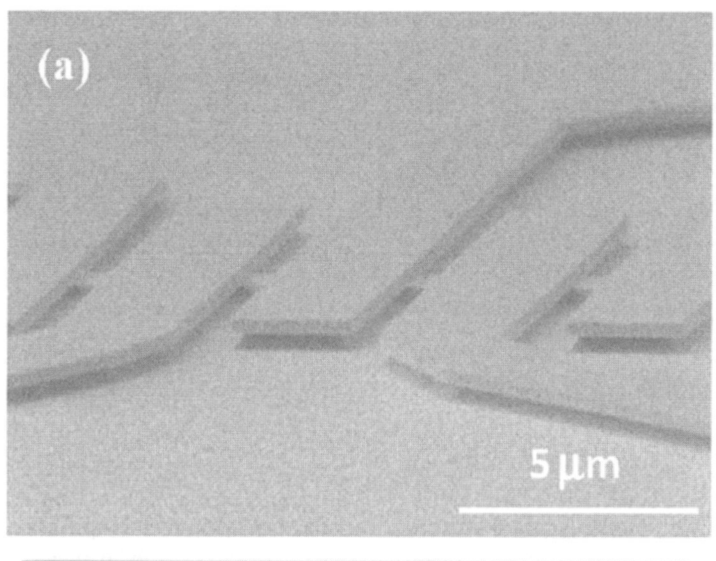

Figure 7. SEM micrographs of the fabricated devices with the suspended graphene nanoribbons [4].

13.2.2. Nanocomposite: Graphene/PVDF

Graphene possesses great electrical properties, but the pure intrinsic form did not give desired performance for a strain gauge sensor. A nanocomposite of graphene was proposed, which combines the electrically high conductivity nanomaterial with flexible polymer to achieve both good electrical conductivity and flexibility. In 2012, V. Eswaraiah et al., reported a nanocomposite of graphene (reduced graphite oxide) with polyvinylidene fluoride (PVDF) to achieve a high strain gauge sensor [5].

Graphene's single layer of sp2-hybridized carbon atoms are arranged in a honey comb lattice producing a high level of electron mobility due to its free π electron of each atom. Strain has been shown to modulate graphene's electronic, magnetic and transport properties by distorting the lattice shape from the regular hexagon as the sp2 bond length cannot change. It is well known that the high surface area and energy alongside the Van der Waals forces and exposed π-π bonds cause graphene powder to tend to aggregate making dispersion in a polymer time-consuming and tedious, thereby reducing the chance of its usage for industrial applications. Thus, synthesis of graphene as thin as possible and dispersed in a homogenous fashion has been a priority in the preparation of electrically conductive graphene composites.

A method of reducing graphite oxide in polymer powered using solar radiation has been demonstrated by Eswaraiah et al. [5]. This procedure simultaneously reduces graphite oxide, melts the polymer and embeds the reduced graphite oxide nanoflakes in the polymer within 5 minutes. The key within the process is the melting point of PVDF, which is 166°C, significantly lower than the temperature of focused solar radiation at 200°C, easily melting the polymer while reducing and exfoliating graphite oxide. The procedure is simple, time efficient and entirely via a dry route with a short list of materials consisting of graphite oxide, PVDF powder, converging lenses and solar radiation.

Both graphite oxide and PVDF are insulating materials. Graphite oxide is mixed with PVDF at a series of different wt% ranging from 0 to 7%. The mixture is then treated with focused solar radiation which simultaneously melts the polymer, reduces and exfoliates PVDF-graphite oxide, drastically reducing the electrical resistivity. Figure 8a) shows the electrical conductivity of the graphene-PVDF composites from different concentrations of graphite oxide. After then, the composites are subjected to a melt press to make thin films out of them. The high and low magnifications of field emission scanning electron microscope (FEI-SEM) images are shown in Figure 9, which clearly demonstrates the uniform dispersion of reduced graphite oxide within the polymer matrix with limited segregation. Advantageous features that this method shows include homogeneity, and an increase of temperature ($\sim$35°C) at the onset decomposition temperature of reduced graphite oxide, thereby increasing the thermal stability of the compositions.

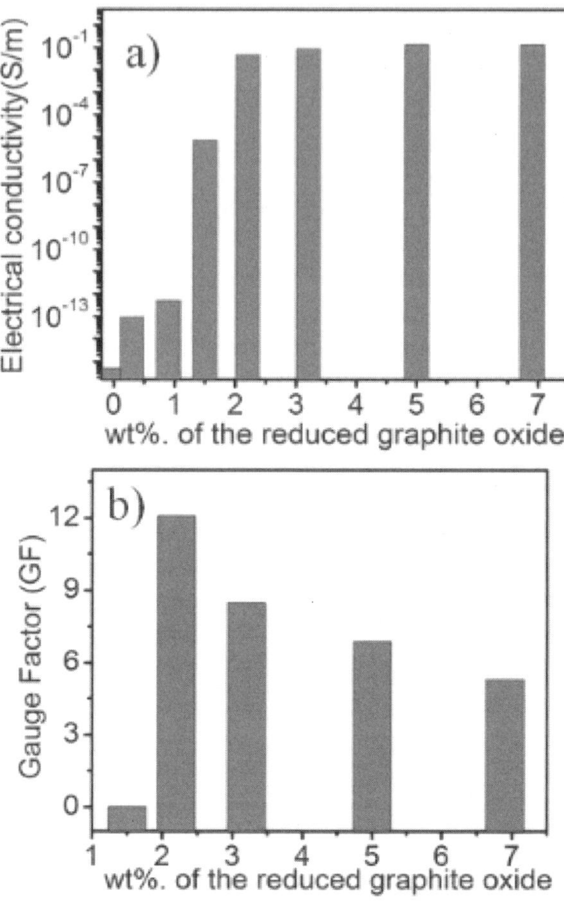

Figure 8. a) Electrical conductivity of various reduced graphite oxide weight fractions in PVDF and b) gauge factors of corresponding composites [5].

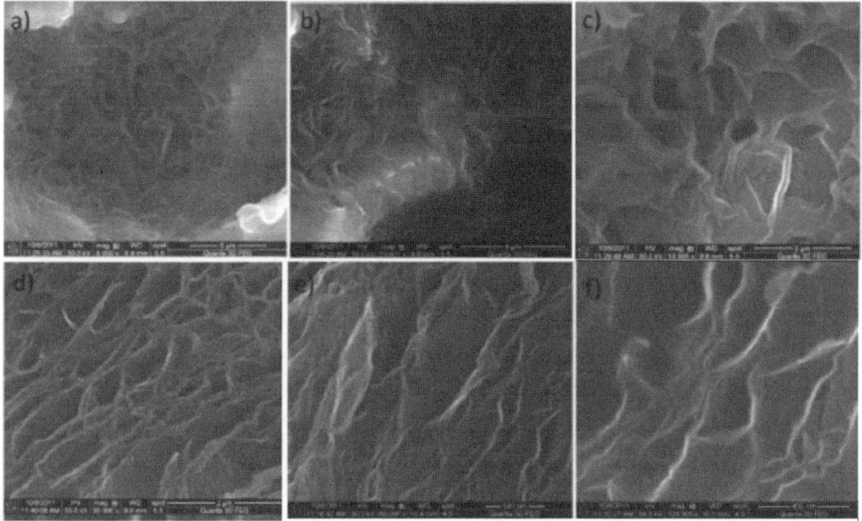

Figure 9. Field emission scanning electron micrographs of 2.2wt% reduced graphite oxide (8000, 13 607, 16 885, 60 000 and 120 000x) [5].

The gauge factors are extracted from each graphite oxide concentration. The extracted result is shown in Figure 8b). The optimal concentration of graphite oxide was determined to be 2.2 wt% to produce the maximum gauge factor of 12.1 with good electrical conductivity. With higher concentrations of graphite oxide in the PVDF matrix, the applied strain is attenuated by adjacent fillers due to the saturation of the conducting fillers. When a sufficient concentration of nanofiller was observed, percolation in electrical conductivity was obtained. The measured relative resistance change versus applied strain is shown in Figure 10c, which is approximately 6 times higher than one found in literature on a single graphite oxide strain sensor with a gauges factor of around 2. Figure 10d-f show the device loading force, strain level, and relative resistance change, respectively.

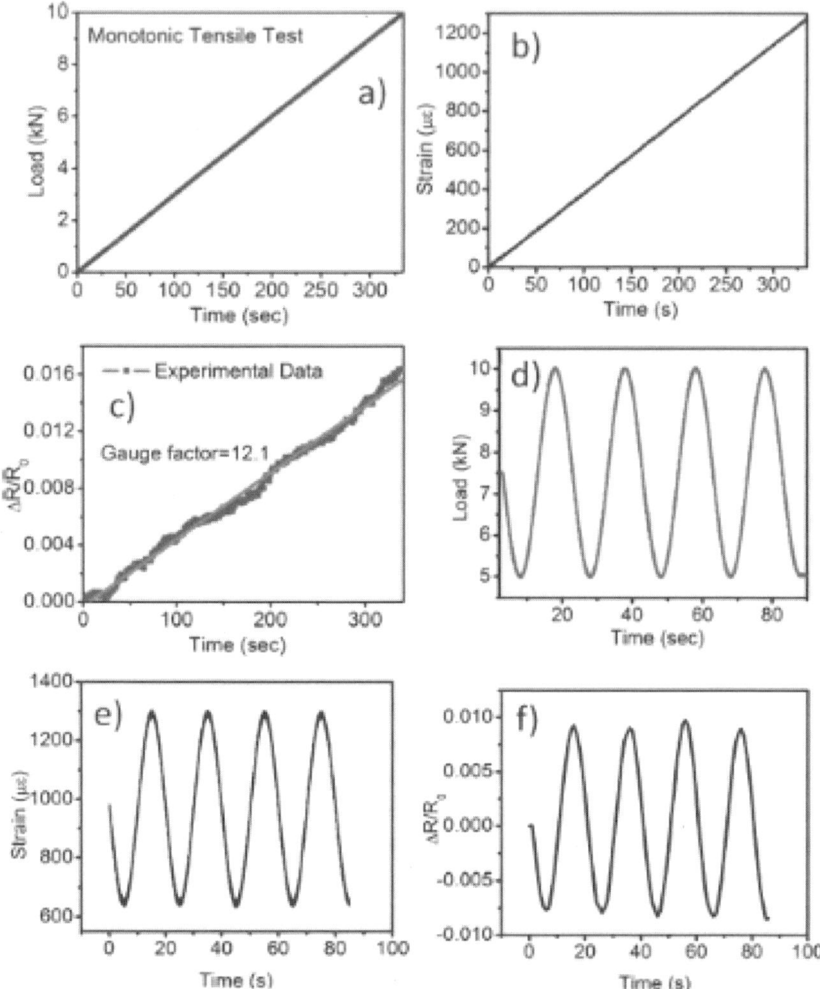

Figure 10. (a-c) monotonic test over 2.2wt% reduced graphite oxide- PVDF film of strain and change in resistance of composite film (d-f). Tensile compressive test from 5-10 kN [5].

This sensor shows the nanocomposite concept to take advantage of both graphene and elastomer materials. Since graphene did not produce the desired electro-mechanical performance, it was embedded in a polymer matrix. As a result, the nanocomposite film has shown much better electro-mechanical properties compared to a pure graphene film.

13.2.3. Carbon Nanotube Strain Sensor

Although carbon nanotubes are currently very expensive to manufacture, they can still produce novel devices which can outperform the current market standards. Additionally, with further research, the manufacturing techniques can be improved and the overall cost may decrease as with every existing technology.

One such carbon nanotube strain gauge sensor was proposed by Takeo Yamada, et al in 2011 [6]. Figure 11 schematically shows the fabrication process of the carbon nanotube strain gauge sensor. First, vertically aligned single wall carbon nanotube (SWCNT) thin films with a height of 1 mm, a thickness of 6 mm, and a length of 16 mm were grown from patterned catalysts using water-assisted chemical vapor deposition. Second, SWCNT films were individually assembled with 1 mm overlap on a 1 mm thick flat elastomeric dog-bone-shaped polydimethylsiloxane (PDMS) substrate, with the alignment of the SWCNT films arranged perpendicular to the strain axis. The film was wet with a droplet of isopropyl alcohol, which resulted in the development of a strong van der Waals contact between SWCNT and PDMS substrate without any additional mechanical pressure. The adhesion strength was measured as ~12 N cm^{-2} and was sufficient to bear large strain.

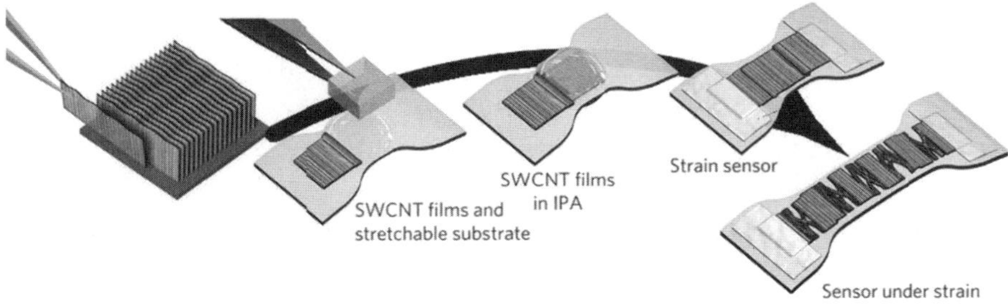

Figure 11. Schematic of the fabrication and working principle of the CNT strain gauge sensors [6].

The Van der Waals forces keep the CNTs aligned and in contact with each other even when the bottom substrate is flexed and stretched. Instead of tearing, as a traditional film would, the carbon nanotube film splits along the nanotube boundaries, where it can easily rejoin. The CNT film becomes porous with these so-called gaps where the nanotubes stretch the length. Figure 12A-B shows the SEM image of homogeneous SWCNT film. An enlarged SEM image of the suspended SWCNT bundle is shown in Figure 12C. In this device, the traditional strain is not measured, as the individual carbon nanotubes are not being strained. However, the device realigns the nanotubes to change the resistivity of the whole film. The so-called islands in the film do not change in resistance when the film is strained. However, in the gaps, the carbon nanotubes are now stretching across instead of up-down, which forces the electric field to span across the length of the CNT instead of just across the width. As the film is stretched more and more, the ratio of gaps to islands increases causing the resistance to increase. Figure 12D shows the distance of the island width and gap width as a function of the different applied strain. The working principle of this architecture is presented using paper, which is shown in Figure 12E.

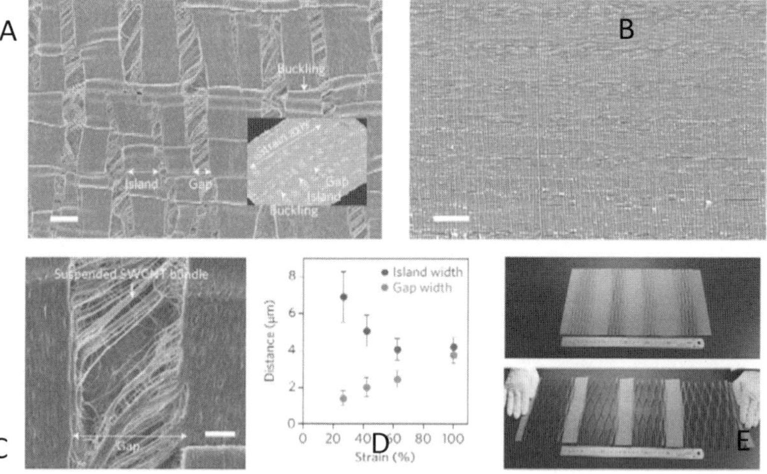

Figure 12. (A-B) SEM images of the island and gap regions showing the CNTs stretching across the gaps. (C) Enlarged SEM image of suspended SWCNT bundles. (D) Graph showing the average widths of the island and gap regions versus the applied strain. {E) Working principle demonstrated with a paper model [6].

Figure 13A shows the cycle tests of the relative change of electrical resistance as a function of strain. The sensor shows good repeatability below 100 % strain, and the electrical response remains nearly unchanged up to 10,000 cycles. However, in high strain, the sensors 2 and 3 show increased relative resistances after cycles by 5 and 10 %, respectively. The gauge factors of these strain gauges are extracted to be 0.82 and 0.02 for an operation strain of less than 40 % and above 60 %, respectively. However, due to the unique operating principle, the strain gauge is capable of measuring a strain as high as 280 %, which is approximately two orders of magnitude higher than that of pure graphene. The result suggests the nanocomposite strain gauge sensor is suitable for wearable applications, which require a high strain range. In addition, since the CNTs themselves undergo very little strain, the sensor is extremely durable with very little drifting over 10,000 cycles.

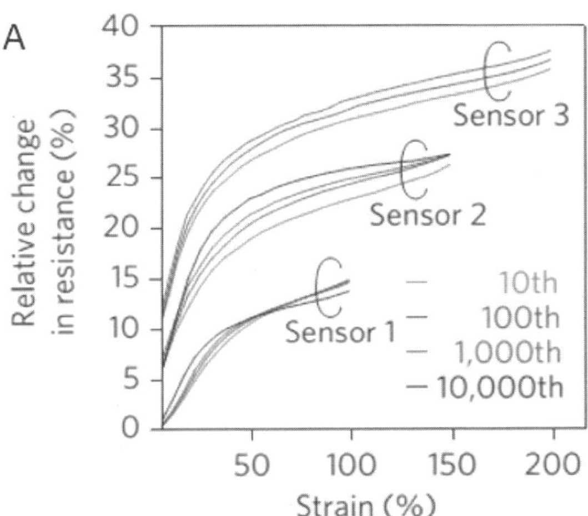

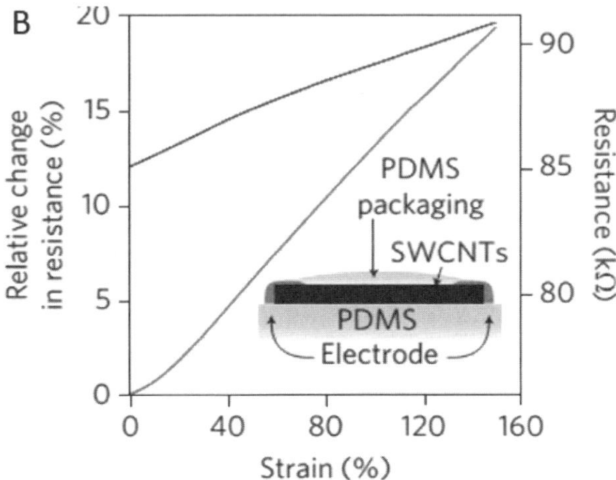

Figure 13. (A) Resistance response to applied strain of three sensors across 10,000 cycles,10 (red), 100 (blue), 1,000 (green) and 10,000 (black) cycles at 5 – 100 % (sensor 1), 5 – 150 % (sensor 2) and 5 – 200% (sensor 3) strain. (B) Difference between the initial (red) and subsequent (blue) cycles showing the break-in required [6].

13.2.4. Printable Silver Nanowire Composite

A cost-effective manufacturing method of nanocomposites was demonstrated by Lee et al., using the direct inkjet printing technique [7]. Figure 14 shows the fabrication process of the silver nanowire composite strain gauge sensor. The traditional plastic ink is replaced with silver nanoparticles dispersed in water. The ink is then printed onto a silicon wafer and then uniformly covered with a PDMS mixture solution. Once PDMS is heated and cured, it is peeled off along with the printed silver nanowire from the Si substrate. The resulting sensor is the flexible nanocomposite of the silver nanowires embedded in PDMS. Therefore, a low cost and high manufacturability nanocomposite strain gauge sensor can be realized using this method.

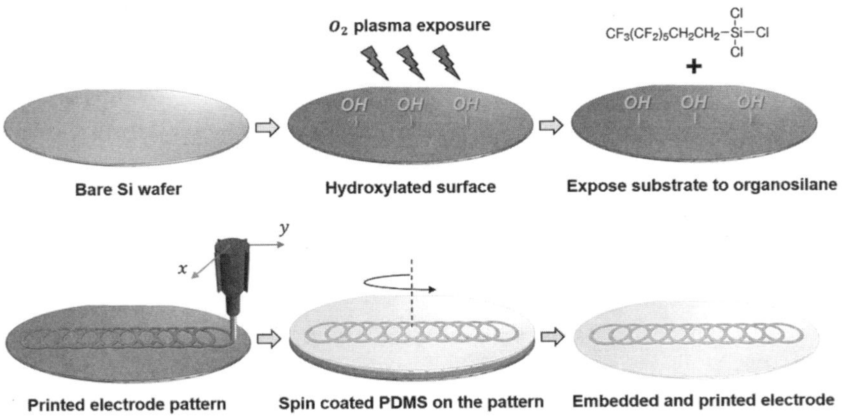

Figure 14. Fabrication of the silver nanowire composite strain gauge sensor [7].

Figure 15 shows the resistance response as a function of applied strain for different silver nanowire patterns. Both circular and diamond patterns show linear responses, however, some hysteresis is observed. These sensors produce a reasonable gauge factor of around 2-3, but their ingenuity lies in their simple and cost-effective fabrication process. The cost of other strain gauge sensors can be very expensive, especially when accounting for equipment and processing cost. However, these sensors can be made as needed, for any application. Their dimensions can be adjusted from sensor to sensor with an easy change of program instead of an expensive lithography mask change.

13.3. Electrospun Nanofiber

One of the unique ways to produce nanomaterials is electrospinning, which is a simple, costeffective way capable of producing nanoporous materials and nanoscale fiber mats. So the process becomes very popular in nanofabrication recent days. Figure 16A shows the trend of the number of publications on nanofibers in the last two decades, rapidly increasing in past several years and exceeding 12,000 in 2014. Figure 16B shows the trend of the number of patents on nanofibers in the last two decades, which also shows significant increase. Around a thousand patents regarding nanofibers are filed every year after 2010, and the numbers are continuously growing. Electrospinning has become a well-known fabrication method to prepare continuous fibers with diameters ranging from tens nanometers to several micrometers.

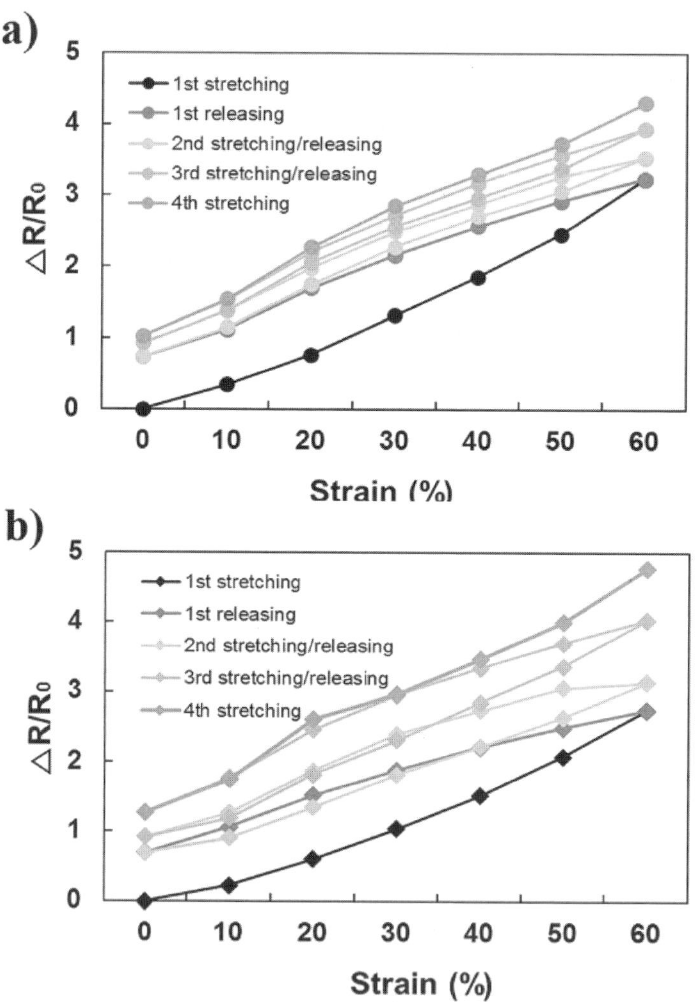

Figure 15. Resistance change versus applied strain for circular (a) and diamond (b) electrode patterns for several test cycles [7].

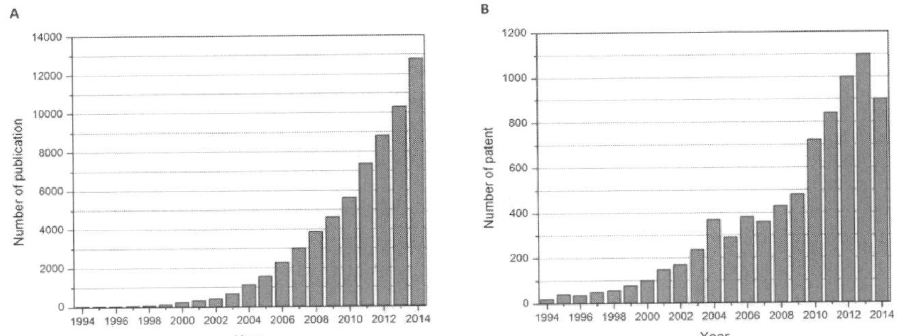

Figure 16. Number of (A) Publications and (B) Patents on "Nanofibers" from 1994 to 2014 (Source: Google Scholar accessed).

Figure 17 shows a conventional electrospinning setup which consists of a mechanical syringe pump, a syringe with a metallic needle, a high voltage source, and a metallic collector plate. The process electrostatically charges a polymer solution to a high electrical potential, dispensing the charged polymer solution with electrostatic force, and collecting produced fibers at a grounded collector. The highly charged solution intrinsically splits into very fine and long fibers due to the electrostatic repulsion force during its flights from the tip of the needle to the collector plate. The electrospinning process is described in detail as followings. First, the polymer solution is loaded into a syringe with a hypodermic needle and pumped out at a constant flow rate using a syringe pump. Second, the needle tip is connected to the positive terminal of the high voltage supply, and a grounded metallic plate is place at certain distance away from the needle tip. As the applied voltage exceeds a critical voltage Vc, the polymer droplet at the needle tip forms a conical shape, the so called Taylor cone, and ejects a jet. The critical voltage Vc can be calculated from the following expression [8].

$$V_n > V_c = \sqrt{4\frac{H^2}{L^2}\left(\ln\frac{2L}{R} - 1.5\right)(0.18\cos\theta\pi R\gamma)} \text{ kV} \qquad (3)$$

where V_C, H, L, R, θ, and γ parameters are the critical voltage in kV, the tip-to-collector distance, the length of needle, the inner radius of needle, the Taylor's cone angle (49.3°), and the surface tension of polymer, respectively. When the electrostatic force exceeds the surface tension of polymer at the need tip, a charged jet of the solution is ejected from the Taylor cone to the metallic collector. The ejected polymer jet undergoes a whipping process travelling toward the grounded metallic plate. As the polymer jet travels, it continues to elongate but shrink in diameter due to its solvent evaporation. The jet experiences a bending instability at a critical point between the needle tip and collector where the jet starts whipping in multiple directions in a chaotic fashion [9]. As a result, a web of non-woven solid polymer fiber is produced on the collector.

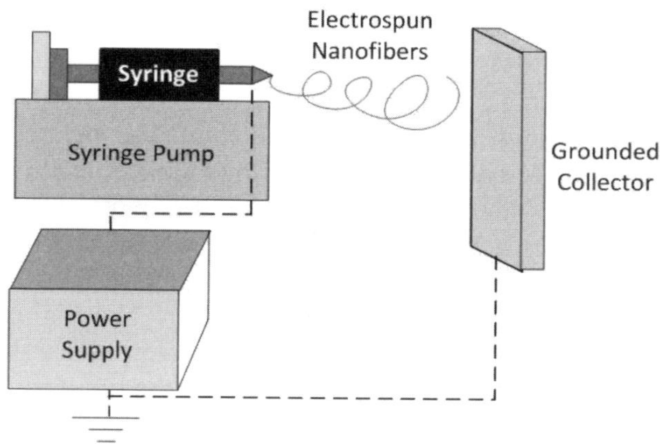

Figure 17. A typical electrospinning setup.

A typical electrospinning setup produces randomly orientated nanofibers due to the inherent whipping process. The challenge of controlling the spatial orientation of electrospun nanofibers has been overcome with some successes. Some researchers have conducted different electrospinning schemes to obtain aligned electrospun nanofibers. Li et al. have demonstrated aligned nanofibers by carefully designing the collector with a small gap [10]. The schematic setup and calculated electric field strength are shown in Figure 18A and 18B, respectively. Lee et al. have studied the effect of the mechanically moving collector, which is shown in Figure 18C. The highly orientated nanofibers can be collected by carefully controlling the speed of the rotating collector [11]. Recently, Jao et al. have demonstrated a 4 collector electrospinning setup with an alternating offset voltage scheme between two electrode pairs to achieve orthogonally aligned nanofibers [12]. The schematic is shown in Figure 18D.

Electrospinning can be used to form nanoscale fibers, microscale spherical beads, or their combination, which result from various conditions of the polymer concentration, electric field strength, tip-to-collector distance, and type of solvent etc. The diameters of nanofibers and spherical beads are in the range of a few tens nanometers to a few micrometers, which can find much usefulness in diverse disciplines due to their large surface-to-volume ratio and nano-/micro porosity [13-14]. Applications of electrospun nanofibers include energy storage devices [15], sensors and actuators [16], and scaffolds for tissue regeneration [17]. Previously, multiple nonlithographic patterning techniques such as mechanical nanofiber patterning [12], laser machine patterning [18], and hydrogel template based patterning [19] have been reported. These techniques are useful for applications which require large nanofibrous structures ranging from few hundred micron to a few millimeter scale. However, patterning nanofibrous structures with a few micrometer resolution still remains as a challenge. In 2009, Shi et al. demonstrated a technique called indirect photolithographical patterning [20] to obtain nanofibrous structures within microscale resolution. First, a nanofiber stack is electrospun on the substrate. Second, a thin layer of photoresist is spin coated on top of the nanofiber stack. Then, the photoresist is patterned by UV photolithography and subsequent reactive ion etching (RIE) of the stack. This technique has enhanced the pattern resolution down to submicron, but it requires an additional process step, excessive process time, and additional manufacturing cost.

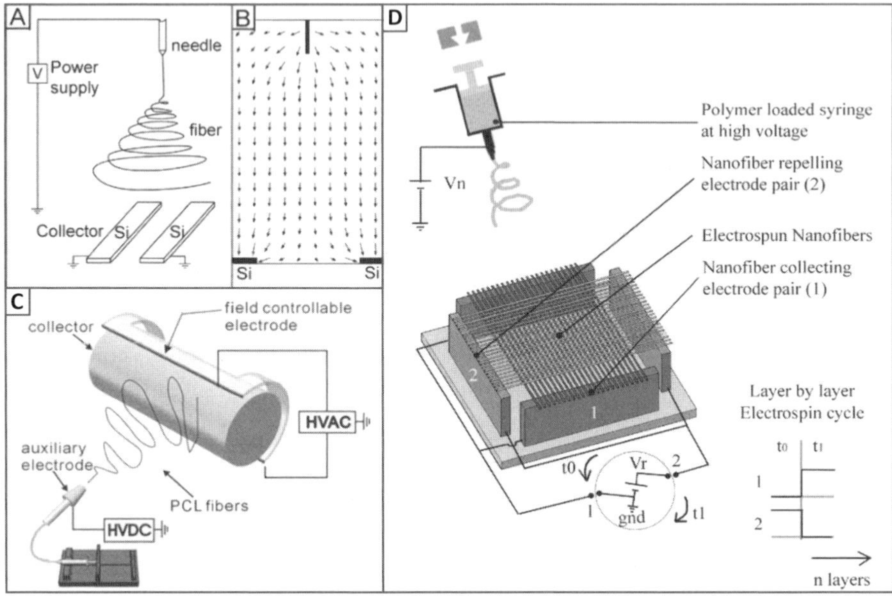

Figure 18. Various electrospinning setups for aligned nanofibers: A. Schematic of electrostatic approach with 2 collectors separated by a gap. B. Calculated electric field strength vectors for 2 collectors [10]. C. Schematic of an electrospinning setup with a rotating collector [11]. D. Schematic of 4 collectors with 2 offset electrodes [12].

Since photolithography is one of the promising technologies to make microscale structures and patterns, researchers have explored the possibility to form nanofibers out of photoresist polymers and directly perform photolithography to the stack of photoresist nanofibers for patterning. Multiple groups have demonstrated photo-patterning of electrospun nanofibers using photosensitive polymers by a direct photo-lithographical approach [21]. This method has enhanced the patterning resolution of a nanofiber stack to a few microns. However, most electro-spun polymer nanofibers are still electrically non-conductive, which limits their usage for electronic applications. Recently, Sharm et al. demonstrated electrically conductive carbon nanofibers (CNFs) der-ived from electrospun photo-patternable polymer (SU-8) nanofibers by high temperature carbonization. Photo-patternable polymer nanofibers offer micrometer precision with the standard CMOS ultra-violet (UV) light photolithography process. After high temperature carbonization, a large surface area and good electrical conductivity CNF microstructure is formed, which becomes a promising candidate for energy storage or high sensitivity sensors. This method opens the door to explore more advanced potential applications using nanofibers.

In the conventional UV patterning process, the photo-patternable vertical thickness of a nanofiber stack is limited due to significant optical scattering and diffraction. The typical diameter of electrospun nanofibers is between 100 nm to 500 nm, in a similar size range of the wavelength of UV light source (λ_{UV} = 365 nm). Experimental work to illustrate the diffraction effect in the nanofibrous stack has been performed by Jao et al., where they have used negative photoresist, SU-8, with a refractive index of 1.68 for electrospun nanofibers in the air environment, resulting in only a limited stack height of 34.6 ± 4.1 μm [22]. Later they have overcome this limitation by replacing the air inside the nanofiber stack with a refractive index matching medium such as canola oil. This process is known as oil-immersion lithography. Figures 19(a-b) show the schematic of oil-immersion lithography. As the air is replaced by the medium which has the same reflective index, the nanofiber stack is optically equivalent to a homogeneous solid film. Figures 19(c,d) show the UV intensity simulation (COMSOL Multiphysics, COMSOL Package Inc.) in the air and oil media. The simulation result shows clearly the reduced scattering effect with the oil medium. Figures 19(e,f) show the fabricated scanning electron microscope (SEM) images of nanofiber stacks with a UV dose of 400 mJ/cm^2. The oil-immersion technique significantly enhances the pattern resolution and patternable aspect ratio. This method enables to fabricate applications requiring high aspect ratio nanoporous microstructures such as 3D artificial tissue scaffold.

13.3.1. Carbon Nanofiber Strain Gauge Sensor

Previously, usage of elastomer/conductor nanocomposites with carbon nanotube, graphene nanoribbon, and silver nanowire has been reported. However, those nanoparticle/ribbon and short segmented nanowires in the elastomer are not highly stretchable, resulting in a limited working span. Electrospun nanofibers are inherently continuous and randomly oriented in deposition, forming a microscale spring. Carbon nanofibers derived from electrospun photopatternable polymers offer high pattern resolution and good electrical conductivity. Encasing CNFs in an elastomer enables to form stretchable conductive polymer with good mechanical support and robust packaging. Figure 20 shows the schematic of the CNF/PDMS nanocomposite for a strain gauge sensor application.

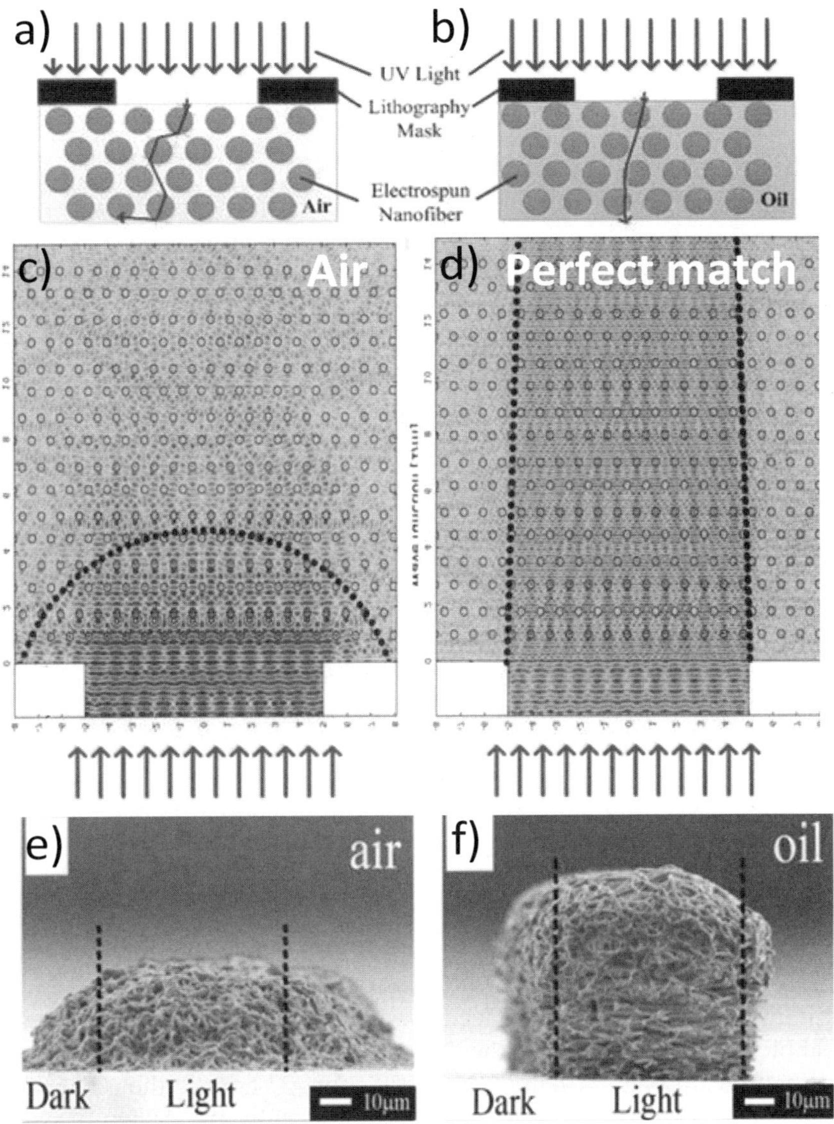

Figure 19. Oil immersion lithography: (a, b) Schematics, (c, d) COMSOL simulation results, and (e, f) SEM images of the fabricated nanofiber microstructures. (a)(c)(e) are obtained with the air medium and (b)(d)(f) are with the oil medium.

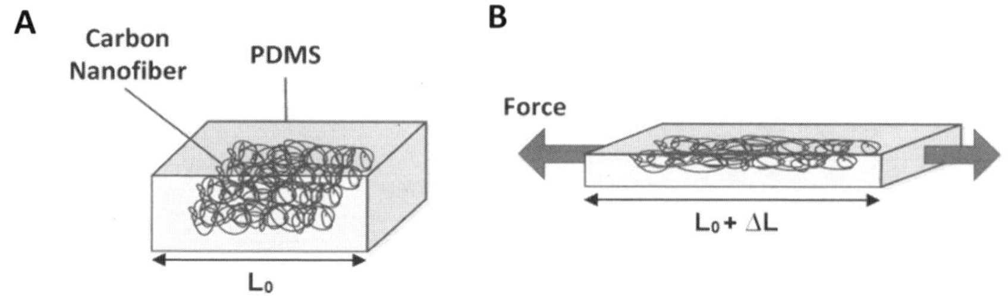

Figure 20. Schematic of the carbon nanofiber(CNF)/PDMS nanocomposite strain sensor.

Figure 21 shows the fabrication process of the composite strain sensor, consisting of CNFs embedded in polydimethysiloxane (PDMS). First, an SU-8 electrospinning solution is prepared by diluting SU-8 2025 (Microchem Inc.) with Dimethylformamide (DMF) to give a concentration of 90 % SU-8 2025. The solution is then carefully loaded into a syringe without introducing air bubbles and capped with a hypodermic needle. A syringe pump is used to maintain a pump rate of 1 ml/hour for steady droplet formation at the needle tip while the needle tip is connected to a high voltage power supply. Then, SU-8 nanofibers are electrospun to have a nanofiber stack thickness of 400 μm using the standard electrospinning setup with an electric field of 1 kV/cm and a distance between the needle tip to collector of 15 cm. Second, SU-8 nanofibers are dried in a vacuum oven for 24 hours to remove excessive solvent and then UV patterned (i-line: 365 nm) in oil medium with 400 mJ/cm^2. Third, high temperature carbonization in a forming gas environment with a steady flow rate of 10 sccm is performed to derive polymer nanofibers to conductive carbon nanofibers. Fourth, the resulting CNF is spin coated with PDMS elastomer solution (Sylgard 184, 1:10 mixture) and the sample is placed in vacuum for an hour to remove any air bubble that is trapped inside the CNF stack so that PDMS can fill the pores. After the PDMS is cured at 80°C for 2 hours, the CNF/PDMS nanocomposite pattern is peeled from the substrate. Then 2 external electrodes are assembled with the aid of carbon paste to increase conductivity and provide some mechanical support. Finally, PDMS solution is applied on the backside of the nanocomposite pattern to hermetically seal the CNF in PDMS producing a flexible packaging.

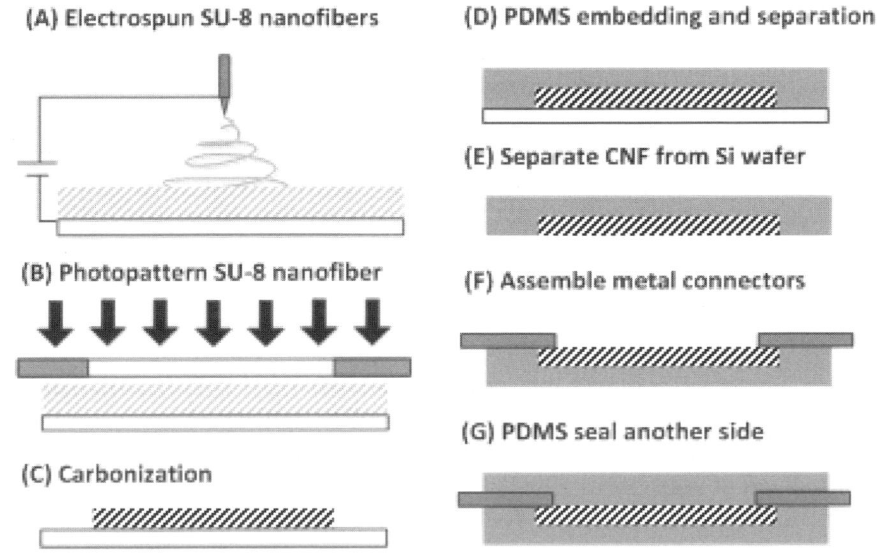

Figure 21. Fabrication process of the CNF/PDMS flexible strain gauge sensor.

The microscope image of the fabricated sensor is shown in Figure 22A. The CNF is completely packaged by PDMS and 2 wires are assembled for external circuitry connection. Figures 22B and 22C show the scanning electron microscope (SEM) images of the photopatterned SU-8 nanofibers before and after high temperature carbonization, respectively. Nanofiber diameter shrinkage during carbonization is observed. The SEM image of the cross-section view of the CNF/PDMS nanocomposite is shown in Figure 22D. The image shows that CNFs are completely embedded in PDMS and sandwiched by PDMS layers which not only serve as electrical insulation layers, but also as mechanical supporting layers.

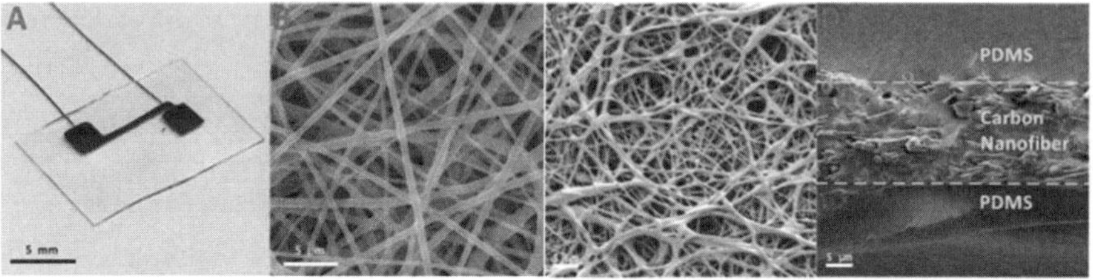

Figure 22. (A) Microscope image of a fabricated CNF/PDMS nanocomposite strain sensor. SEM images of (B) SU-8 nanofibers, (C) Carbon nanofibers, and (D) Cross-section view of the sandwiched CNF/PDMS nanocomposite sensor.

Figure 23 shows the measured I-V curves for the strain sensors using CNFs carbonized at 900°C and 1000°C with the measuring setup shown in inset [15]. The extracted sheet resistances (Rs) of those carbonized at 900°C and 1000°C are 830 kΩ/□ and 50 kΩ/□, respectively. Significant sheet resistance reduction is observed when the sample is carbonized at 1000°C. Figure 24 shows the measured resistance change per strain range, and the result shows a near linear response. A measured maximum gauge factor (GF $= \frac{\Delta R}{R}/\epsilon$) of 23.1 occurs at 50 % strain in room temperature. The strain range can be further increased by optimizing the PDMS thickness. Table 1 shows a comparison of this work with other state-of-the-art strain sensors. This work shows a significant increase in sensitivity (gauge factor) while maintaining high stretchability. The measurement result indicates that the CNF/PDMS nanocomposite has about 12 - 30 times higher the gauge factor than that of the graphene or carbon nanotube based nanocomposite sensor. The result shows that the CNF/PDMS nanocomposite structure has its unique architecture consisting of continuous and tangled electrospun nanofibers, resulting in a high gauge factor with high strain %. The low cost, manufacturable, high sensitivity CNF/PDMS nanocomposite sensor is demonstrated highly useful for smart wearable motion monitoring in fitness, sports, and medicine in the future.

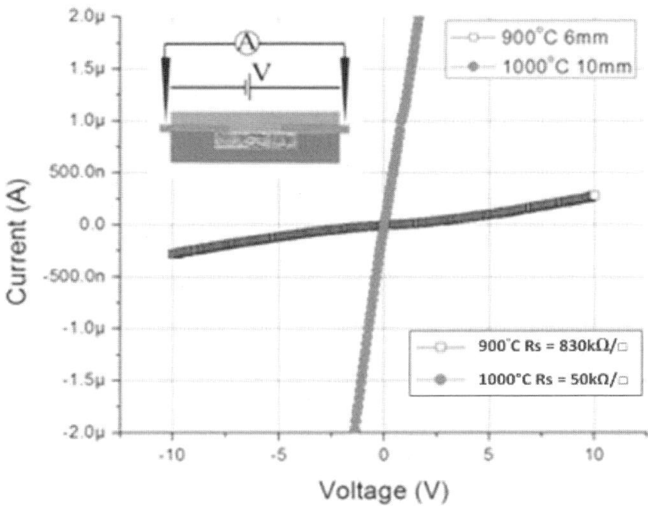

Figure 23. I-V curves of CNF/PDMS strain sensors fabricated at different pyrolysis temperatures. (Inset) Schematic of measurement setup [15].

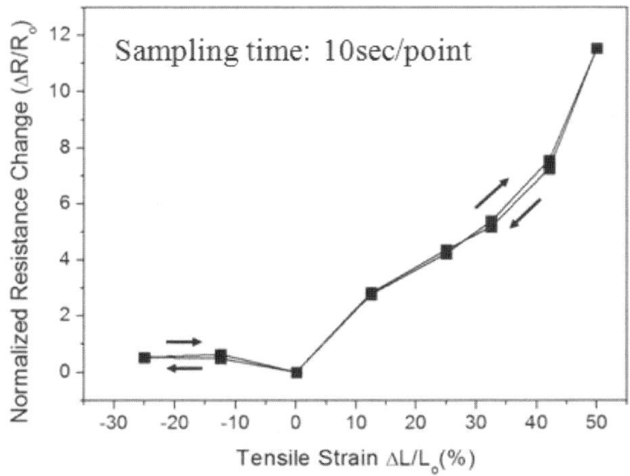

Figure 24. Measured relative resistance change to the applied strain at room temperature.

Table 1. Comparison of stretchability and gauge factor of this work and others.

Composite Material	Maximum Strain	Gauge Factor
CNF in PDMS (This work)	50%	23.1
SWCNT on PDMS [6]	280%	0.82 (<40% strain) 0.06 (>60% strain)
Graphene nanoribbon [5]	3%	1.9
Silver Nanowire [7]	20-60%	2-6
Printed Silver Nanowire [7]	60%	5

13.4. Novel Detection Mechanism

This nanocomposite strain gauge sensor is extremely versatile as a result of the high gauge factor and flexibility. As the society continues to move on the smart wearables technology, this sensor could easily be introduced as a wireless monitoring device on clothing or skin to monitor bio feedback such as on the fingers of a prosthetic arm for motion detection. Rahimi et al. demonstrated also a wireless strain detection mechanism to reduce whole device footprint and completely eliminate readout wire connection, which enables to enhance the compactness of the system.

Figure 25a shows the wireless detection setup with a strain gauge also serving as an antenna whose electrical response changes as a result of variation in strain. The antenna made of nanocomposites is designed to have a resonance frequency in the radio frequency (RF) range. The electrical RF response is detected by the external readout coil. The mechanism is based on the change of inductance (L) and capacitance (C) of the strain sensor. The self-resonant frequency (f) can be expressed as the following equation.

$$f = \frac{1}{2\pi\sqrt{LC}} \qquad (3)$$

Since the L and C changes as the sensor is stretched or compressed laterally, the amount of strain can be determined by reading out the self-resonant frequency shifts. This method could be commercially adapted to a radar system and does not require an embedded power source in the sensor, reducing the sensor size significantly.

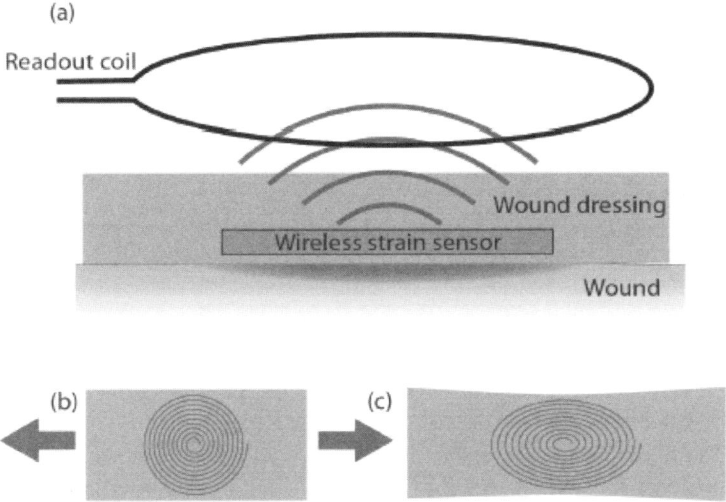

Figure 25. (a) Schematic of the wireless strain sensor embedded in a wound dressing with a readout coil for wireless detection. The proposed strain sensor (b) in resting state, and (c) in deformed state as a result of an applied strain.

Figure 26A shows measured phase responses according to different applied strains. The results show clear shifts of the self resonant frequency by applying different strain levels. The extracted self resonant frequency as a function of strain is plotted in Figure 26B. As a result, the self resonant frequency increases linearly up to 35 % strain. The frequency decreases linearly when the strain sensor is relaxed to its original dimension. The result shows limited hysteresis between 0–35 % strain. An average sensitivity of 150 kHz per % strain is observed.

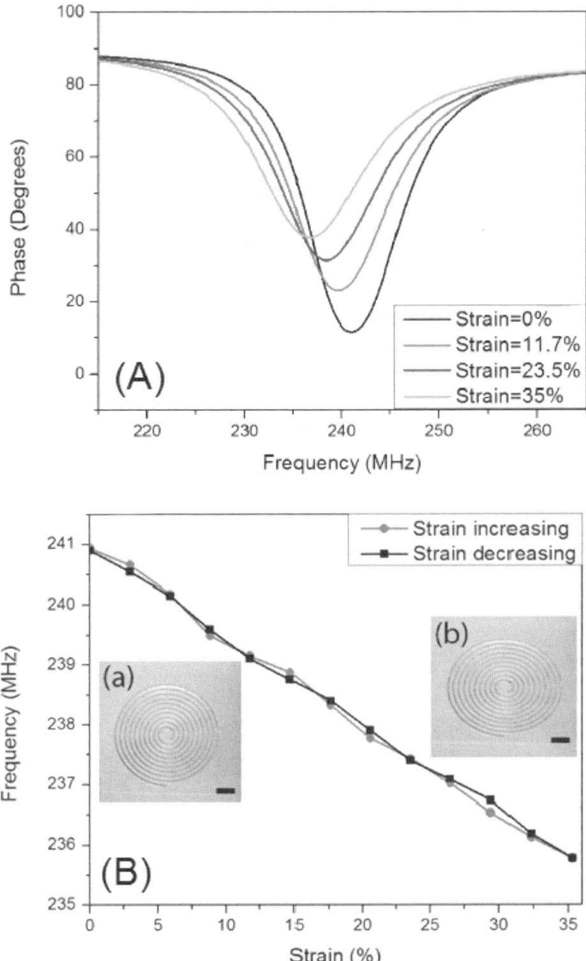

Figure 26. (A) Overlay plot of measured phase-dip curves of the sensor. The dip peak shifts with changes in strain. Changes in the dip amplitude at different strain levels are due to the changing distance between the sensor and the external coils. (B) Sensor self resonant frequency versus strain. Data points indicate the experimental values with increasing and decreasing strain. Insets: photographs of the sensor in its (a) un-stretched and (b) stretched form. Scale bar: 3 mm.

13.5. Conclusion

Nanotechnology offers unique material properties and system compactness. As examples of nanotechnology applications, the concepts of various smart wearable strain gauge sensors including graphene/PVDF, CNT/PDMS, Silver/PDMS, and CNF/PDMS have been reviewed and discussed. Especially, a highly sensitive and flexible strain sensor with the CNF/PDMS nanocomposite is very desirable for advanced wearable applications with a very high strain gauge factor of 23.1. The sensor can be easily integrated with clothes or even directly attached on the human body. Moreover, a wireless sensor system can be realized using the simultaneous strain gauge and antenna concept, which facilitates system size reduction and easy readout.

References

1. F. H. Khan, "Chenical Hazards of Nanoparticles to Human and Environment (A Review)", *Oriental Journal of Chemistry*, vol. 29, no. 4, pp.1399-1408, 2013.
2. "Apply nanotech to up industrial, agri output, "*The Daily Star(Bangladesh)*, 17 April 2012.
3. S.Suresh, "Semiconductor Nanomaterials, Methods and Applications: A Review, " Nanoscience and nanotechnology, 3(3), 62-74, 2013.
4. M. Huang, T. A. Pascal, H. Kim, W. A. Goddard III, and J. R. Greer, "Electronic-Mechanical Coupling in Graphene from in situ Nanoindentation Experiments and Multiscale Atomistic Simulations," *Nano Letters*, vol. 11, pp. 1241-1246, 2011.
5. V. Eswaraiah, K. Balasubramaniam, and S. Ramaprabhu, "One-pot synthesis of conducting graphene–polymer composites and their strain sensing application," *Nanoscale,* vol. 4, pp1258-1262, 2012.
6. T. Yamada, Y. Hayamizu, Y. Yamaoto, Y. Yomogida, A. Izadi-Najafabadi, D.N. Futaba, and K. Hata, "A stretchable carbon nanotube strain sensor for human-motion detection," *Nature Nanotechnology*, vol.6, 2011.
7. H. Lee, B. Seong, H. Moon, and D. Byun, " Directly printed stretchable strain sensor based on ring and diamond shaped silver nanowire electrodes, " *RSC Advance*, vil. 5, pp.28379-28384, 2015.
8. Z.-M. Huang, Y.-Z. Zhang, M. Kotaki, and S. Ramakrishna, "A review on polymer nanofibers by electrospinning and their applications in nanocomposites," *Compos Sci Technol.* 2003;63(15):2223-53.
9. S. Ramakrishna, K. Fujihara, W.-E. Teo, T.-C. Lim, and Z. Ma, "An introduction to electrospinning and nanofibers," *New Jersey: World Scientific*; 2005.
10. D. Li, Y. Wang, and Y. Xia, "Electrospinning of Polymeric and Ceramic Nanofibers as Uniaxially Aligned Arrays," *Nano Letters*, vol. 3, no. 8, pp. 1167-1171, 2003.

11. H. Lee, H. Yoon, and G.H. Kim, "Highly oriented electrospun polycaprolactone micro/nanofibers prepared by a field-controllable electrode and rotating collector," *Apply Physic A*, vol. 97, pp. 559-565, 2009.

12. P.F. Jao, M. Machado, X. Cheng, D.E. Senior, G.J. Kim, D. Ding, W. Sun, and Y.K. Yoon, "Fabrication of Nanoporous Membrane and its Nanolithographic Patterning Using Electrospinning and Stamp-Thru-Mold (ESTM)," *Micro Electro Mechanical Systems (MEMS), 2011 IEEE 24th International Conference on*, pp. 257-260, 2011.

13. J.S. Kim and D.S. Lee, "Thermal properties of eletrospun polyesters," *J. Polymer*, vol. 32, pp.616-618, 2000.

14. H. Fong, W. Liu, C.S. Wang, and R.A. Vaia, "Generation of electrospun fibers of nylon 6 and nylon 6-montmorillonite nanocomposite," *J. Polymer*, vol. 43, pp.775-780, 2002.

15. P.F. Jao, K.T. Kim, G.J. Kim and Y.-K. Yoon, "Fabrication of an all SU-8 electrospun nanofiber based supercapacitor," *J. Micromech. Microeng.*, vol. 23, no. 11, 2013.

16. H. Du, J. Wang, M. Su, P. Yao, Y. Zhang, and N. Yu, "Formaldehyde gas sensor based on SnO2/In2O3 hetero-nanofibers by a modified double jets electrospinning process," *Sensors and Actuators B: Chemical*, vol 166, pp. 746-752, 2012.

17. S.-P. Fang, H. Shang, P.F. Jao, K.T. Kim, G.J. Kim, J.H. Yoon, K. Cho, A.J. Katz, and Y.-K. Yoon, "Mechano-active tissue scaffold system based on a magnetic nanoparticle embedded nanofibrous membrane," *2014 IEEE 27th International Conference on Micro Electro Mechanical System (MEMS)*, pp.917-920, 2014.

18. Y. C. Lim et al., "Micropatterning and characterization of electrospun poly(ε-caprolactone)/gelatin nanofiber tissue scaffolds by femtosecond laser ablation for tissue engineering applications," *Biotechnol. Bioeng.*, vol. 108, no. 1, pp. 116–126, Jan. 2011.

19. H. J. Lee, Y. H. Park, and W.-G. Koh, "Fabrication of nanofiber microarchitectures localized within hydrogel microparticles and their application to protein delivery and cell encapsulation," *Adv. Funct. Mater.*, vol. 23, no. 5, pp. 591–597, Feb. 2013.

20. J. Shi, L. Wang, and Y. Chen, "Microcontacting Printing and Lithographic Patterning of Electrospun Nanofibers," *Langmuir*, vol 25, pp. 6015-6018, 2009.

21. C.S. Sharma, A. Sharma and M. Madou, "Multiscale Carbon Structures Fabricated by Direct MicroPatterning of Electrospun Mats of SU-8 Photoresist Nanofibers," *Langmuir* Vol. 26, No.4, 2010, pp 2218-2222.

22. P.F. Jao, E. Franca, S.P. Fang, J. Yoon, K. Cho, D.E. Senior, G. Kim, B. Wheeler, and Y.K. Yoon, "Fabrication of Carbon Nanofibrous Microelectrode Array (CNF-MEA) Using Nanofiber Immersion Photolithography," *2014 IEEE 27th International Conference on Micro Electro Mechanical System (MEMS)*, 2014.

Pathogen, 103
PEC wet etching, 214
Peptide, 175
Perkinsus marinus, 116
Persistent photoconductivity, 407
Phase noise, 1
Phase-lock-loop, 14
Phosphate buffered saline (PBS), 109
Photochemical assisted binding, 167
Photodetectors, 407
Photodynamic therapy (PDT), 86
Photosensitizers (PS), 86
Phototherapy, 85
Photothermal therapy (PTT), 86
Physical vapor deposition (PVD), 41
Physiological solution, 162
Picric acid (PA), 93
Piezoelectric polarization, 407
Piezoeresitive transductive, 404
Piezoresistivity, 405
Platinum nano-network, 208, 209
Platinum, 39
Point-of-care testing (POCT), 245
Polar c-plane, 201
Polyaniline (PANI), 272

Polyaniline, 247
Polyvinylidene fluoride (PVDF), 446
Power loss, 3
Power-gain cutoff frequency (f_{MAX}), 4
Pressure sensing, 404, 405
Printable silver nanowire composite, 454
Printing circuit board, 1
Propane sultone, 247
Protein-peptide binding, 128
Pt nanourchins. 47
Pt Schottky diodes, 215
Pyrolytic method, 73

Qaudrature-phase, 5
Quality factor, 10

Radar cross section, 3
Radar system test, 22
Radiating sensing, 404
Raman spectra, 229
Reactive ion etching (RIE), 230
Reactive oxygen species (ROS), 86
Receiver, 9
Residue phase, 2
Respiration rates, 2
Ring oscillator, 6
Robustness, 271
Ruthenium chloride complex, 168
Ruthenium, 42